Advances in Volcanology

An Official Book Series of the International Association of Volcanology and Chemistry of the Earth's Interior

Series Editor

Karoly Nemeth, Institute of Natural Resources, Massey University, Palmerston North, New Zealand

Advances in Volcanology is an official book series of the International Association of Volcanology and Chemistry of the Earth's Interior (IAVCEI). The aim of the book series is to publish scientific monographs on a varied array of topics or themes in volcanology. The series aims at building a varied library of reference works, by describing the current understanding of selected themes such as certain types of volcanism or regional aspects. The books in the series are prepared by leading experts actively working in the relevant field. The Advances in Volcanology series contains single and multi-authored books as well as edited volumes. The Series Editor, Dr. Karoly Nemeth, is currently accepting proposals and a proposal document can be obtained from the Publisher, Dr. Annett Buettner (annett.buettner@springer.com).

Thomas C. Pierson · Lee Siebert ·
Kevin M. Scott

Deposits of Volcanic Wet Flows

Identifying Deposits of Lahars, Debris Avalanches, and Water Floods in Volcanic Terrain

Springer

Thomas C. Pierson
U.S. Geological Survey (Retired)
Glenwood, WA, USA

Lee Siebert
Smithsonian Institution (Retired)
Gresham, OR, USA

Kevin M. Scott
U.S. Geological Survey (Retired)
Portland, OR, USA

ISSN 2364-3277 ISSN 2364-3285 (electronic)
Advances in Volcanology
ISBN 978-3-031-66573-8 ISBN 978-3-031-66574-5 (eBook)
https://doi.org/10.1007/978-3-031-66574-5

This Springer imprint is published by the registered company Springer Nature Switzerland AG
The registered company address is: Gewerbestrasse 11, 6330 Cham, Switzerland

Complex sequence of volcanic wet-flow deposits in a fill terrace in the upper Sandy River, about 8 km downstream of the source lava dome on Mount Hood (Oregon, USA)

Preface

The seeds for this book were planted during a field excursion at a 1993 volcano workshop in Guatemala, when volcaniclastic deposits downstream of Santiaguito lava dome had mystified a gathering of experienced volcanologists. The puzzling depositional units appeared to have been laid down by flows of water-mobilized volcanic sediment, but the flow type and significance of the sediments were hotly debated that afternoon at those outcrops. In retrospect, this was not surprising, because few detailed descriptions of water-mobilized, subaerial, volcaniclastic, and epiclastic deposits—deposits emplaced by lahars, debris avalanches, or muddy water floods—had appeared in the literature prior to the 1980s and 1990s (Manville et al. 2009). Primary pyroclastic deposits had been well studied and described over many decades, but deposits of volcanic rock fragments mobilized by or containing liquid water had not. Now, after a long gestation, those seeds are coming to fruition.

Full recognition of volcanic wet flows as significant volcano hazards began to fully emerge following the explosive May 18, 1980, eruption of Mount St. Helens (Washington, USA). Not only were extensive deposits from many processes—tephra fall, pyroclastic density currents, lahars, and high-concentration water floods—emplaced and interbedded during and following that eruption, but the poorly understood process of deposition by a large volcanic debris avalanche was recorded and its deposit left well exposed. Soon after the dust settled, it became clear that all these different kinds of deposits can be uncannily similar in appearance to each other. We three authors, experiencing the Mount St. Helens eruption early in our careers, all realized how important recognizing and differentiating these different kinds of similar-looking deposits would be for accurately assessing volcano hazards. Then, like two big exclamation points, catastrophic damage and loss of life from lahars and volcanic water floods during and following eruptions at Nevado del Ruiz (Colombia) in 1985 and at Mount Pinatubo (Philippines) in 1991 further added to the urgency of sharing what we were learning, eventually with this book.

It is our hope that this volume will be a resource in the identification and interpretation of this full range of unconsolidated volcanic deposits. The book is aimed primarily at advanced undergraduate and graduate students of volcanology, stratigraphy, and sedimentology; professional volcanologists, stratigraphers, and sedimentologists; and geologic mappers working in volcanic terrain to assess volcano hazards. Risks to people living near volcanoes from these different volcanic processes vary greatly, depending on the location, exposure, and preparedness of vulnerable communities. The safety of millions of people worldwide who live on or downstream, downslope, and downwind of active volcanoes could depend on how well we can discern the origin of deposits that most people would simply describe as sand and gravel.

Glenwood, USA — Thomas C. Pierson
Gresham, USA — Lee Siebert
Portland, USA — Kevin M. Scott

Reference

Manville V, Nemeth K, Kano K (2009) Source to sink: a review of three decades of progress in the understanding of volcaniclastic processes, deposits, and hazards. Sediment Geol 220:136–161

Acknowledgements

This volume is an attempt to summarize what is currently known about identification and interpretation of lahar, debris-avalanche, and highly sediment-laden water-flood deposits at volcanoes, and we therefore owe much to many researchers who have worked in this field. Our bibliography attests to the depth of worldwide expertise on this topic. We also owe much to many colleagues who have puzzled over these deposits with us at outcrops and discussed the attributes, quirks, and terminology of these sediments with us—on our home turf in the Cascade Range of the Pacific Northwest and on numerous field trips in other parts of the world. Many of them, particularly the participants in a 1998 international field course on volcanic wet flows in Quito, Ecuador, have offered strong encouragement for writing this book.

Colleagues who gave generously of their time to review and provide input on early versions of parts of this book include Álvaro Amigo, Carlos Borrero, Brittany Brand, Scott Bryan, Gerardo Carrasco Nuñez, Sarah Fagents, Judy Fierstein, Cynthia Gardner, Chris Harpel, Jon Major, Jim O'Connor, John Pitlick, Mark Reid, Willie Scott, Gary Smith, Robert Tilling, Jim Vallance, and Nate Wood. Individual chapter reviews for the current version were provided by subject matter experts Andrew Fountain, Dick Iverson, Paul Kennard, Vern Manville, Jon Procter, Shinji Takarada, and Alexa Van Eaton. Wes Hildreth, James D. L. White, and Jon Major graciously undertook the huge task of providing whole-book reviews of the final manuscript. For all these contributions, we are exceedingly grateful. In addition, we are extremely grateful to our USGS colleague Lisa M. Faust for her many beautifully rendered illustrations in this book, often presented in complex perspective views.

We owe a special debt of gratitude to three colleagues, now deceased, who were pioneers in the field of volcaniclastic sedimentology—Dwight ("Rocky") Crandell, Richard J. ("Dick") Janda, and Harry Glicken. They not only inspired us, but in Dick's case also made it possible for two of us to devote our careers to this topic at the U.S. Geological Survey's Cascades Volcano Observatory.

Finally, we offer profound thanks to our wives—Molly Davis, Cheryl Siebert, and Lisa Scott—who not only tolerated our diverted energy and attention in writing the various drafts of this book, but who also endured our absences at the many long professional meetings, field trips, and international assignments that were required for us to educate ourselves about volcanic sediments.

Contents

About the Authors

Thomas C. Pierson is an internationally recognized authority on lahars and posteruptive sedimentation in volcanically disturbed watersheds. He retired in 2018 after 38 years as a research geologist/hydrologist at the US Geological Survey Cascades Volcano Observatory, serving four of those years as an associate scientist in charge. For many years, he was also an adjunct professor of geology at Portland State University. Investigations over his career focused on understanding the behavior of hazardous wet flows in volcanic and nonvolcanic terrain and on communicating with officials and the public about geologic and hydrologic hazards. Research projects and volcano crisis responses took him beyond the Cascade Range to Alaska, Japan, the Philippines, New Zealand, Mexico, Ecuador, Colombia, and Chile.

Lee Siebert is globally recognized for his work on volcanic debris avalanches. He spent 37 years with the Smithsonian Institution's Global Volcanism Program (GVP), where he researched and analyzed data on the physical characteristics, eruption chronologies, and eruptive characteristics of the world's volcanoes. He served as a GVP director from 2007 to 2010. Field research projects, primarily on volcanic debris-avalanche deposits, took him to volcanoes in Alaska, the Cascade Range, Mexico, Guatemala, El Salvador, Panama, and Indonesia. Following an assignment to the US Geological Survey Cascades Volcano Observatory (CVO) prior to his retirement to the Pacific Northwest in 2011, he continued as a Smithsonian research associate for a dozen years while volunteering at CVO.

Kevin M. Scott is recognized around the world as an expert on large lahars from active and inactive volcanoes. He retired from his position as a research scientist with the US Geological Survey Cascades Volcano Observatory, in which capacity he received the Department of the Interior's Distinguished Service Award, the highest award for service to that agency. He was also a senior international professor at the Chinese Academy of Sciences from 2009 to 2011. His years of working on international volcano crisis responses are documented in his book, *The Voice of This Stone—Learning from Volcanic Disasters Around the World* (Carpe Diem Press, 2019).

Abbreviations

Time

yr	Year
yr BP	Years before present
CE	Christian era (with a date)

Physical Units/Indexes

vol%	Volume percent (solids volume/solids + water volume)
wt%	Weight percent (solids weight/solids + water weight)
VEI	Volcano Explosivity Index

Processes

VDF	Volcanic debris flow
VHF	Volcanic hyperconcentrated flow
VDA	Volcanic debris avalanche
VWF	Volcanic water flood
PDC	Pyroclastic density current
TF	Tephra fall

Deposits Emplaced by Volcanic Wet Flows

Abstract

Gravity-driven flows of material emanating from volcanoes during or following eruptions can range from flowing molten rock to flowing denser-than-air volcanic gas. But the bulk of volcanic flows involve two-phase mixtures of solid rock fragments and a fluid—either liquid water or gas composed of air ± volcanic gases. Deposits of the various particle–fluid flows derive their physical characteristics from mixture properties and from the particle–fluid interactions involved in both the transport and deposition of the mixtures. Interactions can differ substantially among emplacement processes. Flow types vary with the ratio of solid particles to fluid and with fluid composition. Volcanic "wet" flows involve variable mixtures of liquid water + volcanic rock particles (± ice fragments)—lahars, debris avalanches, and water floods. Their deposits exhibit properties differing only slightly from deposits emplaced by "dry" flows involving gas + volcanic rock particles and from deposits emplaced by flowing glacier ice. Until now, wet-flow deposits have not been systematically described in detail. Recognition of the characteristic features of any volcanic or volcaniclastic deposit can lead to identification of its emplacement process, which is critical for accurate volcanic hazard assessment.

1 Overview

Many volcanoes around the world produce far-reaching and potentially deadly flows of solid particles (mainly volcanic rock fragments) mixed with liquid water—*volcanic wet flows*. More precisely, volcanic wet flows are two-phase (solid + fluid) *particle–fluid flows*, driven by gravity, where the fluid between the particles in the flowing mixture is partly or entirely liquid water. The main categories of wet flows at volcanoes include *lahars*, *debris avalanches*, and *water floods* (defined below), which are behaviorally distinct flow types having varying particle concentrations, densities, and sizes. In contrast to these wet flows, the interstitial fluid in pyroclastic density currents (the other main flow hazard at volcanoes) is largely or entirely gas—a combination of exsolving volcanic gas and entrained air.

Lahar-like flows, debris avalanches, and floods occur not only at volcanoes. However, we specifically label the wet flows described in this book as "volcanic" (lahars being implicitly volcanic flows) because the volcanic varieties of these flows differ slightly from, and are generally larger and more destructive than, their nonvolcanic counterparts in mountainous terrain elsewhere.

Volcanic wet flows are partly or wholly mobilized by liquid water but not necessarily directly caused or triggered by water release.

T. C. Pierson et al., *Deposits of Volcanic Wet Flows*, Advances in Volcanology,
https://doi.org/10.1007/978-3-031-66574-5_1

The debris avalanche that occurred at the beginning of the May 18, 1980, eruption of Mount St. Helens (Washington, USA), for example, was triggered by deformation and oversteepening of the volcano's north flank due to intrusion of magma into the upper part of the glaciated stratocone. Indeed, a variety of possible triggers (both related and unrelated to eruptions) can initiate volcanic wet flows. This variety means that volcanic wet flows, in contrast to eruption-triggered pyroclastic density currents or tephra falls, can occur prior to, during, or years after eruptions. Moreover, volcanic wet flows commonly travel much farther downstream than most pyroclastic density currents.

Until recently (Major 2023), the deposits of lahars, volcanic debris avalanches, and volcanic water floods have not been contrasted with each other or with other types of deposits composed of volcanic rock particles, even though many excellent descriptions of process-specific deposits have appeared over the last few decades. There are two main reasons for this. First, recognition of volcanic wet flows as distinct processes has been late in coming. Most in-depth studies of lahars and volcanic debris avalanches (and descriptions of their deposits) have appeared only after the string of late twentieth century eruptions that demonstrated the hazards (and the long downslope reach) of these processes—the May 18, 1980, eruption of Mount St. Helens; the November 13, 1985, eruption of Nevado del Ruiz (Colombia); and the June 15, 1991, eruption of Mount Pinatubo (Philippines) (Manville et al. 2009). Second, the task of deposit differentiation is difficult—see Sect. 7. Furthermore, studies recognizing and describing unusually sediment-rich water-flood deposits and glacial deposits (primarily glacial tills) at volcanoes have been rare.

This book is a guide in the identification and differentiation of deposits emplaced by volcanic wet flows on land (*subareal* deposits), but it also aims to show how wet-flow deposits differ from those of pyroclastic density currents (PDCs), tephra falls (TFs), and glacial deposits composed of volcanic rock fragments. Numerous outcrop photographs and descriptions of features observable or measurable in the field, as well as descriptions of the processes and their impacts on landscapes, form the core of the volume. Our intended audiences include advanced undergraduate- and graduate-level students of volcanology, sedimentology, and stratigraphy; professional volcano scientists, stratigraphers, and sedimentologists; and geologic mappers working in volcanic terrain to assess volcano hazards.

The focus of this book is on the fresh, unaltered, and unconsolidated subaerial deposits encountered on or downstream of volcanoes. More exotic types of localized fragmental deposits linked to autobrecciation of lava flows, such as flow breccias and peperites, are not considered. For chemically altered and lithified volcanic sediments, other references should be consulted, such as Fisher and Schmincke (1984), Cas and Wright (1987), and McPhie et al. (1993). Deposits emplaced in standing bodies of water share some of the characteristics of subaerial deposits, but they also exhibit differences related to the additional effects of quenching, dispersal and sorting by oscillatory currents, density-controlled particle settling, and reworking by submarine gravity-flow processes. For comparative descriptions and interpretations of marine and lacustrine deposits of volcanic rock particles, we refer readers to other sources, including Fisher and Schmincke (1984), Kokelaar and Howells (1984), Stix (1991), Cas and Busby-Spera (1991), McPhie et al. (1993), and White and Riggs (2001).

"Sequencing of chapters in the book moves from this chapter and the other two introductory chapters (Chaps. "Scaling and Timing of Volcanic Wet Flows in Response to Eruptions" and "Deposit Identification and Interpretation") to paired chapters on deposit emplacement processes and deposit characteristics for each type of volcanic wet flow (lahars, debris avalanches, and sediment-rich water flows). These wet-flow chapters are followed by a chapter that contrasts volcanic wet flow deposits with pyroclastic density-current, tephra-fall, and volcano-glacial deposits (Chap. "Other Processes Depositing Volcanic Sediment")." Critical characteristics for identification and differentiation of deposits are tabulated in Chap. "Deposit Identification: Key and Summary Tables" with an identification key and summary tables of sediment characteristics. Finally, Chap. "Annotated Outcrop Photographs: Volcanic Wet-Flow Deposits and Other Unconsolidated Volcanic Deposits" presents a collection of annotated outcrop photographs to aid field identification. Note that the types and locations of specific volcanoes mentioned throughout the book are listed in the Volcano Index.

2 Types of Volcanic Wet Flows

Types of volcanic wet flows differ in their flow behavior mainly due to differences in the concentration of particles in the flowing mixtures, the physical properties and composition of the particles and the fluid, and the distribution of particle sizes within the mixtures. Concentrations in the most concentrated mixtures are limited by the maximum packing density of the particles involved. Mixtures with a wide variety of particle sizes (poorly sorted mixtures) can be packed closer together than uniformly sized (well-sorted) mixtures because finer grains can fit into spaces between coarser grains. Maximum particle concentrations can vary from about 70% of mixture volume (vol%) to about 90 vol% for mixtures containing poorly sorted particles, but commonly the limit is 80–85 vol% (Lambe and Whitman 1969; Frings et al. 2011; Chang et al. 2015). The remaining 15–20% of mixture volume may be filled with water, air, or a combination of both. The most dilute naturally occurring volcanic wet-flow mixtures contain only a few volume percent solids and can be described as muddy water.

Lahars are flowing dense mixtures of water and solid particles sourced from a volcano (which may or may not be erupting) and are fully water saturated but intermediate in mean sediment concentration between normal streamflow and unsaturated debris avalanches (Smith and Lowe 1991; Vallance 2000; Waythomas et al. 2013; Vallance and Iverson 2015; Major 2023) (Chap. "Lahars—Process Overview"). Solid particles are rock or ice fragments (or a combination of both), but lahars often contain wood and other debris, as well. They can flow slowly or rapidly (up to 40 m/s) and are one of the most dangerous types of volcano hazards, commonly affecting communities 100 km or more downstream of the source volcano. Lahar sediment concentrations typically vary simultaneously within different parts of a single lahar, but concentrations also change with time as lahars flow downstream. Depending on localized sediment concentration inside the lahar body, types of flow may include *debris flow* or *hyperconcentrated flow*, or commonly both, plus some muddy *water flow* (Pierson and Scott 1985; Cronin et al. 1999, 2000; Vallance 2000; Vallance and Iverson 2015; Major 2023) (see Chaps. "Lahars—Process Overview" and "Volcanic Water Floods—Process Overview"). Concentrated lahars (or parts of lahars) moving mainly as debris flow are here termed ***volcanic debris flows*** **(VDFs)**, and more dilute lahars (or parts of lahars) moving mainly as hyperconcentrated flow are termed ***volcanic hyperconcentrated flows*** **(VHFs)**.

Volcanic debris avalanches **(VDAs)** are rapid flows of mainly unsaturated mixtures of rock particles and water, sometimes virtually dry, such that the spaces between particles are partly or entirely filled with air during flow (Ui et al. 2000; van Wyk de Vries and Davies 2015) (Chap. "Volcanic Debris Avalanches—Process Overview"). Additionally, volcanic debris avalanches are larger, contain a higher proportion of fine particles (usually including clay minerals), and flow farther than most debris avalanches in nonvolcanic terrain.

Volcanic water floods **(VWFs)** are identified here as a special category of water flood involving unusually high concentrations of suspended fine sediment and pronounced bed-load transport, like many flash floods in arid nonvolcanic terrains (Lekach and Schick 1982; Reid and Frostick 1987; Costa 1988; Reid et al. 1998; Cohen and Laronne 2005) (see Chap. "Volcanic Water Floods—Process Overview"). These muddy flood flows are less concentrated than hyperconcentrated flows but contain up to 100 times more suspended sediment than water floods typically occurring in humid nonvolcanic terrain. Because of their higher load of suspended sediment, volcanic water floods can transport coarser bed-load sediment than most nonvolcanic water floods of equivalent magnitude.

3 Volcanic Rock Fragment Classification

Particle–fluid flows at volcanoes transport different types of rock fragments, and the types of fragments transported can relate to flow origins. Classifications of volcanic rock fragments have been based previously on clast-forming processes (Fisher 1961, 1966; Fisher and Schmincke 1984), on transport and depositional processes (Cas and Wright 1987), or on a combination of both (McPhie et al. 1993; White and Houghton 2006) (Table 1). Composition of dominant component particles and environment of breakage allow further subdivision of particle type. The classification adopted for this book relies on the scheme put forth by White and Houghton (2006). By their definition, volcanic rock fragments are either *volcaniclastic* or *epiclastic*. Volcaniclastic particles and deposits are further subdivided as primary or secondary, depending on emplacement history and timing.

Primary volcaniclastic particles are rock fragments formed from solidifying magma (primary juvenile particles) during eruptions and then directly deposited. Rock fragmentation may be caused by physical stresses imposed during energetic magma ejection or by thermal or gravitational stresses imposed on active lava flows or lava domes. Primary volcaniclastic deposits are derived from juvenile magma but can include older rock fragments scoured from the vent walls during eruptions or entrained from the ground surface by flows (lithic fragments). Deposit emplacement is accomplished by eruption-triggered flows or falls. Sohn and Sohn (2019) argued that if juvenile particle emplacement is modified by a noneruptive geomorphic process (such as the action of wind or rainfall runoff on accumulating tephra particles), such deposits should be distinguished as "reprocessed" primary deposits. This can be a useful distinction, but fundamentally we consider emplacement of juvenile particles at an initial depositional site to be the main determinant for identifying a deposit as primary.

Secondary volcaniclastic particles result from the erosion and transport of primary volcaniclastic particles from original depositional sites to new depositional sites elsewhere. Geomorphic processes reworking and remobilizing primary volcaniclastic deposits include gravity-driven mass flows and transport by flowing water, ice (glaciers), or air (wind).

Lithic fragments are rock particles derived from preexisting rocks (volcanic and nonvolcanic), fragmented by eruptive processes, or already existing as fragments, which are entrained by eruptive processes (PDCs and TFs) and emplaced with pyroclasts in primary volcaniclastic deposits. Fragmentation does not require the action of weathering processes, so lithic fragments are not considered to be epiclastic. Lithic fragments are commonly nonjuvenile volcaniclastic particles or pieces of nonvolcanic country rock, but recycled juvenile particles (i.e., fragments of solidified magma from the same eruption scoured from vent walls) may be included. Most recognizable lithic fragments in primary volcaniclastic deposits are composed of dense or poorly vesicular microcrystalline or crystalline rock; re-entrained pumice or scoria fragments would not be easily recognized as nonjuvenile. Lithic fragments typically make up less than 5 vol% of the particles in a pumiceous or scoriaceous pyroclastic deposit (Fisher and Schmincke 1984).

Epiclastic particles are those eroded directly from preexisting solid rock by weathering processes. Source rock for epiclastic particles may be volcanic (fresh or weathered) or nonvolcanic (local bedrock).

The broader terms, ***volcanic particles*** and ***volcanic sediment***, refer to deposits composed of volcaniclastic or epiclastic particles or a combination of both. It has been estimated that volcanic sediment makes up at least one quarter

Table 1 Categories of rock fragments in volcanic sediments, based on mode of fragmentation or process of deposition or a combination of both

Rock fragment origin	Rock fragment descriptors			
	White and Houghton (2006)	McPhie et al. (1993)	Cas and Wright (1987)	Fisher and Schmincke (1984)
Particles of volcanic rock formed and dispersed by any process (including volcanic epiclastic particles)	Volcanic	Volcaniclastic	Volcaniclastic	Volcaniclastic
Particles (juvenile) ejected from a volcanic vent, regardless of eruption cause or fragment origin, composition, or vesicularity and emplaced at an initial deposition site	Primary volcaniclastic	Pyroclastic	Pyroclastic	Pyroclastic
Particles (juvenile) formed by fragmentation during ejection of magma from a vent or during transport in a pyroclastic density current	Pyroclastic	Pyroclastic	Pyroclastic	Pyroclastic
Particles (juvenile or non-juvenile) formed by steam-driven explosions from a volcanic vent in phreatic or phreatomagmatic eruptions	Pyroclastic	Pyroclastic	Pyroclastic	Hydroclastic
Particles (juvenile) formed by quench-shattering of lava that encountered liquid water or ice	Hyaloclastic	Hyaloclastic	Hyaloclastic	Hydroclastic
Particles (juvenile) fragmented by lava being injected into wet sediment	Peperitic	Peperitic	Peperitic	Hydroclastic
Particles (juvenile) formed by mechanical fragmentation of flowing/deforming lava flows or domes during solidification/cooling	Autoclastic	Autoclastic	Autobreccia	Autoclastic
Particles (juvenile or nonjuvenile) derived from preexisting rocks (volcanic or nonvolcanic) found in primary volcaniclastic deposits *Cognate lithic fragments*—juvenile fragments originating as solidified parts of the erupting magma *Accessory lithic fragments*—pieces of country rock scrubbed from conduit/vent walls by explosive eruptions *Accidental lithic fragments*—fragments entrained from the ground surface by PDCs	Lithic	Lithic	Lithic	Lithic
Primary volcaniclastic particles remobilized from an unconsolidated deposit at its initial site of deposition and transported to another site (one or more times) by geomorphic processes	Secondary volcaniclastic	Resedimented volcaniclastic	Epiclastic	Reworked pyroclastic

(continued)

Table 1 (continued)

Rock fragment origin	Rock fragment descriptors			
	White and Houghton (2006)	McPhie et al. (1993)	Cas and Wright (1987)	Fisher and Schmincke (1984)
Particles formed by the breakage of lithified volcanic or nonvolcanic rock by physical or chemical weathering processes (not by eruptive processes)	Epiclastic	Epiclastic	Epiclastic	Epiclastic

Term in parentheses is not explicitly defined but inferred from discussion in reference. Usage in this book follows that of White and Houghton (2006). Note that lithic particles can include volcanic particles from earlier eruptions or particles of country rock (the host bedrock through which a volcanic conduit ascends)

of all sediment in the world's geologic record (Fisher and Schmincke 1984). Volcanic wet-flow deposits are typically composed of a combination of secondary volcaniclastic and epiclastic particles, but they may also include primary particles (e.g., cryptodome fragments in the 1980 VDA deposit at Mount St. Helens [Glicken 1996]).

4 Classifying Flows Involving Particles: Basic Concepts

Flow behavior is a general term that encompasses the mechanics of flows and their depositional processes. Different types of two-phase flow (wet and dry flow types) have characteristic behaviors that determine, in part, the properties of their deposits. The sections below describe basic concepts about how volcaniclastic particles move when mixed with a fluid.

4.1 Particle–Fluid Interaction

Both wet and dry particle–fluid flows at volcanoes are highly complex mechanically, and sizes of particles involved can vary greatly among different flows. Flow involves complex physical interactions between a granular solid phase and a fluid phase (water, air, or a combination of both)—interactions that also occur in tephra falls (Iverson 1997, 2005; Branney and Kokelaar 1992, 2002; Bonadonna et al. 2015; Dufek et al. 2015; Vallance and Iverson 2015; Lube et al. 2020). Particle motion is mainly determined by a combination of grain–grain interactions (collisions and enduring frictional contacts between particles) and fluid–grain interactions (turbulence, viscous drag, and intergranular pore-fluid pressure). Additional influences on the behavior of volcanic particle–fluid flows include gravity-induced or kinematic size segregation, selective deposition of particles based on size or density, changes in pore-fluid pressure, and dynamic interactions between flows and material on underlying bed surfaces, all of which can increase or decrease mixture concentrations, flow mass, and bulk flow momentum (Iverson 1997; Major 1997, 2000; Major and Iverson 1999; Gray and Ancey 2009; Iverson et al. 2011; Johnson et al. 2012; Iverson and Ouyang 2015; Lube et al. 2020).

Types and degrees of interaction vary and are particularly diverse in volcanic wet flows, where particle concentrations, sizes and size distributions, shapes, and densities vary greatly and where the fluid phase can suspend large quantities of very fine particles effectively becoming part of the fluid phase and increasing the viscosity and density of the fluid. Larger (> 2 mm) particles may or may not move independently of the fluid phase, depending on grain diameter and fluid viscosity and velocity. Thus, how particles and fluid mix and interact is highly dependent on both particle-size distribution and on solids concentration. Beyond these two critical factors, both of which can change during flow,

flow behavior is mainly determined by bulk flow momentum, inertial and frictional grain-interaction forces, fluid forces, and flow path topography (Denlinger and Iverson 2001, 2004; Iverson 1997, 2005; Iverson and Denlinger 2001; Iverson and Vallance 2001).

4.2 Dilute Versus Concentrated Particle–Fluid Flow

A simplistic way of contrasting dilute flows with concentrated flows is that in lowest-concentration flows, the flowing fluid moves the particles; in the highest-concentration flows, the flowing mass of particles moves the intergranular fluid (Middleton and Hampton 1973). Yet it is not quite that simple. The types and magnitudes of dynamic interactions between particles and fluid, as well as flow boundary conditions, control how particle–fluid flows move (i.e., their flow mechanics). Material properties governing these interactions include volumetric solids concentration of particles, internal and basal friction angles of particles, fluid viscosity, fluid compressibility, and hydraulic diffusivity—the rate at which excess internal pore-fluid pressures are dissipated (Iverson 1997, 2005; Iverson and Vallance 2001; Iverson and Denlinger 2001; Iverson et al. 2010).

In concentrated particle–fluid flow, interactions between particles (namely collisions and intergranular friction) commonly play a dominant role in momentum transport and the mechanics of flow, whereas the fluid plays a minor role, depending on the generation and retention of excess pore pressure (Iverson 1997, 2003, 2005; Iverson and Vallance 2001; Iverson and Denlinger 2001). Excess pore pressure retention is largely influenced by the volume concentration of particles and the hydraulic diffusivity (a function of particle-size distribution), which depends largely on permeability and compressibility of the mixture. Partitioning of momentum transfer between interparticle collisions and interparticle friction depends on shear rate, which changes over time in a flow and can loosely be approximated as the flow velocity divided by the flow depth. This partitioning affects a material's bulk resistance to flow, and thus, most particle–fluid flows do not have a fixed rheology[1] (Iverson 2003). At relatively low shear rates, frictional interactions between particles are dominant, whereas at high shear rates, interparticle collisions dominate over frictional contacts.

Fluid forces in dilute particle–fluid flow (viscous shear forces and dynamic pressure forces) dominate momentum transport and flow mechanics, although weak force interactions between solid particles, such as van der Waals and electrostatic forces, also operate in suspensions rich in very fine particles (termed *fines*—see Sect. 4.3 in Chap. "Deposit Identification and Interpretation") and may affect properties of the fluid and of the flow itself (Russel 1978; Middleton and Southard 1984; Iverson and Vallance 2001; Branney and Kokelaar 2002). In wet flows, the concentration of fines is critically important for determining flow type (VDF, VHF, or VWF) and flow behavior (Iverson 1997, 2005; Iverson and Vallance 2001; Vallance and Iverson 2015). Below a certain fines-concentration threshold, generally 11–12 vol% in poorly sorted mixtures of sediment typical of volcanic terrains, fluid suspensions of fine particles in water behave like clear flowing water (a Newtonian fluid) (Pierson 2005). *Newtonian fluids* are those that exhibit a linear proportionality between shear stress and shear rate; the proportionality constant is the fluid viscosity. Above 11–12 vol% fines, a water–sediment suspension commonly takes on yield strength and sometimes also nonlinear viscosity effects. Thus, the mixture becomes a non-Newtonian fluid (hyperconcentrated flow), although it still obeys Newton's laws of motion. As concentration increases further, a flow's ability to transport

[1] Rheology is the study of deformation and flow of matter, which can be applied to gases, liquids, and solids (Barnes et al. 1989). In the study of particle–fluid mixtures that can flow, the term is commonly used to describe a mixture's resistance to deformation at different shear rates (i.e., its viscosity) and any resistance to initiation of deformation (i.e., its shear strength).

additional (and coarser) sediment increases markedly and the flow transforms to debris flow (Hampton 1975; Pierson and Costa 1987; Pierson 2005).

Liquefaction and fluidization are two important mechanisms in volcanic flows, facilitated by particle–fluid mixtures having very poor sorting (containing a wide range of particle sizes) and an abundance of fines. In fines-rich mixtures, the connectivity between intergranular pore spaces is low, which reduces mixture hydraulic diffusivity. If, in addition, the viscosity of the pore fluid is high due to suspension of fines, hydraulic diffusivity is reduced even further within the mixture, and the escape of pore fluid during sediment consolidation and shearing is greatly hindered. Under such conditions, the weight of some or most of the solid particles is carried by the "trapped" (nearly static) or slowly escaping pore fluid, allowing the build-up of non-equilibrium or "excess" pore pressure, p. In other words, as the pore spaces get squeezed by the weight of the overlying deposit, the pore fluid cannot easily escape, and pore pressure increases to high values. High values of p result in greatly decreased frictional contacts between particles, and shear stress is partially (or largely) transferred to the fluid phase—a phenomenon usually termed *liquefaction* when applied to particle–liquid mixtures. Partially liquefied particle–fluid mixtures can flow almost like viscous liquids, even at relatively low shear rates, so long as sufficient fluid is trapped in the pore spaces between particles (Iverson and Vallance 2001). In cases where build-up of high intergranular pore-fluid pressure is rapid, but also where diffusivity of the mixture is high because fluid viscosity is low (i.e., a pore fluid such as dusty air can more easily escape), consolidation of the particle–fluid mixture drives a diffuse upward flow of pore fluid through the granular mass—a process usually termed *fluidization*, which is observed in particle-gas mixtures (Wilson 1980, 1984) and some subaqueous particle-liquid flows (Lowe 1976, 1982). The upward flow of fluid imparts upward viscous drag forces on particles, partially supporting the weight of the granular mass. This reduces intergranular friction and allows the particle mixture to flow more easily. The term *mass flow*—common in the geological literature—is generally applied to a contiguous and finite mass of a high-concentration particle–fluid mixture that is moving as a discrete flow event (Iverson and Vallance 2001).

Terminology of flow processes at volcanoes is evolving. In the past, use of different names for seemingly separate flow types has implied separate and distinct flow processes at work in each type of flow. However, growing understanding of what controls flow behavior suggests that transitions from one flow type to another are not entirely separate and distinct, and furthermore, the transitions are not abrupt. Flow types are found along continua—a continuum from low to high solids concentration and a continuum from a fluid phase containing only gas to one containing only liquid water. Our classification of flow processes (below) recognizes that gradational transitions occur in flow behavior as the concentration of solid particles increases or decreases and as fluid type (liquid or gas) changes. Such changes commonly occur as energy is added or subtracted from flowing solid–fluid mixtures (Branney and Kokelaar 2002; Vallance and Iverson 2015). A working hypothesis of this book is that gradational changes in solids concentration and properties of the fluid phase result in gradational changes in the physical characteristics of a volcanic flow deposit.

4.3 Range of Particle–Fluid Flow Types at Volcanoes

Two-phase gravity flows occurring at volcanoes include not only wet flows—lahars, debris avalanches, and water floods, but also several varieties of pyroclastic density currents.[2] In the following conceptual classification scheme

[2] In the past, concentrated pyroclastic density currents were termed 'pyroclastic flows' and dilute currents were termed 'pyroclastic surges.' These terms have been largely retired from use due to the recognition that (1) pyroclastic currents also occur on a continuum from high to low particle concentrations, and (2) many are concentration-stratified.

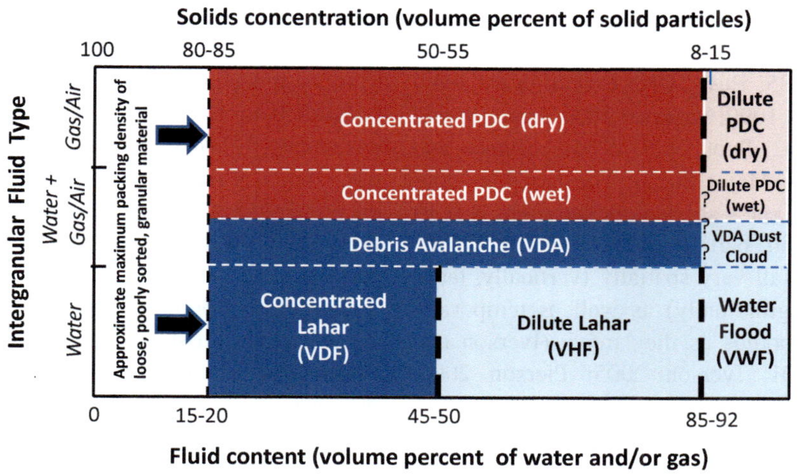

Fig. 1 Conceptual classification of the types of rapid particle–fluid flow that transport rock debris at volcanoes, based on relative proportions of solid particles and fluid and on pore-fluid composition. Red zones represent eruption-triggered flow types (typically hot or warm and transporting mainly primary volcaniclastic particles). Blue zones represent volcanic wet flows, typically composed of mainly secondary volcaniclastic and epiclastic particles. They may or may not be eruption triggered and they are usually cool. Dark colors represent concentrated particle–fluid flows; light colors represent dilute particle–fluid flows. Boundaries between all flow types are gradational. Concentration thresholds between concentrated and dilute flow types differ between particle–water and particle–gas flows. Approximate concentration boundaries between concentrated lahar and dilute lahar and between dilute lahar and water flood are from Pierson (2005). Approximate value for concentration boundary between concentrated and dilute dry pyroclastic density currents (PDCs) are from Branney and Kokelaar (2002) and Lube et al. (2020), although the latter reference recognizes a wide intermediate zone in the transition from concentrated flow to dilute flow. Values for concentration boundaries between concentrated and dilute flows involving mixed gas and liquid pore fluids is unknown. Note that values for solids concentration/fluid content are the average values for the whole vertical section, but many, if not most, of these flow types exhibit vertical concentration gradients when in motion

(Fig. 1), we consider types of volcanic particle–fluid flows, differentiated on the basis of solids concentration and on amounts of gas and liquid water in their fluid phases. Boundaries between flow types are recognized as gradational (e.g., Lube et al. 2020). The types of fluid are shown on the y-axis, and the x-axis shows concentration—both as ratio of solid particle volume to total volume (top axis) and fluid volume to total volume (bottom axis). Hot pyroclastic density currents are essentially dry, with a fluid phase composed of dusty air and exsolved gases, whereas PDCs that have cooled or erupted through water, as well as most debris avalanches, can have intergranular pore spaces occupied by a combination of gas and liquid water (Fisher and Schmincke 1984; Wilson and Houghton 2000; Brown and Andrews 2015). Most water-saturated VDFs (and probably some VHFs) also contain small amounts of air in their pore fluid, but the volumes of interstitial gas bubbles are trivial (< 5 vol%) (Major and Voight 1986). Concentrated lahar mixtures move mainly as debris flows, and dilute lahar mixtures move mainly as hyperconcentrated flows (see Chap. "Lahars—Process Overview").

A fundamental property of wet and dry volcanic flows is that they commonly undergo *flow transformation*[3]—change in flow type—as they add or subtract mass, gain or lose momentum,

[3] Some authors have defined "flow transformation" as a change from laminar to turbulent flow (e.g., Fisher 1983), but we do not use the term in that way.

and change their proportions of fluid and solid components as they move downslope (Pierson and Scott 1985; Best 1992; Cronin et al. 1999, 2000; Pierson 1999; Vallance 2000; Branney and Kokelaar 2002; Scott et al. 2005; Vallance and Iverson 2015; Major 2023). The variables controlling flow type—solids concentration, type of interstitial fluid, particle size, amount of kinetic energy, etc.—all vary spatially (vertically, laterally, and longitudinally) as well as temporally within flow bodies as they move (Iverson and Vallance 2001; Iverson 2005; Pierson 2005; Vallance and Iverson 2015; Major 2023). Thus, any volcanic flow can simultaneously involve multiple flow types in different parts of the flow body. Furthermore, some flows evolve from one primary flow type to another with time and distance traveled as sediment is entrained or deposited or as additional water is ingested. It follows that characteristics of volcanic flow deposits greatly depend on where, when, and how sediment was emplaced along a flow path.

5 Deposits of Particle–Fluid Flows at Volcanoes

Flow type largely determines the properties of volcaniclastic deposits because each flow type has characteristic varieties and degrees of interaction among particles during both flow and deposition. Two important concentration-related determinants of how particles are deposited are (1) hindered gravitational settling of solid particles through the fluid phase (mainly in relatively dilute mixtures) and (2) kinetic sieving (in high concentration mixtures). These are discussed further in individual process chapters. Degree of hindering in settling particles affects stratification in deposits and is controlled by particle properties (size, shape, density, vesicularity, concentration) and by fluid viscosity and turbulence (Dietrich 1982; Major 2003; Pierson 2005; Tomkins et al. 2005; te Slaa et al. 2015). Shear driven kinetic sieving leads to inverse grading by particle size (coarse particles move toward top surface of a flow) and the formation of flow levees (Gray and Ancey 2009; Johnson et al. 2012; Gray 2018). Note that particles suspended in dilute flows commonly have more freedom to move independently of each other and to better sort themselves by size and density than in concentrated flows. However, high-concentration zones of particle movement may still be encountered at the lower flow boundary of dilute flows (e.g., Manville and White 2003; Church and Jakob 2020), so mechanisms operative at the lower flow boundary are most important for deposit characteristics. In general, flow type is largely responsible for deposit features such as presence or absence of stratification, deposit grading and fabric, and sedimentary texture.

Depositional mechanics is not the only determinant of the characteristic of deposits encountered in the field. The structural arrangement of particles also may be altered by gravitational or shear deformation, or by uneven compaction during post-depositional consolidation, by melting out of ice grains, and by pore-fluid escape. Syn- and post-depositional erosion of deposits also alters fresh deposits, as does syn- and post-depositional chemical alteration and recrystallization.

6 Why Interpretation of Volcanic Wet-Flow Deposits Is Important

Rapid flows of two-phase mixtures of solids and fluid, both wet and dry varieties, can pose great risks to people living on or near volcanoes (Tanguy et al. 1998; Witham 2005; Auker et al. 2013). Volcanic wet flows are particularly hazardous due to their ability to travel far downstream of source volcanoes and to the element of surprise they hold when not triggered by eruptions. Lahars can be lethal at more than 100 km from source. For example, in Colombia more than 23,000 people were killed in 1985 by lahars at distances between 70 and 105 km downstream of Nevado del Ruiz volcano in three different river valleys (Lowe et al. 1986; Naranjo et al. 1986; Pierson et al. 1990; Voight 1990; Tanguy et al. 1998). Extremely large lahars can reach even farther—a massive prehistoric lahar

Fig. 2 House in town of Chaitén, Chile, buried by volcanic water-flood deposits within 48 h of the onset of rain following the explosive phase of the 2008–2009 eruption of Chaitén volcano. Rapid channel aggradation and inundation of fan surface on which town was built had not been an anticipated hazard here, even though older deposits on fan showed that it had happened before. USGS photo by T. Pierson, 2010

in Ecuador traveled >300 km from its source (Mothes et al. 1998).

Hazards posed by volcanic wet flows cannot be recognized if their deposits are mistaken for those of other depositional processes. Accurate and effective hazard assessment at volcanoes depends on our ability to recognize past occurrences of different processes based on the deposits preserved in outcrops in the field (Fig. 2). Even though numerical simulations of flow processes have advanced tremendously in their ability to predict flow paths, mass accretion, runout distances, and travel times of particulate flows (e.g., Iverson et al. 1998; Iverson and Denlinger 2001; Denlinger and Iverson 2001, 2004; George and Iverson 2014; Iverson and George 2014; George et al. 2022), the field identification of deposits remains critical for knowing what specific types of flows have reached vulnerable areas in the past and could do so again.

Volcanic wet flows can have direct adverse impacts to vulnerable communities in downstream valleys. But they can also trigger serious indirect hazards (Fig. 3), such as:

- Tsunamis triggered by debris avalanches entering water bodies (e.g., Johnson 1987; Kienle et al. 1987; Smith and Shepherd 1996; Ward and Day 2003; Day 2015).
- Tributary blockage by volcanic wet-flow deposits with consequent tributary valley back-flooding (e.g., Umbal and Rodolfo 1996; Kataoka et al. 2008) and later additional lahars from sudden lake outbreaks (e.g., Umbal and Rodolfo 1996; O'Connor et al. 2001; Manville et al. 1999; Manville 2004).
- Delayed posteruptive sedimentation that causes river channels to aggrade, which can bury valley floors in sediment and induce

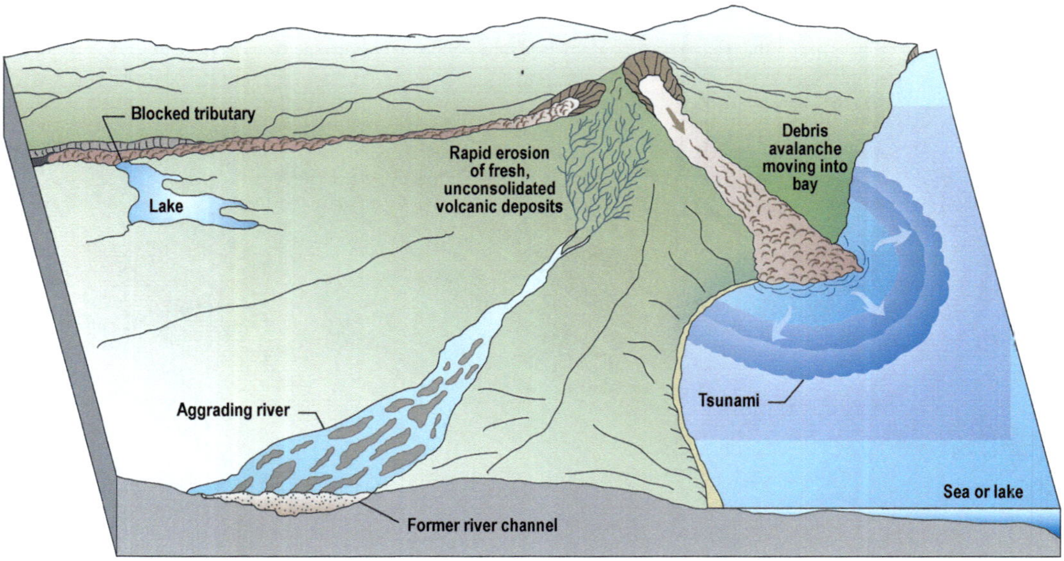

Fig. 3 Schematic representation of sediment deposition by volcanic wet flows and several common indirect hazards associated with those processes

flooding by reducing channel capacity years to decades after eruptive activity (e.g., Kuenzi et al. 1979; White et al. 1997; Hayes et al. 2002; Pierson et al. 2011).

Different direct and indirect volcano hazards can endanger lives and property to different degrees, in different locations, and on different time scales, so our ability to correctly interpret process from deposits is vitally important.

7 Why Interpretation of Volcanic Wet-Flow Deposits Is Difficult

Volcanic wet-flow deposits, even when examined at good outcrops in the field (Fig. 4), can be difficult to differentiate, not only from each other but also from deposits of pyroclastic density currents, pyroclastic falls, and glaciers that can contain the same types and sizes of volcanic particles. Physical differences between all these types of deposits can be subtle. In addition, Orton (1996) identified other reasons that complicate the jobs of volcanologists trying to interpret deposits of volcanic rock particles:

- A wide variety of both eruptive and noneruptive transport and depositional processes can produce fragmental deposits with similar characteristics.
- Depositional processes commonly evolve with time and distance traveled, causing deposits with different characteristics to overlap spatially and grade into each other.
- Textural and compositional variations can have multiple causes.
- Mineral grains and glass fragments are susceptible to chemical alteration in harsh volcanic environments, which can change deposit appearance.
- Localized flow paths and nonuniform rates of sediment supply and transport can result in spatially irregular deposits.
- Deposits can be thin and easily eroded, thus having ephemeral residence times in the geologic record.
- Deposits are vulnerable to disruption and burial by subsequent volcanic or volcano-tectonic activity and processes.

Fig. 4 Sequence of lahar and volcanic water-flood deposits at "Old Maid Flat," downstream of Mount Hood volcano (Oregon, USA) in the upper Sandy River, Oregon. This outcrop shows a variety of sedimentary textures and bedding characteristics. Vertical face is about 6 m tall. Active river floodplain is in foreground. USGS photo by T. Pierson

Until now, deposits of volcanic wet flows have not been systematically characterized, compared, and photographed to allow one to distinguish them from one another and from the other major types of deposits composed of volcanic rock particles. This is a shortfall we attempt to address with this book.

To offer insight into some specific challenges of deposit identification and stratigraphic sequencing involving subtle differences in characteristics, two real-life examples of "head-scratching" moments experienced by volcanologists doing field work are offered below. They have been contributed by experienced research scientists Chris Harpel and Jon Major—colleagues of ours from the U.S. Geological Survey Cascades Volcano Observatory.

7.1 Field Examples

7.1.1 Example 1: On the Flank of El Misti Volcano, Peru (as told by Chris Harpel)

Our team ran across an outcrop that presented us with a key interpretive challenge. A colleague had visited this outcrop a decade earlier and collected a charcoal sample from a paleosol at the base of the relatively narrow exposure (Fig. 5). We dated several other charcoal samples to confirm the age of the paleosol. Overlying the paleosol was a diamict interpreted as a PDC deposit (block-and-ash flow deposit with monomictic dense-clast and ashy matrix componentry). A thin, cross-bedded, fine-grained layer at its base appeared to be a coeval

Fig. 5 Mystery outcrop at El Misti volcano (Peru). Arrows mark locations of two colonial-period potsherds having pottery styles and associated charcoal suggesting different ages—one appearing to be above the contact and the other below. Knowing this, the question in the field was, "Is there a contact between the two?" Upper horizon of the paleosol is the gray band just above the top of the blue pack. USGS photo by C. Harpel

ash-cloud surge deposit. Charcoal from both the surge deposit and top of the PDC deposit, which cropped out several meters away yielded ages consistent with the underlying paleosol. So far everything was making sense.

The interpretive conundrum arose when we returned the following year and noticed that more slope wash had been cleaned away from the outcrop face, which expanded its lateral extent and better revealed the deposits. In this newly exposed outcrop we discovered colonial-age pottery fragments embedded in what we initially thought was the PDC deposit. The pottery age was based on both radiocarbon dating of more charcoal found with the potsherds and an archaeological assessment of the pottery style, all of which seemed to suggest that we were facing two contradictory ages for the PDC unit.

Clearly, we needed to examine the outcrop face more closely. Upon doing so, we noticed that an irregular contact was present in what we thought was the PDC unit. Despite having componentry and a grain-size distribution very similar to the underlying PDC deposit, the overlying ceramic-bearing portion had occasional larger accidental lithics and was a subtly different color than the PDC deposit. Such characteristics led us to interpret the ceramic-bearing unit above the contact as a colonial-age volcanic debris-flow deposit. The contact was an erosional unconformity that represented the sidewall of a gully filled with the debris-flow deposit. As the debris flow eroded this gully into the PDC deposit, it entrained the eroded PDC material giving the two very similar characteristics. The color difference likely resulted from

Fig. 6 Crudely stratified deposit of vitric rhyolite gravel-size clasts in coarse sandy matrix material that was adjacent to an outcrop of the same material that was unstratified and contained charred and uncharred vegetation. Does this outcrop show pulsatory behavior of a dome-collapse lithic pyroclastic density current (PDC) or fluvial reworking of the adjacent massive PDC deposit? Shovel blade is 16 cm wide. USGS photo by T. Pierson

soil that had also been eroded and incorporated into the debris flow. We had not seen this initially because the gully wall had been dusted over by slope wash and we had not been able to see either the subtle color or textural differences between the two units. Our takeaways were: (1) expose as much of an outcrop face as possible and clean it off well, (2) date multiple stratigraphic horizons whenever possible, (3) even slight changes in componentry can be significant, and (4) note color differences, even if subtle.

7.1.2 Example 2: Downstream of Chaitén Volcano, Chile (as told by Jon Major)

In the aftermath of the 2008–2009 eruption of Chaitén volcano, teams of geologists from the United States (USGS) and Chile (SERNAGEOMIN) worked together over two field seasons to decipher the geologic record of volcanic processes that impacted the Chaitén (Blanco) River valley downstream of the volcano. Two important facts to consider were that a torrential rainstorm hit that area just as the initial explosive phase of the eruption was ending and that a PDC (with accompanying ash cloud) was photographed descending the valley in February 2009—a PDC that came within 3 km of entering the still-occupied town of Chaitén at the river's mouth. What appeared in places to be a ~10-m-thick single deposit of polymictic, well-compacted sandy gravel of angular vitric rhyolite (multiple colors, vesicularities, and glass contents) occupied much of the valley floor. It was mainly massive in structure, but in other places it appeared to have crude stratification (Fig. 6). Was this reflecting pulsatory PDC flowage or fluvial reworking? Was it emplaced entirely by the 2009 PDC, but if so, why were some trees and woody debris engulfed by the deposit charred, and others were not? Was the seemingly single flow unit emplaced by two separate flows instead of one, but if two, why was a contact not readily apparent everywhere? And if two flow units, was the lower one emplaced by a lahar or an earlier PDC (there appeared to be only one ash-cloud deposit on the valley side)? The questions had important hazards implications for the town downstream.

*Many outcrops were scraped clean by the research teams to try to answer these questions. These outcrops started at the mouth of the steep narrow channel (scoured clean down to bedrock) that descended from the one narrow breach in the caldera wall about a kilometer upstream. In the end, there were two critical sources of information to help solve the most of the puzzle: (1) the discovery of a single outcrop—the team's "Rosetta Stone" outcrop, which contained all the stratigraphic units (major flow units, flow units from smaller localized flows, and tephra layers) in one vertical sequence and (2) access to detailed satellite imagery that showed circumferential flowage of PDCs in the caldera moat to the caldera breach. In the published analysis of these volcaniclastic sediments (*Major et al. 2013*), it was the stratigraphic relations, detailed textural analysis of the sediment, patient searching for unit contacts, and the vertical distribution and degree of charring on standing tree trunks that led to the conclusion that the lower unit was also a PDC deposit, smaller and cooler than the upper unit. Both originated as partial collapses of the compound rhyolite lava dome during active dome growth. The lahar that impacted the town early in the eruption (*Pierson et al. 2013*) came not from the crater but from the tephra mantle in the surrounding upper watershed that was remobilized by heavy rainfall prior to either of the PDCs.*

Takeaways for the team were: (1) Proximal PDCs and lahars from the same source cannot always be distinguished on the basis of sedimentary texture; (2) outcrops showing at least most of the sequence of events may be essential for interpretation; (3) PDCs can be highly erosive in steep channels and not necessarily hot enough to char vegetation; (4) effects of flows on forest trees can yield important information; (5) post-emplacement fluvial erosion and reworking of sediment can create needed outcrop exposures but also complicate the stratigraphy with new depositional units; (6) proximal sites may contain "too much" information (records of insignificant small local events as well as important major events); and (7) entertaining multiple working hypotheses while scraping and examining outcrops can be helpful. In the end, two short field seasons was not enough to answer all the questions.

References

Auker M, Sparks RSJ, Siebert L, Crosweller HS, Ewert J (2013) A statistical analysis of the global historical volcanic fatalities record. J Appl Volcanol 2(2). https://www.appliedvolc.com/content/2/1/2

Barnes HA, Hutton JF, Walters K (1989) An introduction to rheology, vol 3. Elsevier, Amsterdam

Best JL (1992) Sedimentology and event timing of a catastrophic volcaniclastic mass flow, Volcan Hudson, Southern Chile. Bull Volcanol 54:299–318

Bonadonna C, Costa A, Folch A, Koyaguchi T (2015) Tephra dispersal and sedimentation. In: Sigurdsson H, Houghton B, McNutt SR, Rymer H, Stix J (eds) The encyclopedia of volcanoes, 2nd edn. Elsevier, Amsterdam, pp 587–597

Branney MJ, Kokelaar P (1992) A reappraisal of ignimbrite emplacement: progressive aggradation and changes from particulate to non-particulate flow during emplacement of high-grade ignimbrite. Bull Volcanol 54:504–520

Branney MJ, Kokelaar P (2002) Pyroclastic density currents and the sedimentation of ignimbrites. Geol Soc Lond Mem 27:1–143

Brown RJ, Andrews DM (2015) Deposits of pyroclastic density currents. In: Sigurdsson H, Houghton B, McNutt SR, Rymer H, Stix J (eds) The encyclopedia of volcanoes, 2nd edn. Elsevier, Amsterdam, pp 631–648

Cas R, Busby-Spera C (eds) (1991) Volcaniclastic sedimentation. Sediment Geol 74(1–4):1–362

Cas RAF, Wright JV (1987) Volcanic successions modern and ancient. Allen and Unwin, London, p 528

Chang CS, Wang J-Y, Ge L (2015) Modeling of minimum void ratio for sand–silt mixtures. Eng Geol 196:293–304. https://doi.org/10.1016/j.enggeo.2015.07.015

Church M, Jakob M (2020) What is a debris flood? Water Resour Res 56(8):e2020WR027144. https://doi.org/10.1029/2020WR027144

Cohen H, Laronne JB (2005) High rates of sediment transport by flashfloods in the Southern Judean Desert, Israel. Hydrol Process 19:1687–1702. https://doi.org/10.1002/hyp.5630

Costa JE (1988) Rheologic, geomorphic, and sedimentologic differentiation of water floods, hyperconcentrated flows, and debris flows. In: Baker VR, Kochel RC, Patton PC (eds) Flood geomorphology. Wiley, New York, pp 113–122

Cronin SJ, Neall VE, Lecointre JA, Palmer AS (1999) Dynamic interactions between lahars and stream flow: a case study from Ruapehu volcano, New Zealand. Geol Soc Am Bull 111:28–38

Cronin SJ, LeCointre JA, Palmer AS, Neall VE (2000) Transformation, internal stratification, and depositional processes within a channelized, multi-peaked lahar flow. New Zeal J Geol Geophys 43:117–128

Day SJ (2015) Volcanic tsunamis. In: Sigurdsson H, Houghton B, McNutt SR, Rymer H, Stix J (eds) The encyclopedia of volcanoes, 2nd edn. Elsevier, Amsterdam, pp 993–1009

Denlinger RP, Iverson RM (2001) Flow of variably fluidized granular masses across three-dimensional terrain: 2. Numerical predictions and experimental tests. J Geophys Res B 106:553–566

Denlinger RP, Iverson RM (2004) Granular avalanches across irregular three-dimensional terrain: 1. Theory and computation. J Geophys Res F 109:1–14. https://doi.org/10.1029/2003JF000085

Dietrich WE (1982) Settling velocity of natural particles. Water Resour Res 18(6):1615–1626. https://doi.org/10.1029/WR018i006p01615

Dufek J, Ongaro TE, Roche O (2015) Pyroclastic density currents: processes and models. In: Sigurdsson H, Houghton B, McNutt SR, Rymer H, Stix J (eds) The encyclopedia of volcanoes, 2nd edn. Elsevier, Amsterdam, pp 617–629

Fisher RV (1961) A proposed classification of volcaniclastic sediments and rocks. Geol Soc Am Bull 72:1409–1414

Fisher RV (1966) Rocks composed of volcanic fragments and their classification. Earth-Sci Rev 1:287–298

Fisher RV (1983) Flow transformations in sediment gravity flows. Geology 11:273–274

Fisher RV, Schmincke H-U (1984) Pyroclastic rocks. Springer-Verlag, Berlin, p 472

Frings RM, Schüttrumpf H, Vollmer S (2011) Verification of porosity predictors for fluvial sand-gravel deposits. Water Resour Res 47:W07525. https://doi.org/10.1029/2010WR009690

George DL, Iverson RM (2014) A depth-averaged debris-flow model that includes the effects of evolving dilatancy. II. Numerical predictions and experimental tests. Proc R Soc A 470:20130820. https://doi.org/10.1098/rspa.2013.0820

George DL, Iverson RM, Cannon CM (2022) Modeling the dynamics of lahars that originate as landslides on the west side of Mount Rainier, Washington. U S Geol Surv Open-File Rep 2021-1118. https://doi.org/10.3133/ofr20211118

Glicken H (1996) Rockslide debris avalanche of May 18, 1980, Mount St. Helens Volcano, Washington. U S Geol Surv Open-File Rep 96-677, 90 pp and 5 plates

Gray JMNT (2018) Particle segregation in dense granular flows. Annu Rev Fluid Mech 50:407–433. https://doi.org/10.1146/annurev-fluid-122316-045201

Gray JMNT, Ancey C (2009) Segregation, recirculation and deposition of coarse particles near two-dimensional avalanche fronts. J Fluid Mech 629:387–423. https://doi.org/10.1017/S0022112009006466

Hampton MA (1975) Competence of fine-grained debris flows. J Sediment Petrol 45:834–844

Hayes SK, Montgomery DR, Newhall CG (2002) Fluvial sediment transport and deposition following the 1991 eruption of Mount Pinatubo. Geomorphology 45:211–224

Iverson RM (1997) The physics of debris flows. Rev Geophys 35:245–296

Iverson RM (2003) The debris-flow rheology myth. In: Rickenmann D, Chen CL (eds) Debris-flow hazards mitigation. Mechanics, prediction, and assessment (1). Millpress, Rotterdam, pp 303–314

Iverson RM (2005) Debris-flow mechanics. In: Jakob M, Hungr O (eds) Debris-flow hazards and related phenomena. Praxis-Springer, Berlin, pp 103–134

Iverson RM, Denlinger RP (2001) Flow of variably fluidized granular masses across three-dimensional terrain. 1. Coulomb mixture theory. J Geophys Res B 106:537–552

Iverson RM, George DL (2014) A depth-averaged debris-flow model that includes the effects of evolving dilatancy. I. Physical basis. Proc R Soc A 470(2170):20130819. https://doi.org/10.1098/rspa.2013.0819

Iverson RM, Ouyang C (2015) Entrainment of bed material by earth-surface mass flows: review and reformulation of depth-integrated theory. Rev Geophys 53:27–58

Iverson RM, Vallance JW (2001) New views of granular mass flows. Geology 29:115–118

Iverson RM, Schilling SP, Vallance JW (1998) Objective delineation of lahar inundation hazard zones. Geol Soc Am Bull 110:972–984

Iverson RM, Logan M, LaHusen RG, Berti M (2010) The perfect debris flow? Aggregated results from 28 large-scale experiments. J Geophys Res 115:F03005. https://doi.org/10.1029/2009JF001514

Iverson RM, Reid ME, Logan M, LaHusen RG, Godt JW, Griswold JP (2011) Positive feedback and momentum growth during debris-flow entrainment of wet bed sediment. Nat Geosci 4:116–121. https://doi.org/10.1038/ngeo1040

Johnson RW (1987) Large-scale volcanic cone collapse: the 1888 slope failure of Ritter volcano, and other examples from Papua New Guinea. Bull Volcanol 49:669–679. https://doi.org/10.1007/BF01080358

Johnson CG, Kokelaar BP, Iverson RM, Logan M, LaHusen RG, Gray JMNT (2012) Grain-size segregation and levee formation in geophysical mass flows. J Geophys Res Earth Surf 117:F01032. https://doi.org/10.1029/2011JF002185

Kataoka KS, Urabe A, Manville V, Kajiyama A (2008) Breakout flood from an ignimbrite-dammed valley after the 5 ka Numazawako eruption, northeast Japan. Geol Soc Am Bull 120:1233–1247

Kienle J, Kowalik Z, Murty TS (1987) Tsunamis generated by eruptions from Mount St. Augustine volcano, Alaska. Science 236:1442–1447

Kokelaar BP, Howells MF (eds) (1984) Marginal basin geology: volcanic and associated sedimentary and tectonic processes in modern and ancient marginal basins. Geol Soc Lond Spec Publ 16

Kuenzi WD, Horst OH, McGehee RV (1979) Effect of volcanic activity on fluvial-deltaic sedimentation in a modern arc-trench gap, southwestern Guatemala. Geol Soc Am Bull 90:827–838

Lambe TW, Whitman RV (1969) Soil mechanics. Wiley, New York, p 553

Lekach J, Schick AP (1982) Suspended sediment in desert floods in small catchments. Israel J Earth Sci 31:144–156

Lowe DR (1976) Subaqueous liquified and fluidized sediment flows and their deposits. Sedimentology 23:285–308

Lowe DR (1982) Sediment gravity flows; II. Depositional models with special reference to the deposits of high-density turbidity currents. J Sediment Res 52(1):279–297

Lowe DR, Williams SN, Leigh H, Connor CB, Gemmell JB, Stoiber RE (1986) Lahars initiated by the 13 November 1985 eruption of Nevado del Ruiz, Colombia. Nature 324:51–53

Lube G, Breard ECP, Esposti-Ongaro T, Dufek J, Brand B (2020) Multiphase flow behaviour and hazard prediction of pyroclastic density currents. Nat Rev Earth Environ 1:348–365. https://doi.org/10.1038/s43017-020-0064-8

Major JJ (1997) Depositional processes in large-scale debris-flow experiments. J Geol 105:345–366

Major JJ (2000) Gravity-driven consolidation of granular slurries—implications for debris-flow deposition and deposit characteristics. J Sediment Res 70:64–83

Major JJ (2003) Hindered settling. In: Middleton GV, Church MJ, Coniglio M, Hardie LA, Longstaffe FJ (eds) Encyclopedia of sediments and sedimentary rocks. Kluwer Academic Publishers, Dordrecht, pp 358–360

Major JJ (2023) Subaerial volcaniclastic deposits—influences of initiation mechanisms and transport behaviour on characteristics and distributions. In: Di Capua A, De Rosa R, Kereszturi G, Le Pera E, Rosi M, Watt SFL (eds) Volcanic processes in the sedimentary record: when volcanoes meet the environment. Geol Soc Lond Spec Publ 520:29–100. First published online 12 July 2022. https://doi.org/10.1144/SP520-2021-142

Major JJ, Iverson RM (1999) Debris-flow deposition: effects of pore-fluid pressure and friction concentrated at flow margins. Geol Soc Am Bull 111:1424–1434

Major JJ, Voight B (1986) Sedimentology and clast orientations of the 18 May 1980 southwest-flank lahars, Mount St. Helens, Washington. J Sediment Petrol 56:691–705

Major JJ, Pierson TC, Hoblitt RP, Moreno H (2013) Pyroclastic density currents associated with the 2008–2009 eruption of Chaitén volcano (Chile)—forest disturbances, deposits, and dynamics. Andean Geol 40(2):324–358

Manville VR (2004) Paleohydraulic analysis of the 1953 Tangiwai lahar: New Zealand's worst disaster. Acta Vulcanol 16:137–152

Manville VR, White JDL (2003) Granular mass flows at the base of sediment-laden floods, and the roles of flow competence and flow capacity in the deposition of stratified boulder sands. Sediment Geol 155:157–173

Manville V, White JDL, Houghton BF, Wilson CJN (1999) Paleohydrology and sedimentology of a post-1.8 ka breakout flood from intracaldera Lake Taupo, North Island, New Zealand. Geol Soc Am Bull 111:1435–1447

Manville V, Nemeth K, Kano K (2009) Source to sink: a review of three decades of progress in the understanding of volcaniclastic processes, deposits, and hazards. Sediment Geol 220:136–161

McPhie J, Doyle M, Allen R (1993) Volcanic textures—a guide to the interpretation of textures in volcanic rocks. Centre for Ore Deposit and Exploration Studies, University of Tasmania, Hobart, p 196

Middleton GV, Hampton MA (1973) Sediment gravity flows: mechanics of flow and deposition. In: Middleton GV, Bouma AH (eds) Turbidities and deep water sedimentation. Soc Econ Paleontol Mineral, Pacific Sect, Short Course, pp 1–38

Middleton GV, Southard JB (1984) Mechanics of sediment movement. In: SEPM short course 3, 2nd edn. Providence, 370 pp

Mothes PA, Hall ML, Janda RJ (1998) The enormous Chillos Valley lahar: an ash-flow-generated debris flow from Cotopaxi volcano, Ecuador. Bull Volcanol 59:233–244

Naranjo JL, Sigurdsson H, Carey SN, Fritz WJ (1986) Eruption of the Nevado del Ruiz volcano, Colombia, on 13 November 1985: tephra fall and lahars. Science 233:961–963

O'Connor JE, Hardison III JH, Costa JE (2001) Debris flows from failures of Neoglacial-age moraine dams in the Three Sisters and Mt. Jefferson Wilderness areas, Oregon. U S Geol Surv Prof Pap 1606:1–93

Orton GJ (1996) Volcanic environments. In: Reading HG (ed) Sedimentary environments: processes, facies and stratigraphy. Blackwell Science, London, pp 485–567

Pierson TC (1999) Transformation of water flood to debris flow following the eruption-triggered transient-lake breakout from the crater on March 19, 1982. In: Pierson TC (ed) Hydrologic consequences of hot-rock/snowpack interactions at Mount St. Helens volcano, Washington 1982–84. U S Geol Surv Prof Pap 1586:19–36

Pierson TC (2005) Hyperconcentrated flows—transitional process between water flow and debris flow, chap 8. In: Jakob M, Hungr O (eds) Debris-flow hazards and related phenomena. Praxis-Springer, Berlin, pp 159–202

Pierson TC, Costa JC (1987) A rheologic classification of subaerial sediment-water flows. Geol Soc Am Rev Eng Geol 7:1–12

Pierson TC, Scott KM (1985) Downstream dilution of a lahar: transition from debris flow to hyperconcentrated streamflow. Water Resour Res 21:1511–1524

Pierson TC, Janda RJ, Thouret JC, Borrero CA (1990) Perturbation and melting of snow and ice by the 13 November 1985 eruption of Nevado del Ruiz, Colombia, and consequent mobilization, flow, and deposition of lahars. J Volcanol Geotherm Res 41:17–66

Pierson TC, Pringle PT, Cameron KC (2011) Magnitude and timing of downstream channel aggradation in response to a dome-building eruption at Mount Hood, Oregon. Geol Soc Am Bull 123:3–20

Pierson TC, Major JJ, Amigo Á, Moreno H (2013) Acute sedimentation response to rainfall following the explosive phase of the 2008–2009 eruption of Chaitén volcano, Chile. Bull Volcanol 75:723. https://doi.org/10.1007/s00445-013-0723-4

Reid I, Frostick LE (1987) Flow dynamics and suspended sediment properties in arid zone flash floods. Hydrol Process 1:239–253

Reid I, Laronne JB, Powell DM (1998) Flash-flood and bedload dynamics of desert gravel-bed streams. Hydrol Process 12:543–557

Russel W (1978) The rheology of suspensions of charged rigid spheres. J Fluid Mech 85(2):209–232. https://doi.org/10.1017/S0022112078000609

Scott KM, Vallance JW, Kerle N, Macías JL, Strauch W, Devoli G (2005) Catastrophic precipitation-triggered lahar at Casita volcano, Nicaragua: occurrence, bulking and transformation. Earth Surf Process Landf 30:59–79. https://doi.org/10.1002/esp.1127

Smith GA, Lowe DR (1991) Lahars: volcano-hydrologic events and deposition in the debris flow—hyperconcentrated flow continuum. In: Fisher RV, Smith GA (eds) Sedimentation in volcanic settings. SEPM (Soc Sediment Geol) Spec Publ 45:59–70

Smith MS, Shepherd JB (1996) Tsunami waves generated by volcanic landslides: an assessment of the hazard associated with Kick 'em Jenny. In: McGuire WJ, Jones AP, Neuberg J (eds) Volcano instability on the earth and other planets. Geol Soc Lond Spec Publ 110:115–123

Sohn C, Sohn YK (2019) Distinguishing between primary and secondary volcaniclastic deposits. Sci Rep 9:12425. https://doi.org/10.1038/s41598-019-48933-4

Stix J (1991) Subaqueous, intermediate to silicic-composition explosive volcanism; a review. Earth-Sci Rev 31:21–35

Tanguy J-C, Ribière C, Scarth A, Tjetjep WS (1998) Victims from volcanic eruptions: a revised database. Bull Volcanol 60:137–144

te Slaa S, van Maren DS, He Q, Winterwerp JC (2015) Hindered settling of silt. J Hydraulic Eng 141(9). https://doi.org/10.1061/(ASCE)HY.1943-7900.0001038

Tomkins MR, Baldock TE, Nielsen P (2005) Hindered settling of sand grains. Sedimentology 52(6):1425–1432. https://doi.org/10.1111/j.1365-3091.2005.00750.x

Ui T, Takarada S, Yoshimoto M (2000) Debris avalanches. In: Sigurdsson H, Houghton B, McNutt SR, Rymer H, Stix J (eds) Encyclopedia of volcanoes. Academic Press, San Diego, pp 617–626

Umbal JV, Rodolfo KS (1996) The 1991 lahars of southwestern Mount Pinatubo and evolution of the lahar-dammed Mapanuepe Lake. In: Newhall CG, Punongbayan RS (eds) Fire and mud: eruptions and lahars of Mount Pinatubo, Philippines. Philippine Institute of Volcanology and Seismology, Quezon City, and University of Washington Press, Seattle, pp 951–970

Vallance JW (2000) Lahars. In: Sigurdsson H, Houghton BF, McNutt SR, Rymer H, Stix J (eds) Encyclopedia of volcanoes. Academic Press, San Diego, pp 601–616

Vallance JW, Iverson RM (2015) Lahars and their deposits. In: Sigurdsson H, Houghton B, McNutt SR, Rymer H, Stix J (eds) The encyclopedia of volcanoes, 2nd edn. Elsevier, Amsterdam, pp 649–664. https://doi.org/10.1016/B978-0-12-385938-9.00037-7

van Wyk de Vries B, Davies TR (2015) Landslides, debris avalanches, and volcanic gravitational deformation. In: Sigurdsson H (ed) The encyclopedia of volcanoes, 2nd edn. Elsevier, Amsterdam, pp 665–685. https://doi.org/10.1016/B978-0-12-385938-9.00038-9

Voight B (1990) The 1985 Nevado del Ruiz volcano catastrophe: anatomy and retrospection. J Volcanol Geotherm Res 42(1–2):151–188

Ward SN, Day S (2003) Ritter Island volcano—lateral collapse and the tsunami of 1888. Geophys J Int 154:891–902

Waythomas CF, Pierson TC, Major JJ, Scott WE (2013) Voluminous ice-rich lahars generated during the 2009 eruption of Redoubt Volcano, Alaska. J Volcanol Geotherm Res 259:389–413

White JDL, Houghton BF (2006) Primary volcaniclastic rocks. Geology 34:677–680

White JDL, Riggs NR (eds) (2001) Volcaniclastic sedimentation in lacustrine settings. Spec Publ No.

30. International Association of Sedimentologists, Blackwell Scientific, Oxford, 309 pp

White, JDL, Houghton BF, Hodgson KA, Wilson CJN (1997) Delayed sedimentary response to the A.D. 1886 eruption of Tarawera, New Zealand. Geology 25(5):459–452

Wilson CJN (1980) The role of fluidization in the emplacement of pyroclastic flows: an experimental approach. J Volcanol Geotherm Res 8:231–249

Wilson CJN (1984) The role of fluidization in the emplacement of pyroclastic flow, 2. Experimental results and their interpretation. J Volcanol Geotherm Res 20:55–84

Wilson CJN, Houghton BF (2000) Pyroclast transport and deposition. In: Sigurdsson H, Houghton BF, McNutt SR, Rymer H, Stix J (eds) Encyclopedia of volcanoes. Academic Press, San Diego, pp 545–554

Witham C (2005) Volcanic disasters and incidents: a new database. J Volcanol Geotherm Res 148:191–233

Scaling and Timing of Volcanic Wet Flows in Response to Eruptions

Abstract

The distribution of volcanic and volcaniclastic deposits in both space and time gives important clues to their origins. Distance from volcanic source, depositional setting, and stratigraphic associations are important spatial characteristics. Timing of volcanic flows with respect to associated eruptions, as well as timing of flows with respect to each other, provide additional insight into emplacement processes. Magnitudes of lahars and volcanic water floods are highly influenced by size and type of associated eruptions, which affect the extent, severity, and character of landscape disturbance, including the particle size, bulk volume, and spatial distribution of fresh primary deposits. Magnitudes of volcanic debris avalanches, which may or may not be triggered by eruptions, are highly influenced by height and volume of source volcanic edifices, by the tectonic setting and strength of a source edifice's foundation, and by the structural integrity of source edifices themselves, which can be greatly reduced by hydrothermal alteration of the constituent rock. All three types of volcanic wet flows are capable of mobilizing sediment in volcanic drainage basins at volumes per unit drainage area and per unit time far in excess of volumes of sediment mobilized by most nonvolcanic processes.

Much of a volcano's eruptive history and its potential for producing hazards are recorded in deposits of volcanic sediments that are distributed in both space and time, relative to source-vent location and eruptive activity. Location and timing can be different for volcanic wet flows than for processes emplacing pyroclastic or glacial deposits. This chapter aims to provide the spatial and temporal context for the deposits of mostly unconsolidated volcanic sediments produced by lahars, volcanic debris avalanches, and volcanic water floods.

Natural archives of past eruptions and hazards are commonly preserved both in sequences of lava flows and eruptive deposits on volcanic edifices, as well as in river valleys far downstream of those volcanoes. Wet-flow deposits form an important component of this archive in both settings but are more commonly preserved at sites far from the source vent. For example, the geologic record of the 1991 explosive eruption of Mount Pinatubo (Philippines) was preserved almost entirely in cubic kilometers of pyroclastic density-current (PDC), tephra-fall (TF), volcanic debris-flow (VDF), volcanic hyperconcentrated-flow (VHF), and volcanic water-flood (VWF) deposits. No volcanic debris avalanche (VDA) occurred during this eruption. Different parts of the 1991 eruption story were recorded in deposits emplaced in diverse settings and at varying distances from the vent (Fig. 1).

T. C. Pierson et al., *Deposits of Volcanic Wet Flows*, Advances in Volcanology,
https://doi.org/10.1007/978-3-031-66574-5_2

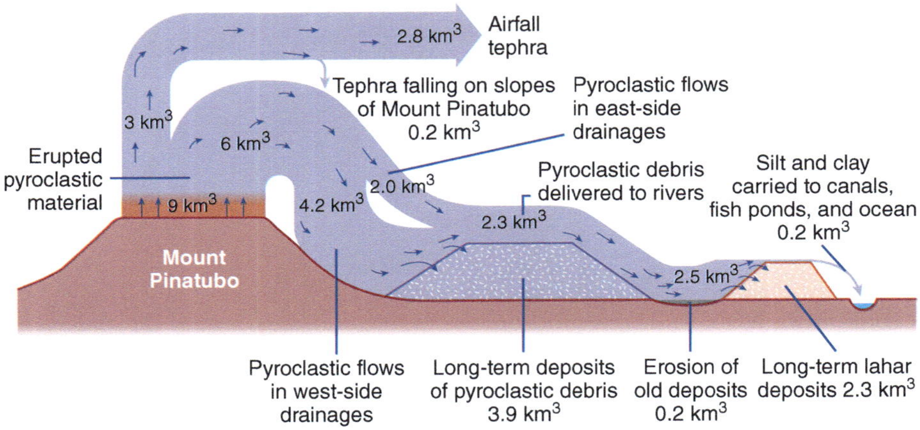

Fig. 1 An eruption recorded mainly by its volcaniclastic deposits—the June 15, 1991, eruption of Mount Pinatubo (Philippines) (Newhall and Punongbayan 1996). This schematic summary diagram, made from rough preliminary volume estimates (Pierson et al. 1992) shows the different types and volumes of primary and secondary volcaniclastic deposits emplaced on the volcano's flanks during and shortly after the eruption. In the years following the eruption, more and more of the pyroclastic deposits were eroded and transported farther downstream (Hayes et al. 2002; Gran et al. 2011). Figure is modified from Pierson et al. (1992); refined values can be found in Newhall and Punongbayan (1996). The volume of material moving off the edifice includes both massive volumes of pyroclastic density-current (PDC) deposits and minor amounts of tephra-fall (TF) deposits emplaced on the volcano flanks. Some of the TF material was incorporated into active PDCs, and some was later eroded and transported by secondary lahars. The estimated volume of older deposits eroded during this time is also shown. The greatest volume of 1991 deposits initially accumulated on the slopes and low-gradient apron of the volcano (shown as essentially flat terrain in the diagram). "Long-term" deposits shown here are those that survived the first flush of erosion in the three months following the eruption, although significant fractions of these deposits were removed by erosion within the next few decades. The bulk of the tephra-fall deposits were emplaced far from the volcano. The "old deposits" noted here were volcaniclastic sediments emplaced during previous eruptive periods. Due to the abundance of primary volcaniclastic material produced by the eruption, it appears that little epiclastic sediment was included in these deposits. This eruption did not produce flank lava flows, although a lava dome later grew within the caldera

Unconsolidated volcanic sediment, including primary and secondary volcaniclastic and epiclastic deposits, may form a significant part of a volcanic edifice itself, but with the passage of time, pyroclastic deposits on many volcanoes (and thin regional tephra blankets farther afield) are vulnerable to erosion and redeposition in downstream river valleys and sedimentary basins. Thus, a more complete and longer-term record of volcanic activity than can be found on edifices may be preserved in volcanic sediments of different ages and origins that were emplaced in river valleys and lake basins (Davies et al. 1978; Mathisen and Vondra 1983; Kataoka and Nakajo 2002; Kataoka 2005). In some cases, volcanic sediments are all that remain to document the timing and extent of a volcano's past impacts on a landscape (Fisher and Schmincke 1984; Smith 1987b, 1988; Orton 1996).

1 Spatial Distribution of Volcanic Wet-Flow Deposits

Relative distance of depositional zones from a volcanic vent is commonly described by the qualitative and subjective terms *proximal*, *medial*, and *distal*, but scaling depends on volcano type and edifice size and geometry (Vessell and Davies 1981; Orton 1996) (Fig. 2). At

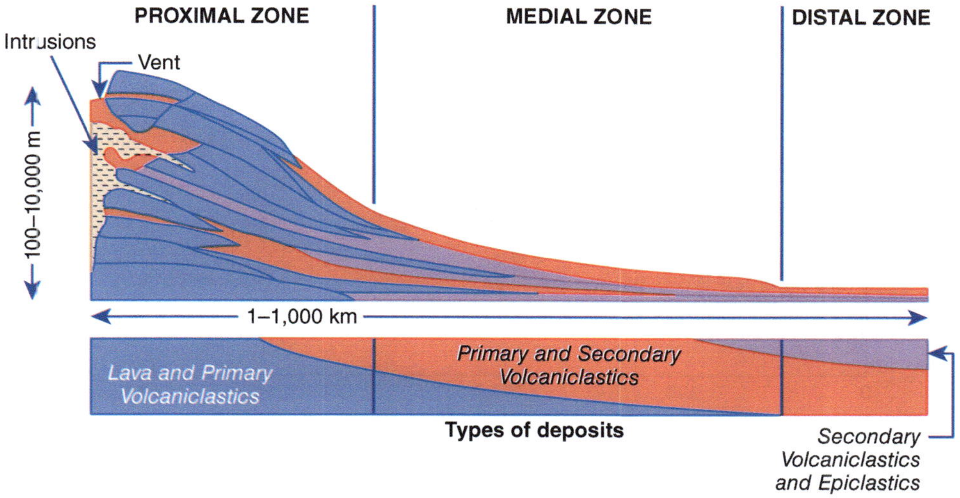

Fig. 2 Simplified schematic vertical half section through a composite volcanic edifice of intermediate composition, showing the relation between volcanic rock and deposit types encountered in the proximal, medial, and distal zones on edifices of composite volcanoes, based on relative distance from vent (modified from Orton 1996). Vent elevation and distance from the vent influence the main types of volcanic rock and volcaniclastic deposits to be found at a particular location and the hazards that could be encountered there. Box at the bottom indicates how relative proportions of different deposit types change with distance away from the vent

stratovolcanoes and dome complexes,[1] **proximal zones** typically encompass the steep mid to upper flanks of edifices. Emplacement of lava flows, coarse TF deposits, autobreccias (from deforming lavas domes and flows), and PDC deposits is common in this zone. Depending on the size of the edifice and the scale of an eruption, proximal zones extend from 0 to about 7 km from the vent. **Medial zones** at these types of volcanoes generally comprise the low-gradient depositional aprons and fringes of edifices and are primarily depositional zones for a mix of primary and secondary volcaniclastic deposits—PDC and TF deposits plus deposits of lahars (more volcanic debris-flow (VDF) deposits than volcanic hyperconcentrated-flow (VHF) deposits), VDAs, and some VWFs. Volcanoglacial deposits are also commonly found in medial zones. Depending on the size of the edifice and the scale of an eruption, medial zones may extend from about 7 km from the vent to about 20 to 70 km. **Distal zones** at stratovolcanoes and dome complexes are found either in river valleys that extend well beyond edifices or in distant terrain that can be broadly blanketed by TF deposits. In river valleys, tephra beds can be found on or interbedded within valley fill terraces, but active floodplains are typically reworked frequently enough by fluvial action that preservation potential there is limited. Depending on the size of the edifice and the scale of an eruption, distal zones extend from 20 to 70 km from the vent to more than 100 km. Distal channel deposits typically are dominated by secondary volcaniclastic (and epiclastic) sediments—mainly dilute lahar deposits and VWF deposits. However, rare long-runout VDA or PDC sediments and some glacial deposits may also be present in distal river valleys. Large eruptions from calderas with little or no relief but with large volumes of water in caldera

[1] For the estimated extents of proximal, medial, and distal zones, measurements from three examples were used: Mount Pinatubo (Philippines), a dome complex, following a very large eruption in 1991; Mount Adams (Washington, USA), a very large stratovolcano that produced mainly lava flows in the past; and Nevado del Ruiz (Colombia), a very large stratovolcano, following very large eruption-triggered lahars in 1985.

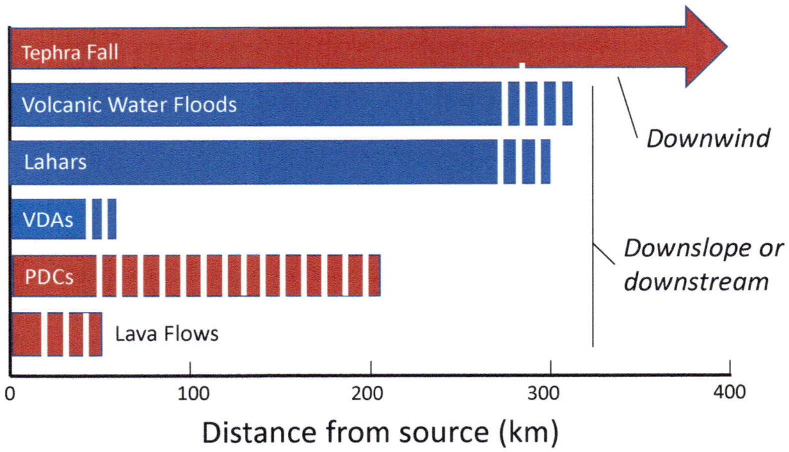

Fig. 3 Distances traveled by volcanic wet flows (blue bars), in comparison to other processes that emplace volcaniclastic deposits on volcanoes. Solid bars represent approximate maximum distances traveled at stratovolcanoes. Dashed bars represent distances recorded at other types of volcanoes. Extraordinary distances traveled by flood basalts are not considered. Except for tephra falls, which can be deposited >1000 km from source (as indicated by arrow), lahars and volcanic floods typically emplace volcaniclastic sediment the farthest from source, and their travel distances are commonly limited by length of river channel prior to dilution and transformation in large rivers, lakes, or the sea

lakes may have proximal zones extending up to 100 km from the vent, with medial and distal zones much farther away (Smith 1987a, b, 1991; Palmer et al. 1993; Manville et al. 2009b).

Deposits of volcanic wet flows at stratocones—arguably where most such deposits are found—are most abundant in the medial to distal zones, generally farther from source than those of PDC or volcano-glacial deposits, and the proportions of VDF, VHF, and VWF deposits change with distance from source (Fig. 3). For example, the Miocene Ellensburg Formation on the east side of the Cascade Range in central Washington State contains about 40% VDF deposits (and no VHF deposits) and 60% VWF deposits at 30–40 km from the inferred volcanic source (Smith 1987b). About 20 km farther downstream, the fraction of VDF deposits drops to 15% and the fraction of VHF units increases to 15% (total lahar content 30%), and VWF deposits increase to 70%.

Wet-flow deposits mobilized during or following a single stratocone eruption commonly produce lahar and VWF aggradational assemblages 20–40 m thick, and >100 m when VDA deposits are included (e.g., Voight et al. 1981, 1983; Pierson et al. 1996, 2011; Rodolfo et al. 1996; Manville et al. 2005). Wet-flow deposits from multiple eruptions can form depositional aprons (ring plains) as much as 700 m thick on the lower flanks of some stratocones (Van Houten 1976; Mathisen and Vondra 1983; Smith 1987a, b). Such massive accumulations of volcanic sediment are the result of extremely high rates of sediment transport by flows of sediment and water, which can be qualified as *sediment yield*—the amount of sediment per unit area removed from a drainage basin during a specified period (usually per year).

2 Temporal Distribution of Volcanic Wet-Flow Deposits

2.1 Timing of Eruptions

Volcanic eruptions are events during which fragments of solidified new magma (juvenile particles) or older volcanic rock (commonly a combination of both) are explosively ejected or new molten magma is extruded onto Earth's surface as lava. Individual eruptions may include relatively short pauses between eruptive phases (see below) (Siebert et al. 2010, 2015). Rapid venting of gas without the ejection of solid material is not considered to be an eruption. Eruptions have discrete starting and ending times, although endings can be difficult to discern (cf. Dzurisin et al. 2015). The production of juvenile rock fragments during eruptions is the primary source of discrete particles comprising volcanic sedimentary deposits, so the timing and duration of volcanic wet-flow occurrence can be closely related to eruption timing.

Documented durations of volcanic eruptions range from minutes (e.g., Kaneko et al. 2016) to more than a century (Siebert et al. 2010). Most historical eruptions have ended in less than three months, few have lasted longer than three years, and the median duration has been 7 weeks (Siebert et al. 2010, 2015). But variations in eruption rates, as well as the character and duration of pauses between eruptive events, have led to varied terminologies. For example, high-resolution categories of time-scale variation (seconds to days) have been applied to Hawaiian and Strombolian eruptions (Taddeucci et al. 2015). We restrict the term *eruption* in this book to an event involving the nearly continuous ejection or extrusion of magma, whereby pauses in magma venting do not exceed three months (Simkin et al. 1981; Fisher and Schmincke 1984; Siebert et al. 2010). More specific terms describing durations of eruption pauses may be applied: (1) *Eruptive pulses* are periods of active ejection or extrusion of magma (e.g., Romero et al. 2016) or intervals of increased eruption rate during ongoing ejection or extrusion (e.g., Lautze et al. 2004); (2) *eruptive phases* are groupings of eruptive pulses based on timing or eruptive style (Fisher and Schmincke 1984); (3) *eruptive periods* are clusters of eruptions spanning decades to centuries (e.g., Clynne et al. 2008); and (4) *eruptive stages* are clusters of eruption periods spanning thousands to millions of years (e.g., Orton 1996; Clynne et al. 2008; Barry et al. 2010). Mount St. Helens has had well-dated eruptions over the last 1500 years that clearly show groupings of eruptions separated by pauses of varying length (Fig. 4), as well as eruption stages of longer duration (Clynne et al. 2005, 2008). Some volcanoes exhibit cyclic patterns of eruptions, as well as cyclic patterns of edifice growth and collapse and subsequent accelerated sedimentation (e.g., Zernack et al. 2009).

2.2 Timing of Flows with Respect to Eruptions

Timing of volcanic wet flows relative to eruptive activity can be divided into three time periods: (1) the *syneruptive* period—the time interval defined by active production of volcaniclastic particles by eruptive processes (eruptive pulses); (2) the *posteruptive* period—the time interval following placement of volcaniclastic particles on the landscape by eruptions and during which time landscapes actively respond to volcanic disturbance with altered basin hydrology, remobilization of deposits, unstable river channels, and elevated sediment yields; and (3) the *intereruptive* period—the period during which landscape conditions and sediment yield are considered to be at normal background levels (Rodolfo et al. 1989; Manville 2002; Manville and Wilson 2004; Németh and Cronin 2007; Manville et al. 2009a, b; Pierson and Major 2014). Earlier authors defined *syneruptive* to encompass both (1) and (2) above (e.g., Smith 1991) or split the posteruptive period into a lahar-dominated fan-building phase and a flood-dominated channel braiding phase (Vessell and Davies 1981), but we find the simple three-part division to be most useful. Volcanic wet flows can occur during any of the three time periods defined above, although they are much more frequent during the syneruptive and early posteruptive periods (Kuenzi et al. 1979; Vessell and Davies 1981; Smith 1987a, b; Palmer et al. 1993; Hayes et al. 2002; Zernack 2021) (Fig. 5). It is also true that lahars are the dominant flow type immediately flowing eruptions and volcanic water floods become dominant later in posteruptive periods (Major et al. 2000; Gran and Montgomery 2005). Syneruptive and posteruptive lahars and

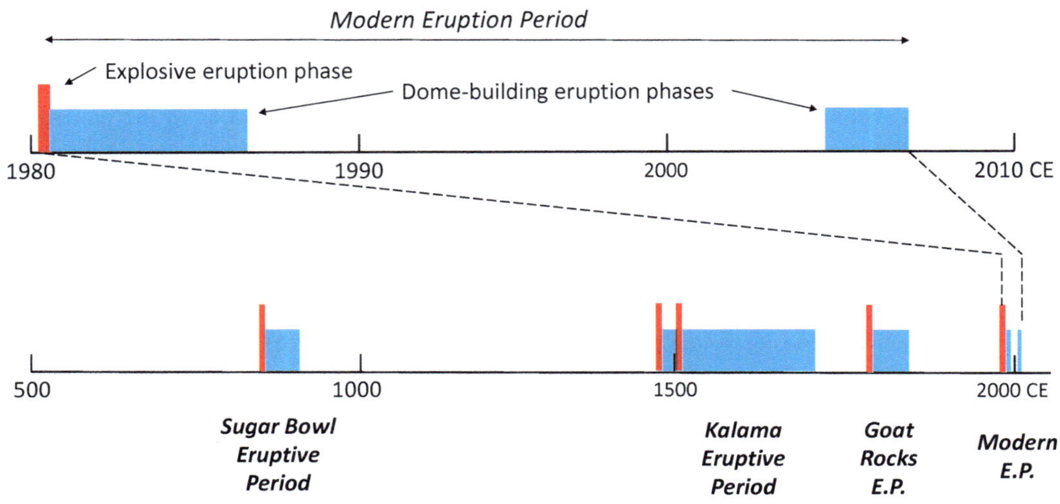

Fig. 4 Temporal distribution of eruption periods—an example from Mount St. Helens (Washington, USA) over about the last 1500 years. Timeline (expanded for 1980–2010) shows eruptive periods recorded since 500 CE (adapted from Clynne et al. 2005). These eruptive periods are part of the Spirit Lake eruptive stage, which has lasted about 4000 years. Explosive phases (red) are the major producers of highly erodible volcaniclastic particles. Periods dominated by lava-dome or lava-flow extrusion (blue) may be continuous or intermittent. The explosive phase of the modern eruption period lasted from May to October 1980 and included six separate tephra-producing eruptions (Lipman and Mullineaux 1981). The two dome-building eruptions were from late 1980 to 1986 (intermittent lava-extruding events) and from late 2004 to early 2008 (continuous plug extrusion) (Swanson and Holcomb 1990; Dzurisin et al. 2015)

water floods also tend to be larger than those during intereruptive periods.

The lengths of syneruptive, posteruptive, and intereruptive periods depend on eruption magnitude, types of eruptive processes, and degree of landscape disturbance. For small eruptions, posteruptive periods can last from several months to a few years, depending on the extent and degree to which the landscape is affected (Vessell and Davies 1981; Major et al. 2000; Manville et al. 2009a). Posteruptive water-flood sedimentation more typically lasts for decades (Kuenzi et al. 1979; Major et al. 2000; Hayes et al. 2002; Gran and Montgomery 2005; Pierson et al. 2011), but geomorphic/landscape recovery and stabilization following very large explosive eruptions may require centuries to millennia (Manville and Wilson 2004; Manville et al. 2009b; Pierson and Major 2014).

Lahars, debris avalanches, and water floods can all be triggered directly by eruptions, and when this happens, they are syneruptive flows. However, such flows are posteruptive events when triggered by meteorologic or geomorphic processes (such as heavy rainfall or failures of natural dams) after eruptive activity has ceased. The magnitudes of posteruptive flows are highly influenced by the degree and recency of eruption disturbance to the landscape—vegetation destruction or burial, the volume of available loose sediment, and the timing and volume of water applied to or released from an affected landscape (Pierson and Major 2014). The more time available to landscapes for stabilization following eruptive disturbance, the smaller and less frequent the posteruptive flows—an effect influenced by regrowth of stabilizing vegetation, compaction and cementation of loose deposits, and prior removal of erodible sediment.

Occurrence of posteruptive wet flows is one type of geomorphic response to volcanically disturbed landscapes, but other related types of measurable responses include increased sediment yields from river basins, changes in riverbed elevations (aggradation and degradation), and changes in river channel geometry (channel

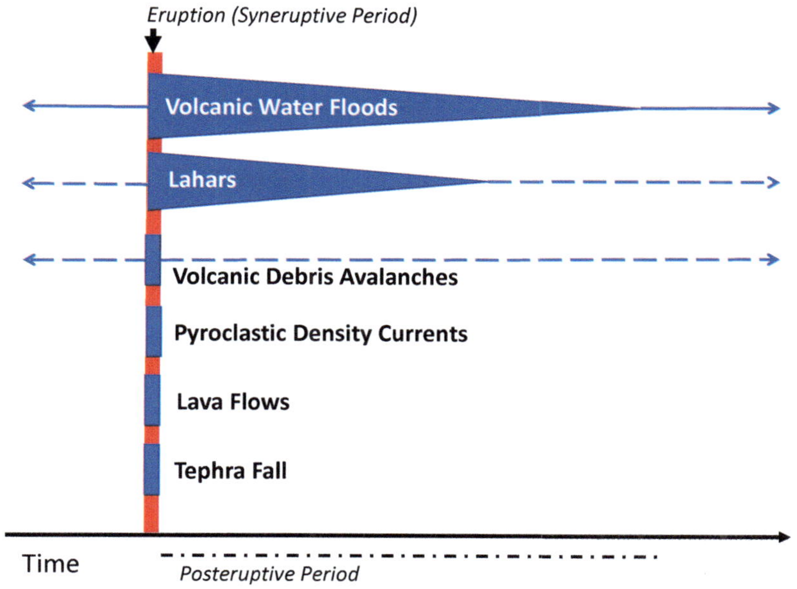

Fig. 5 Potential timing of volcanic wet flows (in blue) relative to eruption timing and compared to timing of primary eruptive processes (in red). Volcanic wet flows, depending on process, are most likely to occur during syneruptive periods (during or immediately after eruptions), generally decreasing in magnitude and frequency during the posteruptive period (years to decades after eruptions). However, volcanic wet flows can occur anytime, as indicated by the thin horizontal arrows. In contrast, eruptive processes occur only during syneruptive periods. Frequency and potential magnitude of floods and lahars are schematically depicted by the thickness of horizontal arrows. Solid arrows indicate that floods occur at the same average frequency (in one area and ignoring climate change) regardless of eruptive activity, but they are larger and carry more sediment in the syneruptive and posteruptive periods. Thin dashed arrows indicate the continuing possibility of occurrence but with lower frequency and magnitude. Some debris avalanches and lahars occur thousands of years after initial emplacement of the source material (e.g., Vallance and Scott 1997; Pareschi et al. 2000; Scott et al. 2001, 2005)

form, width, depth, and slope). Timing of wet-flow occurrence relative to eruption occurrence typically involves a *reaction time*, a *destabilization time*, and a *relaxation time*, relative to timing of an eruption (Graf 1977; Pierson and Major 2014) (Fig. 6). During destabilization times, lahars are likely to be concentrated and coarse grained, and heavy rainfall (and rapid runoff) tends to produce mostly lahars, not water floods. During relaxation times (periods when less sediment is available for erosion and vegetation cover is reestablishing), dilute lahars and water floods become more common and concentrated lahars less so. Once drainage basin sediment and water fluxes have returned to preeruption background levels or to a new background level (return to an intereruptive period), dilute water floods are the dominant sediment-transporting process. In general, syneruptive and posteruptive periods tend to be short relative to longer intereruptive periods at a given volcano or volcanic region (Smith 1991). However, in volcanic terrains subjected to relatively frequent eruptions (and volcanic disturbance), an intereruptive period may not often be reached and the landscape may continue in a prolonged or perpetual state of recovery (Major et al. 2021).

2.3 Timing of Flows with Respect to Each Other

Interpretation of the timing of sediment sequences emplaced by volcanic wet flows (and PDCs), more often found within river valleys

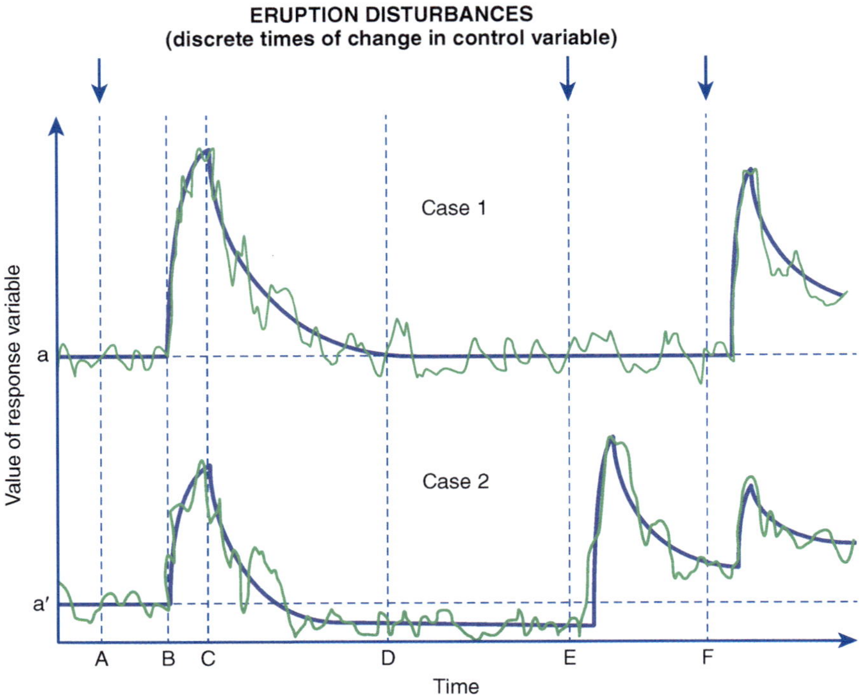

Fig. 6 Schematic representation of timing of geomorphic response, shown as a change to a geomorphic response variable (such as rate of sediment production or channel-bed elevation) in a fluvial system, which is adjusting to disturbances imposed by explosive volcanic eruptions, assumed here to have very short durations (modified from Pierson and Major 2014, and originally modified from Graf 1977). The thick blue lines represent the mean condition, while the finer green lines represent short-term fluctuations. In a simple example represented by case 1, where eruptions occur at time A and F, the response variable at first does not change following the first eruption (lag time AB). At time B, the variable begins deviating from a preeruption steady-state value (value *a*), reaches a maximum deviation value (at time C), and then more gradually returns to a value like that in the preeruption condition, reaching that point at time D. For the interval DF, one can say that the response variable has recovered its preeruption steady-state condition. Note, however, that not all response variables may recover at the same time. Interval AB is the **reaction time**. Interval BC is the **destabilization time**, when the value of the response variable is moving away from the value characterizing preeruption stability. Interval CD is the **relaxation time**, when the variable moves back toward and asymptotically approaches the normative preeruption condition. Interval AD also represents the **posteruptive period** for this disturbed fluvial system, while DF represents the **intereruptive period**, the time of steady-state equilibrium. In the more complex case 2 example, the drainage basin is affected by three more closely spaced eruptions at times A, E, and F. In this example, the response variable does not ever return to its original preeruption value (value *a′*), and at time F another eruption occurs before the variable has reached a steady-state value. Syneruptive periods are not instantaneous points in time, however, as assumed in this figure. Eruption durations range from hours to years, so the start of reactions times should more realistically be taken from the time when an eruption creates a new sediment source, such as when an explosive pulse (within a longer duration eruption) emplaces a fine-grained tephra layer

than on broad depositional surfaces, can be challenging. Tephra-fall deposits adhere nicely to the straightforward stratigraphic law of superposition—the proposition that in undeformed sequences of strata, the oldest deposit will be on the bottom and the youngest on top (Fig. 7a). However, volcanic wet-flow and PDC deposits emplaced within incised valley fills might appear to be in the reverse order (Fig. 7b). The key to interpretation is in figuring out whether the exposed units are a true cross-section of the terrace fill or thin coatings plastered on the

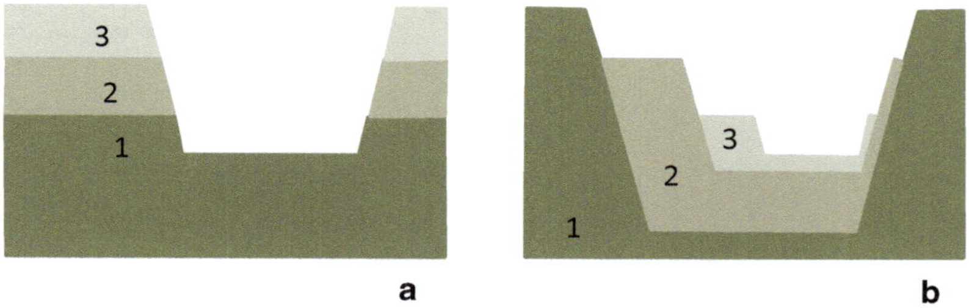

Fig. 7 Schematic diagram of two different ways that volcanic flow deposits can be preserved in river valleys while still obeying the stratigraphic law of superposition: **a** successive flows emplace deposits on top of each other, so that the oldest (unit 1) is at the bottom of the sequence and the youngest (unit 3) is on top. Incision following deposition of unit 3 exposes the sequence. **b** Successively smaller flows emplace deposits within valleys incised within previous deposits, so that the oldest deposit is preserved at a higher level than the youngest

side of a cut (as on the right side of Fig. 7b). Digging into outcrop faces may be required for a 3-dimensional perspective. Further complexity is commonly added when multiple episodes of incision and deposition are caused by flows of different magnitude (Fig. 8).

3 Effect of Eruption Size and Type on Wet-Flow Type and Magnitude

Lahar and water flood magnitudes are generally a function of the size and type of the contributing eruption (or eruptions), the degree and extent of eruption-induced landscape disturbance, and the particle size and distribution of primary deposits. Size of explosive eruptions is commonly rated with the Volcano Explosivity Index (VEI) (Newhall and Self 1982). Debris-avalanche magnitude is a function of factors controlling the size of edifice collapses initiating the avalanches—edifice geometry and rock strength (determined by degree of fracturing and hydrothermal alteration), basement characteristics, magmatic or tectonic forces, and degree of hydrothermal pressurization. Volcanic edifice collapses are among the largest landslides on the planet, sometimes displacing and fragmenting cubic kilometers of volcanic rock.

Lahars and water floods in volcanic fluvial systems, measured as peak flow rate (discharge) or volume, are commonly 2–4 orders of magnitude larger than their nonvolcanic counterparts (Pierson 1995, 1998). A minimum threshold eruption magnitude (probably about VEI 3) may be required for lahars and floods to produce significant sedimentation responses from rainfall runoff or rapid snowmelt (Vessell and Davies 1981), but on the other end of the spectrum, extremely large eruptions (VEI 7–8) can produce large calderas that can impound immense volumes of water. Breakout floods from large intracaldera lakes have produced floods that are among the largest postglacial deluges on Earth, which have had enormous sedimentation responses (Manville 2002, 2010; Manville et al. 1999, 2009b). Explosive eruptions of VEI 4–6 from stratovolcanoes consistently produce lahars and floods large enough and frequent enough to trigger major sedimentation responses (e.g., Newhall and Punongbayan 1996; Major et al. 2000; Hayes et al. 2002; Mothes et al. 2004; Pierson et al. 2013). Even the VEI 3 eruption of Nevado del Ruiz (Colombia) in 1985 produced catastrophic lahars in three of the four rivers draining the volcano (Lowe et al. 1986; Pierson et al. 1990), primarily because that small eruption emplaced its limited volume of hot PDC deposits directly on a broad ice cap, focusing and maximizing meltwater production. In some cases, even non-explosive dome-building eruptions can generate sufficiently large volumes of volcaniclastic material through dome-collapse block-and-ash flows to trigger large sedimentation responses, such as the 1781–1793

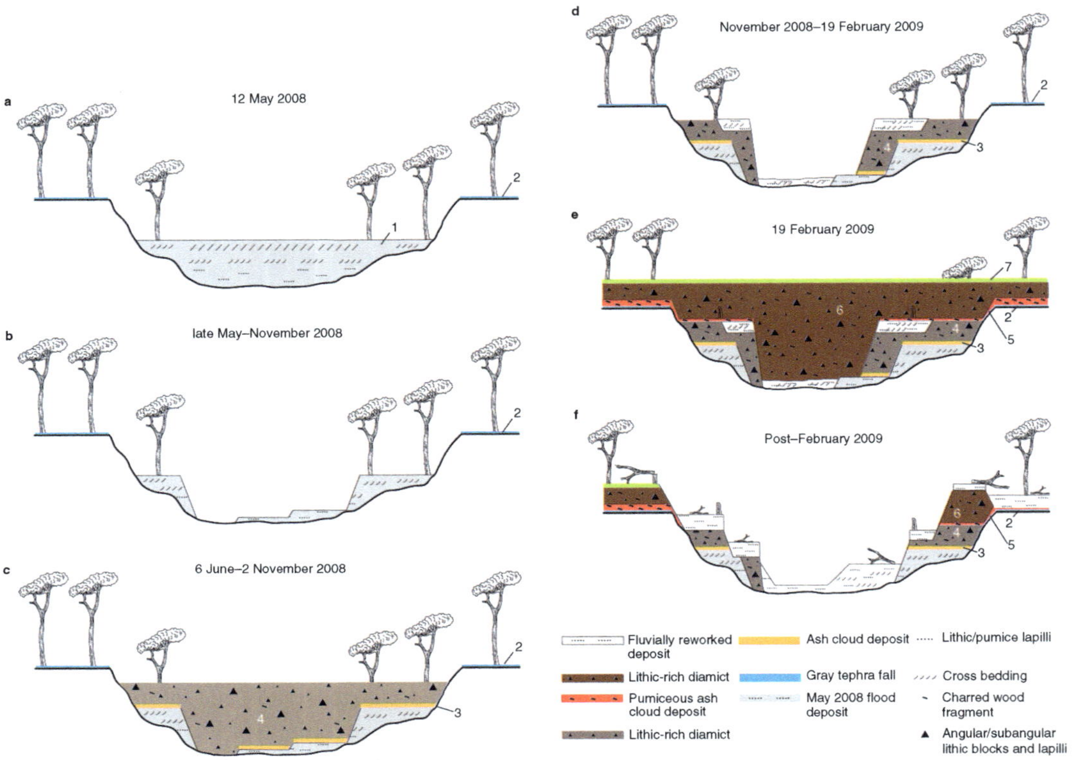

Fig. 8 Complex stratigraphic relations in deposits from a single eruption, emplaced mainly by pyroclastic density currents (PDCs) and tephra fall (TF). Example shows timing sequence (cross-sections **a–f**) of volcaniclastic deposits emplaced by the 2008–2009 eruption of Chaitén volcano (Chile), which is complicated by intervening episodes of erosion and redeposition of secondary deposits. These volcanic sediments were emplaced in the Chaitén River (aka Blanco River) valley between May 2008 and January 2010. Order of deposition of the primary deposits is indicated by numbers. Unit 1 is an ambiguous deposit—a channel-confined sandy gravel with angular, fresh, glassy clasts. It is partly unstratified and partly faintly bedded, suggesting emplacement by a low-temperature PDC near the beginning of the eruption that either had multiple surges or mixed with water during transport. Unit 2 is a TF deposit emplaced during the explosive phase of the eruption in May 2008. Unit 3 is an ash-cloud surge (type of PDC) deposit. Unit 4 is a dome-collapse block-and-ash flow (type of PDC) deposit. Unit 5 is a possible ash-cloud fallout (type of TF) deposit. Unit 6 is another dome-collapse block-and-ash flow deposit. Unit 7 is an ash-cloud fallout deposit. White units are secondary volcaniclastic deposits, mainly fluvial, composed of reworked primary volcaniclastic sediment. Woody debris is both charred and uncharred. Modified from Major et al. (2013)

dome-building eruption of Mount Hood (Oregon, USA) (Pierson et al. 2011).

Sediment production and transport rates are commonly extremely high over broad areas in volcanically disturbed terrain, as they can be (at least locally) in nonvolcanic terrain affected by some other natural disturbances such as wildfires, earthquakes, or hurricanes. Importantly, however, large explosive eruptions emplace enormous quantities of new, readily erodible sediment particles (10^8–10^{11} m^3) on the landscape surface in addition to destroying vegetation and focusing powerful erosive forces on the terrain. Moreover, such new reserves of erodible sediment can be mobilized long after the original landscape disturbance—for years, decades, or even centuries following volcanic disturbance (Pierson and Major 2014) (Fig. 9).

Large volumes of loosely packed and highly erodible volcaniclastic particles on the landscape are important but not sufficient for massive sedimentation responses (e.g., Segerstrom

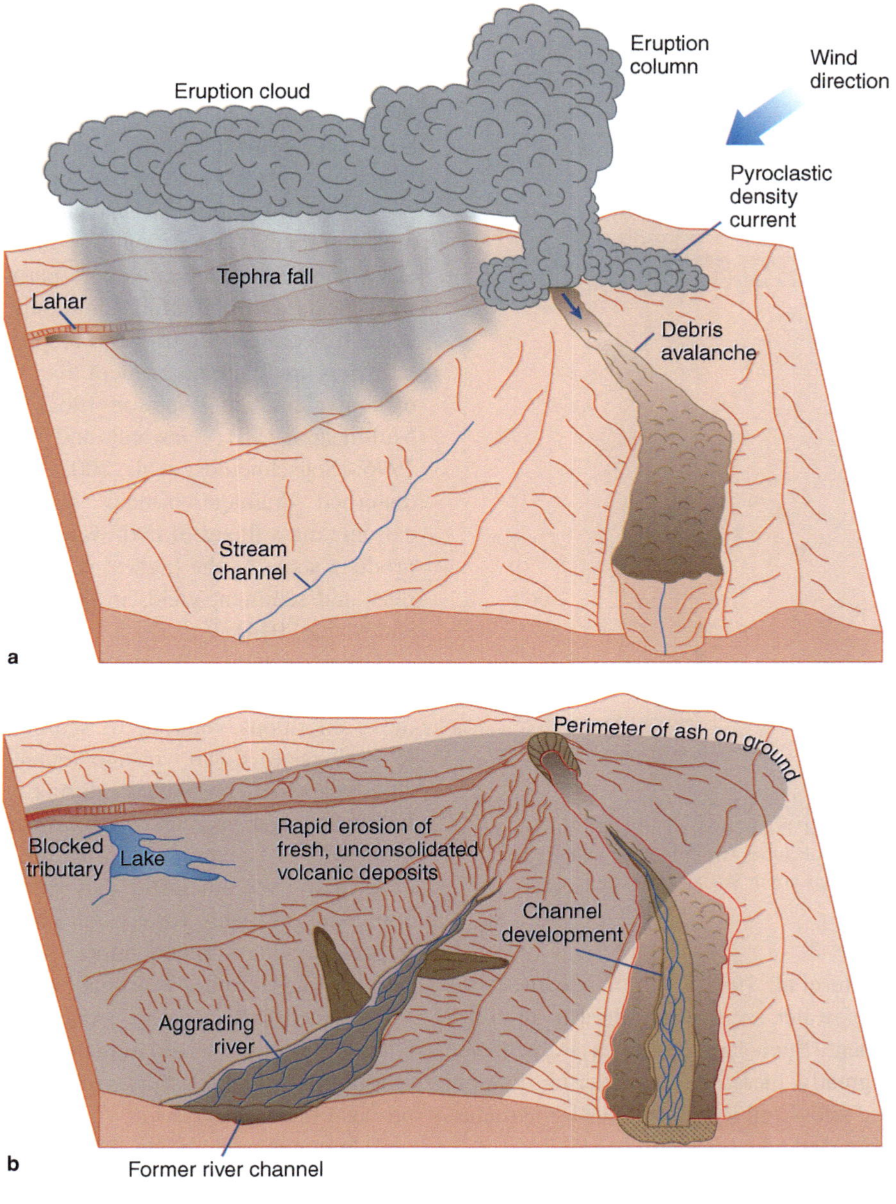

Fig. 9 Schematic diagram illustrating explosive eruption processes and subsequent landscape responses (from Pierson and Major 2014): **a** volcanic processes from or related to an explosive eruption, including tephra fall, pyroclastic density currents, a debris avalanche, and a lahar; **b** landscape changes and responses to disturbances caused by eruption processes shown in upper diagram include rapid erosion of fresh tephra by rilling and landsliding, erosion of channel fill and development of new channel networks, channel aggradation and formation of braided channel patterns; and impoundment of channel-margin lakes. Shaded area denotes deposition zone of tephra-fall deposits

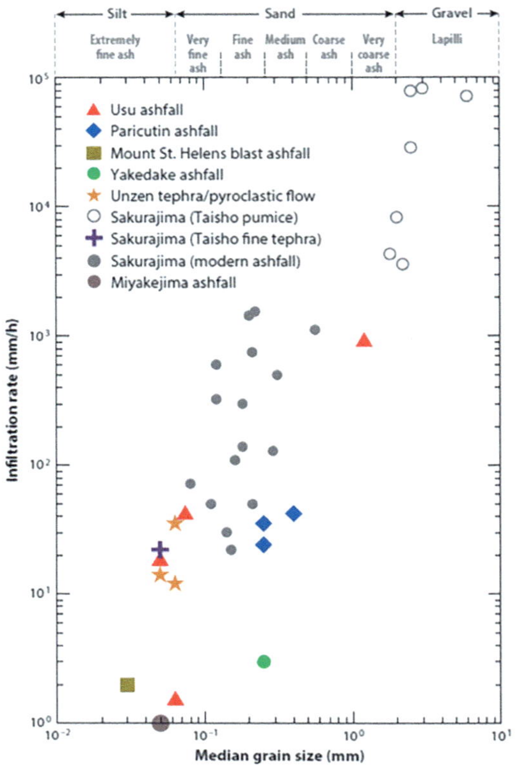

Fig. 10 Relationships between infiltration rate and grain size of tephra-fall and pyroclastic density-current deposits at different volcanoes (from Pierson and Major 2014). Volcano locations are in Volcano Index

1950; White et al. 1997). This is because large volumes of water are also required to mobilize the sediment. Such large volumes are achieved through greatly increased rainfall runoff, rapid melting of snow and ice on edifices by pyroclastic deposits, and lake breakout floods.

Heavy rainfall is the most common mechanism for mobilizing sediment over large areas in volcanically disturbed terrain. Top layers of many newly emplaced primary volcaniclastic deposits are very fine-grained to extremely fine-grained, including surfaces of both TF deposits mantling steep hillslopes and PDC deposits filling valley bottoms. Such fine-grained surface layers can decrease infiltration rates of rainfall by up to four orders of magnitude (Fig. 10) (Pierson and Major 2014). The slowing of infiltration, combined with the destruction of vegetation that intercepts rain (Fig. 11), greatly increases rates of rainfall runoff, because much less water soaks into the ground and much more runs off into streams. Water also runs off bare hillslopes faster than on vegetated hillslopes, which decreases the time lag between rainfall peaks and flood peaks. This more voluminous and more rapid runoff quickly and efficiently erodes and entrains fresh tephra particles. Thus, early posteruptive rainfall commonly mobilizes enough sediment to trigger lahars, which can transport much more sediment than water floods of equivalent size (Waldron 1967; Lipman and Mullineaux 1981; Newhall and Punongbayan 1996; Segschneider et al. 2002). Volcanically disturbed drainage basins, when generating extreme mass fluxes of both water and sediment, produce some of the highest sediment transport rates and sediment yields in the world (Pierson and Major 2014) (Fig. 12).

Lahars generated by rainfall and rapid snowmelt, commonly highly concentrated, typically move enormous volumes of sediment (relative to basin size) down volcano-sourced rivers at stratovolcanoes, particularly during eruptions or within the first year or two following an eruption. This is because meters to tens of meters of primary volcaniclastic deposit thickness is commonly available for erosion in the eruption-impacted zones—much more sediment than can be transported by normal fluvial erosion. Where such volcanoes are situated in monsoon climates, multiple lahars may be generated daily in different channels for weeks or months to produce the high sediment yields (e.g., Newhall and Punongbayan 1996; Thouret et al. 1998; Lavigne et al. 2000a, b; Lavigne and Thouret 2002; Lavigne 2004; de Bélizal et al. 2013). In other cases, a large single lahar (or only a few lahars) may be responsible for most of the sediment yield for an entire season (e.g., Major et al. 2000). However, where volumes of water generated or released are large relative to the volume of sediment available for erosion, lahars may be dilute or altogether absent, due to the dilution effect of so much water moving down river channels. Examples of this effect may be found with large breakout floods from intracaldera or

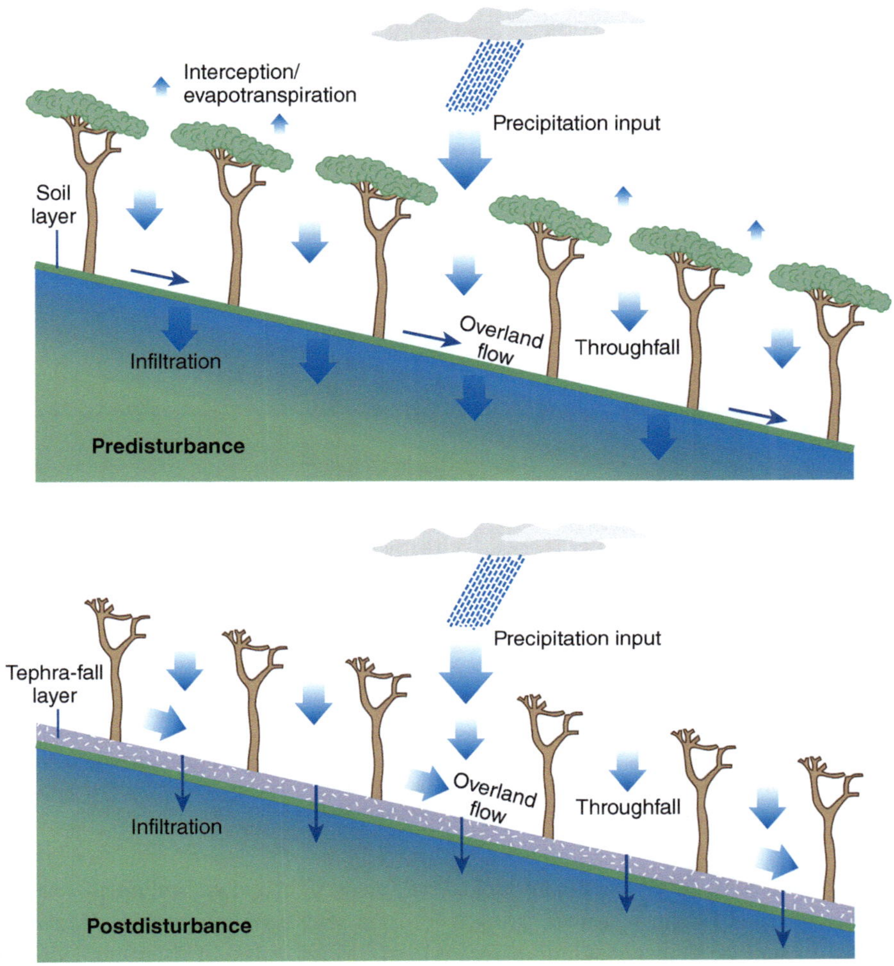

Fig. 11 Schematic depiction of effects of severe damage to forest vegetation by tephra fall or dilute pyroclastic density currents (reduced canopy interception and evapotranspiration) and the resulting hydrologic effects that can increase the volume, flow rate, and erosive capacity of surface runoff during rainstorms (from Pierson and Major 2014). This effect leads to larger and more frequent lahars and water floods in drainage basins affected in this way. Width of arrows depicts relative magnitude of water transfer

subglacial lakes (e.g., Guðmundsson et al. 1997; Manville et al. 1999; Maria et al. 2000; Manville 2010) or with runoff or snowmelt from drainage basins on stratocones, where lesser thicknesses of deposits are exposed to prolonged, high-intensity rainfall or catastrophic rates of snowmelt (Waythomas et al. 2013; Pierson and Major 2014). In these cases, large water floods produce most of the high sediment yields.

At stratovolcanoes in humid environments, magnitude and frequency of lahars and water floods, as well as sediment yield, commonly peak shortly after landscape-disturbing eruptions and then decrease exponentially within a few years. Yields decrease either directly to preeruption levels or to a higher-than-normal plateau level (Pierson and Major 2014). For example, sediment production at Mount St. Helens within the first two years following the May 18, 1980, explosive eruption was as much as 500 times higher than normal background levels, but thereafter it declined nonlinearly for about 7 more years and then plateaued at nearly 100 times above background levels for

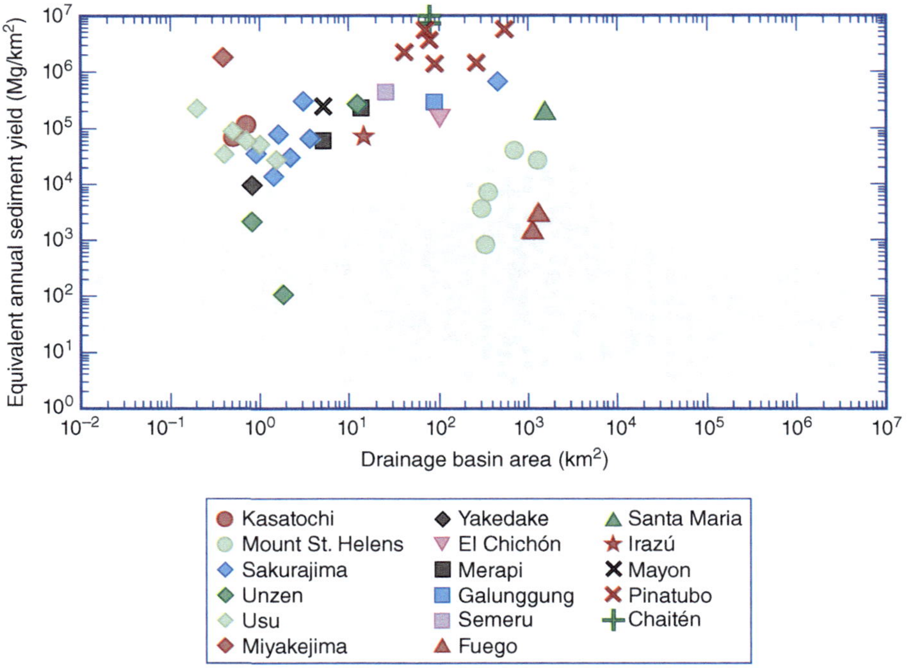

Fig. 12 Average annual sediment yield as a function of drainage basin area from volcanically disturbed basins (volcanoes listed) and from non-volcanic river basins worldwide, modified from Pierson and Major (2014). Colored data points represent data from drainage basins affected by recent volcanic eruptions at those volcanoes. Gray data points represent data from basins in volcanic terrain where no recent eruptions have occurred and from basins in nonvolcanic terrain. Volcano locations are in Volcano Index

two decades (Major et al. 2000). Sediment yield plateaus occur where sediment reservoirs are so large that they cannot be depleted within a few years or decades (Fig. 13). High posteruptive sediment yields can be inferred without direct measurement from channel patterns and valley or fan morphologies that are altered by the greatly increased storage of sediment in downstream river reaches (e.g., Davies et al. 1978; Kuenzi et al. 1979; Newhall and Punongbayan 1996; Hayes et al. 2002; Manville et al. 2005; Pierson et al. 2011, 2013) (Fig. 14). Erosion and transport of sediment by lahars and water floods in arid environments is likely to be limited and only episodic (e.g., Segerstrom 1950).

At calderas, huge breakout floods can occur if large intracaldera lakes form in the craters and if the lakes destabilize the crater walls impounding the lakes (Manville 2010). Such floods can erode and transport enormous volumes of volcaniclastic sediment if abundant sediment is available for erosion and entrainment (e.g., McGimsey et al. 1994; Waythomas et al. 1996; Manville et al. 1999; Manville and Wilson 2004). Timing of such sedimentary responses at calderas may differ greatly from those at stratovolcanoes. There may be long intervals (decades to millennia) between time of eruption and time of lake breakout at calderas, owing mainly to the time required for intracaldera lakes to fill to critical depths and destabilize or overtop caldera walls (Waythomas et al. 1996; White et al. 1997; Manville et al. 1999; Manville and Wilson 2004; Kataoka 2011). Furthermore, periods of higher-than-normal sediment transport and geomorphic responses following caldera eruptions may persist for decades to millennia, due to the extraordinary volumes of volcaniclastic sediment emplaced on landscapes (White et al. 1997; Manville and Wilson 2004; Manville et al. 2009b).

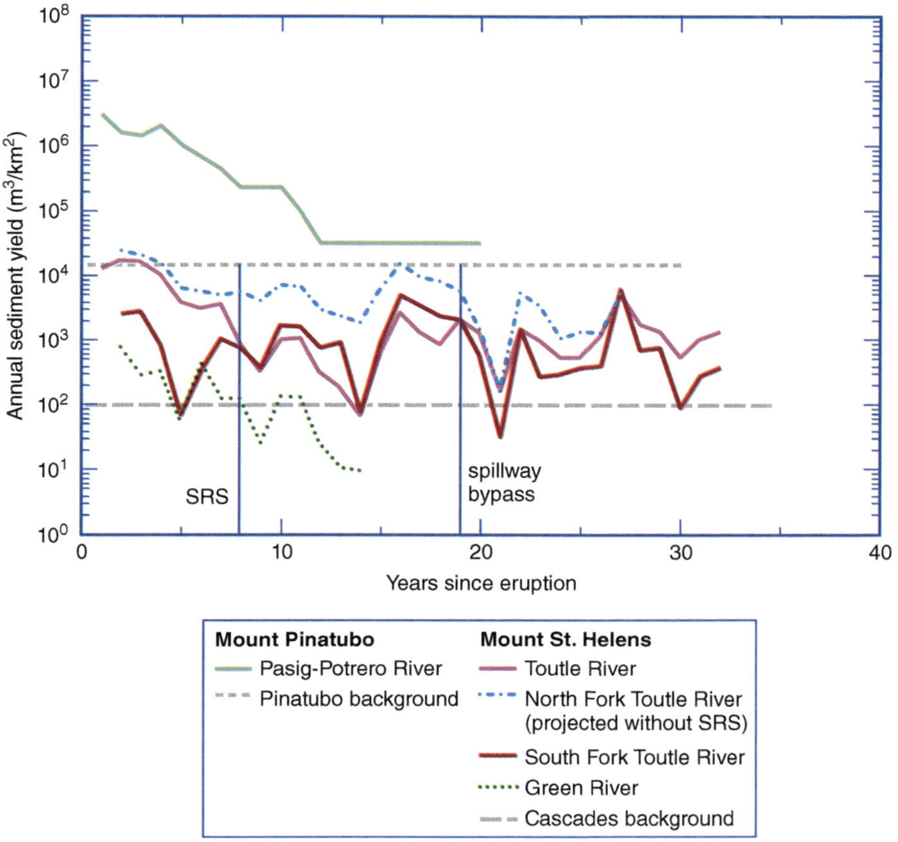

Fig. 13 Computed posteruptive trends in sediment yield for four rivers at Mount St. Helens (Washington, USA)—Toutle River and its tributaries (disturbed by the May 18, 1980, eruption) and one river at Mount Pinatubo (Philippines) (disturbed by the June 15, 1991, eruption). For each, year zero is the year of the eruption (from Pierson and Major 2014). Yields for all but one of the rivers plateaued above background levels for several decades. The vertical line labeled "SRS" indicates the time the Sediment Retention Structure (SRS)—a dam to hold back bedload sediment but not water and suspended sediment—was completed on North Fork Toutle River. The SRS created a large catch basin to trap sand and gravel washing down the river. The vertical line labeled "spillway bypass" indicates the time that the catch basin filled to the level of the SRS spillway, so that most of the transported sand and gravel could then bypass the sediment dam and once again move downriver. The sediment yield curve for the North Fork Toutle River combined the rate of sediment fill in the catch basin with stream gage measurements (stream discharge and sediment concentration) to estimate what sediment yield would have been without the presence of the SRS. The actual preeruption sediment yield for these specific rivers is not known, but the average yield for rivers in the Cascade Range is shown as "Cascades background"

Regardless the magnitude, timing, and duration of syneruptive and posteruptive sediment transport by volcanic wet flows, the downstream transport of excessive volumes of volcaniclastic particles for tens to hundreds of kilometers can seriously impact downstream communities. Those impacts, often inflicted by flow events without warning, can be both costly and extremely hazardous (e.g., Voight 1990; Newhall and Punongbayan 1996; Segschneider et al. 2002; Manville et al. 2009b; Pierson et al. 2014).

Fig. 14 Excessive in-channel sediment storage in the Rayas River that drains Chaitén volcano (Chile), about 25 km downstream of the volcano. This braided channel pattern results from excessive volumes of sediment washing downstream from the heavily tephra-mantled source terrain near the volcano, which produced the tephra during the explosive phase of an eruption in May 2008. The excess sediment caused channel aggradation here (note inundated areas of forest within river braidplain). USGS photo by T. Pierson, 2010

References

Barry TL, Self S, Kelley SP, Reidel S, Hooper P, Widdowson M (2010) New $^{40}Ar/^{39}Ar$ dating of the Grande Ronde lavas, Columbia River Basalts, USA: implications for duration of flood basalt eruption episodes. Lithos 118:213–222

Clynne MA, Ramsey DW, Wolfe EW (2005) Pre-1980 eruptive history of Mount St. Helens, Washington. U S Geol Surv Fact Sheet 2005–3045, 4 pp

Clynne MA, Calvert AT, Wolfe EW, Evarts RC, Fleck RJ, Lanphere MA (2008) The Pleistocene eruptive history of Mount St. Helens, Washington—from 300,000 to 12,800 years ago. In: Sherrod D, Scott WE, Stauffer PH (eds) A volcano rekindled; the renewed eruption of Mount St. Helens, 2004–2006. U S Geol Surv Prof Pap 1750:593–627

Davies DK, Vessell RK, Miles RC, Foley GG, Bonis SB (1978) Fluvial transport and sediment modifications in an active volcanic region. In: Miall AD (ed) Fluvial sedimentology. Can Soc Pet Geol Mem 5:61–84

de Bélizal E, Lavigne F, Hadmoko DS, Degeai J-P, Dipayana GA, Mutaqin BW, Marfai MA, Coquet M, Le Mauff B, Robin A-K, Vidal C, Cholik N, Aisyah N (2013) Rain-triggered lahars following the 2010 eruption of Merapi volcano, Indonesia: a major risk. J Volcanol Geotherm Res 261:330–347. https://doi.org/10.1016/j.jvolgeores.2013.01.010

Dzurisin D, Moran SC, Lisowski M, Schilling SP, Anderson KR, Werner C (2015) The 2004–2008 dome-building eruption at Mount St. Helens, Washington: epilogue. Bull Volcanol 77:89. https://doi.org/10.1007/s00445-015-0973-4

Fisher RV, Schmincke H-U (1984) Pyroclastic rocks. Springer-Verlag, Berlin, p 472

Graf WL (1977) The rate law in fluvial geomorphology. Am J Sci 277:178–191

Gran KB, Montgomery DR (2005) Spatial and temporal patterns in fluvial recovery following volcanic eruptions: channel response to basin-wide sediment loading at Mount Pinatubo, Philippines. Geol Soc Am Bull 117:195–211

Gran KB, Montgomery DR, Halbur JC (2011) Long-term elevated post-eruption sedimentation at Mount Pinatubo, Philippines. Geology 39(4):367–370. https://doi.org/10.1130/G31682.1

Guðmundsson MT, Sigmundsson F, Björnsson H (1997) Ice-volcano interaction of the 1996 Gjálp subglacial eruption, Vatnajökull, Iceland. Nature 389:954–957

Hayes SK, Montgomery DR, Newhall CG (2002) Fluvial sediment transport and deposition following the 1991 eruption of Mount Pinatubo. Geomorphology 45:211–224

Kaneko T, Maeno F, Nakada S (2016) 2014 Mount Ontake eruption: characteristics of the phreatic eruption as inferred from aerial observations. Earth Planet Space 68:72. https://doi.org/10.1186/s40623-016-0452-y

Kataoka K (2005) Distal fluvio-lacustrine volcaniclastic resedimentation in response to an explosive silicic eruption. The Pliocene Mushono tephra bed, central Japan. Geol Soc Am Bull 117:3–17

Kataoka KS (2011) Geomorphic and sedimentary evidence of a gigantic outburst flood from Towada caldera after the 15 ka Towada-Hachinohe ignimbrite eruption, northeast Japan. Geomorphology 125:11–26

Kataoka K, Nakajo T (2002) Volcaniclastic resedimentation in distal fluvial basins induced by large-volume explosive volcanism. The Ebisutoge—Fukuda tephra, Plio-Pleistocene boundary, central Japan. Sedimentology 49:319–334

Kuenzi WD, Horst OH, McGehee RV (1979) Effect of volcanic activity on fluvial-deltaic sedimentation in a modern arc-trench gap, southwestern Guatemala. Geol Soc Am Bull 90:827–838

Lautze NC, Harris AJ, Bailey JE, Ripepe M, Calvari S, Dehn J, Rowland SK, Evans-Jones K (2004) Pulsed lava effusion at Mount Etna during 2001. J Volcanol Geotherm Res 137(1–3):231–246. https://doi.org/10.1016/j.jvolgeores.2004.05.018

Lavigne F (2004) Rate of sediment yield following small-scale volcanic eruptions: a quantitative assessment at the Merapi and Semeru stratovolcanoes, Java, Indonesia. Earth Surf Process Landf 29:1045–1058

Lavigne F, Thouret J-C (2002) Sediment transportation and deposition by rain-triggered lahars at Merapi Volcano, Central Java, Indonesia. Geomorphology 49:45–69

Lavigne F, Thouret J-C, Voight B, Suwa H, Sumaryono A (2000a) Lahars at Merapi volcano, Central Java: an overview. J Volcanol Geotherm Res 100:423–456

Lavigne F, Thouret J-C, Voight B, Young K, LaHusen R, Marso J, Suwa H, Sumaryono A, Sayudi DS, Dejean M (2000b) Instrumental lahar monitoring at Merapi Volcano, Central Java, Indonesia. J Volcanol Geotherm Res 100:457–478

Lipman PW, Mullineaux DR (eds) (1981) The 1980 eruptions of Mount St. Helens, Washington. U S Geol Surv Prof Pap 1250:844 pp. https://doi.org/10.3133/pp1250

Lowe DR, Williams SN, Leigh H, Connor CB, Gemmell JB, Stoiber RE (1986) Lahars initiated by the 13 November 1985 eruption of Nevado del Ruiz, Colombia. Nature 324:51–53

Major JJ, Pierson TC, Dinehart RL, Costa JE (2000) Sediment yield following severe volcanic disturbance—a two-decade perspective from Mount St. Helens. Geology 28:819–822

Major JJ, Pierson TC, Hoblitt RP, Moreno H (2013) Pyroclastic density currents associated with the 2008–2009 eruption of Chaitén volcano (Chile)—forest disturbances, deposits, and dynamics. Andean Geol 40(2):324–358

Major JJ, Spicer KR, Mosbrucker AR (2021) Effective hydrological events in an evolving mid-latitude mountain river system following cataclysmic disturbance—a saga of multiple influences. Water Resour Res 57(2):e2019WR026851. https://doi.org/10.1029/2019WR026851

Manville VR (2002) Sedimentary and geomorphic responses to ignimbrite emplacement: readjustment of the Waikato River after the A.D. 181 Taupo eruption, New Zealand. J Geol 110:519–541

Manville V (2010) An overview of break-out floods from intracaldera lake. Global Planet Change 70(1–4):14–23

Manville V, Wilson CJN (2004) The 26.5 ka Oruanui eruption, New Zealand: a review of the roles of volcanism and climate in the post-eruptive sedimentary response. New Zeal J Geol Geophys 47:525–547

Manville V, White JDL, Houghton BF, Wilson CJN (1999) Paleohydrology and sedimentology of a post-1.8 ka breakout flood from intracaldera Lake Taupo, North Island, New Zealand. Geol Soc Am Bull 111:1435–1447

Manville V, Newton EH, White JDL (2005) Fluvial responses to volcanism: resedimentation of the 1800a Taupo ignimbrite eruption in the Rangitaiki River catchment, North Island, New Zealand. Geomorphology 65:49–70

Manville V, Nemeth K, Kano K (2009a) Source to sink: a review of three decades of progress in the understanding of volcaniclastic processes, deposits, and hazards. Sediment Geol 220:136–161

Manville V, Segschneider B, Newton E, White JDL, Houghton BF, Wilson CJN (2009b) Environmental impact of the 1.8 ka Taupo eruption, New Zealand: landscape responses to a large-scale explosive rhyolite eruption. Sediment Geol 220:318–336

Maria A, Carey S, Sigurdsson H, Kincaid C, Helgadóttir G (2000) Source and dispersal of jökulhlaup sediments discharged to the sea following the 1996 Vatnajökull eruption. Geol Soc Am Bull 112:1507–1521

Mathisen ME, Vondra CF (1983) The fluvial and pyroclastic deposits of the Cagayan Basin, Northern Luzon, Philippines—an example of non-marine volcaniclastic sedimentation in an interarc basin. Sedimentology 30:369–392

McGimsey RG, Waythomas CF, Neal CA (1994) High stand and catastrophic draining of intracaldera Surprise Lake, Aniakchak volcano, Alaska. In: Geologic studies in Alaska by the U.S. geological survey, 1993. U S Geol Surv Bull 2017:59–71. https://doi.org/10.3133/70180217

Mothes P, Hall ML, Andrade D, Samaniego P, Pierson TC, Ruiz AG, Yepes H (2004) Character, stratigraphy and magnitude of historical lahars of Cotopaxi volcano (Ecuador). Acta Vulcanol 16:85–108

Németh K, Cronin SJ (2007) Syn- and post-eruptive erosion, gully formation, and morphological evolution of a tephra ring in tropical climate erupted in 1913 in West Ambrym, Vanuatu. Geomorphology 86:115–130

Newhall CG, Punongbayan RS (1996) Fire and mud: eruptions of Mount Pinatubo, Philippines. Philippine Institute of Volcanology and Seismology, Quezon City, and University of Washington Press, Seattle, 1126 pp

Newhall CG, Self S (1982) The volcanic explosivity index (VEI) an estimate of explosive magnitude for historical volcanism. J Geophys Res Oceans 87(C2):1231–1238. https://doi.org/10.1029/JC087iC02p01231

Orton GJ (1996) Volcanic environments. In: Reading HG (ed) Sedimentary environments: processes, facies and stratigraphy. Blackwell Science, London, pp 485–567

Palmer BA, Purves AM, Donoghue SL (1993) Controls on accumulation of a volcaniclastic fan, Ruapehu composite volcano, New Zealand. Bull Volcanol 55:176–189. https://doi.org/10.1007/BF00301515

Pareschi MT, Favalli M, Giannini F, Sulpezio R, Zanchetta G, Santacroce R (2000) May 5, 1998, debris flows in circum-Vesuvian areas (southern Italy): insights for hazard assessment. Geology 28:639–642

Pierson TC (1995) Flow characteristics of large eruption-triggered debris flows at snow-clad volcanoes: constraints for debris-flow models. J Volcanol Geotherm Res 66:283–294

Pierson TC (1998) An empirical method for estimating travel times for wet volcanic mass flows. Bull Volcanol 60:98–109

Pierson TC, Major JJ (2014) Hydrogeomorphic effects of explosive volcanic eruptions on drainage basins. Annu Rev Earth Planet Sci 42:469–507. https://doi.org/10.1146/annurev-earth-060313-054913

Pierson TC, Janda RJ, Thouret JC, Borrero CA (1990) Perturbation and melting of snow and ice by the 13 November 1985 eruption of Nevado del Ruiz, Colombia, and consequent mobilization, flow, and deposition of lahars. J Volcanol Geotherm Res 41:17–66

Pierson TC, Janda RJ, Umbal JV, Daag AS (1992) Immediate and long-term hazards from lahars and excess sedimentation in rivers draining Mt. Pinatubo, Philippines. U S Geol Surv Water-Resour Invest Rep 92-4039, 35 pp

Pierson TC, Daag AS, Delos Reyes PJ, Regalado MTM, Solidum RU, Tubianosa BS (1996) Flow and deposition of posteruption hot lahars on the east side of Mount Pinatubo, July–October 1991. In: Newhall CG, Punongbayan RS (eds) Fire and mud: eruptions and lahars of Mount Pinatubo, Philippines. Philippine Institute of Volcanology and Seismology, Quezon City, and University of Washington Press, Seattle, pp 921–950

Pierson TC, Pringle PT, Cameron KC (2011) Magnitude and timing of downstream channel aggradation in response to a dome-building eruption at Mount Hood, Oregon. Geol Soc Am Bull 123:3–20

Pierson TC, Major JJ, Amigo Á, Moreno H (2013) Acute sedimentation response to rainfall following the explosive phase of the 2008–2009 eruption of Chaitén volcano, Chile. Bull Volcanol 75:723. https://doi.org/10.1007/s00445-013-0723-4

Pierson TC, Wood NJ, Driedger CL (2014) Reducing risk from lahar hazards: concepts, case studies, and roles for scientists. J Appl Volcanol 3:16. http://www.appliedvolc.com/content/3/1/16

Rodolfo KS, Arguden AT, Solidum RA, Umbal JV (1989) Anatomy and behaviour of a post-eruptive rain lahar triggered by a typhoon on Mayon Volcano, Philippines. Bull Int Assoc Eng Geol 40:55–66

Rodolfo KS, Umbal JV, Alonso RA, Remotigue CT, Paladio-Melosantos ML, Salvador JHG, Evangelista D, Miller Y (1996) Two years of lahars on the western flank of Mount Pinatubo: initiation, flow processes, deposits, and attendant geomorphic and hydraulic changes. In: Newhall CG, Punongbayan RS (eds) Fire and mud: eruptions and lahars of Mount Pinatubo, Philippines. Philippine Institute of Volcanology and Seismology, Quezon City, and University of Washington Press, Seattle, pp 989–1013

Romero JE, Morgavi D, Arzilli F, Daga R, Caselli A, Reckziegel F, Viramonte J, Díaz-Alvarado J, Polacci M, Burton M, Perugini D (2016) Eruption dynamics of the 22–23 April 2015 Calbuco Volcano (Southern Chile): analyses of tephra fall deposits. J Volcanol Geotherm Res 317:15–29

Scott KM, Macías JL, Naranjo JA, Rodríguez S, McGeehin JP (2001) Catastrophic debris flows transformed from landslides in volcanic terrains: mobility, hazard assessment, and mitigation strategies. U S Geol Surv Prof Pap 1630:59 pp

Scott KM, Vallance JW, Kerle N, Macías JL, Strauch W, Devoli G (2005) Catastrophic precipitation-triggered lahar at Casita volcano, Nicaragua: occurrence, bulking and transformation. Earth Surf Process Landf 30:59–79. https://doi.org/10.1002/esp.1127

Segerstrom K (1950) Erosion studies at Parícutin volcano, State of Michoacan, Mexico. In: Geologic investigations in the Parícutin area, Mexico. U S Geol Surv Bull 956-A:1–164

Segschneider B, Landis CA, Manville V, White JDL, Wilson CJN (2002) Environmental response to a large, explosive rhyolite eruption: sedimentology of post-1.8 ka pumice-rich Taupo volcaniclastics in the Hawke's Bay region, New Zealand. Sediment Geol 150:275–299

Siebert L, Simkin T, Kimberly P (2010) Volcanoes of the world, 3rd edn. University of California Press, Berkeley, p 568

Siebert L, Cottrell E, Venzke E, Andrews B (2015) Earth's volcanoes and their eruptions: an overview. In: Sigurdsson H, Houghton B, McNutt S, Rymer H, Stix J (eds) The encyclopedia of volcanoes, 2nd edn. Elsevier, Amsterdam, pp 239–254

Simkin T, Siebert L, McClelland L, Bridge D, Newhall C, Latter JH (1981) Volcanoes of the world: a regional directory, gazetteer, and chronology of volcanism during the last 10,000 years. Hutchinson Ross, Stroudsberg, PA, 232 pp

Smith GA (1987a) The influence of explosive volcanism on fluvial sedimentation: the Deschutes Formation (Neogene) in central Oregon. J Sediment Petrol 57:613–629

Smith GA (1987b) Sedimentology of volcanism-induced aggradation in fluvial basins: examples from the Pacific Northwest, U.S.A. In: Ethridge FG, Flores RM, Harvey MD (eds) Recent developments in fluvial sedimentology. Soc Econ Paleontol Mineral Spec Publ 39:217–228

Smith GA (1988) Neogene synvolcanic and syntectonic sedimentation in central Washington. Geol Soc Am Bull 100:1479–1492

Smith GA (1991) Facies sequences and geometries in continental volcaniclastic sediments. In: Fisher RV, Smith GA (eds) Sedimentation in volcanic settings. SEPM (Soc Sediment Geol) Spec Publ 45:109–121

Swanson DA, Holcomb RT (1990) Regularities in growth of the Mount St. Helens dacite dome, 1980–1986. In: Fink JH (ed) Lava flows and domes. The international association of volcanology and chemistry of the earth's interior, vol 2. Springer, Berlin, Heidelberg. https://doi.org/10.1007/978-3-642-74379-5_1

Taddeucci J, Edmonds M, Houghton B, James MR, Vergniolle S (2015) Hawaiian and Strombolian eruptions, chap 27. In: Sigurdsson H, Houghton B, McNutt SR, Rymer H, Stix J (eds) The encyclopedia of volcanoes, 2nd edn. Elsevier, Amsterdam, pp 485–503. https://doi.org/10.1016/B978-0-12-385938-9.00027-4

Thouret J-C, Abdurachman KE, Bourdier J-L, Bronto S (1998) Origin, characteristics, and behavior of lahars following the 1990 eruption of Kelud volcano, eastern Java (Indonesia). Bull Volcanol 59:460–480

Vallance JW, Scott KM (1997) The Osceola Mudflow from Mount Rainier: sedimentology and hazard implications of a huge clay-rich debris flow. Geol Soc Am Bull 109:143–163. https://doi.org/10.1130/0016-7606(1997)109<0143:TOMFMR>2.3.CO;2

Van Houten FB (1976) Late Cenozoic volcaniclastic deposits, Andean foredeep, Colombia. Geol Soc Am Bull 87:481–495

Vessell RD, Davies KK (1981) Nonmarine sedimentation in an active fore arc basin. In: Ethridge FG, Flores RM (eds) Nonmarine depositional environments: models for exploration. Soc Econ Paleontol Mineral Spec Publ 31:34–45

Voight B (1990) The 1985 Nevado del Ruiz volcano catastrophe: anatomy and retrospection. J Volcanol Geotherm Res 42(1–2):151–188

Voight B, Glicken H, Janda RJ, Douglass PM (1981) Catastrophic rockslide avalanche of May 18. In: Lipman PW, Mullineaux DR (eds) The 1980 eruptions of Mount St. Helens, Washington. U S Geol Surv Prof Pap 1250:347–377

Voight B, Janda RJ, Glicken H, Douglass PM (1983) Nature and mechanics of the Mount St. Helens rockslide-avalanche of 18 May 1980. Geotechnique 33:243–273. https://doi.org/10.1680/geot.1983.33.3.243

Waldron HH (1967) Debris flow and erosion control problems caused by the ash eruptions of Irazu volcano, Costa Rica. U S Geol Surv Bull 1241-I:37 pp

Waythomas CF, Walder JS, McGimsey RG, Neal CA (1996) A catastrophic flood caused by drainage of a caldera lake at Aniakchak volcano, Alaska, and implications for volcanic hazards assessment. Geol Soc Am Bull 108(7):861–871

Waythomas CF, Pierson TC, Major JJ, Scott WE (2013) Voluminous ice-rich lahars generated during the 2009 eruption of Redoubt Volcano, Alaska. J Volcanol Geotherm Res 259:389–413

White JDL, Houghton BF, Hodgson KA, Wilson CJN (1997) Delayed sedimentary response to the A.D. 1886 eruption of Tarawera, New Zealand. Geology 25(5):459–452

Zernack AV (2021) Volcanic debris-avalanche deposits in the context of volcaniclastic ring plain successions—a case study from Mt. Taranaki. In: Roverato M, Dufresne A, Procter J (eds) Volcanic debris avalanches. Advances in volcanology, vol 6. Springer, Cham, pp 211–254. https://doi.org/10.1007/978-3-030-57411-6_9

Zernack AV, Procter JN, Cronin SJ (2009) Sedimentary signatures of cyclic growth and destruction of stratovolcanoes: a case study from Mt. Taranaki, New Zealand. Sed Geol 220(3–4):288–305. https://doi.org/10.1016/j.sedgeo.2009.04.024

Deposit Identification and Interpretation

Abstract

Volcanic and volcaniclastic fragmental deposits can be identified and differentiated through examination of their sediment-body characteristics, the physical properties of their constituent particle mixtures, and any impacts to terrain caused by the processes that emplace them. These identification criteria are observable and measurable at different scales: landscape scale (kilometers to meters), outcrop scale (meters to millimeters), and particle scale (millimeters and smaller). Landscape-scale criteria (seen on the ground surface and in vertical sections) involve the location, geometry, internal architecture, and lithofacies associations of the sediment body as a whole. Outcrop-scale criteria include the size, size distribution, composition, spatial arrangement, and physical characteristics of the deposit's easily visible coarser particles. Particle-scale criteria include the size, size distribution, composition, spatial arrangement, and physical characteristics of the deposit's finer particles, which are typically evaluated in the laboratory. Terrain impacts mainly include erosional and depositional structures or patterns, physical damage to trees and other vegetation, and occasionally chemical damage to vegetation. This chapter illustrates and discusses the identification criteria used in this book for most types of relatively fresh volcanic and volcaniclastic deposits. Identification criteria for lithified, chemically altered, profoundly weathered, or unusual types of volcanic deposits are not covered.

1 Examining Deposits at Different Scales

Sequences of volcanic sediments comprise particles emplaced by multiple depositional events over time, typically involving several different processes. Once a sediment body has been emplaced, criteria useful for identification and interpretation can be applied at three different scales: the landscape, outcrop, and particle scales (generalized in Table 1). It is seldom sufficient to examine a deposit at just one scale. These criteria involve the size, composition, spatial arrangement, and physical characteristics of the deposit's constituent particles, as well as the location, geometry, and internal architecture of the sediment body as a whole.

Landscape-scale criteria are typically evaluated from remote sensing imagery or by synthesizing observations from multiple field exposures. Outcrop-scale characteristics are examined in the field, typically at vertical exposures of the deposit in road cuts, quarry walls,

T. C. Pierson et al., *Deposits of Volcanic Wet Flows*, Advances in Volcanology,
https://doi.org/10.1007/978-3-031-66574-5_3

Table 1 Scales at which all volcaniclastic deposits can be evaluated for identification and interpretation

Landscape-scale evaluation criteria	Outcrop-scale evaluation criteria	Particle-scale evaluation criteria
Shape of sediment body as a whole and nature of contacts between the sediment body and other units	Geometry of bedding and bed forms (or lack thereof), and nature of contacts between beds	Particle composition and degree/type of chemical alteration
Internal structure of subunits (lithofacies) and characteristics of their bounding surfaces	Degree of uniformity in spatial distribution of particles of different size or composition Degree/type of coloration of unit and subunits	Shape and size of particles. Shape and size of voids between particles (bubble vesicles or intergranular spaces)
Spatial arrangement or orientation of lithofacies	Grading Spatial orientation of coarse particles with respect to each other and to deposit substrate (fabric of deposit)	Statistical parameters of particle-size distribution
Erosional impacts to landscape by transport and depositional processes	Sedimentary structures and internal deformation	Particle roundness, surface roughness, and degree of fracturing
Types and degree of damage to vegetation by transport and depositional processes	Consistency: degree of hardness from compaction or cementation	Characteristics of any cementing agent between clastic particles, coating particles, or lining voids Presence/absence of matrix between clasts

Examples of sediment and deposit properties used as evaluation criteria shown here are defined and discussed in the following sections or riverbanks. Particle-scale properties can be estimated in the field with a hand lens, but quantification requires bulk sampling and laboratory analysis.

2 Landscape-Scale Evaluation Criteria

Landscape-scale features of volcanic wet-flow deposits are physical characteristics too large to be evaluated from a single outcrop. These deposits can be up to kilometers in length or width and hundreds of meters in thickness. Other fundamental features are the geometric shape and position of the sediment body on the landscape, commonly characterized by its planform length-to-width aspect ratio, thickness-to-area aspect ratio,[1] shape and nature of the exterior and interior bounding surfaces (contacts), distance from source, and topographic position with respect to underlying relief. Additionally, large sedimentary structures on deposit surfaces can be found at this scale, including channel bars and berms, compression ridges, grabens, lateral shears, explosion pits, pull-aparts, boulder clusters, hummocks, and flow levees. Together, the landscape-scale physical characteristics of the sediment body itself define its *architecture*. Large-scale but indirect clues to a deposit's origin can also be discerned from physical impacts to the landscape caused by the flow or other process that emplaced the deposit—the type (physical or chemical) and severity of damage to vegetation and the degree and extent of soil and bedrock erosion along flow paths.

Inside a depositional body of sediment, three-dimensional spatial variations in sedimentary textures and structures define mappable zones of sediment having similar physical and chemical characteristics—the different *lithofacies* within the body (Miall 1996). Lithofacies is a landscape-scale construct that requires descriptive characterizations based on evaluations of sediment textures and structures from multiple outcrops.

[1] Thickness-to-area aspect ratios for volcanic deposits are usually determined by dividing mean thickness of the deposit by the diameter of a circle equivalent in area to the area of the deposit (Neuendorf et al. 2005).

2.1 Deposit Architecture and Morphology

2.1.1 Sediment-Body Shape

The three-dimensional shape of a deposit can be defined in terms of *aspect ratios*, which highlight the relative differences between two geometric attributes: the length to width ratio (L/W) and the thickness to planform area ratio (H/A). Shape can also be described by comparing the relative values of the three orthogonal axes of the sediment body. End-member geometric forms include ones in which length, width, and thickness are approximately equivalent (an *equant* form); ones in which thickness is much less than length and width (a sheet shape, or oblate form); and ones in which both thickness and width are much less than length (a ribbon or rod shape, or prolate form). Complexity is added when axes are curvilinear and when thickness and/or width vary with distance.

Equant forms of deposit bodies are rare. Sheet-like deposits typically result from tephra fall and from flow deposits emplaced on low-gradient undissected terrain (on debris fans or aprons). Ribbon-shaped deposits require confinement by channel or valley sides or by flow levees, and they are more typically found on steeper slopes within incised topography. In addition, concentrated-flow deposits tend to terminate more abruptly in the downstream/downslope direction than dilute-flow deposits.

2.1.2 Structures and Bounding Surfaces

Sediment bodies are circumscribed by unit bounding surfaces—the contacts between the sediment body and the material surrounding it, as well as its upper free surface (if not buried by other deposits). The term *unconformity* is typically applied to geologic units of significantly

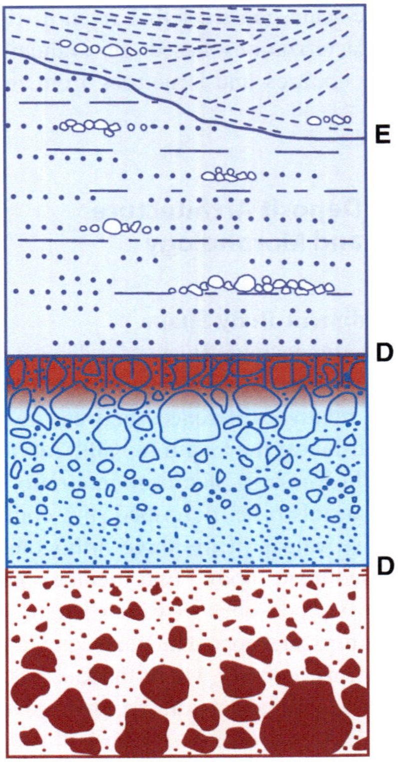

Fig. 1 Examples of unconformable bounding surfaces in lahar and water-flood deposits that could be encountered in vertical outcrop faces. An erosional unconformity (E) is where the time gap is marked by truncation of strata by scour. Disconformities (D) are found where a time gap between units is marked by a bedding-parallel erosional surface. Disconformities may be recognized by a buried soil at the top of the unit beneath the disconformity (indicated by the reddish zone below the upper D), by an abrupt change in sediment type or texture (indicated by the change in sediment texture at the lower D), or by a uniform change in degree of particle weathering (commonly displayed in reddish or yellowish orange oxidation of iron-bearing minerals in clasts below a contact but not above it)

different ages, but the terminology is useful in describing types of contacts between more contemporaneous flow deposits in sequences of volcaniclastic sediments. At outcrops, the challenge is often in trying to differentiate contacts between different flow units and bedding variations within single flow units.

Unconformable contacts represent periods of nondeposition, even if short. Nondeposition may or may not involve erosion (Fig. 1). Pauses in deposition can be brief, perhaps only minutes long, but duration of nondeposition usually cannot be determined unless other evidence is present, such as datable material above and below the contact. Longer hiatuses may be inferred where units on either side of a contact exhibit different degrees of weathering, organic layers are sandwiched between the units, or a paleosol (buried soil) is present at the top of the lower unit. In some cases, however, such indicators of long hiatuses can be eroded off (bed-parallel erosion) by the flow that deposited the upper unit.

Angular unconformities are clearly erosional, formed in sediments where scour by an overriding flow truncates strata in the underlying deposit. *Disconformities* are unconformable surfaces between parallel sets of strata and may or may not represent erosion. They can sometimes be recognized by an abrupt change in sediment type or texture above and below a contact. Contemporaneous unconformities also exist within individual sediment bodies, separating different lithofacies or groupings of bed forms within lithofacies (Miall 1988, 1996; Collinson 1996).

Upper free surfaces of sediment bodies have additional features that can be helpful in process identification. Examples of such features include the overall flatness or convexity of surfaces, both in the main part of a sediment body and at its margins; the relative roughness/smoothness of surfaces, and patterns of abrupt or gradual changes in positive or negative surface relief; and the mean size, type, and spatial distribution of roughness elements (e.g., pits, grabens, cracks, boulders, hummocks, compression ridges, levees, woody debris, etc.).

2.2 Lithofacies

A *lithofacies*[2] is a spatially definable (laterally and vertically) and mappable subdivision of a

[2] Term always written in plural form, whether referring to one or several types.

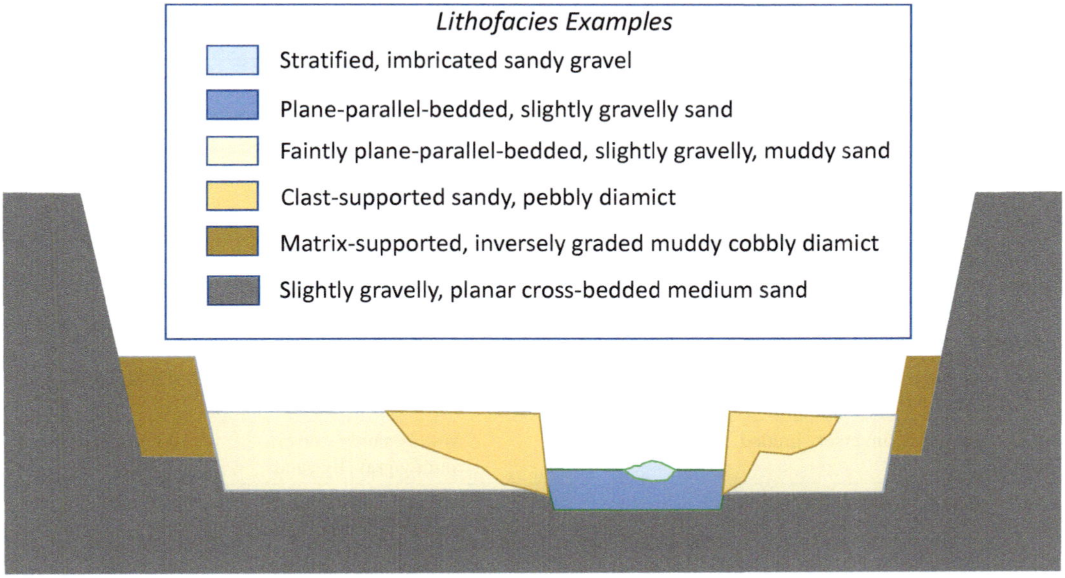

Fig. 2 Examples of lithofacies (zones sediment with consistent textural and structural properties) identified in a schematic valley cross-section of a river valley draining a volcano (see Table 2 for the codes that can be used to note sediment textures and lithofacies in field notes). In this example, the deposits record the passage of a large lahar. Examination of multiple such exposures allows determination of the extent of landscape-scale characteristics

stratigraphic unit, distinguished by the dominance of a consistent material composition, sedimentary texture,[3] and structural elements—characteristics that are produced under a particular set of depositional conditions and discerned from outcrop- and particle-scale features and properties (Miall 1978, 1996; Reading and Levell 1996; Neuendorf et al. 2005) (Fig. 2). A lithofacies may extend laterally over tens or hundreds of meters and is thus classified as a landscape-scale feature, but the outcrop-scale and particle-scale properties needed to define a lithofacies are typically collected at much smaller exposures in multiple locations. Once a lithofacies is defined, however, interpretation of depositional process can seldom be done when examining that lithofacies in isolation. The association of one lithofacies with surrounding lithofacies is critically important. Thus, *lithofacies associations* are the combinations of lithofacies that, when examined together, allow both the delineation of the depositional elements forming a sediment body and the interpretation of specific depositional processes and environments (Smith 1987). Another term in common use, *depositional facies*, is more or less synonymous with lithofacies association. Whereas individual lithofacies may not be process-specific, lithofacies associations can be used to infer the process or processes that emplace volcaniclastic deposits (e.g., Mathisen and Vondra 1983; Scott 1988a, b; Glicken 1986, 1996; Palmer et al. 1991; Smith 1987, 1991; Vallance and Scott 1997).

Systematic classification and coding of lithofacies (e.g., Smith 1987; Miall 1996) are now standard in sedimentology and are useful for volcaniclastic deposits (Table 2, left column).

[3]A deposit's *sedimentary texture* has been defined by the particle size (absolute size and size distribution), shape, roundness, and fabric (Neuendorf et al. 2005). The size and spatial distribution of voids in a deposit also aids in defining texture.

Table 2 Simplified codes for sedimentary lithofacies encountered in volcaniclastic deposits (left side of table) and codes for sedimentary textures of those same deposits (right side of table)

Lithofacies description	Facies code	Texture description	Texture code
FRACTURED ROCK—deposits composed of angular blocks rock; minor interblock and intrablock matrix is crushed rock			
Breccia blocks (< 4.1 m diameter)	Rm	Slightly sandy gravel; slightly muddy sandy gravel	sG; (m)sG
Megabreccia (blocks ≥ 4.1 m in diameter)	Rb	Slightly sandy gravel; slightly muddy sandy gravel	sG; (m)sG
DIAMICT—poorly sorted, mainly unstratified mixture of particles ranging in size from clay to boulders			
Matrix-supported, massive diamict (< 10% stratified)	Dmm	Muddy sandy gravel; muddy gravelly sand	msG; mgS
Matrix-supported, inversely graded diamict	Dmg(i)	Muddy sandy gravel; muddy gravelly sand	msG; mgS
Matrix-supported, normally graded diamict	Dmg(n)	Muddy sandy gravel; muddy gravelly sand	msG; mgS
Matrix-supported, weakly stratified diamict (> 10% stratified)	Dms	Muddy sandy gravel; muddy gravelly sand	msG; mgS
Clast-supported massive diamict (< 10% stratified)	Dcm	Muddy sandy gravel; muddy gravelly sand	msG; mgS
Clast-supported, graded diamict	Dcg	Muddy sandy gravel	msG; (m)sG
Clast-supported, weakly stratified diamict (> 10% stratified)	Dcs	Muddy sandy gravel	msG; (m)sG
GRAVEL—clastic particles larger than 2 mm and finer than 4096 mm in diameter (includes cobbles and boulders)			
Openwork gravel (having voids between gravel particles)	Go	Gravel; slightly sandy gravel	G; (s)G
Massive gravel (< 10% stratified)	Gm	Gravel; sandy gravel	G; sG
Imbricated gravel (stratified with long and intermediate axes of tabular particles dipping upstream)	Gi	Gravel; sandy gravel	G; sG
Trough cross-bedded gravel (stratified)	Gt	Gravel; sandy gravel	G; sG
Planar cross-bedded gravel (stratified; low-angle cross beds < 15°)	Gp	Gravel; sandy gravel	G; sG
Horizontally bedded gravel (stratified)	Gh	Gravel; sandy gravel	G; sG
SAND—particles larger than 0.063 mm and finer than 2 mm in diameter			
Massive sand (< 10% stratified)	Sm	Sand; slightly gravelly sand; gravelly sand	S; (g)S; gS
Horizontally bedded sand	Sh	Sand; slightly gravelly sand; gravelly sand	S; (g)S; gS
Trough cross-bedded sand (stratified)	St	Sand; slightly gravelly sand; gravelly sand	S; (g)S; gS
Planar cross-bedded sand (stratified)	Sp	Sand; slightly gravelly sand; gravelly sand	S; (g)S; gS
Sand (stratified) with scour structures	Ss	Sand; slightly gravelly sand; gravelly sand	S; (g)S; gS
Ripple cross-laminated sand (stratified)	Sr	Sand	S
Plane-parallel laminated sand (stratified)	Sl	Sand	S

(continued)

Table 2 (continued)

Lithofacies description	Facies code	Texture description	Texture code
MUD—particles finer than 0.063 mm in diameter (silt and clay), also called fines			
Massive mud or sandy mud (massive)	Fm	Mud; slightly sandy mud; sandy mud	M, (s)M; sM
Laminated mud (stratified) with fine plane-parallel or wavy laminations	Fl	Mud; slightly sandy mud; sandy mud	M, (s)M; sM
Ripple cross-laminated mud	Fr	Mud; slightly sandy mud; sandy mud	M, (s)M; sM

See Table 5 for size class limits that define gravel (G or g), sand (S or s), and mud (M or m) textures and for equivalent pyroclastic particle terminology. Lithofacies codes here are modified from Miall (1978, 1996) and Smith (1987). Note that these two types of codes can be combined for field notation, with lithofacies modifiers to the right of the main sediment type (R, D, G, S, or F/M) and texture modifiers to the left; importance or significance of the modifier decreases away from the capital letter sediment type. Textural codes, based on frequency of different grain sizes in the deposit, are from Folk (1954). Lithofacies codes use F (fines) instead of M (mud) to denote particles finer than 0.063 mm, so either one can be used when combining codes. Small amounts of a textural component are indicated with parentheses, for example, (m)sG for slightly muddy sandy gravel. In deposits containing gravel-size clasts, the term *matrix* commonly refers to particles in the sediment mixture finer than 2 mm in diameter—a contrast readily seen with the naked eye in outcrops. More broadly, however, *matrix* is defined as finer clastic particles filling the interstices between larger particles at any scale (Neuendorf et al. 2005)

Codes used to describe sediment texture without genetic interpretation (Folk 1954) (Table 2, right column) resemble lithofacies codes and can be used in conjunction with them. Both lithofacies and texture codes are built around a capital letter that represents the dominant particle size or type. In lithofacies codes, descriptive modifier abbreviations (one or more small-case letters) are placed to the right of the capital letter, whereas in texture codes the modifying abbreviations are placed to the left. This convention allows the two systems to be combined, and additional modifying abbreviations can be applied as needed. For example, a sub-horizontal plane bedded sand unit, "Sh" (the lithofacies code), which texturally can be described as a muddy gravelly sand, "mgS" (the texture code), forms the combined code "mgSh." Such combined codes can be quite useful in describing and differentiating sequences of volcaniclastic deposits (e.g., Mathisen and Vondra 1983; Smith 1987), and their use can speed up notetaking in the field.

2.3 Impacts to Terrain

Both volcanic wet flows and pyroclastic flows and falls damage vegetation and alter terrain through erosion and deposition of material—impacts that provide clues to the type of sediment-emplacement process causing the damage, as well as cause larger and more frequent lahars and floods than would otherwise occur.

2.3.1 Forest Vegetation Damage

Flowing and falling volcaniclastic particles damage forest vegetation in impact zones in different ways and to different degrees (Table 3), and vegetation damage can vary from minimal disturbance to complete destruction (Swanson and Major 2005; Ayris and Delmelle 2012; Swanson et al. 2013). Variables determining the type and degree of damage to vegetation are related to the kind of sediment emplacement process and distance from the source vent. In flows, key variables include the magnitude, velocity,

Table 3 Range of damaging effects to forest vegetation within and bordering paths/areas of volcanic wet flows, pyroclastic density currents, and tephra falls

Vegetation damage	Physical effects of particle–fluid flows	Physical effects of tephra falls	Thermal and chemical effects of flows and falls
Extreme	Trees felled and removed. Almost all trees sheared off at ground level and carried away (many broken into pieces). Remaining upright trees stripped of bark, branches, and leaves. Understory vegetation stripped away, leaving bare mineral soil	Trees stripped of all leaves and most branches. Understory vegetation completely buried and may be burned	*Thermal*: Charring/ignition of above-ground trees and understory vegetation. Wood (including logs) completely charred if buried in deposits *Chemical*: Total leaf kill
Severe	Most trees toppled. Some trunks broken off above ground level; others tipped over but still rooted and not broken. Branches largely broken off. Standing trees have severely abraded bark and wood on up-current side. Understory vegetation partially buried, flattened, or stripped away	Trees stripped of all or most leaves by abrasion from coarse tephra, or heavy breakage of branches by loading by fine tephra. Understory vegetation bowed and partially buried	*Thermal*: Intense scorching by flows with thin char layer on wood facing upslope; bark remains on downslope side. Extensive charring of wood buried in deposits. Exposed leaves singed and killed *Chemical*: Total leaf kill
Moderate	Some trees toppled but most still standing (young trees bent over) and sustain abrasional damage. Some bark on up-current sides of standing trees abraded in separated patches (near ground level and near peak-flow level). Understory vegetation abraded, flattened, and broken	Trees stripped of some leaves; minor damage to branches. Understory vegetation damaged in proportion to tephra thickness	*Thermal*: Intense blistering and desiccation of bark and outer wood (with localized light charring) of standing trees. Slight to moderate charring of wood when buried in deposits. Total leaf kill *Chemical*: Significant leaf kill
Light	All trees still standing with minor damage to bark on side facing upslope. Some small trees bent over. Varied abrasional damage to understory vegetation, with partial burial in deposits	Little or no physical damage to leaves. Leaves may be coated in fine ash that washes off later during rainfall	*Thermal*: No charring of standing or buried wood; patchy leaf kill *Chemical*: Patchy leaf kill

Interpreted from Waitt (1981), Hupp et al. (1987), Abdurachman et al. (2000), Pierson (2005a), Druitt et al. (2002), Loughlin et al. (2002), and Swanson et al. (2013)

concentration, duration, temperature, particle-size distribution, and pH of flowing particle–fluid mixtures, as well as position on a valley floor relative to the main axis of flow. Damage generally decreases with distance away from the source vent and from the channel axis, but even vegetation outside the actual flow limits can be splashed or heated by passing flows. In the case of falling tephra, variables affecting degree of damage include the size, roughness, density, and temperature of tephra particles and the fall duration, as well as pH of acidic gas condensates adsorbed on particle surfaces.

Degree of tree mortality in affected forest stands depends not only on immediate abrasion and blunt-force injury to plant tissues, but also on the delayed effects of burial of root systems and the physical, thermal, or chemical damage to leaves. In addition, apparent tree mortality may not be a good indicator of long-term damage. Moderate heat or chemical impacts can kill the affected leaves but not the trees, with new leaves budding out after a few months or years. Some tree species suffering death of aboveground plant tissues may still be able to regenerate by resprouting from roots in subsequent growing seasons.

2.3.2 Erosional Effects

All volcanic particle–fluid flows can erode landscapes. Severity varies with flow volume and depth, slope gradient (velocity), size and concentration of transported particles, and the type of substrate over which a flow moves. Substrates on flanks of stratovolcanoes are commonly soft and readily erodible where not paved by lava rock or compacted glacial deposits. Erosion by flows during and following single eruptions can cause pronounced bedrock abrasion (striations or grooves) and soil erosion (Pierson et al. 1990; Sparks et al. 1997), deep erosion of glaciers on volcano summits (Brugman and Meier 1981; Pierson et al. 1990; Waythomas et al. 2013), and incision of channels on slopes and in valley floors (Kieffer and Sturtevant 1988; Scott et al. 1996). Different types of flows leave different erosional signatures, which are discussed in following chapters. Tephra fall is unlikely to cause erosion on hillslopes, except in the cases of large ballistic fragments cratering into soft substrates on the ground or oversteepened tephra layers sloughing off hillsides.

3 Outcrop-Scale Evaluation Criteria

Properties of the particle mixtures comprising deposits are considered in this section at the scale of a typical field outcrop. Dominant particle lithology, sedimentary texture, and the presence or absence of internal structures are key assessment criteria. In addition, degree of cohesion in a deposit (due to consolidation, clay content, or cementation) along with indicators of high emplacement temperature, can greatly aid deposit identification.

Sedimentary texture of a deposit can be subjectively characterized by its overall appearance at an outcrop—the average size of its particles (mean, median, and mode), the size variability of those particles (sorting), and average spatial orientation of gravel-size clasts (fabric). The size, type, and abundance of voids between particles also contribute to sedimentary texture. However, measurements on outcrop faces, combined with bulk sampling and laboratory analysis at the particle scale, are required for an accurate assessment of sedimentary texture. Note that the term "texture" also can be applied to surfaces of individual particles—*surface texture*. This includes the presence or absence of pits, striations, or gouges on particle surfaces resulting from abrasion and collisions during transport, commonly best seen on gravel-size clasts in outcrops.

Internal structures include stratification (bedding) or lack thereof, bed forms, fluid-escape structures and accompanying soft-sediment deformation, and shear deformation structures. Another aspect of deposit structure is the spatial relationship between gravel-size clasts and matrix material (particles finer than 2 mm in diameter). When gravel-size clasts in the mixture are separated from each other and appear to be floating in the matrix, the structure is described as *matrix-supported*, and conversely when gravel-size particles in the sediment mixture are in contact with each other and matrix fills the gaps in-between, the structure can be described as *clast-supported* or *framework-supported*. Gravel clasts in contact with each other but without much or any matrix between clasts form an *openwork* structure.

Over time, bulk sedimentary and particle-surface textures evolve in response to chemical and physical processes, which for volcaniclastic sediments include compaction, devitrification, solution, precipitation, hydration, and diagenetic or hydrothermal alteration (crystallization of new minerals). Each process modifies and overprints the original textures on the way to an unconsolidated deposit being lithified to solid rock, with additional modification ultimately added by metamorphism and tectonic deformation (McPhie et al. 1993).

3.1 Stratification (Bedding)

Stratification results from the direct deposition of granular material in layers (strata) of varying thickness and character (Neuendorf et al. 2005), and it partly defines the internal structure of a deposit. Such layers are commonly distinguished by abrupt changes in sedimentary texture, cohesion, or color. Pseudostratification (pseudobedding) is layering in deposits that has the appearance of stratification, but which has been produced by processes other than sedimentary deposition during deposit accumulation. These processes can include (1) shearing along parallel shear planes formed within a deposit by shear stress imposed by an overpassing flow and (2) layering of localized unconformities created by alternating deposition and erosion that forms a vertical sequence of erosional surfaces during passage of a flow. In the second mechanism, boundary surfaces may or may not cut across depositional layers. Pseudostratification also can result from postdepositional processes—oxidation, cementation, or leaching and translocation of chemical compounds and minute particles due to the movement and action of groundwater, particularly above zones of relative impermeability within a deposit sequence. Stratification (type, orientation, and spacing of strata) is one of the key criteria for distinguishing different emplacement mechanisms.

Strata (layers of sediment) < 1 cm thick are referred to as *laminae* or *laminations*, whereas thicker layers are termed *beds*. Depositional units (the beds deposited during a single emplacement event) may lack any discernable internal bedding if particles are well-sorted and of similar size and composition. Boundaries between depositional units likewise may not be obvious. Field geologists often know, however, that bedding is commonly subtle and that contacts between packets of similar sediment can appear on outcrop faces after a few weeks or months of etching by rain or wind. Important variables in determining the character of the bedding are: (1) particle concentration of the flow just above the bed during deposition; (2) steadiness of the flow; (3) fluid properties (density and viscosity); (4) particle properties (size, shape, sorting, and density); and (5) flow properties (velocity and depth). Contacts between beds are sharpest when changes in any of these variables are abrupt.

Bedding-plane boundaries can be parallel, subparallel, or at a sharp angle to the overall bedding orientation of the deposit (Fig. 3). Subhorizontal bedding parallel to a channel bed is *plane-parallel bedding*.[4] Bedding at an angle with respect to the plane of the channel bed is termed *cross-bedding*, and it can be low-angled (< 15°) or high-angled (> 15° up to about 30°). Bedding orientation, the relative sharpness or diffuseness of boundaries between beds, average thickness of bedding layers, and relative differences between mean particle sizes in juxtaposed beds are all important aspects to note in outcrops.

[4] Plane-parallel bedding typically is within a few degrees of horizontal and is sometimes called "horizontal bedding."

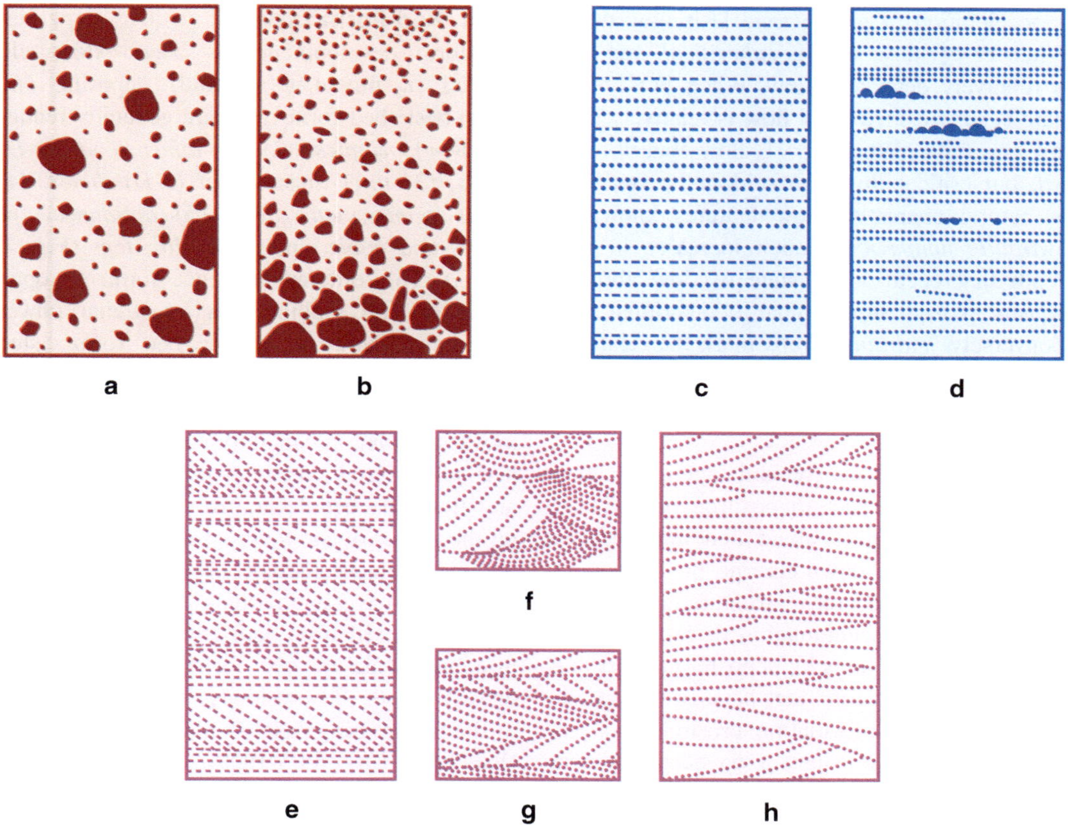

Fig. 3 Basic types of bedding commonly encountered in depositional units of volcanic sediments, as viewed in vertical outcrop faces. Definitions are expanded in text. **a** Massive (unstratified) diamict, ungraded. **b** Massive diamict, normally graded. **c** Plane-parallel bedding in relatively well-sorted sandy deposit. **d** Plane-parallel bedding in poorly sorted sand with gravel lenses. **e** Cross-bedded sand with high-angle (> 15°) planar tabular cross-beds, formed by flow moving left to right and produced by two-dimensional ripple or dune bed forms (outcrop face is parallel flow direction). **f** High-angle trough cross-bedding in sandy deposit produced by three-dimensional dunes (outcrop face is transverse to flow direction). **g** High-angle trough cross-bedding in sandy deposit produced by three-dimensional dunes (outcrop face is oblique to flow direction). **h** Low-angle cross-bedding (< 15°) dipping both upstream and downstream (outcrop face is parallel to flow direction). Diagrams **f** and **g** are modified from McKee and Weir (1953)

3.2 Bed Forms

Bed forms are localized, three-dimensional (3-D), depositional or erosional topographic structures (pits, troughs, mounds, wedges, or ridges) that develop on bedding surfaces at the interface between a static bed and a particle–fluid flow that is depositing bed load. The type of bedding preserved in deposits (and observed in two-dimensional [2-D] vertical outcrop faces in those deposits) is determined by the bed form(s) preserved by burial in aggrading sediment.

Bed-form type is indicative of the *state of flow*, which is determined primarily by the relative values of flow depth and flow velocity at the time of deposition and defined by the Froude number (Fr) of the depositing open-channel flow (Chow 1959; Harms and Fahnestock 1965; Simons et al. 1965; McBride et al. 1975; Costello and Southard 1981). In its simplest form, $Fr = V/(gD)^{1/2}$, where V is flow velocity, g is acceleration of gravity, and D is flow depth. When $Fr < 1$, flow state is *subcritical*, and flow is relatively deep and slow. When $Fr > 1$, flow state is *supercritical*, and flow is relatively shallow and rapid. When $Fr = 1$, flow state is *critical*, but critical flow can seldom be sustained in natural channels. Flow features known as *hydraulic jumps* are sudden increases in flow depth that occur when state of flow abruptly transitions from supercritical to subcritical flow. Conversely, *hydraulic drops* mark the relatively abrupt decrease in flow depth that occurs as state of flow transitions from subcritical to supercritical flow.

Some bed forms remain stationary during flow with respect to position on the bed, while others migrate along bedding surfaces, advancing either down-current or up-current. Preservation of bed-form structures in deposits requires the net accumulation of sediment and burial of the structures by deposition. Most outcrops reveal bed forms that have been partially truncated by post-formation erosion, and patterns of localized erosion surfaces can be just as important as the pattern within preserved beds for process identification.

Bed forms are particularly common and diverse in water-flood deposits. They scale with flow depth and are classified as microforms or mesoforms, as opposed to larger (landscape-scale) sedimentary structures (macroforms) that scale with channel dimensions (Miall 1996). Bed forms are particularly important for analyzing the flow conditions of water floods (Simons and Richardson 1962; Harms and Fahnestock 1965; Langbein and Leopold 1968; Allen 1970; Smith 1971a, b; Miall 1977, 1996; Blodgett and Stanley 1980; Hand and Bartberger 1988; Ashley 1990; Collinson 1996; Slootman and Cartigny 2020). However, the preservation of bed forms is not necessarily diagnostic of deposition by water floods because they also occur in lahar, pyroclastic density-current (PDC), and eolian deposits at volcanoes. Dimensions of bed forms can be orders of magnitude larger than the particles that form them, and they can occupy a significant fraction of the depth of the flow that creates them.

3.2.1 Bed-Form Types

Bed-form development has been extensively studied in fluvial (water-laid) deposits, and seven basic bed-form types are generally recognized—(1) *ripples*, (2) *two-dimensional (2-D) dunes*, (3) *three-dimensional (3-D) dunes*, (4) *upper-regime plane beds*, (5) *antidunes*, (6) *chutes and pools*, and (7) *cyclic steps* (Figs. 4 and 5). This ordered sequence forms in open channels with mobile beds as stream power increases. There is another bed form listed in some classifications termed *lower-regime plane bed* that occurs when flow just begins to move bed load, but it is typically not found in high-relief, high-energy volcanic settings. In any flow, the bed form that develops depends on the power of the flow to move sediment and on the sediment particle properties that resist movement—average particle size, shape, density, size distribution, and cohesion. The structure and scale of bed forms also vary with density, depth, flow velocity, and flow duration of the transporting fluid (Schmincke et al. 1973; Wan 1985; Cas and Wright 1987). Ripples and dunes are indicative of subcritical turbulent flow (*lower flow regime*). Dunes wash out and *upper-regime plane beds* form at the transition from subcritical flow, through critical flow, to supercritical turbulent flow (*upper flow regime*). The first bed forms to develop in upper-regime flow are *antidunes*. As stream power continues to increase, upstream-breaking antidune waves at the flow surface evolve into hydraulic jumps. Small, hydraulic jumps form *chute-and-pool* bed forms and larger hydraulic jumps form *cyclic-step* bed forms (Simons et al. 1965; Costello and Southard 1981; Ashley 1990; Miall 1996; Fielding 2006; Kostic et al. 2010; Cartigny et al. 2014; Slootman and Cartigny 2020; Slootman et al. 2021).

The development, characteristics, and behavior of the different bed-form types have been documented from numerous experimental flume studies and field observations over the last half century (e.g., Simons and Richardson 1962; Simons et al. 1963, 1965; Harms and Fahnestock 1965; McKee et al. 1967; Costello

2-D Dunes (with superimposed ripples)

Tabular-planar cross-bedding (high angle, dipping downstream when viewed transverse to flow direction). Ripples and dunes have similar forms, but ripples are < 5 cm tall and dunes are > 5 cm tall.

Lower flow regime

a

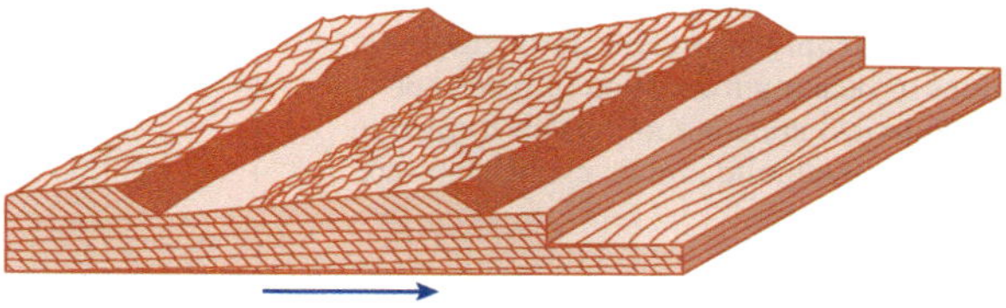

3-D Dunes

Wavy cross-bedding (high and low angle, dipping downstream when viewed oblique and transverse to flow direction); trough cross-bedding when viewed parallel to flow direction.

Lower flow regime

b

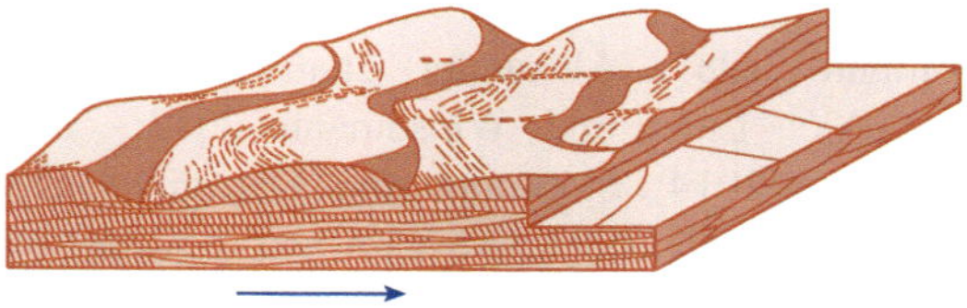

Upper Regime Plane Beds

Plane-parallel bedding, typically with both sand and gravel.

Critical flow, transition to upper flow regime

c

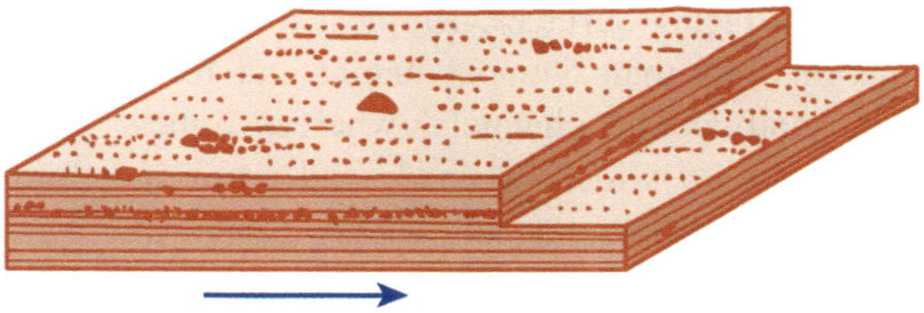

Antidunes

Wavy cross-bedding (low angle, dipping both upstream and downstream); through-going lines of bedding truncation (erosion set boundaries) are common.

Upper flow regime

d

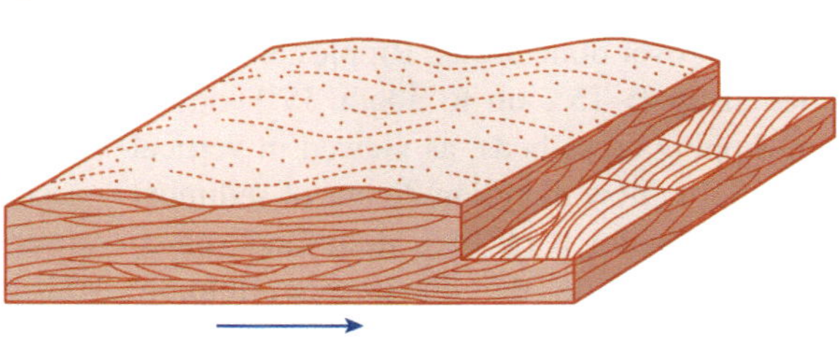

Fig. 4 Block diagrams of the five basic bed-form patterns found in the seven bed-form types. Ripples share a basic pattern with dunes as do chutes and pools with cyclic steps. Size of bed forms determines bed form type in these two cases. **a** Ripples and 2-D dunes. **b** 3-D dunes. **c** Upper regime plane beds. **d** Antidunes. **e** Chute-and pool structures and cyclic steps. Diagrams show bed-form structures in three dimensions and the two-dimensional bedding patterns in the vertical sections that would be seen in outcrops. Bed-form amplitudes can vary vertically up to about a meter in most volcanically disturbed fluvial systems; wave lengths can vary horizontally (downstream) from centimeters to meters. Arrows indicate flow direction of water flow in each frame. Patterns **a** and **b** are modified from Ashley (1990). Pattern **e** is modified from Ashley (1990) and Slootman and Cartigny (2020)

Chute-and-Pool and Cyclic-Step Bed Forms Wavy cross-bedding (mainly low angle and dipping mainly upstream); through-going lines of bedding truncation (erosion set boundaries) are common. Both formed by trains of hydraulic jumps; cyclic steps are larger structures. ***Upper flow regime***	e 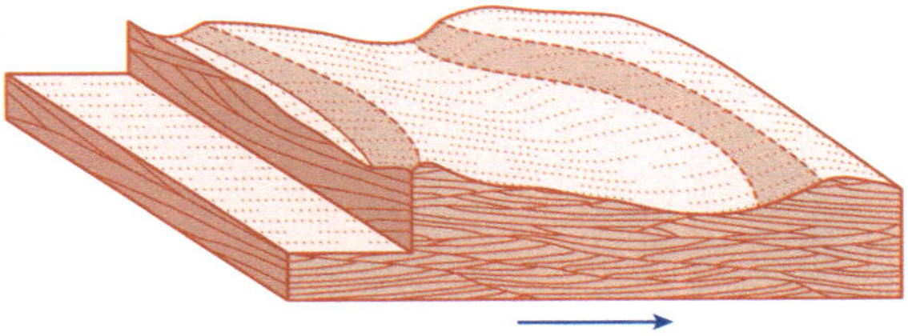

Fig. 4 (continued)

and Southard 1981; Alexander et al. 2001; Fielding 2006; Lang and Winsemann 2013; Cartigny et al. 2014; Slootman and Cartigny 2020; Slootman et al. 2021).

Ripple and dune bed forms are downstream-migrating, periodically spaced, cross-sectionally asymmetric ridges of particles that form transverse to the direction of flow. They have gently sloping upstream-facing surfaces and steeper downstream-facing surfaces. Ripples in alluvial settings are less than 5 cm in height and form only in sand-bedded channels, whereas dunes are larger, more widely spaced, and form in sand- and gravel-bed channels. Dunes can form as so-called 2-D or 3-D structures,[5] sometimes with small dunes superimposed on large dunes. The water surface is typically slightly elevated and more turbulent (upwelling boils) over dune troughs than over dune crests. Ripples and dunes produce steep (> 15°, up to about 35°), downstream-dipping cross-beds, as viewed in vertical cuts parallel to flow direction. Scour of sediment from the upstream back of a dune and subsequent deposition on the downstream-dipping, angle-of-repose slip face of the dune causes downstream migration of these bed forms. Bundles of steep downstream-dipping beds observed in outcrops are commonly referred to as *foresets*.

Plane-parallel bed forms are topographically featureless, sub-horizontal, plane-parallel bedding surfaces (Fig. 4c) that form under two sets of conditions. *Lower regime plane beds* begin to form with incipient particle movement on low-gradient channel beds in relatively slow subcritical flow. This occurs in sand-bed channels before ripples begin to form in sand <0.7 mm in diameter, before dunes start to form in sand >0.7 mm in diameter, and at the transition between ripples and dunes in beds of 0.6–0.7 mm sand (Costello and Southard 1981). *Upper regime plane beds* form in subcritical but near-critical flow close to the transition point to supercritical flow after dunes wash out and before antidunes begin to form (Costello and Southard 1981; Southard and Boguchwal 1990; Grant 1997; Cartigny et al. 2014). Further acceleration of flow deforms the plane bed into a series of antidunes and in-phase surface waves (Grant 1997).

Upper regime plane beds observed in outcrop are commonly composed of plane-parallel bedding that often show abrupt changes in grain size between adjacent beds and may include thin parallel lenses of coarser gravel or single large clasts sandwiched between beds of finer gravelly sand. During formation, bed-load transport over upper regime plane beds is intense.

[5] 2-D dunes are not truly two-dimensional—they have the dimensions of width, length, and height. But the length and width are relatively constant for a particular dune, and they are traditionally called 2-D dunes.

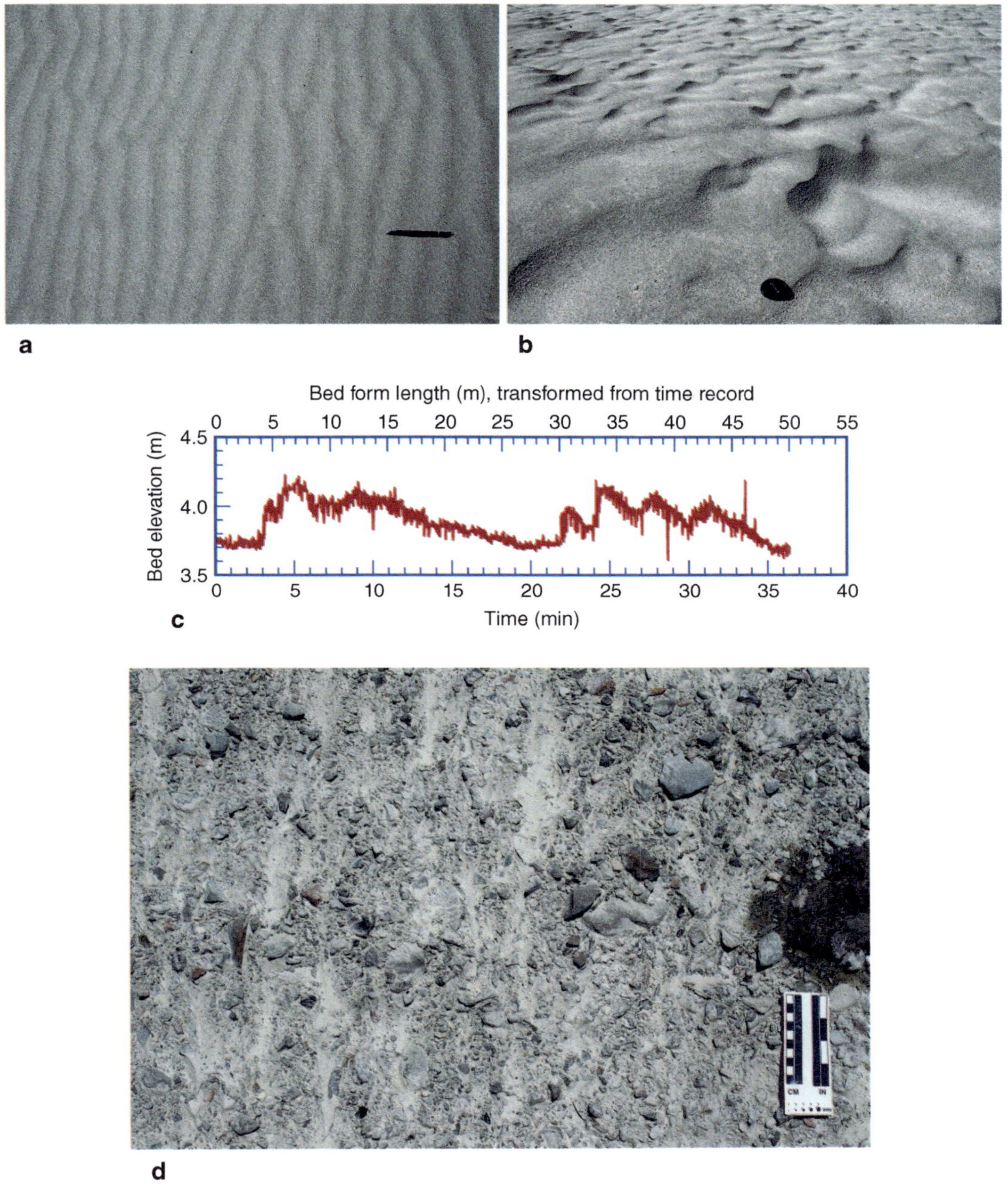

Fig. 5 Examples of common bed forms in sand and gravel-bedded fluvial channels at volcanoes. **a** 2-D ripples (1–2 cm high) viewed vertically on top of a sandbar; pencil points downstream. **b** 3-D dunes (10–15 cm high) on the surface of a sandbar, viewed obliquely; flow was from upper left to lower right. Lens cap for scale. **c** Sonar record of large, downstream-migrating, gravel dunes (20–50 cm high) passing a point on the bed of the North Fork Toutle River, 49 km downstream of Mount St. Helens (Washington, USA) during the recession limb of a volcanic water flood on December 6, 1989 (modified from Dinehart 1992). Note the smaller dunes superimposed on the back of the second large dune. Mean bed-load particle diameter was 3 cm. **d** Flat plane bed on top surface of a broad longitudinal bar, viewed vertically from above. Flow direction was from top to bottom. Coatings of fines on bed surface indicate the extremely high suspended-sediment concentration of the flow. USGS photos by T. Pierson

Antidune bed forms are positive-relief ridges of sediment that form in evenly spaced sets transverse to flow direction (Fig. 4d). They develop directly beneath and in phase with sets of stationary, smoothly undulating, free-surface water waves (wave trains) when accelerating critical flow in a channel starts to become supercritical. As waves form, flow moving upward to the wave crest slows just enough to become subcritical, but it accelerates slightly on the back side of the wave to become supercritical again in the wave trough. Thus, state of flow oscillates smoothly (not abruptly) back and forth between supercritical and subcritical flow in a set of standing waves (average flow state remains critical, i.e., $Fr \approx 1$). Standing waves can also be described as *undular hydraulic jumps* (Grant 1997; Dietterich et al. 2022). In contrast to antidunes, dunes formed in subcritical flow are asymmetrical in transverse cross-section and out of phase with their associated free-surface waves.

With a further increase in flow velocity (and stream power), however, antidunes and their surface wave steepen, begin migrating upstream, become asymmetric and out of phase, and begin breaking upstream as more abrupt hydraulic jumps. Increased scour of the bed accompanying breaking antidune waves leads to washing out of the antidunes and a return to plane-bed conditions (Grant 1997). Cyclic washing out and reforming of antidunes is common. Antidunes have relatively short wavelengths (Kostic et al. 2010). Spacing between dune crests in rivers of the size typically draining volcanoes is in the range of 2–6 m (Simons et al. 1965), and under certain conditions wavelength spacing can be used to calculate flow depth (Kennedy 1963).

Antidune bedding structures are not always well preserved in natural sediments due to an imbalance between aggradation rate and erosion rate (Fig. 6). Flume and field studies (Lang and Winsemann 2013; Cartigny et al. 2014) show that migrating antidunes produce low-angle ($< 15°$) upstream-dipping parallel bedding (lamina sets) that are truncated by sub-horizontal erosion surfaces. Bundles of upstream-dipping beds are also called *backsets*. Bedding tends to be sinusoidal rather than planar. Lamina sets are more continuous when formed by stable antidunes and more lenticular when formed by unstable antidunes. Lenticular lamina sets tend to be structureless (massive) at the base and more distinctly bedded in the upper part. If aggradation rate is low, erosion surfaces (erosion-set boundaries) are closely spaced, and pseudostratification develops that resembles plane-parallel bedding (Fig. 6, bottom frame). If aggradation rate is high, lamina sets are thicker, and backsets are steeper (up to 15°).

Chute-and-pool bed forms develop at stream powers beyond those forming antidunes (Slootman and Cartigny 2020; Slootman et al. 2021). On an undulating channel bed, supercritical flow accelerates down a steep "chute," encounters a flat section of channel and transitions abruptly to subcritical flow (at a highly turbulent and erosive hydraulic jump) where flow slows, deepens, and forms a "pool" in a scoured depression. The pool then overflows a broad ridge on the downstream side of the pool that is formed by sediment thrown into suspension by the hydraulic jump. Flow accelerates again down the steepening backside of the ridge, repeating the sequence. Thus, downstream of each hydraulic jump, the lower-regime flow in the "pool" begins the next chute-and-pool couplet. Accelerating flow leaving a pool may add complexity by forming antidunes in the next chute before reaching the next pool.

Preservation of chute-and-pool bed-forms requires extremely high aggradation rates (Cartigny et al. 2014), so the occurrence of chute-and-pool structures is rare in bodies of fluvial sediment but more common in PDC deposits (Schmincke et al. 1973; Cole 1991). Bedding produced by synthetically aggrading chute-and-pool structures in a laboratory flume (Cartigny et al. 2014) can be described as cross-bedding involving both downstream-dipping foreset beds and steeper upstream-dipping backset beds. Beds are both planar and sinusoidal in form, separated by mildly concave-upward erosional surfaces (troughs) (Fig. 4e). Cross-bedding is generally less than 15° but backsets can locally be up to 20° in alluvial deposits

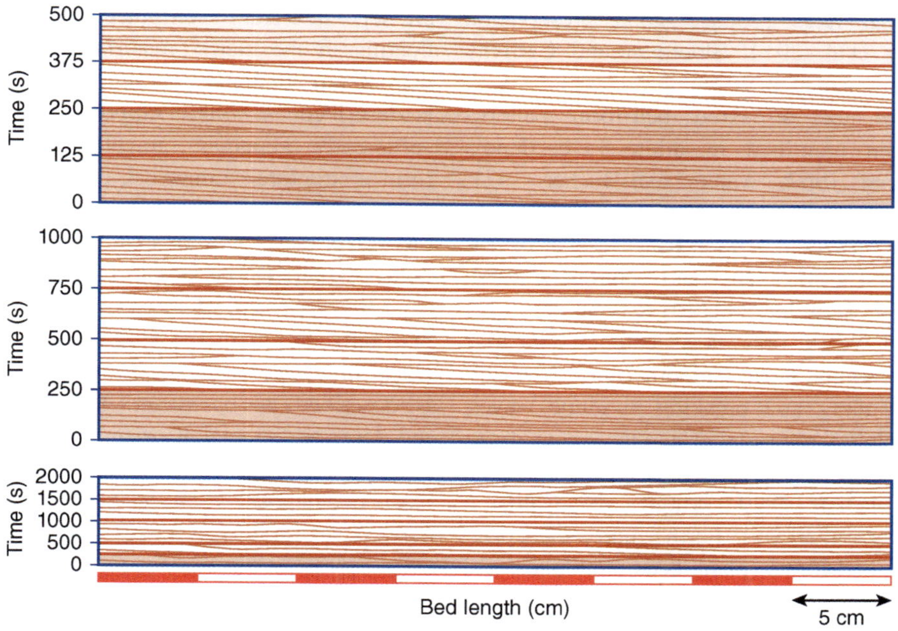

Fig. 6 Effect of different bed aggradation rates on antidune bed-form preservation in flume experiments; **flow direction is from right to left** (modified from Cartigny et al. 2014). The beds produced under a high aggradation rate in the top frame show well-defined upstream-dipping lamina sets between widely spaced sub-horizontal erosional surfaces, which were produced from scouring by migrating stable antidunes. Middle and lower frames show progressive obscuring of cross-bedding in the deposits as aggradation rates decrease and scouring increases. In the bottom frame, where the aggradation rate was lowest, the erosion surfaces are so closely spaced that the low-angle cross-bedding produced by the antidunes is difficult to see at all, and the deposit has pseudostratification formed by parallel erosion surfaces that resembles plane-parallel bedding. Shaded zones indicate relative deposit thickness after 250 s in each run

(Lang and Winsemann 2013) and up to 35° in PDC deposits (Cole 1991). Chute-and-pool bedding is strongly lenticular and like that produced by cyclic steps.

Cyclic-step bed forms are similar in form to chute-and-pool structures and likewise produced by alternating supercritical and subcritical flow in steep channels (Slootman and Cartigny 2020), but cyclic-step structures are deeper and have longer wavelengths than chute-and-pool structures. Channels under these conditions commonly exhibit trains of slowly upstream-migrating 'steps', which are broader versions of the transverse sediment ridges forming downstream of the pools in chute-and-pool couplets. Steps are separated by relatively deep transverse trenches. Steps have wide, gently upstream-dipping surfaces ("stair treads") and narrow, steeply downslope-dipping faces ("stair risers"), where flow accelerates to the next hydraulic jump forming at a trench. The trenches are produced by the intense erosive power of the hydraulic jumps that form at the downstream ends of the troughs. Cyclic steps have deeper transverse trenches and steeper chutes than chute-and-pool structures. Cross-bedding produced by cyclic steps and chute-and-pool structures are similar in outcrop and dominated by gently upstream-dipping beds truncated by numerous erosional surfaces (erosion set boundaries), which are more or less planar (Cartigny et al. 2014; Slootman and Cartigny 2020). Backset laminae can be unstratified at their bases. Overall, cyclic step deposits consist of elongated lenses that are concave-upward and truncated by each other at low angles (Cartigny et al. 2014). The degree of preservation of chute-and-pool and cyclic step bed forms in the depositional record depends mainly on aggradation rates—high rates produce good preservation (Cartigny et al. 2014; Slootman and Cartigny 2020; Slootman et al. 2021).

All of the discussed bed-form patterns have been identified in dilute PDC deposits (Moore 1967; Waters and Fisher 1971; Schmincke et al. 1973; Wohletz and Sheridan 1979; Hoblitt et al. 1981; Rowley et al. 1981, 1985; Waitt 1981; Fisher and Schmincke 1984; Cole 1991; Douillet et al. 2013; Sohn and Sohn 2019). However, only flow features indicative of upper-regime flow (smooth rapid flow surfaces, standing and breaking antidune waves, and hydraulic jumps and drops) have been documented in the field for lahars (e.g., Pierson et al. 1996; Dinehart 1999; Pringle and Cameron 1999; Pierson 2005b). Furthermore, the corresponding bed forms (plane beds, antidunes, and chute-and-pool/cyclic-step bedforms) are seldom preserved in lahar deposits unless bedding is well developed (as it is in some dilute lahar deposits).

3.2.2 Preservation of Bed Forms

Bed forms, where preserved, can reveal much about flow conditions in particle–fluid flows, although high aggradation rates are required for good preservation. Little to no aggradation will result in the continual and near-complete erosion of a bed structure by the next migrating bed form. Low aggradation rates result in sets of closely spaced parallel or subparallel erosional surfaces that closely resemble plane-bed deposition (Smith 1971b; McBride et al. 1975; Cartigny et al. 2014) (Fig. 6). Close examination of bedding planes in the field is needed to detect the difference.

3.3 Gravel-Size Clast Framework

Gravel-size clasts are strong visual elements in outcrops of volcanic sediment, and such clasts may or may not form a structural framework within volcaniclastic deposits, depending on the proportion of coarse clasts available for transport and on the transporting process. Where gravel-size clasts in sediment mixtures are separated from each other and appear to be floating in the matrix material (particles finer than 2 mm in diameter), the deposit structure is described as *matrix supported.* Conversely, sufficiently abundant gravel-size clasts can form a structural framework of gravel clasts in contact with each other. The spaces between the gravel clasts may or may not be filled with matrix material. The structure of these deposits is *clast supported.* If matrix material is partly or totally absent in the spaces between gravel clasts in contact with each other, the structure is described as *openwork.* Fabric and grading describe the spatial orientation and vertical trends in average size, respectively, of gravel clasts in a deposit.

3.4 Fabric

Dimensional fabric in a sedimentary deposit is the measure of the degree of similarity in orientation of its elongate or planar particles within three-dimensional space (Neuendorf et al. 2005) (Fig. 7). In other words, it is the degree of anisotropy in the spatial orientation of dimensionally non-equant particles in a deposit (Friedman and Sanders 1978). Fabric can develop during flow and deposition of a particle–fluid mixture, or it can be imposed later by deformation, including syn-depositional soft-sediment deformation (in a lahar deposit) or welding and compaction (in a PDC deposit). Fabric may be measured with respect to a compass direction and true horizontal, with respect to orientation of bedding planes, or with respect to the direction of flow. A weak or random fabric is one in which there is little or no similarity of particle orientation. A strong fabric is one in which a large proportion of particles are consistently similar in orientation to each other. *Imbrication* (Fig. 8) is a type of fabric where disc- or blade-shaped particles (particles with a distinct plane defined by the long and intermediate axes, i.e., the AB plane) dip upstream at an angle to the bedding plane (Neuendorf et al. 2005).

Fabric can be magnetic as well as dimensional in volcanic sediments. Magnetic fabric

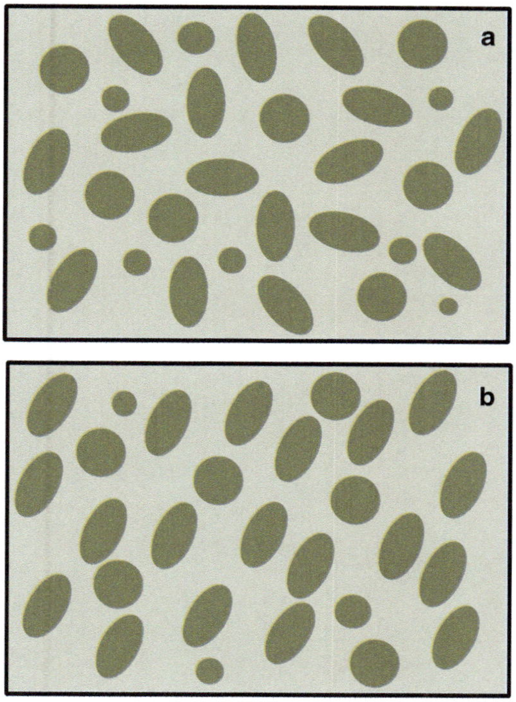

Fig. 7 Schematic representation of deposit fabric viewed in outcrop. **a** Random fabric, in which the elongate particles show no preferred orientation. **b** Strongly developed fabric, in which virtually all the elongate particles have the same orientation. Fabric can be observed on surfaces of deposits or in vertical cuts

develops in hot pyroclastic deposits as ferromagnetic mineral particles realign their magnetic axes with Earth's magnetic field as the deposit slowly cools to below the Currie point, thus locking in the same magnetic orientation. Magnetic fabric can be measured in the field (e.g., Hoblitt and Kellogg 1979) and has been useful for distinguishing PDC deposits emplaced at high temperature from other volcanic sediments. Magnetic fabrics are strong and uniform in the simple case of a sufficiently hot PDC coming to rest and its deposit cooling in place. However, more complex fabrics should be expected in cases where PDCs entrain cold clasts, where rapid cooling of some clasts precedes deposition, or where the flows are sourced from partially cooled lava domes or remobilized PDC deposits (e.g., Rader et al. 2015; Risica et al. 2022). Another technique for assessing magnetic fabric, *anisotropy of magnetic susceptibility* (AMS), does not require deposition at above Currie point temperatures (e.g., Biró et al. 2015).

Dimensional fabric strength can be used to discriminate between different depositional processes. Fabric analysis based on particle shape has traditionally involved field measurement and orientation in space of the three major axes of a clast: length (a-axis), width (b-axis), and thickness (c-axis), with analysis of the axial orientations using the eigenvalue method (Scheidegger 1965; Mark 1973). More recently, the use of outcrop photographs analyzed with image analysis software is more common (e.g., Moreno Chávez et al. 2018). However, even a subjective evaluation of fabric strength can be useful to the field geologist in discriminating between deposits of different flow types.

3.5 Grading

Grading is the progressive change in average particle size with vertical position within an individual bed (Friedman and Sanders 1978; Neuendorf et al. 2005). In concentrated particle-flow deposits, it is the larger particles in a mixture that most strongly reflect the anisotropic spatial distribution of clasts, whereas smaller particles are more likely to be evenly distributed vertically (Sparks et al. 1973). For this reason, the term *coarse-tail grading* is sometimes used, referring to the coarse-particle tail of the particle frequency distribution plot. A *normally graded* bed is one where average particle size becomes finer upward (Fig. 9a); an *inversely (or reversely) graded* bed is one where average particle size coarsens upward (Fig. 9b). An ungraded bed displays no particular vertical trend in particle size. Some individual beds may display both normal and inverse grading (termed *symmetric grading*)—an inversely graded base and a normally graded top (or vice versa), with perhaps an ungraded zone in the middle (Fig. 9c, d). In this figure, frames a–d show grading based only on size, assuming that the density of particles in the mixture is uniform. But in volcanic

Fig. 8 Subtly imbricated gravel on the surface of a water-flood bar in the Sandy River, about 70 km downstream of Mount Hood (Oregon, USA); camera case for scale in foreground. View is transverse to flow direction, which was from right to left. Note that the larger clasts that are relative flat or elongate are dipping about 15–45° to the right (upstream); equant clasts show little orientation preference. USGS photo by T. Pierson

settings, particles of greatly different densities may be part of the sediment mixture—volcaniclastic particles of variable vesicularity as well as dense lithic particles. In such cases, density-segregation grading may be observed (Fig. 9e).

Grading in flow and fall deposits develops dynamically as the result of incremental accretion of particles of progressively finer or coarser size on the ground or bed surface. Lahars and PDCs do this through progressive incremental deposition from the base of a moving fluid as particles are accreted at the flow boundary layer (Vallance 1994, 2000; Vallance and Scott 1997; Branney and Kokelaar 2002; Vallance and Iverson 2015) (see Sect. 6.1 in Chap. "Lahars—Process Overview"). Normal grading occurs by progressive accretion of increasingly finer particles at the flow boundary, while inverse grading forms by the progressive accretion of increasingly coarser particles.

Normal grading may also result from the preferential gravitational settling of coarse or dense gravel-size clasts in matrix-rich lahars that are slowly moving or static but still liquefied (Vallance and Scott 1997; Branney and Kokelaar 2002). Inverse grading may also develop within a concentrated lahar or PDC from *kinetic sieving* and *squeeze expulsion* (Rosato et al. 1987; Savage 1987; Savage and Lun 1988; Gray 2018). Size segregation by both these mechanisms is strongest in mixtures having relatively low-viscosity intergranular fluid phases. In kinetic sieving, smaller grains preferentially fall into momentarily open gaps between larger

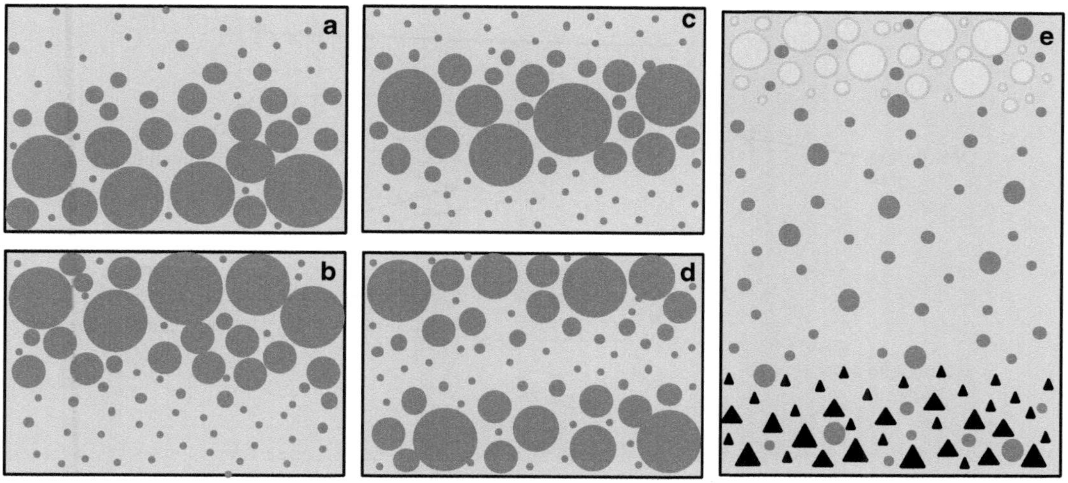

Fig. 9 Types of grading encountered in volcaniclastic deposits, shown in schematic vertical outcrop faces: **a** normal grading, **b** inverse (or reverse) grading, **c** symmetric grading, inverse to normal (going upward), **d** symmetric grading, normal to inverse (going upward), **e** density-segregation grading, normal for the black dense lithic fragments and inverse for the gray low-density pumice fragments (with no grading in the other volcaniclastic particles)

grains as the mixture is dilated and jostled during flow. This results in smaller grains being displaced downward, because there is a higher probability of a small grain finding a gap to fall into than there is of a large grain finding a large enough gap to drop downward. Squeeze expulsion causes a net upward movement of large grains, because even though fluctuating contact forces between grains can induce upward or downward movement with equal probability, the spaces below large grains are more likely to be already filled with small grains, so net upward movement is facilitated. Upward displacement of large grains is probably further enhanced by the complementary processes of (1) "rolling and climbing" upward of larger grains (Haff and Werner 1986; Vallance 1994; Iverson 1997), whereby upward-increasing velocity gradients impart forward rotation more to larger grains than to smaller ones because they stick up farther into the flow, causing them to roll forward and upward over smaller grains.

3.6 Fluid-Escape Structures and Deformation

3.6.1 Water-Escape Structures

Water-escape structures and soft-sediment deformation commonly form in rapidly emplaced and rapidly consolidating lahar deposits and other saturated sediments shortly after deposition (Lowe and LoPiccolo 1974; Lowe 1975; Fisher and Schmincke 1984; Johnson 1986; Cas and Wright 1987; Hodgson 1993; Cronin et al. 2000) (Fig. 10). This occurs because such flow deposits are, to some degree, inflated when they are emplaced, and they compact under their own weight soon thereafter. Fresh lahar deposits can compact about 10% in volume during post-depositional consolidation (Pierson 1981; Major 1996, 2000; Major and Iverson 1999; Cronin et al. 2000). The intergranular fluid in such deposits—liquid water mixed with suspended silt and clay in the case of lahars—is pressurized

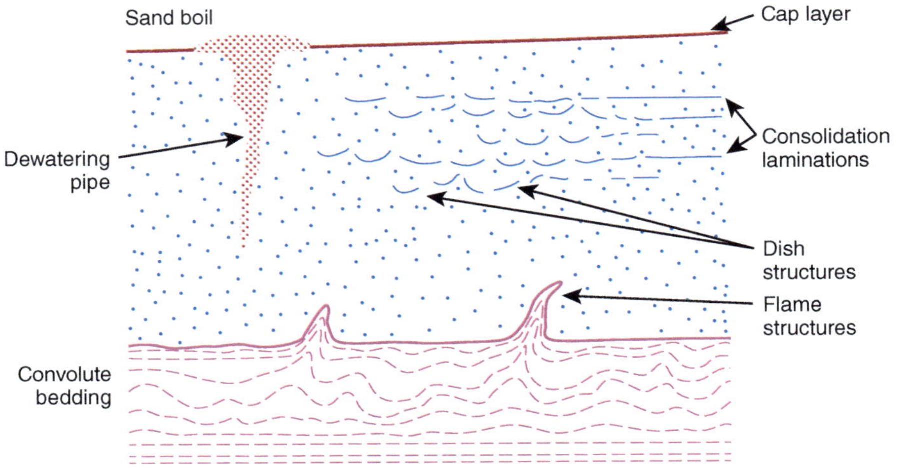

Fig. 10 Different types of water-escape structures and soft-sediment deformation occurring within liquefied zones in consolidating water-saturated deposits. They are commonly seen in poorly to moderately sorted silty sand deposits

during consolidation, and the rate of compaction, the permeability of the material, and the viscosity of the fluid determine how quickly the pressurized pore fluid can escape (Lowe 1975). Permeability is largely a function of particle-size distribution (mean size and degree of sorting). If deposits are well sorted and the rate of emplacement is slow, intergranular fluid escapes at the rate consolidation applies pressure, and no fluid-escape structures will form. If deposits are poorly sorted, positive pore pressures will build up within a consolidating deposit, and fluid-escape structures can form.

Freshly emplaced water-saturated deposits produce several types of fluid-escape structures, as well as soft-sediment deformation. Where water expulsion is not constrained by subhorizontal layering of low-permeability zones, pressurized pore water is expelled upward, either diffusely through the sediment or concentrated in vertical or subvertical conduits termed *dewatering pipes* (Lowe 1975; Cronin et al. 2000; Pierson 2005b) (Fig. 10). The upward flowing pore water in these pipes can be sufficiently rapid to fluidize the sediment along these pathways and to entrain clay, silt, and even fine to medium sand from the matrix of the deposit, leaving conduits that are preserved as zones of openwork, fines-depleted, coarse particles (Crandell 1987; Arguden and Rodolfo 1990; Smith and Lowe 1991). Some slight tilting in the downstream direction is not uncommon and probably results from post-depositional creep of the deposit. Dewatering pipes originate within a compacting deposit above its base, extend upward as narrow (1–5 cm), high-permeability conduits, and then flare out within 10–20 cm of the top of the deposit. Rounded mounds of sand (sand boils) may be found above the flare-out zones on the upper surface of the deposit (Fig. 10). When the upward escaping pore water reaches the deposit surface, it can flow out and emplace the entrained fines in thin sheets of mud—*cap layers*—that fill shallow depressions on the deposit surface (Fig. 11). Cap layers can also be formed by direct suspension fall-out of fines from slowly moving waning flow in channel embayments or from a later dilute flow overtopping a lahar deposit. Cap layers are commonly preserved between successive flow deposits (Cronin et al. 2000).

During consolidation, *consolidation laminations* (Lowe 1975) can form where upward escaping pore water is blocked by low-permeability zones or layers. Pressurized pore fluid is then diverted into thin, generally bedding-parallel, sheet-like flow paths beneath or between less permeable beds. Clay- and silt-size particles

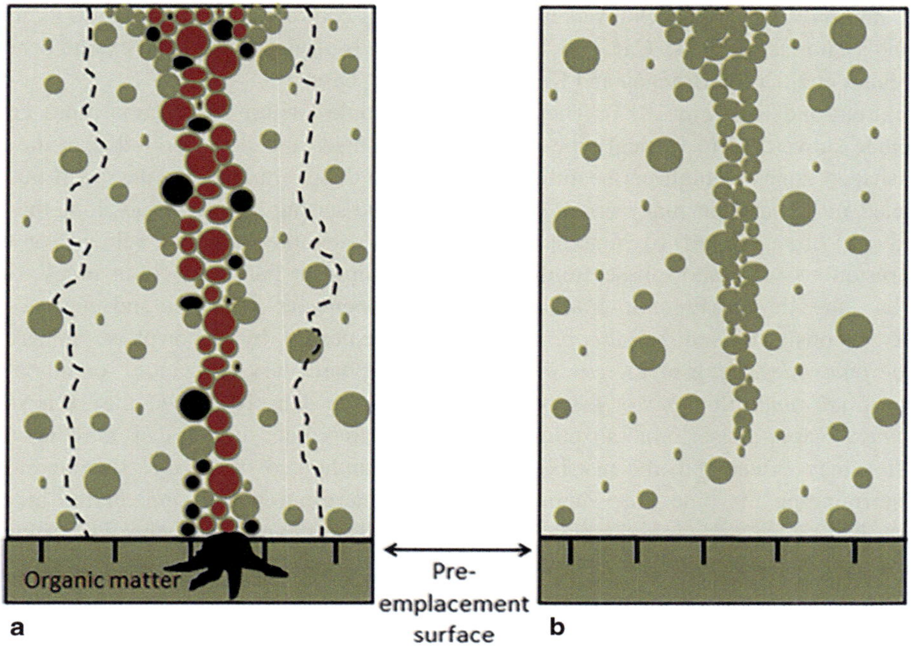

Fig. 11 Schematic diagrams of vertical outcrop faces showing two types of gas-escape pipes in pyroclastic density-current (PDC) deposits. **a** Gas-escape pipe with a localized source of heat-generated volatiles (e.g., pyrolyzed buried tree stump) below the deposit. Gas-escape pipes commonly have highly irregular boundaries, oxide-encrustation, and cementation of coarse particles in throat of pipe, and surrounding aureoles of vapor-phase crystallization in the PDC deposit matrix (aureole bounded by dashed lines in frame **a**). **b** Gas escape pipe of the rootless-fumarole type. This pipe originates within the deposit

transported sub-horizontally with the migrating pore fluid eventually plug these sandy flow paths to form thin bedding-parallel laminations that extend laterally a meter or two (Fig. 10). Consolidation laminations may cut across some bedding planes, however.

Continued diffuse upward migration of water and entrained sediment during deposit consolidation, if focused in narrow zones, can disrupt sub-horizontal consolidation layers to form segments with upturned edges—concave-upward *dish structures* (Lowe and LoPiccolo 1974; Lowe 1975; Hodgson 1993; Neuendorf et al. 2005) (Fig. 10). Dish structures are commonly 5–15 cm across and occur at the level of formerly continuous consolidation laminations.

Convolute bedding and *flame structures* (Fig. 10) form where intergranular fluid pressures rise faster than pore fluid can escape—a condition called undrained loading, such as when a still-saturated deposit from a lahar surge is suddenly covered by another surge deposit (Lowe 1975). If internal pore pressure in the underlying deposit approaches that of the weight of the overbearing load, that deposit can liquefy and begin to flow or deform under the weight of the imposed load. This can greatly deform the initially sub-horizontal beds within the deposit. Protrusion of knobs or fingers of liquefied sediment injected into the overlying strata are commonly termed flame structures.

3.6.2 Gas-Escape Structures

Gas-escape pipes and fissures form post-depositionally in PDC deposits by rapid escape of high-temperature air and exsolved acidic gas, along with entrained fine ash particles (Fig. 11). They are widespread in PDC deposits, both welded[6] and nonwelded varieties (Walker 1971; Fisher and Schmincke 1984; Cas and Wright 1987; Freundt et al. 2000; Branney and Kokelaar 2002; Hildreth and Fierstein 2012). These features include subvertical to vertical pipe-, pod-, or sheet-shaped zones of high permeability that are depleted in ash and, in many cases, locally cemented (and often colored) by mineral oxides and encrustations of vapor-phase sublimates within the pipe throat (Freundt et al. 2000). They are variously termed *gas-escape pipes, elutriation pipes, degassing pipes, gas segregation pipes, fossil fumarole pipes* or *fissures*, and *lapilli pipes*. Vapor-phase mineralization and cementation may extend into the matrix of the deposit surrounding the pipe—the vapor-phase aureole.

Gas-escape pipes generally are subcircular to irregular in cross-section and funnel shaped near deposit tops; they have been recognized as small as 10 cm long and 1 cm in diameter (Fisher and Schmincke 1984) and as large as 20 m long and tens of centimeters in diameter (flaring out to nearly 2 m at the top) (Hildreth and Fierstein 2012). Pipes are highly irregular in shape, commonly sinuous, and near-vertical in orientation, with single-conduit or branching forms (sometimes splitting and rejoining). They may be deformed into sigmoidal shapes, however, by late-stage shearing or burial compaction. Steam and volatiles produced from heated water or buried organic matter in the underlying substrate escape upward to form pipes through the entire thickness of the deposit (Fig. 11a). Otherwise, if not localized by fracture systems, pipes tend to develop only in the upper parts of cooling units, commonly termed *rootless fumaroles* when actively degassing (Fig. 11b). Buried stream courses may not only localize fumaroles but also cause small phreatic explosions (Banks and Hoblitt 1996). Fissure fumaroles, formed along thermal contraction or compaction joints, can be hundreds of meters long and extend nearly to the deposit base.

Particles within gas-escape pipes are generally coarser and better sorted than the material around them, although the degree of fines-depletion and sorting by elutriation along the conduits depends on the flux and velocity of escaping gas. Remnant particles within pipes commonly have openwork structures and may be oxidized and cemented by vapor-phase mineralization. Vapor-phase crystallization commonly penetrates the densely packed matrix beyond pipe walls to create cylindrical aureoles that are more indurated than the surrounding material. Encrustation and cementation may extend upward into overlying tephra-fall deposits, producing erosion-resistant rims or mounds.

3.7 Shear Deformation

Internal shear is distributed to some degree within the vertical flow profile of all flowing particle–fluid mixtures. In dense, highly concentrated mixtures, it is limited to a narrow, highly localized shear zone at or near the base of the flow. High shear at that boundary during deposition by deep flows can produce sub-horizontal pseudostratification and clast breakage (Scott 1988a; Glicken 1996, 1998; Siebert 1996). But shear is more widely distributed throughout the vertical profile in very dilute mixtures yet remains more concentrated near the lower flow boundary.

Gravel-size clasts within sediment bodies undergoing shear may display either plastic or brittle deformation (Fig. 12). In plastic materials such as pervasively fractured but coherent debris blocks or in clods of soil or floodplain silts entrained in a deposit, shear deformation is recorded in stretching, smearing, or folding of the material (Siebert 1996). In high strength material such as large rock fragments or highly coherent debris-avalanche blocks, shear results in offset faulting and linear trains or elongate clusters of smaller angular rock fragments having the same composition (Takarada et al. 1999).

[6] *Welded* PDC deposits are those hardened to solid rock by the combined action of heat and mechanical compaction. High heat above the glass transition temperature (> 550 °C, depending on glass composition) and retained within a deposit over time, allows glassy particles to fuse together (Fisher and Schmincke 1984).

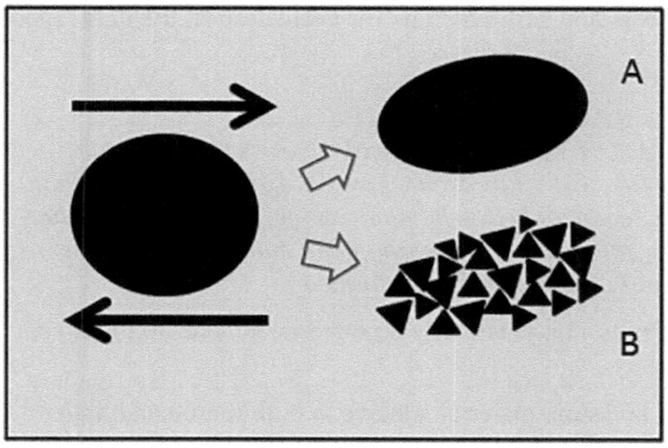

Fig. 12 Schematic diagram of a clast within a volcanic flow being subjected to high shear stress. Solid arrows indicate the sense of shear. Depending on the strength and composition of the clast, shear deformation is either plastic (**a**) or brittle (**b**). If shear displacement is limited, clast fragments remain in proximity, but with extensive shear displacement, fragments are highly dispersed along a shear zone

3.8 Material Consistency

Unconsolidated volcaniclastic particles are deposited initially as loose mixtures of grains, but once deposited, that initial looseness typically changes. Material *consistency* describes the degree of coherence or relative resistance to rupture or deformation of unconsolidated sediment or soil material in the deposit, as a function of cohesion among particles in the material (Neuendorf et al. 2005) (Table 4). Consistency is mainly due to (1) type and the amount of clay present (with smectite clays being responsible for some of the strongest effects); (2) cohesion of particles due to increased surface contact among particles resulting from compaction; or (3) the presence of a cementing agent. In practical terms, consistency is the way in which the material behaves in your hands or how it responds when dug or scraped with a tool. Consistency is affected by the moisture content of the material, and moisture content should be noted whenever consistency is described (e.g., wet, damp, or dry).

Granular deposits that are extremely hard (well indurated) can no longer be considered unconsolidated, but when occurring only within specific zones or layers in a sequence of mainly unconsolidated sediment, such localized induration can be helpful in deciphering the depositional process. Induration results from the action of heat, pressure, or a cementing material (Neuendorf et al. 2005), and degree of induration is independent of sediment moisture content. With the passage of time, any deposit will tend to become more indurated because downward-percolating surface water or circulating groundwater carries dissolved minerals (e.g., calcium carbonate, silica, iron oxides) that can precipitate in void spaces between particles in the deposit and cement particles together.

3.9 Evidence of Heat

Evidence of prior heating of a deposit or of some particles within a deposit to high temperatures (> 100 °C) may include (1) characteristic fracture patterns in clasts, (2) the presence of charcoal, (3) red or pink coloration on individual clasts or the bulk deposit, (4) uniform thermoremanent magnetic orientation of clasts, (5) phreatic explosion pit craters, and (6) gas-escape structures (already described in Sect. 3.6.2). Some of these features may be found in deposits that were emplaced at ambient temperature, however. This may be the case if an eroded, once-hot PDC deposit provided the source sediment for a secondary lahar deposit.

3.9.1 Fracture Patterns in Clasts

Rapid cooling or heating of volcanic rock produces differential thermal stresses, owing to the thermal contraction or expansion that starts first at the particle's surface and moves inward (Ryan and Sammis 1978; Reiter et al. 1987). When tensile stresses exceed the tensile strength of the rock, fractures form. Fractures tend to be either perpendicular or parallel to the outer surface of a volcanic clast, depending on whether the clast is heated or cooled. Rapid cooling occurs in primary volcaniclastic particles emplaced by

Table 4 Descriptors of consistency of unconsolidated sediments and soil materials for evaluation in the field (modified from Taylor and Pohlen 1970)

Material consistency	Explanation
Consistency when wet (*stickiness and plasticity*)	*Moisture content at or slightly above field capacity (i.e., moisture remaining after free water has drained away). For field evaluation of stickiness, press sediment between thumb and finger. For field evaluation of plasticity, roll material between thumb and finger and observe whether a thin rod of sediment can be formed*
Nonsticky	After release of pressure, practically no material adheres to thumb or finger
Slightly sticky	After release of pressure, material adheres to both thumb and finger but comes off one or the other rather cleanly; material not appreciably stretched when digits separated
Sticky	After release of pressure, material adheres to both thumb and finger and tends to stretch somewhat when digits separated; material pulls apart rather than pulling free from either digit
Very sticky	After release of pressure, material adheres strongly to both thumb and finger and is decidedly stretched when digits are separated
Nonplastic	No rod can be formed by rolling material between hands
Slightly plastic	Thin rod ("wire") is barely formable and can very easily be broken
Plastic	Thin rod is formable but when broken cannot be reformed
Very plastic	Thin rod is easily formable and not easily broken; if broken, it can be reformed several times
Consistency when moist (*firmness*)	*Moisture content approximately midway between air dry and field capacity; loose sediment will cohere again when pressed together. For field evaluation, attempt to crush material by hand*
Loose	Noncoherent
Very friable	Material crushes under very gentle pressure but coheres when pressed together
Friable	Material crushes easily under gentle to moderate pressure between thumb and forefinger; coheres when pressed together
Firm	Material crushes under moderate pressure between thumb and forefinger
Very firm	Material crushable between thumb and forefinger
Extremely firm	Material cannot be crushed between thumb and forefinger; can still be broken apart by hand
Consistency when dry (*hardness*)	*Sample has "air dry" moisture content; crushed material will not cohere again when pressed together. For field evaluation, break sample apart by hand*
Loose	Noncoherent
Soft	Material is very weakly coherent and fragile; breaks into individual grains under very slight pressure
Slightly hard	Material is weakly resistant to pressure; easily broken between thumb and forefinger
Hard	Material is moderately resistant to pressure; barely breakable between thumb and forefinger
Very hard	Material is very resistant to pressure; can be broken by hand only with difficulty; not breakable between thumb and forefinger

This table is a subjective guide for field assessment of material consistency only. Note that a descriptor that is sometimes used, "compact," should refer to high packing density, rather than firmness or hardness

Fig. 13 Prismatic jointing (radial cooling fractures) in hot volcanic clasts: **a** prismatic jointing in hand specimen of dacite (dome fragment) from Mount St. Helens (Washington, USA). **b** Prism-shaped fragment of a quenched and broken hyaloclast from the Columbia River Gorge (Oregon, USA). Pen is 13 cm long. USGS photos by T. Pierson

pyroclastic density currents or tephra fall. Such fracturing is more pronounced in cobble- and boulder-size pyroclasts, because ash particles and finer lapilli reach thermal equilibrium too quickly for large differential stresses to develop.

A volcanic clast that is initially wholly or partly molten when formed, then subjected to rapid cooling from the outside in after being formed, develops radial fractures. This happens because the fractures resulting from thermal contraction during cooling propagate inward approximately normal to the cooling front, which is usually normal to surfaces of equal tensile stress and which is also approximately normal to the clast's surface (Williams and McBirney 1979; McPhie et al. 1993; Francis 1993). The fracture pattern thus radiates out from the clast's core (Fig. 13a). If the temperature differential is small, such as when the clast is surrounded by other hot particles in a thick PDC deposits and cooling is gradual, stress differentials are relatively low, and fracturing may not occur or may be limited. The inwardly intersecting fracture planes in gravel-size clasts can produce prism-shaped fragments (Fig. 13b). Consequently, such radial fractures are also called *prismatic fractures* or *prismatic joints*.

Gravel-size clasts having outer rinds that are very rapidly cooled and solidified while the clast core is still molten and partially gas-charged are a special case. Vesiculation in the plastic core and the resultant volumetric expansion causes the prismatic fractures at the clast surface to widen noticeably (Williams and McBirney 1979; Fisher and Schmincke 1984; Wright et al. 2007) (Fig. 14a). A pattern of open fractures on clast surfaces distinguishes these clasts as *breadcrust* clasts, which originate either as hot ballistic projectiles that achieve rapid cooling during flight (breadcrust bombs) or as very hot fragments carried and rapidly cooled in a dome-collapse avalanche (Fig. 14b, c). Breadcrust clasts and bombs are more common in lava fragments of gas-rich intermediate to silicic composition, whereas basaltic lava fragments have less obvious surface cracking (Fisher and Schmincke 1984). Contact of hot clasts with surface water (possibly from rapid snowmelt) causes even more rapid cooling and may produce more finely fractured surfaces and the formation of cauliflower-like surface textures—*cauliflower bombs* (Fig. 14d) (Fisher and Schmincke 1984). Extremely rapid cooling from direct immersion of near-molten rock in cold water or snow

Fig. 14 Breadcrust and cauliflower clast surface textures. **a** Cross-section of small breadcrust bomb, with dense chilled outer rind and highly vesiculated core from Llaima volcano (Chile). **b** Expanded radial fractures in an unweathered dacite breadcrust clast from Mount St. Helens (Washington, USA). **c** Breadcrusted, prismatically jointed block from Mount St. Helens where physical weathering (mainly freeze–thaw action) has expanded some radial fractures and is breaking the block apart. Trekking pole shaft diameter 15 mm. **d** Cauliflower bomb with lumpy, finely fractured surface (unknown location). USGS photos **a**, **b**, **c** by T. Pierson; **d** from Smithsonian Institution archive (public domain)

results in quenching and extensive fine-scale fragmentation and rock disintegration; such particles are termed *hyaloclasts* (Fisher and Schmincke 1984).

Initially cool large clasts entrained by a hot PDC would be subjected to a temperature differential opposite of the case explained above, resulting in a rapidly heated outer surface relative to a cooler core. In this case, the outer layer of a clast expands differentially and creates compressive circumferential stresses and radial tensile stresses that produce fractures that are parallel to the clast surface. Heating of cold clasts in hot deposits is rare, but it can create concentric exfoliation fractures parallel to the clast surface, which may explain the observed spalling of accidental clasts in pyroclastic-flow deposits (Williams and McBirney 1979) (Fig. 15). Interpretation of exfoliating clasts as having been initially cool and then entrained

Fig. 15 Volcanic clast with exfoliation fractures, Hoyo Negro crater, Cumbre Vieja volcano (Canary Islands, Spain). It is unclear, however, whether this fracturing was caused by reheating or by other mechanisms. Note hand lens on cord for scale (length of cord loop is about 45 cm). USGS photo by T. Pierson

by a hot flow requires that the clasts be fresh, unweathered rock and that the clasts not be on the deposit surface. This is because other processes can cause the outer surfaces of surface clasts to concentrically expand and exfoliate, such as low-temperature spheroidal weathering of old clasts (Fletcher et al. 2006) or rapid heating of clast surfaces by wildfire (Zimmerman et al. 1994).

The relative abundance and type of thermal fracturing of gravel-size clasts is important for distinguishing between flows that were initially hot and flows that incorporated material from cool but once-hot deposits. Clasts with extensive prismatic jointing are quite fragile and tend not to survive transport intact (Francis 1993). Because clasts in concentrated PDC deposits (and some hot lahar deposits) cool mainly after they are deposited, they can contain a high proportion of intact, prismatically jointed breadcrust clasts, particularly on deposit surfaces where cooling is most rapid (Francis 1993). For example, on the surface of a PDC deposit from the 1968 eruption of Mayon volcano, most of the clasts >4 cm diameter (78%) were breadcrusted (Moore and Melson 1969). Secondary lahars, volcanic debris avalanches (VDAs), and even alpine glaciers can entrain prismatically jointed clasts from other deposits, but their deposits are unlikely to contain many intact examples of such clasts because of stresses imposed during transport (Francis et al. 1985; Francis 1993).

3.9.2 Charcoal In or Beneath Deposits

Charcoal forms when organic material, most commonly living woody material, is heated to at least 230–250 °C (depending on the heating

Fig. 16 Minor charring on a log carried by the February 19, 2009, pyroclastic density current downstream of Chaitén volcano in the Chaitén River valley (Chile), described in Major et al. (2013). Camera lens cap (6 cm diameter) for scale. USGS photo by T. Pierson

time) in the absence of air (Lockwood and Lipman 1980; Scott and Jones 1991). Under static conditions, the decomposition reaction (*pyrolysis*) will commonly leave black, tarry distillation by-products on the rock or soil material immediately surrounding the charcoal itself. Data from hot PDC deposits from Mount St. Helens that took months to cool (Banks and Hoblitt 1996) indicate that deposits heated to only 150 °C contain wood but not charcoal, while those heated to about 300 °C contain only charcoal. Once formed, charcoal is highly inert chemically and cannot burn unless it is heated to >500 °C (Lockwood and Lipman 1980).

The presence of charcoal fragments or partial charred wood in a deposit is unambiguous evidence of heating to temperatures greater than approximately 250 °C, but the heating may not necessarily have occurred in the deposit that contains it. Minor charring of the outside of wood can be subtle and difficult to distinguish from the darkening of wood that naturally occurs in the early stages of decomposition in wet deposits (Fig. 16). Even minimal charring of wood in relatively fresh deposits will give off a burnt smell when scraped.

A hot PDC origin of a deposit is confirmed if almost all the organic material present is charcoal and the tarry distillation by-products mentioned above can be found surrounding at least some of the charcoal fragments (proof the wood fragments turned to charcoal in situ after the flow had come to rest). Many PDC deposits contain large pieces of fully carbonized wood (charcoal), such as whole tree trunks, but because PDCs cool as they flow, distal deposits may contain partially charred or uncharred wood. On the other hand, lahars initiated by the rain mobilization of old PDC deposits may also contain charcoal, but in this case the distillation by-products would not be present and the amount and size of charcoal fragments are likely to be quite small due to the grinding action of rock particle mixtures during lahar flow. In addition, charcoal fragments and uncharred wood may be juxtaposed throughout such lahar deposits. Hot lahars may contain some charcoal that was formed in situ by organic matter being situated next to large hot clasts, but charring only would occur immediately next to such clasts. Elsewhere in the deposit, wood would be uncharred because the average temperature of the deposit would not have exceeded 100 °C. Glacial deposits could contain some small amounts of charcoal fragments, but those would have to be accidental fragments of preexisting charcoal from forest fires or in PDC deposits on land overrun by the glacier.

3.9.3 Red or Pink Coloration

Pyroclastic particle surfaces, exposed to air and water-rich magmatic gases for at least hours while still very hot, acquire faint red, pink, or orange coloration due to oxidation of iron in magnetite and pyrrhotite[7] to hematite (Hoshyaripour et al. 2014; Matsumoto and Nakamura 2017; Matsumoto 2020). Such iron oxidation imparts these hues to pumice fragments in hot eruption columns (Hoshyaripour et al. 2014; Matsumoto 2020) and to the surfaces of particularly hot PDC deposits (Crandell 1987; Freundt et al. 2000). Reddish coloration is especially apparent on whitish pumice and black scoria deposits. These reddish colors usually differ from the brownish yellow/orange colors typically imparted to unconsolidated deposits by iron hydroxides formed by low-temperature weathering reactions (Hurst 1977; Nagano et al. 2002). Subtle differences in color can be documented using published rock and soil color charts (e.g., Munsell Soil Color Charts 1975). Interiors of pumice clasts may show more color than the exteriors, and larger clasts may develop darker colors (Rowley et al. 1981).

Reddish coloration of PDC deposits is commonly restricted to a surface layer up to several meters thick (Crandell 1987), but thickness of such colored zones can increase with time in thick, hot deposits. For example, coloration developed to a depth of 2–5 cm at one location on the surface of the June 12, 1980, pumiceous pyroclastic-flow deposit at Mount St. Helens one day after emplacement, but this pinkish layer thickened to 1.5 m over the next 2 months (Rowley et al. 1981). The emplacement temperature of this flow ranged from 360 to 600 °C (Banks and Hoblitt 1981). Notably, reddish coloration may be imparted to the top of any previously unoxidized deposit if overridden by a lava flow or hot PDC (Fig. 17).

3.9.4 Thermoremanent Magnetic Orientation of Clasts

Thermoremanent magnetization (TRM) of a volcaniclastic deposit occurs when rocks containing iron or iron–titanium oxide minerals within the deposit have been heated above the Curie Temperature (about 570 °C for magnetite) and then cooled following deposition in the presence of Earth's magnetic field to below the blocking temperature (200–300 °C), which locks in the magnetic orientation at the time of cooling. Thus, all clasts within such a deposit should have the same magnetic orientation. TRM provides a powerful means for discriminating between deposits emplaced at high temperature and those emplaced below about 200 °C. Field measurement of the directional consistency or lack thereof of TRM in a sample of clasts from a coarse volcanic deposit of unknown origin can be done using a portable field magnetometer to reveal whether the deposit was emplaced while hot (e.g., Aramaki and Akimoto 1957; Crandell and Mullineaux 1973; Wright 1978). Three possibilities exist for volcanic flow deposits:

(1) If a flow was initially emplaced at a temperature above about 570 °C and the clasts within it cooled without post-depositional reorientation, all the clasts containing Fe or Fe–Ti oxide minerals, whether large or small, should exhibit a single paleomagnetic orientation parallel to the local geomagnetic field at the time of emplacement. In this case, emplacement by a hot PDC can be inferred, because the probability of

[7] Pyrrhotite is an iron-bearing sulfide mineral typically present in arc magmas.

Fig. 17 Faint oxidation of volcaniclastic sediment by an overlying lava flow on the lower north flank of Llaima volcano (Chile). Note that the basal rubbly zone of the lava flow was also oxidized. USGS photo by T. Pierson

dipole orientations of cool clasts aligning themselves by chance in only one direction is virtually nil.

(2) If large clasts are much hotter than small clasts as a deposit is emplaced (i.e., hot large clasts transported within cooler matrix material), the large clasts may acquire magnetic orientation parallel to the then-ambient geomagnetic field but the smaller clasts would be oriented randomly. Such a mixed result suggests origins such as an unevenly heated or cooled (possibly remobilized) block-and-ash flow (Fisher and Schmincke 1984; Risica et al. 2022), a warm lahar transporting very hot boulders (Dinehart 1999), or a debris avalanche transporting parts of an edifice-intruding cryptodome (Glicken 1996).

(3) If the flow was relatively cool and emplaced at a temperature entirely below about 200 °C, the minimum blocking temperature, all clasts will be oriented randomly.

A more involved but also more accurate application of thermoremanent magnetism involves progressive thermal demagnetization of clast samples in the laboratory, which allows estimation of the temperatures at which the clasts were emplaced (Hoblitt and Kellogg 1979). Emplacement temperatures to within ±25 °C can be estimated by this technique.

3.9.5 Phreatic Explosion Pits

Phreatic explosion pits are craters on deposit surfaces, generally circular in plan and up to 100 m in diameter and 20 m in depth (Fig. 18). Particularly large examples can be hundreds of meters in diameter and nearly 40 m deep (Rowley et al. 1981; Hildreth and Fierstein 2012). They form when hot PDC deposits cover springs, ponds, or streams; heating by the deposit pressurizes the confined liquid water and triggers upward directed steam explosions. They can also form in otherwise cool flow deposits, such as debris-avalanche deposits, that are transporting extremely hot and large fragments of

Fig. 18 Phreatic-explosion pit craters in pumiceous pyroclastic density-current deposits near shore of Spirit Lake, 6 km north-northeast of the active vent at Mount St. Helens (Washington, USA), summer 1980. The lake surface in the background is largely covered by floating logs. USGS photo archive (public domain)

volcanic rock. When migrating subsurface water encounters such point sources of heat, similar flashing to steam can occur. Rings of ejecta around such explosion pits may exhibit dune-like bed forms and bomb sags.

4 Particle-Scale Evaluation Criteria

Particle-scale properties of unconsolidated sediment may be visible at outcrop scale in the field, but particle-scale analysis of bulk samples is required for determination of those properties, including the statistical parameters of a sample's particle-size distribution, namely measures of average particle size, sorting, skewness, and kurtosis (Inman 1952; Folk 1974). The determination of the sedimentary texture of a deposit, however, requires analysis of both particle-scale properties (size distribution, roundness, and shape of particles) and an outcrop-scale property, fabric (spatial distribution and orientation of grains), determine the *sedimentary texture* of a deposit (Neuendorf et al. 2005). Particle-scale data can also be applied to more sophisticated techniques, such as involving geospatial and multivariate analyses, to discriminate among different deposit types (Bellotti et al. 2010).

4.1 Particle Composition

Composition of particles in sediment bodies can characterize a deposit and provide evidence for emplacement mechanism and source. The meaning of *particle composition* here encompasses not only the specific types of mineral and rock fragments present, but also the relative percentages of crystals, glass, and rock fragments in the juvenile component and the relative percentages of juvenile volcaniclastic particles, cognate and accessory lithic fragments from a source volcanic vent, and accidental lithic fragments scoured from a flow path. However, the proportions of different types of particles can change with distance from a source volcano, so a sampling strategy should include multiple samples taken at different distances from source.

Volcanic particles are commonly composed of some combination of individual mineral grains (crystals or crystal fragments), dense polycrystalline rock fragments (volcanic lithic fragments), vesicular glassy fragments (pumice or scoria), and glass shards. Additional but less common components may include accretionary lapilli, spatter, bombs, carbonized wood, or even human cultural debris (fragments of pottery, bricks, roof tiles, etc.). In volcanic wet-flow and PDC deposits, the componentry of a particle mixture typically reflects both rocks from the volcanic source and the fragments of country rock entrained along its flow path. The rate of increase in the proportion of volcanic or nonvolcanic accidental particles in a deposit can reflect the volumetric rate of sediment entrainment by the flow (e.g., Pierson et al. 1990).

Heterogeneity of volcanic rock particle compositions in deposit mixtures has commonly been described by volcanologists with use of the informal terms "monolithologic" (composed of a single rock type) and "heterolithologic" (composed of multiple rock types), although these two terms seem not to have official definitions (Neuendorf et al. 2005). Sedimentologists usually opt for slightly more nuanced terms—*monomictic*, *oligomictic*, and *polymictic*—to describe the degree of particle mixing in unconsolidated deposits (Friedman and Sanders 1978; Neuendorf et al. 2005). But the definitions of these sedimentologic terms vary slightly among references, have never (to our knowledge) been applied to volcanic sediments, and depend on a distinction between "rock type" and "rock species" that is not clearly defined. Therefore, in applying the sedimentologic terms to volcanic sediments in this volume, we interpret *rock type* to be a particular classification of volcanic rock as defined by the IUGS petrological classification of volcanic rocks (Le Bas and Streckeisen 1991). We then interpret *rock species* to be a particular subcategory of that rock type having specific visual and (if known) chemical characteristics. Visual characteristics could include groundmass color and crystallinity; abundance and size of bubble vesicles; and abundance, size, color, and mineralogy of phenocrysts. For example, the rock type encountered in a deposit might be *andesite*, and the rock species might be *medium gray, slightly vesicular, high-silica andesite with large (up to 5 mm) white plagioclase phenocrysts*. A rock species with such specificity would probably be the product of a particular eruption. Thus, we apply the sedimentologic terms for volcanic deposit heterogeneity as follows:

1. *Monomictic* deposits—sediment mixtures almost entirely composed of a single rock species—a rock type (e.g., andesite) having specific visual and chemical characteristics that set it apart from other rocks of that type.
2. *Oligomictic* deposits—sediment mixtures almost entirely composed of a single rock type but containing a few different species of the same rock type.
3. *Polymictic* deposits—sediment mixtures composed of two or more different rock types in quantities of more than a few percent of total components.

Note that monomictic and oligomictic deposits may contain up to a few percent of entrained accidental lithic fragments of diverse compositions without requiring reclassification of the descriptor.

4.2 Particle Roundness, Shape, Fracturing, and Surface Texture

The roundness, shape, character of fracturing, and surface texture of sedimentary particles reflect what has happened to particles after they were formed. Different flow processes modify

Fig. 19 Roundness categories subjectively applied to cobble-size clasts of nonvesiculated volcanic rock (Friedman and Sanders 1978). Categories from upper left to lower right are **a** very angular, **b** angular, **c** subangular, **d** subrounded, **e** rounded, and **f** well rounded. USGS photo by T. Pierson

particles differently during transport (e.g., Komorowski et al. 1991; Caballero and Capra 2011).

Roundness is the degree of angularity of a particle's edges and corners, which is used as a measure of abrasion of clastic particles during transport. Note that old volcanic rock fragments can be rounded by in situ chemical and physical weathering, as well as by abrasion. Roundness is independent of particle sphericity. A method of quantifying roundness was developed by Wadell (1932), but normally a subjective evaluation of roundness based on 6 main roundness classes is sufficient for field descriptions of deposits (Fig. 19). For a given particle–fluid flow, four principal factors appear to affect degree of particle rounding: degree of particle dispersion in the flow, distance traveled, particle size, and particle hardness. These factors affect the frequency, duration, and intensity (force) of collisions and grinding physical contacts. Particle hardness determines resistance to abrasion or breakage during collisions. Rounding is greater in flows that are more concentrated, that travel greater distances, and that contain weathered or soft rock particles, with large gravel-size clasts showing a greater degree of rounding than small clasts and matrix grains. It follows that particle size, composition, and degree of weathering or hydrothermal alteration should be noted together with degree of roundness in field descriptions.

The degree of rounding of large gravel-size clasts (large cobbles and boulders) can be particularly helpful in interpreting the origin of some diamicts. For example, we have observed that freshly broken, boulder-size, unweathered rock fragments of dacite or andesite (e.g., from dome collapse) can change from very angular

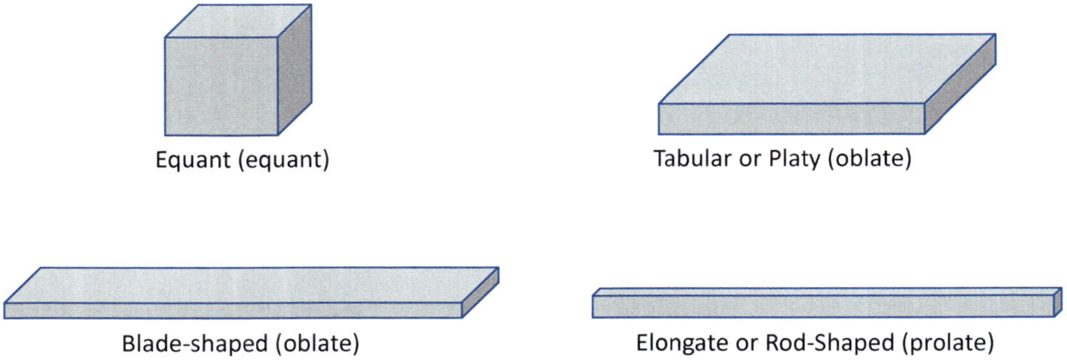

Fig. 20 Common names for idealized particle shapes that may be found in volcanic particle–fluid flows, with geometric form name in parentheses (modified from Folk 1974)

to sub-angular or subrounded after transport of 10–20 km by a concentrated lahar (volcanic debris flow), whereas clasts of similar size and composition in more inflated PDCs or volcanic debris avalanches typically show less rounding after transport of equivalent distances. Pebble-size and smaller fragments of the same rocks in those deposits, however, typically remain angular in all flows that contain fresh, hard, rock particles. On the other hand, flows transporting delicate fresh pumice fragments only a kilometer or two will cause the pumice clasts of all sizes to become well rounded (Manga et al. 2011; Kueppers et al. 2012).

Not all clasts in a particle–fluid flow deposit are necessarily primary particles. Flows may entrain particles from streambeds or slope surfaces that had already been rounded to some degree by other processes. The juxtaposition of differently rounded gravel-size clasts (e.g., angular clasts and well-rounded clasts) in the same deposit indicates entrainment of exotic clasts.

Particle shape is determined by (1) initial shape when particle was first formed, (2) size, (3) hardness, (4) size of component grains or crystals if particle is a fragment of rock, (5) inherited planes of weakness; and (6) mechanical stresses during transport (from transport process type, magnitude, and duration). Shape terminology (Fig. 20) is based on the relative lengths of the particle's three main axes (Folk 1974).

Particle fracturing reflects the type, duration, and intensity of stress experienced by particles during transport and sometimes prior to mobilization (e.g., in a cooling lava dome prior to collapse or within a stratocone during magma intrusion). Fracture orientation, density, and degree of fracture opening observed in particles of different sizes may be evidence of exposure to high heat or rapid cooling (Sect. 3.9.1), or fractures (and freshly broken clasts) may reflect the intensity of collision, shear, or compressive stresses within a moving flow. A fractured but unbroken gravel-size clast within a deposit generally suggests minimal exposure to high shear stress following fracture formation, although many pervasively fractured clasts and debris-avalanche blocks formed by the extreme normal (compressive) stresses during edifice collapse tend to disaggregate with continued downslope flowage of VDAs (e.g., Glicken 1996).

Surface texture refers to the overall roughness or smoothness of particle surfaces, including any pits, striations, chatter marks, or coatings that may be observed (Friedman and Sanders 1978). Physical processes causing variations in surface texture (e.g., collisional impacts and abrasion) act with more force and typically leave more evidence of such interactions on surfaces of cobbles and boulders than on finer gravel- and sand-size grains.

4.3 Particle-Size Distribution and Sedimentary Texture

The frequency distribution of different particle sizes within a mixture of particles is a fundamental physical property that is related both to properties of the source material and to transport and depositional processes. Measurement and characterization of particle-size distribution (granulometry) is exceedingly important in interpreting the origin of particle–fluid flow deposits. It is primarily the differences in the shapes of distribution curves, quantified by the statistical parameters of those curves, that allows different flow types to be identified from deposits (Fig. 21).

Individual particles in volcanic settings can range from boulders tens of meters in diameter to tiny flakes of clay less than a micron in diameter. With such a range in particle size, statistically representative samples can usually only be obtained by using a combination of sampling techniques and a well-designed sampling strategy (Kellerhals and Bray 1971; Mosley and Tindale 1985; Ibbeken and Schleyer 1986; Wolcott and Church 1991; Fripp and Diplas 1993). Where deposits do not contain clasts larger than 16 or 32 mm, it is practical to do a particle-size analysis entirely from samples collected in the field. In this case, the size of a field sample to be collected for particle size analysis depends on the weight of the largest clast in the deposit. To be representative of a deposit, the heaviest clast should weigh less than 5% of the weight of the entire bulk sample, which accounts for different particle densities (Mosley and Tindale 1985). Any description of particle-size characteristics of a deposit should specify (1) whether there is an upper or lower particle size limit for the deposit sample being analyzed and (2) which techniques were used to obtain the size distribution.

The mix of low-density vesicular rock fragments with normal-density crystals, glass shards, and lithic fragments results in sediment textures that are unique to volcanic sediments. In water, relatively large, low-density pumice or scoria grains have hydrodynamic behaviors like much smaller non-vesicular rock fragments and thus are deposited together in deposits that are more poorly sorted than sediments in otherwise similar nonvolcanic depositional environments (Fisher 1965; Whitham and Sparks 1986; Manville et al. 1998, 2002; White et al. 2001; Segschneider et al. 2002). In addition, the density of pumice grains can vary with time as water soaks into the pore spaces (White et al. 2001).

Particle-size classification systems are based on particle diameter (D), which must be measured in different ways, depending on size fraction. Size-fraction terminology differs, however, between sedimentologists and volcanologists, and the differing particle-size terms are shown (Table 5). The sedimentology terms are purely descriptive and refer to the sizes of solid particles regardless of their origin; these terms are used preferentially throughout this book unless size gradations are limited to volcanic particles The volcanology terms imply an eruptive or pyroclastic origin. All sedimentary particles larger than 2 mm are broadly classified as *gravel* (Neuendorf et al. 2005), with terms *granule*, *pebble*, *boulder*, and *block* describing progressively larger subdivisions of gravel. The terms *sand*, *silt*, and *clay*[8] describe the broad categories of finer grainer particles making up the matrix (or matrix material) occupying the spaces between gravel-size clasts in volcaniclastic deposits. Silt and clay fractions are sometimes combined and termed *mud* or *fines*.

Particle sizes in deposits can be measured in various ways. Sand-size (0.063–2 mm) and pebble-size (2–32 mm) particles usually are measured in the laboratory by sieving—passing samples of particle mixtures through a stack of

[8] Clay-size particles can be composed of various clay minerals or simply very fine fragments of other minerals. Clay minerals have physical properties that, when suspended in liquid water, can strongly affect the viscosity and shear strength of fluid suspensions and the behavior of volcanic wet flows.

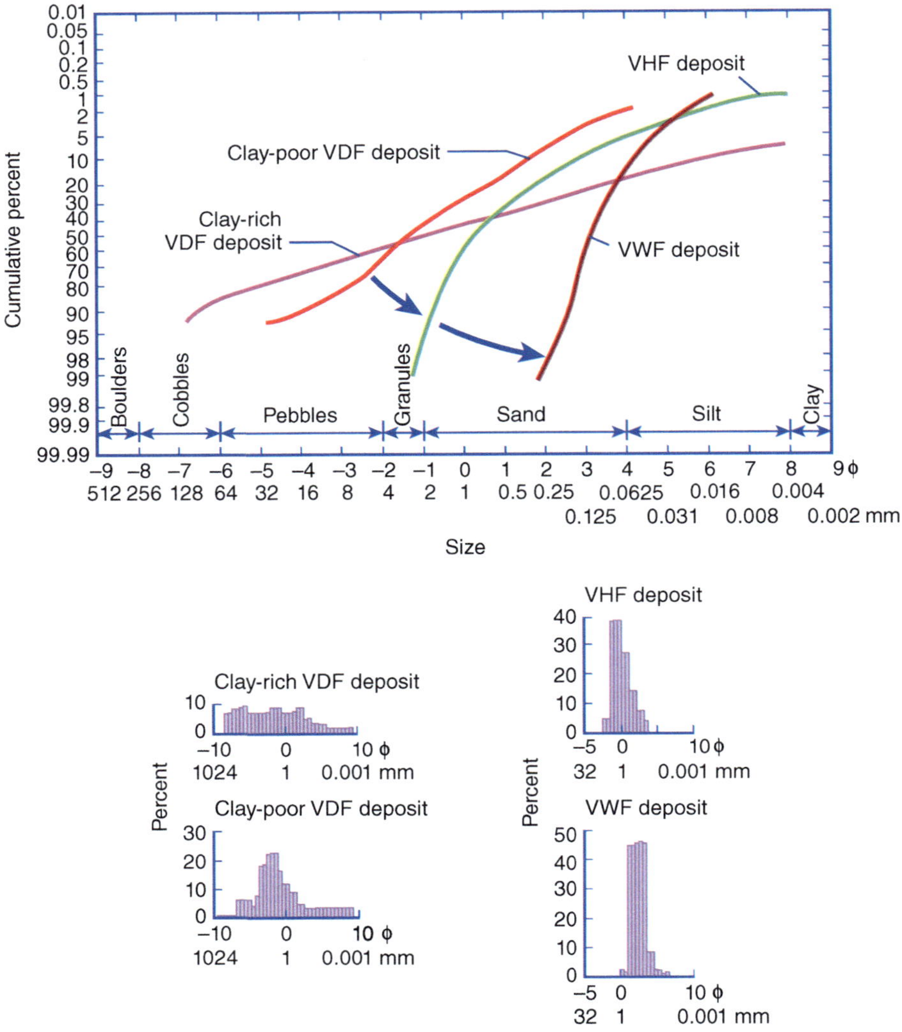

Fig. 21 Differences in typical particle-size distributions for samples of deposits of volcanic debris flows (VDFs), volcanic hyperconcentrated flows (VHFs), and volcanic water floods (VWFs), shown as both cumulative frequency plots and frequency-distribution histograms (modified from Scott et al. 1995). Arrows indicate how coarse tails of distributions steepen as sorting values decrease. Note that all distributions are fine-tailed, but this may owe to sampling limitations that inadequately account for the cobble- and boulder-size clasts in the deposit

sieves with openings of known dimensions—or by using image analysis software that utilizes two-dimensional cross-sectional images and stereology with statistical computations to estimate particle distributions in three dimensions (e.g., Fernlund et al. 2007; Jutzeler et al. 2012). Physical measurement techniques for clasts too large to sieve include (1) direct measurement of the length of the clast's major axes ($D = (abc)^{1/3}$ or, when less accuracy is needed, $D \approx$ b-axis length) and (2) direct measurement of particle volume (usually by fluid displacement) and then assigning the calculated diameter of a sphere of the same volume and density as the particle. Silt and clay fractions cannot be separated by mechanical sieving, so other techniques must be

Table 5 Size classes and terminology for sedimentary particles of unspecified origin coarser than 0.063 mm and for particles of similar size having a primary volcanic origin

Particle-size categories		Names for particles in sedimentology		Names for particles in volcanology	
Diameter (phi)	Diameter (mm)	Unconsolidated	Lithified	Unconsolidated	Lithified
		Mud (silt and clay)[a]	Mudstone, shale	Extremely fine ash	Extremely fine tuff
4	0.063				
		Very fine sand	Very fine sandstone	Very fine ash	Very fine tuff
3	0.125				
		Fine sand	Fine sandstone	Fine ash	Fine tuff
2	0.25				
		Medium sand	Medium sandstone	Medium ash	Medium tuff
1	0.5				
		Coarse sand	Coarse sandstone	Coarse ash	Coarse tuff
0	1				
		Very coarse sand	Very coarse sandstone	Very coarse ash	Very coarse tuff
− 1	2				
		Granule (gravel)	Grit, granule conglomerate	Fine lapilli	Fine lapilli-tuff
− 2	4				
		Fine pebble (gravel)	Pebble conglomerate	Medium lapilli	Medium lapilli-tuff
− 3	8				
		Medium pebble (gravel)	Pebble conglomerate	Medium lapilli	Medium lapilli-tuff
− 4	16				
		Coarse pebble (gravel)	Pebble conglomerate	Coarse lapilli	Coarse lapilli-tuff
− 5	32				
		Very coarse pebble (gravel)	Pebble conglomerate	Coarse lapilli	Coarse lapilli-tuff
− 6	64				
		Fine cobble (gravel)	Cobble conglomerate	Block, bomb	Breccia
− 7	128				
		Coarse cobble (gravel)	Cobble conglomerate	Block, bomb	Breccia
− 8	256				
		Fine boulder (gravel)	Boulder conglomerate	Block, bomb	Breccia

(continued)

Table 5 (continued)

Particle-size categories		Names for particles in sedimentology		Names for particles in volcanology	
Diameter (phi)	Diameter (mm)	Unconsolidated	Lithified	Unconsolidated	Lithified
– 9	512 (0.5 m)				
		Medium boulder (gravel)	Boulder conglomerate	Block, bomb	Breccia
– 10	1024 (1 m)				
		Coarse boulder (gravel)	Boulder conglomerate	Block, bomb	Breccia
– 11	2048 (2 m)				
		Very coarse boulder (gravel)	Boulder conglomerate	Block, bomb	Breccia
– 12	4096 (4.1 m)				
		Block[b] (megagravel)	Mega-conglomerate	Block, bomb	Breccia

The sedimentologic classification scheme is the Udden–Wentworth classification (Udden 1914; Wentworth 1922) that is used for epiclastic (and other) sedimentary particles; it has been extended to 4 m by Blair and MacPherson (1999). Proposed terms for clasts >4 m are suggested in Terry and Goff (2014). The volcanology classification scheme shown here is the White and Houghton (2006) classification for primary volcaniclastic particles

[a] The boundary between silt and clay is 8 phi or 0.004 mm; see Blair and MacPherson (1999) for subdivisions of silt

[b] Sedimentary blocks are described up to – 16 phi or 65.5 m in diameter in Blair and MacPherson (1999), but larger blocks can exist. Sedimentary and volcanic blocks differ from debris-avalanche blocks, as discussed in Chap. "Volcanic Debris-Avalanche Deposits"

employed. The hydraulic-equivalent diameter of silt and clay particles has long been measured using physical methods based on particle settling velocities in distilled water (pipette and hydrometer sedimentation methods), but now various noncontact methods to measure the size distribution of fine particles in water suspensions include X-ray beam attenuation ("sedigraph"), dynamic light scattering, and laser diffraction (Coates and Hulse 1985; Vitton and Sadler 1997; Di Stefano et al. 2010).

The widely used Udden–Wentworth size classification scheme shown in Table 5 (Udden 1898, 1914; Wentworth 1922) uses a base-2 logarithmic scale in which each successive size class is twice as large as the preceding one. For example, in the fine gravel range, size classes are bounded by 2.00, 4.00, 8.00, 16.00 mm, and so on. A later addition to the scheme (Krumbein 1934) uses the logarithms themselves in the φ (phi) scale to delineate size classes. The phi scale has 1.00 mm as its origin, and the dimensionless φ units were originally defined as:

$$\varphi = -\log_2 \frac{D}{D_0}$$

where φ is the particle size in phi-units, D is the particle size in millimeters, and D_0 is the particle size at the scale origin (1.00 mm). This equation can also be written in base-10 form:

$$\varphi = -\log_{10} D / \log_{10} 2$$

Quick conversions from one scale to the other can be done using a calculator (the second equation), nomographs (Fig. 22), or tables. Complete conversion tables at quarter-phi intervals can be found in other texts (e.g., Gale and Hoare 1991, pp. 58–59). Other size classification schemes are used in the fields of engineering geology, pedology, and volcanology. Differences between some of these other schemes and the Udden–Wentworth scale are that (a) they use 2 μm for

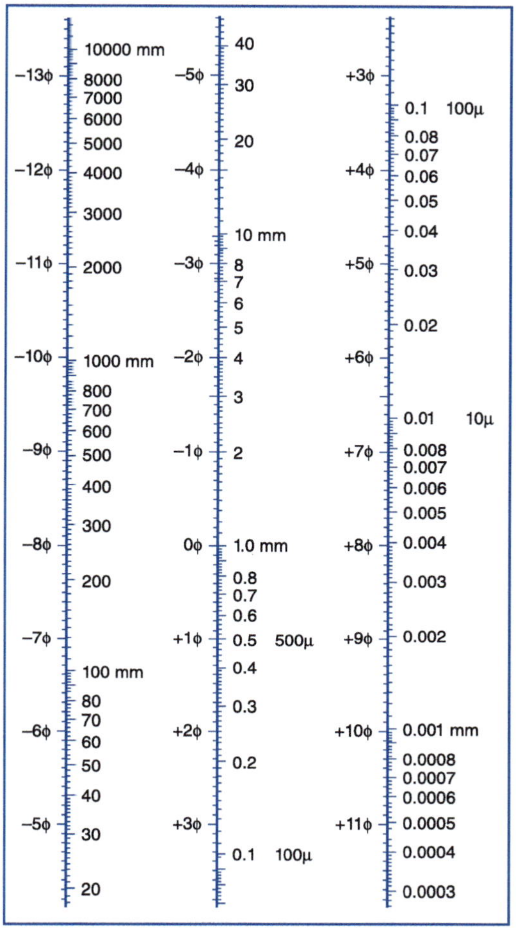

Fig. 22 Nomograph for quickly converting particle diameter in φ (phi) units to millimeters (modified from Friedman and Sanders 1978)

the silt–clay boundary instead of 4 μm, and (b) they restrict the term "gravel" to pebble-size clasts, whereas common usage today includes all particles larger than 2 mm as gravel.

The size terminology for volcaniclastic particles (White and Houghton 2006) is also shown in Table 5. We agree with White and Houghton (2006) that a classification scheme that marries the terminology traditionally used by volcanologists for primary volcaniclastic particles (e.g., Fisher 1961) with the particle-size terminology used by sedimentologists makes the most sense. Resedimented volcaniclastic particles could be described using either classification. Although the term "clast" technically refers to particle of any size (Neuendorf et al. 2005), general usage by many sedimentologists and volcanologists in field descriptions of deposits informally differentiates "clasts" (particles >2 mm in diameter—gravel or lapilli and blocks) from "matrix" (particles <2 mm—sand and mud or ash). The term "larger clasts" is informally used in this book to describe clasts that are easily visible in outcrops—a size threshold that can vary with one's distance from an outcrop.

Folk (1954) developed a way to roughly characterize the *sedimentary texture* of a sedimentary or volcaniclastic deposit based on its proportions of gravel, sand, and mud (silt+clay) (Fig. 23). Correct assignment of textural terms requires a full particle-size analysis from deposit samples, but with practice a geologist can describe deposits in the field using this terminology. For example, slightly gravelly sandy mud has a different look and feel than muddy sandy gravel.

To quantitatively assess and compare grain size distributions among different volcaniclastic deposits, it is necessary to determine statistical parameters for those distributions (obtained from representative samples, as noted above). Parameters commonly computed are *mean particle size*, *sorting*, *skewness*, and *kurtosis* (Folk 1974). Some statistical methods for analysis of grain-size distributions of sediment mixtures assume that the distributions are approximately log-normal and have a single mode. These two assumptions are often violated. For example, volcaniclastic sediment size distributions are commonly polymodal and asymmetrical. Where highly accurate assessment of complex distributions is needed, sophisticated statistical methods can be applied (e.g., Wohletz et al. 1989; Weltje and Prins 2007; Wu et al. 2020). For most applications, however, the methods for determining statistical parameters first defined by Folk and Ward (1957) are adequate and are determined graphically from cumulative particle-size

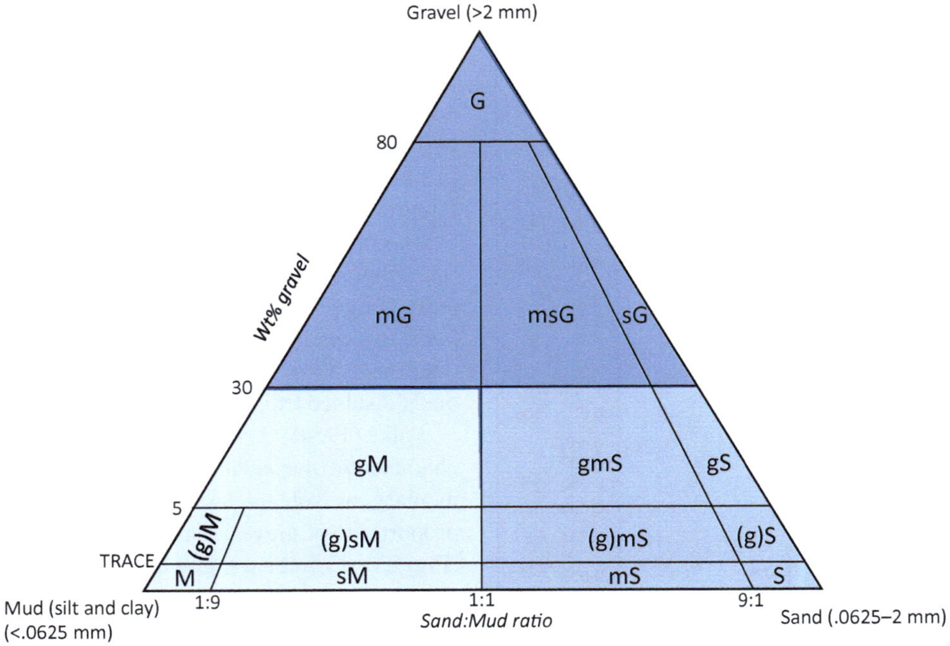

Fig. 23 Sedimentary texture nomenclature for describing various mixtures of differently sized sediment particles, depending on the proportions of size classes (modified from Folk 1954). Abbreviations are G = gravel, S = sand, and M = mud (silt + clay). When capitalized, that size is the major component; lower case components decrease in abundance and importance in moving to left away from the capital letter. Components shown in parentheses are trace components (for example, "slightly gravelly"). Trace amounts are assumed to be < 0.01 wt%

distribution curves: (1) Graphic Mean (M_z) for defining mean particle size; (2) Inclusive Graphic Standard Deviation (σ_I) for defining degree of sorting (uniformity of particle size); (3) Inclusive Graphic Skewness (Sk_I) for defining the symmetry/asymmetry of a distribution curve, with positive values indicating a larger distribution tail of fine particles; and (4) Graphic Kurtosis (K_G) for measuring the peakedness of a distribution curve, where normal Gaussian curves have a K_G value of 1.00, curves with values > 1.00 are excessively peaked (leptokurtic), and those with values < 1.00 are deficiently peaked (platykurtic) relative to a normal distribution (Folk 1974).

Degree of sorting—a measure of the uniformity of particle size within a sample (Fig. 24)—is an especially useful parameter for comparing different types of volcanic sediments, but volcanologists and sedimentologists largely differ in how this parameter is calculated and how it is evaluated, which can be a problem when comparing sorting values published by volcanologists and sedimentologists. Volcanologists commonly define sorting (a measure of the degree of deviation from the median particle size) using the Inman Method (Inman 1952), which uses only the 16th and 84th percentile particle diameters (from a cumulative weight-percent plot of the particle-size distribution) (Fisher and Schmincke 1984; Cas and Wright 1987). In recent years, sedimentologists have largely defined sorting using the method of Folk and Ward (1957), which uses the 5th, 16th, 50th, 84th, and 95th percentile particle diameters from the cumulative plot, more accurately characterizing multimodal and strongly nonnormal size-distribution curves found in

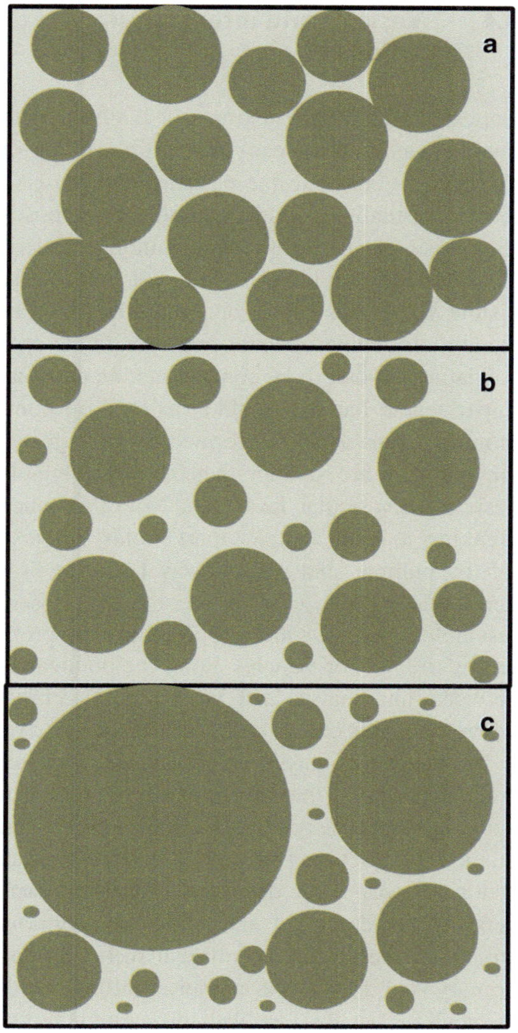

Fig. 24 Schematic depiction of degree of sorting within a sediment sample. **a** Sample is relatively well sorted; most of the particles are similar in size. **b** Sample is moderately sorted. **c** Sample is relatively poorly sorted; the particle sizes vary widely

many volcaniclastic sediments (Folk 1966). It thus includes particle-size variability in the tails of the distributions that the Inman method ignores. Any of the statistical parameters for particle-size distributions obtained by these two methods can be somewhat similar but are not directly comparable (Folk 1966). Be aware that

Table 6 Terms used to describe degrees of sorting in volcanological and sedimentological studies

Terms largely used in volcanology (Cas and Wright 1987)	Sorting values (σ_{ϕ})	Terms largely used in sedimentology (Folk 1974)	Sorting values (σ_I)
Very well sorted	0–1.0ϕ	Very well sorted	<0.35ϕ
		Well sorted	0.35–0.5ϕ
		Moderately well sorted	0.5–0.71ϕ
		Moderately sorted	0.71–1.0ϕ
Well sorted	1.0–2.0ϕ	Poorly sorted	1.0–2.0ϕ
Poorly sorted	2.0–4.0ϕ	Very poorly sorted	2.0–4.0ϕ
Very poorly sorted	>4.0ϕ	Extremely poorly sorted	>4.0ϕ

The sedimentology terms are used in this book

volcanologists and sedimentologists also evaluate the degree of sorting in sediment mixtures differently (Table 6). In this book we use sorting values computed by the Folk and Ward method for obtaining the sorting parameter (σ_I) when those values are available and Inman values (σ_{ϕ}) otherwise. We also use the Folk and Ward criteria (modified in Folk 1974) for evaluating the degree of sorting.

Determination of particle-size distribution is normally measured based on the weight of particles in each size class. Therefore, the density of the particles is important. The density of quartz (2.65 g/cm^3) is sometimes taken as the approximate average for most volcanic silicate rocks, but dense-rock densities vary depending on composition (Table 7). Many volcanic rocks contain vesicles, so actual densities of some rocks are less than the dense-rock values shown in the table. Highly vesiculated volcanic rocks (pumice and scoria) can have densities less than 1 g/cm^3, a property which greatly affects how such particles are transported and deposited. Mixtures containing pumice, crystals, and lithic

Table 7 Densities of common minerals and dense (nonvesicular) rock fragments that may be found in volcaniclastic deposits, including entrained nonvolcanic particles (modified from Clark 1966; Deer et al. 1967)

Minerals/rocks	Density (g/cm^3)
Quartz (pure)	2.65
Orthoclase (feldspar)	2.55
Plagioclase (feldspar)	2.62–2.76
Biotite (mica)	2.7–3.3
Cummingtonite–Grunerite (amphibole)	3.10–3.60
Hornblende (amphibole)	3.02–3.45
Enstitite–Orthoferrosilite (pyroxene)	3.21–3.96
Diopside–Hedenbergite (pyroxene)	3.22–3.56
Augite–Ferroaugite	2.96–3.52
Olivine	3.22–4.39
Hematite	5.23
Ilmenite	4.70–4.78
Magnetite	5.2
Rhyolite/Granite	2.52–2.81
Dacite/Granodiorite	2.67–2.79
Andesite/Diorite	2.72–2.96
Basalt/Gabbro	2.85–3.12
Sandstone	2.17–2.70
Limestone, Marble, Dolomite	2.37–2.80
Schist	2.70–3.03
Gneiss	2.66–3.06

Densities of vesicular particles (pumice and scoria) vary widely, from less than 0.5 g/cm^3 up to the values of their nonvesicular rock types. Density ranges in individual mineral species are due to variations in chemical formulas. Chemical weathering also causes variation in density

fragments can be very poorly sorted in terms of grain diameters, but they may be better sorted in terms of their hydrodynamic properties (Oehmig and Wallrabe-Adams 1993; Manville et al. 2002, 2009) (Fig. 25).

4.4 Vesicles and Intergranular Voids

Deposits with muddy textures or diamicts with muddy sand matrix can trap entrained air or gas bubbles between fine particles and preserve them as small (~ 1 mm) vesicles in deposits (Fig. 26a). Such voids are commonly spherical in shape, randomly distributed, and easily visible with a hand lens, but some may be flattened if air bubbles rise upward through the consolidating deposit to be concentrated beneath an impermeable barrier. Bubble vesicles are common in lahar deposits, formed from ingested air, but they are smaller than most gas-bubble vesicles in vesicular lava rock. Note that when breaking a hand sample apart, voids are also left by pulling sand grains away from the broken surface. Such grain impression can closely resemble vesicles. One way around this problem is to look for vesicles along the boundaries between fine matrix and large smooth clasts, where no grains can be pulled out of the matrix.

Some deposits may be dominated by cobble- or boulder-size clasts that have irregular intergranular voids between clasts where liquefied matrix has drained out, has been washed out by rain or stream action, or was never deposited (Fig. 26b). Such deposits are described as having an *openwork structure*. One way to determine if such voids are primary is to dig into the interior of the deposit. If the voids persist, they are likely primary.

Other voids encountered in volcaniclastic deposits include open gaps between fractures in debris blocks, and elongate voids formed as molds around woody debris (tree branches or logs), prior to decay and removal of the wood by groundwater solution. The empty cast of a log or branch indicates that the wood was not fully carbonized. Charcoal is chemically inert and will not decay.

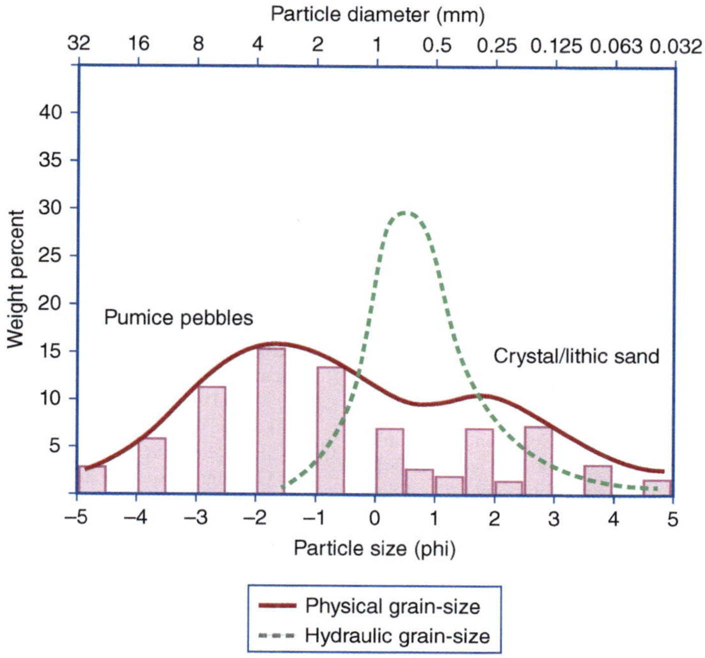

Fig. 25 Comparison of hydraulic and physical grain-size distributions in particle mixtures containing pumice fragments. Pumice particles form the modal peak on left based on particle weight and smaller dense-rock fragments form the modal peak on right. This is a common physical particle-size distribution in volcanic terrain. Both populations of particles behave similarly, however, during sediment transport by water flow—the hydraulic grain-size distribution (modified from Manville et al. 2002, 2009)

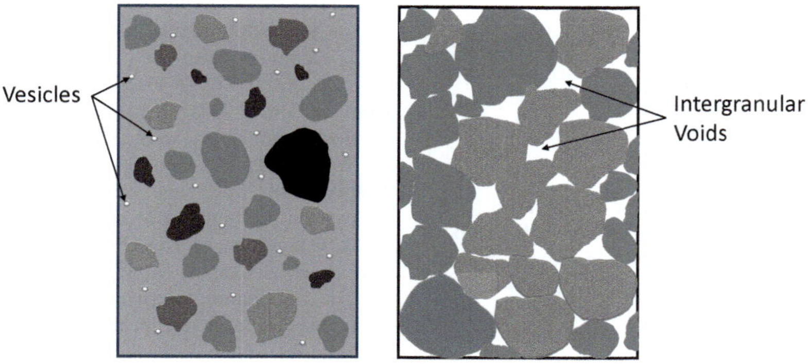

Fig. 26 Contrasting vesicles and intergranular voids in volcaniclastic deposits: **a** gas-bubble vesicles in deposit matrix material and **b** intergranular voids in a deposit of larger clasts without matrix material in the spaces between clasts. White spaces represent air-filled pockets in both deposits. Gas-bubble vesicles are scale-dependent (usually about 1–2 mm in diameter), but intergranular voids occur at a variety of scales

References

Abdurachman EK, Bourdier J-L, Voight B (2000) Nuées ardentes of 22 November 1994 at Merapi volcano, Java, Indonesia. J Volcanol Geotherm Res 100(1–4):345–361. https://doi.org/10.1016/S0377-0273(00)00144-X

Alexander J, Bridge JS, Cheel RJ, Leclair SF (2001) Bedforms and associated sedimentary structures formed under supercritical water flows over aggrading sand beds. Sedimentology 48:133–152. https://doi.org/10.1046/j.1365-3091.2001.00357.x

Allen JRL (1970) Physical processes of sedimentation: an introduction. Elsevier, New York, p 248

Aramaki S, Akimoto SI (1957) Temperature estimation of pyroclastic deposits by natural remanent magnetism. Am J Sci 255:619–627

Arguden AT, Rodolfo KS (1990) Sedimentologic and dynamic differences between hot and cold laharic debris flows of Mayon volcano, Philippines. Geol Soc Am Bull 102:865–876

Ashley GM (1990) Classification of large-scale subaqueous bedforms: a new look at an old problem. J Sediment Petrol 60:160–172

Ayris PM, Delmelle P (2012) The immediate environmental effects of tephra emission. Bull Volcanol 74:1905–1936

Banks NG, Hoblitt RP (1981) Summary of temperature studies of 1980 deposits. In: Lipman PW, Mullineaux DR (eds) The 1980 eruptions of Mount St. Helens, Washington. U S Geol Surv Prof Pap 1250:295–313

Banks NG, Hoblitt RP (1996) Direct temperature measurements of deposits, Mount St. Helens, Washington, 1980–1981. U S Geol Surv Prof Pap 1387:1–76. https://doi.org/10.3133/pp1387

Bellotti F, Capra L, Sarocchi D, D'Antonio M (2010) Geostatistics and multivariate analysis as a tool to characterize volcaniclastic deposits: application to Nevado de Toluca volcano, Mexico. J Volcanol Geotherm Res 191(1–2):117–128. https://doi.org/10.1016/j.jvolgeores.2010.01.005

Biró T, Karátson D, Márton E, Józsa S, Bradák B (2015) Paleoflow directions of a subaqueous lahar deposit around the Miocene Keserűs Hill lava dome complex (North Hungary) as constrained by photo-statistics and anisotropy of magnetic susceptibility (AMS). J Volcanol Geotherm Res 302:141–149

Blair TC, McPherson JG (1999) Grain-size and textural classification of coarse sedimentary particles. J Sediment Res 69:6–19

Blodgett RH, Stanley KO (1980) Stratification, bedforms, and discharge relations of the Platte braided river system, Nebraska. J Sediment Res 50(1):139–148. https://doi.org/10.1306/212F7987-2B24-11D7-8648000102C1865D

Branney MJ, Kokelaar P (2002) Pyroclastic density currents and the sedimentation of ignimbrites. Geol Soc Lond Mem 27:1–143

Brugman MM, Meier MF (1981) Response of glaciers to the eruptions of Mount St. Helens. In: Lipman PW, Mullineaux DR (eds) The 1980 eruptions of Mount St. Helens, Washington. U S Geol Surv Prof Pap 1250:743–756

Caballero L, Capra L (2011) Textural analysis of particles from El Zaguán debris avalanche deposit, Nevado de Toluca volcano, Mexico: evidence of flow behavior during emplacement. J Volcanol Geotherm Res 200:75–82

Cartigny MJB, Ventra D, Postma G, Van den Berg JH (2014) Morphodynamics and sedimentary structures of bedforms under supercritical-flow conditions: new insights from flume experiments. Sedimentology 61:712–748. https://doi.org/10.1111/sed.12076

Cas RAF, Wright JV (1987) Volcanic successions modern and ancient. Allen and Unwin, London, p 528

Chow VT (1959) Open-channel hydraulics. McGraw-Hill, New York, 680 pp

Clark SP Jr (ed) (1966) Handbook of physical constants. In: Geological society of America memoir, vol 97, pp 1–587

Coates GF, Hulse CA (1985) A comparison of four methods of size analysis of fine-grained sediments. New Zeal J Geol Geophys 28:369–380. https://doi.org/10.1080/00288306.1985.10422234

Cole PD (1991) Migration direction of sand-wave structures in pyroclastic-surge deposits: implications for depositional processes. Geology 19(11):1108–1111. https://doi.org/10.1130/0091-7613(1991)019<1108:MDOSWS>2.3.CO;2

Collinson JD (1996) Alluvial sediments. In: Reading HG (ed) Sedimentary environments: processes, facies and stratigraphy, 3rd edn. Blackwell Science, Oxford, pp 37–82

Costello WR, Southard JB (1981) Flume experiments on lower-flow-regime bed forms in coarse sand. J Sediment Res 51(3):849–864. https://doi.org/10.1306/212F7DC4-2B24-11D7-8648000102C1865D

Crandell DR (1987) Deposits of pre-1980 pyroclastic flows and lahars from Mount St. Helens volcano, Washington. U S Geol Surv Prof Pap 1444:1–91

Crandell DR, Mullineaux DR (1973) Pine Creek volcanic assemblage at Mount St. Helens, Washington. U S Geol Surv Bull 1383-A:23 pp

Cronin SJ, LeCointre JA, Palmer AS, Neall VE (2000) Transformation, internal stratification, and depositional processes within a channelized, multi-peaked lahar flow. New Zeal J Geol Geophys 43:117–128

Deer WA, Howie RA, Zussman J (1967) An introduction to the rock-forming minerals. Wiley, New York, p 528

Di Stefano C, Ferros V, Mirabile S (2010) Comparison between grain-size analyses using laser diffraction and sedimentation methods. Biosyst Eng 106(2):205–215. https://doi.org/10.1016/j.biosystemseng.2010.03.013

Dietterich HR, Grant GE, Fasth B, Major JJ, Cashman KV (2022) Can lava flow like water? Assessing applications of critical flow theory to channelized basaltic lava flows. J Geophys Res Earth Surf 127(9):e2022JF006666. https://doi.org/10.1029/2022JF006666

Dinehart RL (1992) Evolution of coarse gravel bed forms; field measurements at flood stage. Water Resour Res 28:2667–2689

Dinehart RL (1999) Sediment transport in the hyperconcentrated phase of the March 19, 1982, lahar. In: Pierson TC (ed) Hydrologic consequences of hot-rock/snowpack interactions at Mount St. Helens volcano, Washington, 1982–84. U S Geol Surv Prof Pap 1586:37–52

Douillet GA, Pacheco DA, Kueppers U, Letort J, Tsang-Hin-Sun E, Bustillos J, Hall M, Ramón P, Dingwell DB (2013) Dune bedforms produced by dilute pyroclastic density currents from the August 2006 eruption of Tungurahua volcano, Ecuador. Bull Volcanol 75:62. https://doi.org/10.1007/s00445-013-0762-x

Druitt TH, Calder ES, Cole PD, Hoblitt RP, Loughlin SC, Norton GE, Ritchie LJ, Sparks RSJ, Voight B (2002) Small-volume, highly mobile pyroclastic flows formed by rapid sedimentation from pyroclastic surges at Soufrière Hills volcano, Montserrat: an important volcanic hazard. In: Druitt TH, Kokelaar BP (eds) The eruption of Soufrière Hills volcano, Montserrat, from 1995 to 1999. Geol Soc Lond Mem 21:263–279

Fernlund J, Zimmerman R, Kragic D (2007) Influence of volume/mass on grain-size curves and conversion of image-analysis size to sieve size. Eng Geol 90:124–137. https://doi.org/10.1016/j.enggeo.2006.12.007

Fielding CR (2006) Upper flow regime sheets, lenses and scour fills: extending the range of architectural elements for fluvial sediment bodies. Sediment Geol 190:227–240. https://doi.org/10.1016/j.sedgeo.2006.05.009

Fisher RV (1961) A proposed classification of volcaniclastic sediments and rocks. Geol Soc Am Bull 72:1409–1414

Fisher RV (1965) Settling velocity of glass shards. Deep Sea Res 12:345–353

Fisher RV, Schmincke H-U (1984) Pyroclastic rocks. Springer-Verlag, Berlin, p 472

Fletcher RC, Buss HL, Brantley SL (2006) A spheroidal weathering model coupling porewater chemistry to soil thicknesses during steady-state denudation. Earth Planet Sci Lett 244:444–457

Folk RL (1954) The distinction between grain size and mineral composition in sedimentary rock nomenclature. J Geol 62:344–359

Folk RL (1966) A review of grain-size parameters. Sedimentology 6:73–93

Folk RL (1974) Petrology of sedimentary rocks. Hemphill, Austin, 184 pp

Folk RL, Ward WC (1957) Brazos River bar: a study in the significance of grain size parameters. J Sediment Petrol 27:3–26

Francis PW (1993) Volcanoes—a planetary perspective. Clarendon Press (Oxford University Press), New York, p 443

Francis PW, Gardeweg M, Ramirez CF, Rothery DA (1985) Catastrophic debris avalanche deposit of Socompa volcano, northern Chile. Geology 13:600–603. https://doi.org/10.1130/0091-7613(1985)13<600:CDADOS>2.0.CO;2

Freundt A, Wilson CJN, Carey SN (2000) Ignimbrites and block-and-ash deposits. In: Sigurdsson H, Houghton BF, NcNutt SR, Rymer H, Stix J (eds) Encyclopedia of volcanoes. Academic Press, San Diego, pp 581–599

Friedman GM, Sanders JE (1978) Principles of sedimentology. Wiley, New York, p 792

Fripp JB, Diplas P (1993) Surface sampling in gravel streams. J Hydraul Eng 119(4):473–490

Gale SJ, Hoare PG (1991) Quaternary sediments. Belhaven Press, London, p 323

Glicken H (1986) Rockslide-debris avalanche of May 18, 1980, Mount St. Helens volcano, Washington. Dissertation, University of California, Santa Barbara, 303 pp

Glicken H (1996) Rockslide debris avalanche of May 18, 1980, Mount St. Helens volcano, Washington. U S Geol Surv Open-File Rep 96-677, 90 pp and 5 plates

Glicken H (1998) Rockslide-debris avalanche of May 18, 1980, Mount St. Helens volcano, Washington. Bull Geol Surv Jpn 49:55–106

Grant GE (1997) Critical flow constrains flow hydraulics in mobile-bed streams: a new hypothesis. Water Resour Res 33(2):349–358

Gray JMNT (2018) Particle segregation in dense granular flows. Annu Rev Fluid Mech 50:407–433. https://doi.org/10.1146/annurev-fluid-122316-045201

Haff PK, Werner T (1986) Computer simulation of the mechanical sorting of grains. Powder Technol 48(3):238–245. https://doi.org/10.1016/0032-5910(86)80048-1

Hand BM, Bartberger CE (1988) Leeside sediment fallout patterns and the stability of angular bedforms. J Sediment Res 58(1):33–43. https://doi.org/10.1306/212F8D05-2B24-11D7-8648000102C1865D

Harms JC, Fahnestock RK (1965) Stratification, bed forms, and flow phenomena (with an example from the Rio Grande). In: Primary sedimentary structures and their hydrodynamic interpretation—a symposium. Spec Publ Soc Econ Paleontol Mineral 84–115

Hildreth W, Fierstein J (2012) The Novarupta-Katmai eruption of 1912—largest eruption of the twentieth century: centennial perspectives. U S Geol Surv Prof Pap 1791:259 pp

Hoblitt RP, Kellogg KS (1979) Emplacement temperatures of unsorted and unstratified deposits of volcanic rock debris as determined by paleomagnetic techniques. Geol Soc Am Bull 90(1):633–642

Hoblitt RP, Miller CD, Vallance JW (1981) Origin and stratigraphy of the deposit produced by the May 18 directed blast. In: Lipman PW, Mullineaux DR (eds) The 1980 eruptions of Mount St. Helens, Washington. U S Geol Surv Prof Pap 1250:401–419

Hodgson KA (1993) Late Quaternary lahars from Mt. Ruapehu in the Whangaehu River valley, North Island, New Zealand. Dissertation, Massey University, 242 pp

Hoshyaripour G, Hort M, Langmann B, Delmelle P (2014) Volcanic controls on ash iron solubility: new insights from high-temperature gas–ash interaction modeling. J Volcanol Geotherm Res 286:67–77. https://doi.org/10.1016/j.jvolgeores.2014.09.005

Hupp CR, Osterkamp WR, Thornton J (1987) Dendrogeomorphic evidence and dating of recent debris flows on Mount Shasta, northern California. U S Geol Surv Prof Pap 1396B

Hurst VJ (1977) Visual estimation of iron in saprolite. Geol Soc Am Bull 88(2):174–176

Ibbeken H, Schleyer R (1986) Photo-sieving: a method for grain-size analysis of coarse-grained, unconsolidated bedding surfaces. Earth Surf Process Landf 11:59–77. https://doi.org/10.1002/esp.3290110108

Inman DL (1952) Measures of describing the size distribution of sediments. J Sediment Petrol 22:125–145

Iverson RM (1997) The physics of debris flows. Rev Geophys 35:245–296

Johnson SY (1986) Water-escape structures in coarse-grained, volcaniclastic, fluvial deposits of the Ellensburg Formation, south-central Washington. J Sediment Petrol 56:905–910

Jutzeler M, Proussevitch AA, Allen SR (2012) Grain-size distribution of volcaniclastic rocks 1: a new technique based on functional stereology. J Volcanol Geotherm Res 239:1–11

Kellerhals R, Bray DI (1971) Sampling procedures for coarse fluvial sediments. J Hydraul Div Proc Am Soc Civil Eng 97(HY8):1165–1179

Kennedy JF (1963) The mechanics of dunes and antidunes in erodible-bed channels. J Fluid Mech 16(4):521–544

Kieffer SW, Sturtevant B (1988) Erosional furrows formed during the lateral blast at Mount St. Helens, May 18, 1980. J Geophys Res 93(B12):14793–14816

Komorowski J-C, Glicken HX, Sheridan MF (1991) Secondary electron imagery of microcracks and hackly fracture surfaces in sand-size clasts from the 1980 Mount St. Helens debris-avalanche deposit: implications for particle-particle interactions. Geology 19:261–264. https://doi.org/10.1130/0091-7613(1991)019<0261:SEIOMA>2.3.CO;2

Kostic S, Sequeiros O, Spinewine B, Parker G (2010) Cyclic steps: a phenomenon of supercritical shallow flow from the high mountains to the bottom of the ocean. J Hydro-Environ Res 3:167–172. https://doi.org/10.1016/j.jher.2009.10.002

Krumbein WC (1934) Size frequency distributions of sediments. J Sediment Petrol 4:65–77

Kueppers U, Putz C, Spieler O, Dingwell DB (2012) Abrasion in pyroclastic density currents: insights from tumbling experiments. Phys Chem Earth 45–46:33–39. https://doi.org/10.1016/j.pce.2011.09.002

Lang J, Winsemann J (2013) Lateral and vertical facies relationships of bedforms deposited by aggrading supercritical flows: from cyclic steps to humpback dunes. Sediment Geol 296:36–54. https://doi.org/10.1016/j.sedgeo.2013.08.005

Langbein WB, Leopold LB (1968) River channel bars and dunes—theory of kinematic waves. U S Geol Surv Prof Pap 422-L:20 pp

Le Bas MJ, Streckeisen AL (1991) The IUGS systematics of igneous rocks. J Geol Soc Lond 148:825–833

Lockwood JP, Lipman PW (1980) Recovery of datable charcoal beneath young lavas; lessons from Hawaii. Bull Volcanol 43:609–615

Loughlin SC, Calder ES, Clarke A, Cole PD, Luckett R, Mangan MT, Pyle DM, Sparks RSJ, Voight B, Watts RB (2002) Pyroclastic flows and surges generated by the 25 June 1997 dome collapse, Soufrière Hills volcano, Montserrat. Geol Soc Lond Mem 21(1):191–209

Lowe DR (1975) Water escape structures in coarse-grained sediments. Sedimentology 22:157–204

Lowe DR, LoPiccolo RD (1974) The characteristics and origins of dish and pillar structures. J Sediment Petrol 44:484–501

Major JJ (1996) Experimental studies of deposition by debris flows: process, characteristics of deposits and effects of pore-fluid pressure. Dissertation, University of Washington, 341 pp

Major JJ (2000) Gravity-driven consolidation of granular slurries—implications for debris-flow deposition and deposit characteristics. J Sediment Res 70:64–83

Major JJ, Iverson RM (1999) Debris-flow deposition: effects of pore-fluid pressure and friction concentrated at flow margins. Geol Soc Am Bull 111:1424–1434

Major JJ, Pierson TC, Hoblitt RP, Moreno H (2013) Pyroclastic density currents associated with the 2008–2009 eruption of Chaitén volcano (Chile)—forest disturbances, deposits, and dynamics. Andean Geol 40(2):324–358

Manga M, Patel A, Dufek J (2011) Rounding of pumice clasts during transport: field measurements and laboratory studies. Bull Volcanol 73:321–333. https://doi.org/10.1007/s00445-010-0411-6

Manville V, Hodgson KA, White JDL (1998) Rheological properties of a remobilised tephra layer associated with the 1995 eruption of Ruapehu volcano, New Zealand. New Zeal J Geol Geophys 41:157–164

Manville V, Segschneider B, White JDL (2002) Hydrodynamic behavior of Taupo 1800a pumice: implications for the sedimentology of remobilized pyroclasts. Sedimentology 49:955–976

Manville V, Nemeth K, Kano K (2009) Source to sink: a review of three decades of progress in the understanding of volcaniclastic processes, deposits, and hazards. Sediment Geol 220:136–161

Mark DM (1973) Analysis of axial orientation data, including till fabrics. Geol Soc Am Bull 84:1369–1374

Mathisen ME, Vondra CF (1983) The fluvial and pyroclastic deposits of the Cagayan Basin, Northern Luzon, Philippines—an example of non-marine volcaniclastic sedimentation in an interarc basin. Sedimentology 30:369–392

Matsumoto K (2020) Pyrrhotite oxidation as a proxy for thermal structure of eruption clouds: a comparative study of Plinian and Vulcanian eruptions of Asama and Sakurajima volcanoes, Japan. J Volcanol Geotherm Res 392:106758. https://doi.org/10.1016/j.jvolgeores.2019.106758

Matsumoto K, Nakamura M (2017) Syn-eruptive breakdown of pyrrhotite: a record of magma fragmentation, air entrainment, and oxidation. Contrib Mineral Petrol 172:83. https://doi.org/10.1007/s00410-017-1403-8

McBride EF, Shepherd RG, Crawley RA (1975) Origin of parallel, near-horizontal laminae by migration of bed forms in a small flume. J Sediment Petrol 45:132–139

McKee ED, Weir GW (1953) Terminology for stratification and cross-stratification in sedimentary rocks. Geol Soc Am Bull 64(4):381–390. https://doi.org/10.1130/0016-7606(1953)64[381:TFSACI]2.0.CO;2

McKee ED, Crosby EJ, Berryhill HL Jr (1967) Flood deposits, Bijou Creek, Colorado, June 1965. J Sediment Petrol 37:829–851

McPhie J, Doyle M, Allen R (1993) Volcanic textures—a guide to the interpretation of textures in volcanic rocks. Centre for Ore Deposit and Exploration Studies, University of Tasmania, Hobart, p 196

Miall AD (1977) A review of the braided river depositional environment. Earth Sci Rev 13:1–62

Miall AD (1978) Lithofacies types and vertical profile models in braided river deposits: a summary. In: Miall AD (ed) Fluvial sedimentology. Can Soc Petrol Geol Mem 5:597–639

Miall AD (1988) Reservoir heterogeneities in fluvial sandstones: lessons from outcrop studies. Bull Am Assoc Pet Geol 72:682–697

Miall AD (1996) The geology of fluvial deposits. Springer, Berlin, p 582

Moore JG (1967) Base surge in recent volcanic eruptions. Bull Volcanol 30:337–363

Moore JG, Melson WG (1969) Nuees ardentes of the 1968 eruption of Mayon volcano, Philippines. Bull Volcanol 33:600–620

Moreno Chávez G, Castillo Rivera F, Sarocchi D, Borselli L, Rodríguez-Sedano LA (2018) FabricS: a user-friendly, complete and robust software for particle shape-fabric analysis. Comput Geosci 115:20–30

Mosley MP, Tindale DS (1985) Sediment variability and bed material sampling in gravel-bed rivers. Earth Surf Process Landf 10:465–482

Munsell Soil Color Charts (1975) Munsell color. Kollmorgen Corp, Baltimore

Nagano T, Isobe H, Nakashima S, Ashizaki M (2002) Characterization of iron hydroxides in a weathered rock surface by visible microspectroscopy. Appl Spectrosc 56(5):651–657

Neuendorf KKE, Mehl JP, Jackson JA (2005) Glossary of geology, 5th edn. American Geological Institute, Alexandria, VA, 779 pp

Oehmig R, Wallrabe-Adams H-J (1993) Hydrodynamic properties and grain-size characteristics of volcaniclastics deposits on the Mid-Atlantic Ridge north of Iceland (Kolbeinsey Ridge). J Sediment Res 63(1):140–151. https://doi.org/10.1306/D4267AAF-2B26-11D7-8648000102C1865D

Palmer BA, Alloway BV, Neall VE (1991) Volcanic debris-avalanche deposits in New Zealand—lithofacies organization in unconfined, wet-avalanche flows. In: Fisher RV, Smith GA (eds) Sedimentation in volcanic settings. SEPM Spec Publ 45:89–98

Pierson TC (1981) Dominant particle support mechanisms in debris flows at Mt. Thomas, New Zealand, and implication for flow mobility. Sedimentology 28:49–60

Pierson TC (2005a) Distinguishing between debris flows and floods from field evidence in small watersheds. U S Geol Surv Fact Sheet 2004-3142, 4 pp

Pierson TC (2005b) Hyperconcentrated flows—transitional process between water flow and debris flow, chap 8. In: Jakob M, Hungr O (eds) Debris-flow hazards and related phenomena. Praxis-Springer, Berlin, pp 159–202

Pierson TC, Janda RJ, Thouret JC, Borrero CA (1990) Perturbation and melting of snow and ice by the 13 November 1985 eruption of Nevado del Ruiz, Colombia, and consequent mobilization, flow, and deposition of lahars. J Volcanol Geotherm Res 41:17–66

Pierson TC, Daag AS, Delos Reyes PJ, Regalado MTM, Solidum RU, Tubianosa BS (1996) Flow and deposition of posteruption hot lahars on the east side of Mount Pinatubo, July–October 1991. In: Newhall CG, Punongbayan RS (eds) Fire and mud: eruptions and lahars of Mount Pinatubo, Philippines. Philippine Institute of Volcanology and Seismology, Quezon City, University of Washington Press, Seattle, pp 921–950

Pringle PT, Cameron KA (1999) Eruption-triggered lahar on May 14, 1984 In: Pierson TC (ed) Hydrologic consequences of hot-rock/snowpack interactions at Mount St. Helens volcano, Washington 1982–84. U S Geol Surv Prof Pap 1586:81–103

Rader E, Geist D, Geissman J, Dufek J, Harpp K (2015) Hot clasts and cold blasts: thermal heterogeneity in boiling-over pyroclastic density currents. In: Ort MH, Porreca M, Geissman JW (eds) The use of palaeomagnetism and rock magnetism to understand volcanic processes. Geol Soc Lond Spec Publ 396:1–11. https://doi.org/10.1144/SP396.16

Reading HG, Levell BK (1996) Controls on the sedimentary rock record. In: Reading HG (ed) Sedimentary environments: processes, facies and stratigraphy: Blackwell Science, Oxford, pp 5–36

Reiter M, Barroll MW, Minier J, Clarkson G (1987) Thermo-mechanical model for incremental fracturing in cooling lava flows. Tectonophysics 142:241–260

Risica G, Rosi M, Pistolesi M, Speranza F, Branney MJ (2022) Deposit-derived block-and-ash flows: the hazard posed by perched temporary tephra accumulations on volcanoes; 2018 Fuego disaster, Guatemala. J Geophys Res Solid Earth 127:e2021JB023699. https://doi.org/10.1029/2021JB023699

Rosato A, Strandburg KJ, Prinz F, Swendsen RH (1987) Why the Brazil nuts are on top: size segregation of particulate matter by shaking. Phys Rev Lett 25:1038–1040

Rowley PD, Kuntz MA, Macleod NS (1981) Pyroclastic-flow deposits. In: Lipman PW, Mullineaux DR (eds) The 1980 eruptions of Mount St. Helens, Washington. U S Geol Surv Prof Pap 1250:489–512

Rowley PD, MacLeod NS, Kuntz MA, Kaplan AM (1985) Proximal bedded deposits related to pyroclastic flows of May 18, 1980, Mount St. Helens, Washington. Geol Soc Am Bull 96:1373–1383

Ryan MP, Sammis CG (1978) Cyclic fracture mechanisms in cooling basalt. Geol Soc Am Bull 89:1295–1308

Savage SB (1987) Interparticle percolation and segregation in granular materials: a review. In: Selvadurai APS (ed) Developments in engineering mechanics. Elsevier, Amsterdam, pp 347–363

Savage SB, Lun CKK (1988) Particle size segregation in inclined chute flow of dry cohesionless granular solids. J Fluid Mech 189:311–335

Scheidegger AE (1965) On the statistics of the orientation of bedding planes, grain axes, and similar sedimentological data. U S Geol Surv Prof Pap 525:164–167

Schmincke H-U, Fisher RV, Waters AC (1973) Antidune and chute and pool structures in the base surge deposits of the Laacher See area, Germany. Sedimentology 20:553–574

Scott KM (1988a) Origins, behavior, and sedimentology of lahars and lahar-runout flows in the Toutle-Cowlitz River system. U S Geol Surv Prof Pap 1447-A:74 pp

Scott KM (1988b) Origin, behavior, and sedimentology of prehistoric catastrophic lahars at Mount St. Helens, Washington. In: Clifton HE (ed) Sedimentologic consequences of convulsive geologic events. Geol Soc Am Spec Pap 229:23–36. https://doi.org/10.1130/SPE229-p23

Scott AC, Jones TP (1991) Fossil charcoal: a plant-fossil record preserved by fire. Geol Today 7(6):214–216

Scott KM, Vallance JW, Pringle PP (1995) Sedimentology, behavior, and hazards of debris flows at Mount Rainier, Washington. U S Geol Surv Prof Pap 1547:56 pp

Scott WE, Hoblitt RP, Torres RC, Self S, Martinez MML, Nillos T (1996) Pyroclastic flows of the June 15, 1991, climactic eruption of Mount Pinatubo. In: Newhall CG, Punongbayan RS (eds) Fire and mud: eruptions and lahars of Mount Pinatubo, Philippines. Philippine Institute of Volcanology and Seismology, Quezon City, University of Washington Press, Seattle, pp 545–557

Segschneider B, Landis CA, Manville V, White JDL, Wilson CJN (2002) Environmental response to a large, explosive rhyolite eruption: sedimentology of post-1.8 ka pumice-rich Taupo volcaniclastics in the Hawke's Bay region, New Zealand. Sediment Geol 150:275–299

Siebert L (1996) Hazards of large volcanic debris avalanches and associated eruptive phenomena. In: Scarpa R, Tilling RI (eds) Monitoring and mitigation of volcano hazards. Springer-Verlag, Berlin, pp 541–572

Simons DB, Richardson EV (1962) Resistance to flow in alluvial channels. Am Soc Civil Eng Trans 127(1):927–1006

Simons DB, Richardson EV, Haushild WL (1963) Studies of flow in alluvial channels, some effects of fine sediment on flow phenomena. U S Geol Surv Water-Supply Pap 1498-G, 46 pp

Simons DB, Richardson EV, Nordin CF (1965) Sedimentary structures generated by flow in alluvial channels. Soc Econ Paleontol Mineral Spec Publ 12:34–52

Slootman A, Cartigny MJB (2020) Cyclic steps: review and aggradation-based classification. Earth-Sci Rev 201:102949, 30 pp. https://doi.org/10.1016/j.earscirev.2019.102949

Slootman A, Vellinga AJ, Moscariello A, Cartigny MJB (2021) The depositional signature of high-aggradation chute-and-pool bedforms. The build-and-fill structure. Sedimentology 68(4):1640–1673. https://doi.org/10.1111/sed.12843

Smith ND (1971a) Transverse bars and braiding in the lower Platte River, Nebraska. Geol Soc Am Bull 82:3407–3420

Smith ND (1971b) Pseudo-planar stratification produced by very low amplitude sand waves. J Sediment Petrol 41:69–73

Smith GA (1987) The influence of explosive volcanism on fluvial sedimentation: the Deschutes Formation (Neogene) in central Oregon. J Sediment Petrol 57:613–629

Smith GA (1991) Facies sequences and geometries in continental volcaniclastic sediments. In: Fisher RV, Smith GA (eds) Sedimentation in volcanic settings. SEPM (Soc Sediment Geol) Spec Publ 45:109–121

Smith GA, Lowe DR (1991) Lahars: volcano-hydrologic events and deposition in the debris flow—hyperconcentrated flow continuum. In: Fisher RV, Smith GA (eds) Sedimentation in volcanic settings. SEPM (Soc Sediment Geol) Spec Publ 45:59–70

Sohn C, Sohn YK (2019) Distinguishing between primary and secondary volcaniclastic deposits. Sci Rep 9:12425. https://doi.org/10.1038/s41598-019-48933-4

Southard JB, Boguchwal LA (1990) Bed configurations in steady unidirectional water flows. Part 2. Synthesis of flume data. J Sediment Petrol 60:658–679

Sparks RSJ, Self S, Walker GPL (1973) Products of ignimbrite eruptions. Geology 1:115–118

Sparks RSJ, Gardeweg MC, Calder ES, Matthews SJ (1997) Erosion by pyroclastic flows on Lascar volcano, Chile. Bull Volcanol 58:557–565

Swanson FJ, Major JJ (2005) Physical events, environments, and geological—ecological interactions at Mount St. Helens: March 1980–2004. In: Dale VH, Swanson FJ, Crisafulli CM (eds) Ecological responses to the 1980 eruption of Mount St. Helens. Springer, New York, NY. https://doi.org/10.1007/0-387-28150-9_3

Swanson FJ, Jones JA, Crisafulli CM, Lara A (2013) Effects of volcanic and hydrologic processes on forest vegetation: Chaitén volcano, Chile. Andean Geol 40:359–391

Takarada S, Ui T, Yamamoto Y (1999) Depositional features and transportation mechanism of valley-filling Iwasegawa and Kaida debris avalanches, Japan. Bull Volcanol 60:508–522. https://doi.org/10.1007/s004450050248

Taylor NH, Pohlen IJ (1970) Soil survey method. New Zeal Soil Bur Bull 25:242 pp

Terry JP, Goff J (2014) Megaclasts: proposed revised nomenclature at the coarse end of the Udden–Wentworth grain-size scale for sedimentary particles. J Sediment Res 84:192–197

Udden JA (1898) The mechanical composition of wind deposits. Augustana Library Publ No 1, Rock Island, Illinois, 69 pp

Udden JA (1914) Mechanical composition of clastic sediments. Geol Soc Am Bull 25:655–744

Vallance JW (1994) Experimental and field studies related to the behavior of granular mass flows and the characteristics of their deposits. Dissertation, Michigan Technological University, 197 pp

Vallance JW (2000) Lahars. In: Sigurdsson H, Houghton BF, McNutt SR, Rymer H, Stix J (eds) Encyclopedia of volcanoes. Academic Press, San Diego, pp 601–616

Vallance JW, Iverson RM (2015) Lahars and their deposits. In: Sigurdsson H, Houghton B, McNutt SR, Rymer H, Stix J (eds) The encyclopedia of volcanoes, 2nd edn. Elsevier, Amsterdam, pp 649–664. https://doi.org/10.1016/B978-0-12-385938-9.00037-7

Vallance JW, Scott KM (1997) The Osceola Mudflow from Mount Rainier: sedimentology and hazard implications of a huge clay-rich debris flow. Geol Soc Am Bull 109:143–163. https://doi.org/10.1130/0016-7606(1997)109<0143:TOMFMR>2.3.CO;2

Vitton S, Sadler L (1997) Particle-size analysis of soils using laser light scattering and X-ray absorption technology. Geotech Test J 20(1):63–73. https://doi.org/10.1520/GTJ11421J

Wadell HA (1932) Volume, shape, and roundness of rock particles. J Geol 40:443–451

Waitt RB (1981) Devastating pyroclastic density flow and attendant air fall of May 18—stratigraphy and sedimentology of deposits. In: Lipman PW, Mullineaux DR (eds) The 1980 eruptions of Mount St. Helens, Washington. U S Geol Surv Prof Pap 1250:439–458

Walker GPL (1971) Grain-size characteristics of pyroclastic deposits. J Geol 79:696–714

Wan Z (1985) Bed material movement in hyperconcentrated flow. J Hydraulic Eng 111:987–1002

Waters AC, Fisher RV (1971) Base surges and their deposits: Capelinhos and Taal volcanoes. J Geophys Res 76:5596–5614

Waythomas CF, Pierson TC, Major JJ, Scott WE (2013) Voluminous ice-rich lahars generated during the 2009 eruption of Redoubt volcano, Alaska. J Volcanol Geotherm Res 259:389–413

Weltje GJ, Prins MA (2007) Genetically meaningful decomposition of grain-size distributions. Sediment Geol 202(3):409–424

Wentworth CK (1922) A scale of grade and class terms for clastic sediment. J Geol 30:377–392

White JDL, Houghton BF (2006) Primary volcaniclastic rocks. Geology 34:677–680

White JDL, Manville VR, Wilson CJN, Houghton BF, Riggs N, Ort M (2001) Settling and deposition of 181 A.D. Taupo pumice in lacustrine and associated environments. In: White JDL, Riggs NR (eds) Volcanogenic sedimentation in lacustrine settings. Int Assoc Sedimentol Spec Publ 31:141–150

Whitham AG, Sparks RSJ (1986) Pumice. Bull Volcanol 48:209–233

Williams H, McBirney AR (1979) Volcanology. Freeman, Cooper, & Co, San Francisco, p 397

Wohletz KH, Sheridan MF (1979) A model of pyroclastic surge. In: Chapin CE, Elston WE (eds) Ash-flow tuffs. Geol Soc Am Spec Pap 180:177–194

Wohletz KH, Sheridan MF, Brown WK (1989) Particle size distributions and the sequential fragmentation/transport theory applied to volcanic ash. J Geophys Res 94(B11):15703–15721

Wolcott J, Church M (1991) Strategies for sampling spatially heterogeneous phenomena: the example of river gravels. J Sed Petrol 61(4):534–543

Wright JV (1978) Remanent magnetism of poorly sorted deposits from the Minoan eruption of Santorini. Bull Volcanol 41:131–135

Wright HMN, Cashman KV, Rosi M, Cione R (2007) Breadcrust bombs as indicators of Vulcanian eruption dynamics at Guagua Pichincha volcano, Ecuador. Bull Volcanol 69:281–300. https://doi.org/10.1007/s00445-006-0073-6

Wu L, Krijgsman W, Liu J, Li C, Wang R, Xiao W (2020) CFLab: a MATLAB GUI program for decomposing sediment grain size distribution using Weibull functions. Sediment Geol 398:105590

Zimmerman SG, Evenson EB, Gosse JC, Erskine CP (1994) Extensive boulder erosion resulting from a range fire on the type-Pinedale moraines, Fremont Lake, Wyoming. Quat Res 42:255–265

Lahars—Process Overview

Abstract

Lahars are highly hazardous volcanic wet flows, owing to their potentially large volumes, rapid velocities, long runouts, frequent occurrence during and following eruptions, and unpredictability. However, lahars are not a specific flow type. They are, instead, time-discrete flow events like flash floods, which involve up to three distinct types of flow—debris flow (~ 50 to ~ 80 volume percent solid particles), hyperconcentrated flow (~ 10 to ~ 50 volume percent solid particles), and water flow (0 to ~ 10 volume percent solid particles). These concentration boundaries vary slightly for different particle-size distributions of sediment mixtures. All three flow types can occur simultaneously in different parts of the same lahar, and therefore lahars can evolve in unique ways in time and space as they flow downstream. Debris flow, hyperconcentrated flow, and water flow have fundamentally different flow behaviors, bulk properties, particle transport mechanics, and modes of deposition. These differences can lead to distinctly different kinds of deposits, as well as different lithofacies zones within any single lahar deposit.

1 Definitions and Context

Lahars are flowing dense mixtures of water and solid particles from a volcanic source, fully water-saturated but intermediate in mean sediment concentration between normal streamflow and unsaturated debris avalanches (Smith and Lowe 1991; Vallance 2000; Waythomas et al. 2013; Vallance and Iverson 2015; Major 2023). Solid particles are mainly rock fragments but may include ice, wood, and other debris (Cronin et al. 1996; Kilgour et al. 2010; Lube et al. 2009; Waythomas et al. 2013; Waythomas 2014). Mean solids concentrations in lahars vary from about 10% solids by volume (vol%) in the most dilute lahars up to about 80 vol% in the most concentrated flows (see Chap. "Deposits Emplaced by Volcanic Wet Flows"). However, average sediment concentration typically varies with time and distance from source within any lahar as it flows downslope or downstream, and it can even vary significantly (vertically, laterally, and longitudinally) at any point in time within some flowing lahars (Vallance and Iverson 2015; Major 2023). Concentration variability is due to the actions of particle erosion and entrainment, selective particle deposition,

T. C. Pierson et al., *Deposits of Volcanic Wet Flows*, Advances in Volcanology,
https://doi.org/10.1007/978-3-031-66574-5_4

Fig. 1 Aerial view of site of the former town of Armero (Colombia) after it was swept over by a very large, rapidly flowing lahar—a volcanic debris flow (VDF)—from the Azufrado River on November 13, 1985 (Lowe et al. 1986; Naranjo et al. 1986; Pierson et al. 1990; Thouret 1990; Voight 1990; Thouret et al. 2007). The lahar was triggered by a small explosive eruption from Nevado del Ruiz volcano that rapid melted surface snow and ice on the broad summit ice cap with hot pyroclastic density currents. About 23,000 people perished in three simultaneous lahars that flowed from about 40 to 100 km into several communities around the volcano. The victims were caught by surprise late at night, with no warning system in place and no time to escape (Voight 1990). USGS photo by R. J. Janda, December 9, 1985

incorporation of overtaken river water, and gravity segregation. *Primary lahars* are those triggered directly or indirectly by volcanic eruptions (syneruptive); *secondary lahars* are those triggered by noneruptive processes. Lahars are significant volcanic hazards, owing to their potentially large volumes, rapid velocities, long runouts, frequent occurrence during and following eruptions, and unpredictability (Fig. 1).

Lahars, viewed longitudinally as downstream-moving bodies of flowing material, have spatially identifiable components: the *flow front* (the point at which flow depth and discharge in a channel abruptly begin to increase), the *flow head* (the frontal segment of the flow body in which discharge is increasing), the main body of the lahar that typically includes surging or pulsing behavior, the *flow peak* (the point of highest discharge), and the *flow tail* (the segment in which discharge is steadily decreasing). Lahars commonly exhibit multiple downstream-moving surges or waves that may or may not be periodic. Lahar durations (viewed from a fixed point along a channel) can vary from minutes to tens

Fig. 2 A very large lahar moving downstream. Aerial view of the North Fork Toutle River lahar, a large concentrated syneruptive lahar (volcanic debris flow) that was triggered about 5 h after onset of the May 18, 1980, eruption of Mount St. Helens (Washington, USA) by localized melting of entrained snow and ice, dewatering, and subsequent liquefaction of parts of the freshly emplaced debris-avalanche deposit in the upper reaches of the drainage basin (Fairchild 1985, 1987; Voight et al. 1981, 1983; Glicken 1996). Photograph of the flowing lahar was taken just as an accumulation of logs floating on the lahar surface (entrained from a timber company log yard) pushed State Route 504 highway bridge off its support piers. Location is just downstream of the confluence of the North Fork Toutle and South Fork Toutle Rivers, about 54 km downstream of the volcano. Bridge deck was buried and never found. Photo by Roger Werth, © 1980 The Daily News (Longview, Washington), used with permission

of hours from the arrival of the flow front to the passing of the tail (i.e., return to pre-lahar discharge). A plot of flow surface elevation versus time is a lahar's *stage hydrograph*.[1]

Lahars are thus time-discrete "flow events" (like flash floods) that are now recognized to commonly involve two or three distinct types of flow, depending on variation of solids concentration within the flow body (Vallance 2000; Vallance and Iverson 2015). Concentrated lahars move mainly as *debris flow* (concentrated particle–fluid flow), while dilute lahars move mainly as *hyperconcentrated flow* (dilute particle–fluid flow) (see Chap. "Deposits Emplaced by Volcanic Wet Flows"). Heads and tails of both concentrated and dilute lahars commonly also include zones of muddy streamflow (water flow). Because all three flow types can occur simultaneously in different parts of the same lahar, lahars can evolve in unique ways in time and space as they flow downstream. Debris flow, hyperconcentrated flow, and water flow

[1] Stage hydrographs are two-dimensional plots of flow surface elevation of lahars or water floods passing a fixed points along a channel. Stage hydrographs can be used to estimate volumetric flow rate (discharge) hydrographs where channel cross-section dimensions are known (and assumed to be relatively stable). If sediment samples are collected during lahar passage, mass flux can be estimated.

have fundamentally different flow behaviors, bulk properties, particle transport mechanics, and modes of deposition. These differences can lead to distinctly different kinds of deposits or different lithofacies zones within a single lahar deposit. Deposits emplaced mainly by *volcanic debris flow* (VDF) are treated separately in this book from deposits emplaced mainly by *volcanic hyperconcentrated flow* (VHF). Both debris flows and hyperconcentrated flows occur in nonvolcanic terrain as well as at volcanoes, but VDFs and VHFs tend to be larger, more mobile, and more hazardous than their nonvolcanic counterparts (Fig. 2).

Lahar terminology in the literature remains complex. Since the recognition of hyperconcentrated flow and muddy streamflow (in addition to debris flow) as components of many lahars (Pierson and Scott 1985; Scott 1988a; Rodolfo 1989; Cronin et al. 1997, 1999, 2000; Vallance 2000; Major 2023), the definition of a lahar has been evolving. The term 'lahar' is Indonesian (Javanese) in origin, long used by villagers in that part of the world to describe the flowing mixtures of rock debris, mud, and water from volcanoes that regularly threaten their villages. Its first use in a Western publication was in 1854 to describe flows from Kelut volcano (reported in Schmidt 1934, p. 101), and the first precise definition was provided by Van Bemmelen (1949)—*a mudflow containing debris and angular blocks of chiefly volcanic origin*. English translations of terms describing lahars by early twentieth century geologists and engineers (writing mainly in German) also include 'sand flow,' 'mud stream,' and 'mudflow' (Scrivenor 1929; Schmidt 1934; Van Bemmelen 1949). According to Neall (1976), first mention of 'lahar' in an English-language article was by Escher (1922) for a flow triggered by the 1912 Novarupta-Katmai eruption, which ironically was a large pyroclastic density current (the Valley of Ten Thousand Smokes [VTTS] ignimbrite) and not a lahar (E. W. Hildreth, U. S. Geological Survey, written commun., 2021). In addition, the term 'lahar' had been restricted to only VDFs (also termed 'mudflows') for many years (e.g., Crandell 1971; Fisher and Schmincke 1984; Smith 1986; Bates and Jackson 1987; Neuendorf et al. 2005), but a broader definition was adopted by the 1988 Geological Society of American Penrose Conference on Volcanic Influences on Terrestrial Sedimentation (Smith and Fritz 1989): *a general term for a rapidly flowing mixture of rock debris and water (other than normal streamflow) from a volcano*. Although this broader definition makes room for hyperconcentrated flow and muddy streamflow, it also permits inclusion of debris avalanches. In this book we adopt the spirit of the 1988 definition in our definition stated in the first sentence of this chapter, but we acknowledge that lahars may include other types of solid components.

2 Properties of Volcanic Debris Flows and Volcanic Hyperconcentrated Flows

2.1 Volcanic Debris Flows (VDFs)

VDFs have concentrations of solid particles ranging from approximately 50 to 80 vol% (73–92 wt%), with bulk densities in the range ~ 1800 to ~ 2300 kg/m^3 for mixtures with particle densities of 2.6–2.7 g/cm^3 (Iverson 1997; Pierson 2005) (Table 1). Many VDF particle–fluid mixtures resemble wet concrete in consistency, both while they are flowing and after they have come to rest. Concentrations above 70 vol% are usually restricted to zones of dynamically segregated cobble- and boulder-size clasts. Closest possible packing in extremely poorly sorted debris mixtures is achieved somewhere between 70 and 90 vol% (Iverson and LaHusen 1993), thus some moving VDF lahar mixtures may approach the densest possible static particle-packing configurations.

A defining feature of all debris flows is the full and continuous (relative to flow duration) suspension of at least some gravel-size (≥ 4 mm) particles in the mixtures during flow, which is due to both complex,

Table 1 Sediment concentrations in samples from flowing volcanic debris flows (VDFs)

Volcanic debris flows (VDFs)	Sampling (in main body of lahar)	Sediment concentration (vol%)	Sorting coefficient (inclusive standard deviation in ϕ units)	Silt + clay fraction (wt%)	References
North Fork Toutle River, Mount St. Helens, May 18, 1980 (8:45 PM)	Dip sample (1), 83 km downstream of crater (65–70 km from source); particles up to ~ 4 cm diameter sampled	66	nd	nd	Dinehart et al. (1981)
Small channel on, SE flank Mount St. Helens, 1981–1983	Dip samples (26), 2 km from source; particles <9 cm diameter	53–70	2.60–3.94	8.5–19.4	Pierson (1986)
Sacobia River, Mount Pinatubo, Aug 1991	Dip samples (2), 17 and 28 km from caldera rim, sampled with 5 gal. buckets	55–70	3.96, 3.32	12.9, 16.2	Pierson et al. (1996)
Abacan River, Mount Pinatubo, Sep 1991	Dip samples (2), 18 km from caldera rim; sampled with 5 gal. buckets	55, 51 (flow near transition between VDF and VHF)	2.57, 3.33	15.6, 14.4	Pierson et al. (1996)
Whangaehu River, Mount Ruapehu, Sep and Oct 1995	Dip samples (2), 24 and 42 km from crater rim; sampled with 5 L jars	54, 52 (flow near transition between VDF and VHF)	3.7, 2.8	12, 8	Cronin et al. (1999)
Curah Lengkong River, Semeru volcano, Nov 1999–Nov 2000	Dip samples (15), 9.5 km from summit; sampled with "large" plastic jars	48–69	nd	nd	Lavigne and Suwa (2004)
Bucao River, Mount Pinatubo, 1991 and 1992 monsoon seasons	Dip samples (10), sampled using buckets	39–86 Mean 57 (pumice-rich mixtures, 45–70 °C)	nd	nd	Rodolfo et al. (1996)

Two VDFs were triggered by eruptions; four were triggered by heavy posteruption rainfall. Sample sizes generally 5–10 kg in bulk and collected by hand from riverbanks; clasts larger than medium pebbles not sampled (or not present). Number of dip samples taken at each location are shown in parentheses. Determination of flow type as debris flow was based on presence of gravel-size particles (> 4 mm) in sustained suspension at flow surface. Abbreviation "nd" = "not determined." See Volcano Index for volcano locations and details shear-rate-dependent particle–particle and particle–fluid interactions during flow and the material properties of the particle–fluid mixture (Iverson 1997, 2005; Major and Iverson 1999; Estep and Dufek 2013). Furthermore, the capacity of debris flows to suspend progressively larger clasts increases with increasing sediment concentration (Yang et al. 2014). Turbulence is not necessary for suspension of gravel-size clasts. Although experimental mixtures of fine particles and water can exhibit fixed and uniform rheologies with composition- and concentration-determined yield strengths, poorly sorted VDFs do not (Iverson 1997, 2003, 2005).

Two types of VDFs are generally recognized based on content of clay-size particles[2] in their deposits. *Clay-poor VDF deposits* have less than 5 wt% (1.9 vol%) clay-size particles in their matrix fraction (sand + silt + clay) (Vallance and Iverson 2015), and the combined fines (silt + clay) fraction of the matrix is generally <30 wt% (11 vol%) (Pierson 1986; Scott 1988b). The distinction is important for flow behavior because clay-poor VDF lahars are susceptible to mixing with overtaken river water in channels, and they commonly exhibit relatively rapid downstream attenuation in discharge with selective deposition over limited travel distances (Pierson and Scott 1985; Scott 1988a; Scott et al. 1995). This occurs because the particle mixtures in clay-poor VDFs have sufficiently high hydraulic diffusivities to allow rapid drainage of excess pore fluid, and rapid loss of pore fluid can speed deposition and limit travel distance (Major 2000). *Clay-rich VDF deposits* contain between about 5 and 40 wt% clay-size particles (commonly including clay minerals), and the fines fraction of the matrix can be as much as 64 wt% (Vallance and Scott 1997; Vallance and Iverson 2015). In contrast to clay-poor mixtures, clay-rich VDF mixtures have low hydraulic diffusivities and can maintain high pore-fluid pressures and high mobility for long periods, flowing longer distances (> 100 km) and spreading out more widely on flat valley floors than clay-poor VDFs (Mizuyama et al. 1992; Scott et al. 1995; Vallance and Scott 1997). Clay-rich VDFs also tend to have larger volumes due to their typical initiation as volcanic edifice collapses (rich in hydrothermally altered rock), and most are capable of resisting dilution by river water.

The adjectives *noncohesive* and *cohesive* have been widely applied as synonyms for 'clay-poor' and 'clay-rich' when discussing lahars and their deposits (e.g., Scott 1988a; Scott et al. 1995; Vallance and Scott 1997; Mothes et al. 1998; Lavigne and Thouret 2002; Bursik and Reid 2004). Indeed, 'cohesive' is descriptive of the nature of deposits of many clay-rich lahars—the deposits can be hard as bricks when dry. However, cohesion (in the mechanical sense) plays a negligible role in the balance of forces acting on most of the particles in flowing lahar mixtures (Iverson 1997). For this reason, the descriptors 'clay-poor' and 'clay-rich' are used in this book. In clay-poor lahars, some of the coarsest particles in suspension tend to be concentrated at the surface, particularly toward flow heads and lateral margins, owing to kinetic sieving with highest velocities at flow surfaces (Vallance and Iverson 2015). Gravelly flow fronts and lateral margins can be close to 100% coarse solid particles (Pierson 1986; Suwa and Yamakoshi 1999; Major and Iverson 1999). Size distributions within bodies of clay-rich lahars, where kinetic sieving appears to be hindered by matrix viscosity, tend to be more uniform and cobble- and boulder-size clasts are not concentrated on lahar surfaces (Vallance and Iverson 2015).

2.2 Volcanic Hyperconcentrated Flows (VHFs)

Hyperconcentrated flow, as a process, involves a viscous fluid phase and a granular solid phase as in debris flow, but VHFs (unlike VDFs) cannot

[2] It is assumed that VDF deposits have clay contents the same as or slightly less than VDF flows, because intergranular pore fluid with suspended fines is deposited with the coarse particles. Some loss of fines may occur as deposits dewater during consolidation. One study (Pierson et al. 1996) that measured both showed similar clay contents in dip samples of flows and their deposits.

suspend coarse sand and fine gravel in the fluid phase for long durations without flow turbulence (Pierson 2005). In addition to turbulence, a minimum concentration of suspended *fines*[3]—a fines-concentration threshold—is required for long-duration suspension and transformation of water flow to hyperconcentrated flow (Beverage and Culbertson 1964; Gerson 1977; Qian et al. 1981; Wan and Song 1987; Rickenmann 1991; Xu 2002, 2003). Those suspended fines, usually including silt and clay plus some fine sand, are necessary to maintain the viscosity and yield strength in the fluid phase needed to slow the settling of coarser particles and allow turbulence to keep them in dynamic suspension (Qian et al. 1981; Cao and Qian 1990; Rickenmann 1991; Dinehart 1999; Xu 2002, 2003; Yang et al. 2014). In dilute VHFs, macroturbulent vortices can pull particles off the bed and into suspension, and the upward fluid drag provided, balanced by hindered settling, helps keep the coarser grains suspended (Wilson 1985; Komar 1988). Upward fluid drag is also imparted by convection cells, common in VHFs, where downward settling of some coarse particles can generate upwelling along cell margins to lift other particles (Druitt 1995; Cronin et al. 1999; Major 2003; Yang et al. 2014). Without turbulence (i.e., in VHF-concentration mixtures that are only slowly moving or static), the gravel and coarse sand fractions settle out of suspension in seconds (gravel) to hours (silt and clay) at settling velocities controlled by particle size, shape, and density and by fluid viscosity and flow depth. Where flow is turbulent, VHFs can transport extraordinarily large volumes of sediment, particularly sand (e.g., Wilcox et al. 2014). Total suspended-load transport rates in VHFs can be much greater than in water flow at the same discharge. For example, during the 1982 VHF in the Toutle River at Mount St. Helens (Washington, USA), the quantity of sediment in suspension was between 15 and 40 times more than during volcanic water flows (VWFs) of similar magnitude at the same gauging stations (Pierson and Scott 1985; Dinehart 1999).

Given the necessity of a fines-concentration threshold for VHFs to develop, it might seem that knowing the proportion of fines in a deposit could determine whether the deposit had a VHF origin or not. But this is not a valid assumption for several reasons: both the type of clay (clay size vs. clay minerals) and the proportions of silt versus clay in suspension affect the proportion of suspended fines required for hyperconcentrated flow. In addition, the proportion of fines found in a deposit usually does not reflect the proportion of fines in the flow that emplaced it.

2.2.1 Effect of Clay Type in Transformation to VHF

Clay-mineral particles have strong electrical charges on their surfaces, and clay mineral particles are more effective than clay-size particles of other minerals in increasing the viscosity of clay–water suspensions. Mixtures of smectite clay and water behave as viscous hyperconcentrated fluids with as little as 1 vol% solids (Hampton 1975), but in many volcanic terrains with little or no such clay, such as at Mount St. Helens, a minimum fines concentration of about 8 vol% (mostly silt-size particles) appears to be necessary for the transformation from water flow to hyperconcentrated flow (VWF to VHF) (Dinehart 1999; Pierson 2005). At fines concentrations below 6 vol% in rivers draining Mount St. Helens, transformation to VHF does not occur, and high-discharge flow events occur as muddy VWFs, not VHFs (Dinehart 1999). But in terrains where the fines fraction contains more clay (and more clay minerals, particularly montmorillonite/smectite), lower fines-concentration thresholds to hyperconcentrated flow are found (Hampton 1975; Beverage and Culberton 1964; Xu 2002).

[3] "Fines" are considered by North American researchers to be particles <0.063 mm, i.e. the full range of silt- and clay-size grains. In China, however, where much of the work on hyperconcentrated flow has been done, "fines" are defined as particles <0.01 mm, i.e., fine silt and clay. The North American convention is used in this book. However, the suspended fine particles in VHFs include some fine sand.

2.2.2 Range in Suspended-Sediment Concentration Values in VHFs

Once a muddy flood suspends between 1 and 8 vol% silt and clay (with some fine sand mixed in), the threshold to hyperconcentrated flow is crossed, and flow turbulence assists the weak yield strength of the viscous fines suspension in abruptly increasing the capacity of the flow to suspend and transport coarser and much larger quantities of sediment (Dinehart 1999; Cronin et al. 1999; Pierson 2005). The total amount of sediment in suspension after this flow transformation greatly exceeds the amount of fines alone. Total suspended-sediment concentrations in VHF can vary widely, depending on the particle-size distribution of the sediment available for transport (Table 3).

Thus, total suspended-solids concentrations for VHFs in most volcanic terrains range from approximately 10 to 50 vol% (Fig. 3; Table 2). These are terrains where the entrained sediment is poorly sorted. A particular value is mainly determined by the mineralogical makeup and the proportion of the fines in the flow, as previously discussed, and by suspended-particle densities (Pierson 2005). Bounding values for concentration ranges shift downward in any of three instances: as clay- and silt-size particle contents increase, as the percentage of clay minerals increases, or as the percentage of low-density pumice grains in the mixture increases.

Notably, the proportion of fines in a VHF deposit does not necessarily reflect the fines content of the flow that emplaced it. This is

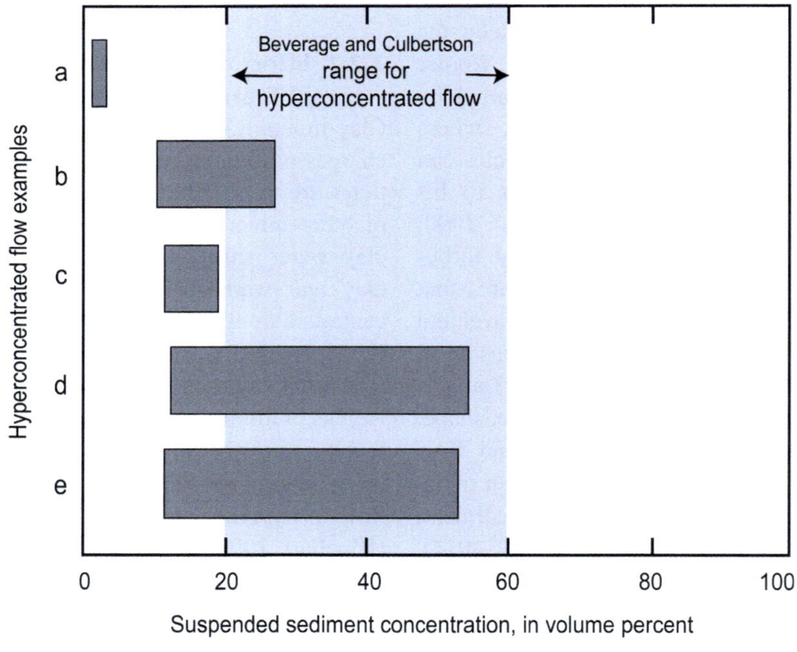

Fig. 3 Approximate ranges in total suspended-sediment concentrations for hyperconcentrated flows having different particle-size distributions, compared to the range originally defined by Beverage and Culbertson (1964) (from Pierson 2005). **a** Mixture of water and smectite clay at 1–3 vol% solids concentration range (Hampton 1975), **b** mixture of water and kaolinite clay at 10–27 vol% solids concentration range (Hampton 1975), **c** mixture of water and silt with some clay and fine sand from Yellow River in China at 11–19 vol% solids concentration range (Qian et al. 1981), **d** mixture of water and poorly sorted volcaniclastic sediment from Mount Ruapehu (New Zealand) at 12–54 vol% solids concentration range (Cronin et al. 1997, 1999, 2000), and **e** mixture of water and poorly sorted volcaniclastic sediment from Mount St. Helens (Washington, USA) at 11–53 vol% solids concentration range (Pierson and Scott 1985)

Table 2 Measured sediment concentrations in volcanic hyperconcentrated flows (VHFs); number of samples collected shown in parentheses

Volcanic hyperconcentrated flow (VHF)	Sampling point (in main body or tail of lahar)	Total sediment conc. (vol%)	Sorting coefficient (inclusive standard deviation in φ units)	Silt + clay fraction (wt%)	References
Small channels on SE flank Mount St. Helens, 1981–1983	Dip samples (8), 2 km from source; all particles <9 cm diameter	37–51	2.1–2.6	13–30	Pierson (1986)
Abacan River, Mount Pinatubo, Sep 1991	Dip samples (2), 18 km from caldera rim; hand-dip samples with 5 gal. buckets	28	3.3	19	Pierson et al. (1996)
Whangaehu River, Mount Ruapehu, Sep and Oct 1995	Dip samples (5), 24 and 42 km from crater rim; sampled with 5 L wide-mouth jars	7	2.7–3.0	19–50	Cronin et al. (1999)
Toutle River, Mount St. Helens, Mar 19–20, 1982	Dip sample (1), 73 km from crater source; sampled with USGS wide-mouth water-sample bottle	42	nd	22	Dinehart (1999)
Curah Lengkong River, Semeru volcano, Nov 1999–Nov 2000	Dip samples (15), 9.5 km from summit; sampled with large plastic jars	31–42	nd	nd	Lavigne and Suwa (2004)
Sacobia River, Mount Pinatubo, Sep 23, 1993	Dip depth-integrated sample (1), 25 km from caldera rim, sampled with 1-qt. zipper-locking plastic bag	30	nd	34	Pierson (2005)
Santo Tomás Rivers, Mount Pinatubo, 1991–1992	Dip samples (multiple from 3 lahars), 40 km downstream from caldera rim with 5 gal. bucket	18–31 (pumice-rich mixtures)	nd	nd	Rodolfo et al. (1996)
White Salmon River, draining Mount Adams volcanic field—hyperconcentrated flow from rapid breaching of dam and erosion of reservoir sediment fill, Oct 2011	Pump samples, depth-integrated samples from cableway and dip samples (126 total); sampled 2.3 and 3.0 km downstream of breached dam	15–32	nd	50 When measured at peak total concentration of 32 vol%	Wilcox et al. (2014)

The first seven examples are from lahars; the eighth was a nonvolcanic flow (dam-breach flood) transporting volcanic sediment that accumulated behind the dam. Dip-sample sizes generally 1–5 kg in bulk were collected by hand from riverbanks; suspended clasts larger than medium pebbles (usually pumice) were absent in flows. Determination of flow type as VHF based on the absence of gravel-size particles in sustained suspension at flow surface. See Volcano Index for volcano locations and details

Table 3 Common initiation mechanisms of lahars

Basic initiation mechanism	Contributing processes	Examples
Ejected hot rock causing rapid melting of snow and ice	Rapid melting of snow/ice by pyroclastic density currents (PDCs) and lava dome explosions or avalanches through conductive heat transfer; mixing by turbulent scouring action, and snow avalanching. Limited melting by flowing lava unless autobrecciation occurs. Lava tends to burrow into snow, forming insulated lava tubes	*PDCs*: Nevado del Ruiz—1985, PDCs on snow (Naranjo et al. 1986; Pierson et al. 1990); Redoubt Volcano—1989–1990, PDCs on glacier ice (Dorava and Meyer 1994) *Dome explosions*: Mount St. Helens—1982, 1984 (Pierson and Scott 1985; Pierson 1999; Pringle and Cameron 1999) *Eruptions through ice-filled craters*: Redoubt Volcano—2009 (Waythomas et al. 2013) *Lava flows on snow*: Villarrica volcano—1971, 1980–84 (Major and Newhall 1989); Sangay volcano—1974 (Hall 1978); Hekla volcano—1947 (Kjartansson 1951)
Eruptions through crater lakes	Ejection of crater lake water with pyroclastic particles onto the outer slopes of a volcano	Kelut volcano—1919 (Scrivenor 1929); Mount Ruapehu—1975, 1995 (Nairn et al. 1979; Cronin et al. 1996; Kilgour et al. 2010)
Volcano edifice failure (flank collapse)[4]	Landslides, leading to slide-induced fragmentation of water-saturated rock debris, from upper portions of volcanoes Failure of flanks of an edifice, triggered by seismic shaking or heavy rainfall	*Deep failure*: Mount Rainier—prehistoric (Crandell 1971; Vallance and Scott 1997; Vallance et al. 1995); Casita volcano—1998 (Scott et al. 2005) *Shallow failure*: Ontake volcano—1984 (Oyagi 1987; Voight and Sousa 1994); San Vicente volcano—1995, 2001, and others (Major et al. 2004)
Lake-breakout outburst floods	Sudden release of water from lakes on or adjacent to the volcano due to natural dam failure	*Debris-dammed lakes*: Mount Pinatubo—1991–92 (Arboleda and Martinez 1996; Umbal and Rodolfo 1996); Mount St. Helens—prehistoric (Scott 1988b); Nevado de Colima—prehistoric (Capra and Macías 2002) *Moraine-dammed lakes*: Three Sisters volcano, Mount Jefferson—twentieth century (O'Connor et al. 2001) *Failed crater lakes*: Mount Ruapehu—1953, 2007 (O'Shea 1954; Manville and Cronin 2007; Manville 2010)

(continued)

[4] Only edifice failures cutting deep into a volcano and involving weak hydrothermally altered rock (deep failure) can produce clay-rich volcanic debris flows (VDFs) (Vallance and Iverson 2015), but not all deep failures involve hydrothermally altered rock (Reid et al. 2010).

Table 3 (continued)

Basic initiation mechanism	Contributing processes	Examples
Glacier outburst floods (jökulhlaups)	Sudden release of water stored within or beneath a glacier	Mount Rainier (Richardson 1968; Walder and Driedger 1994a, b); Iceland—various examples (Maizels 1989; Russell and Knudsen 1999; Marren et al. 2009)
Heavy rainfall	Intense rainfall on loose, fine-grained pyroclastic particles produces surface runoff that easily erodes large sediment volumes	*Syneruptive*: Mount Pinatubo—1991 (Major et al. 1996; Oswalt et al. 1996) *Posteruptive*: Chaitén volcano—2008 (Pierson et al. 2013); Mount Merapi—1994, 2010 (Lavigne and Thouret 2002; Bélizal et al. 2013); Irazú volcano—1963–65 (Waldron 1967); Mount Pinatubo—1991–93 (Arboleda and Martinez 1996; Martinez et al. 1996; Umbal and Rodolfo 1996); Mount St. Helens—1980–2000 (Major et al. 2000)

The first two mechanisms occur only during eruptions and produce primary lahars. The last four mechanisms may or may not occur during eruptions and produce primary or secondary lahars. Examples given are only a small sample of relevant cases. See Volcano Index for volcano locations and details

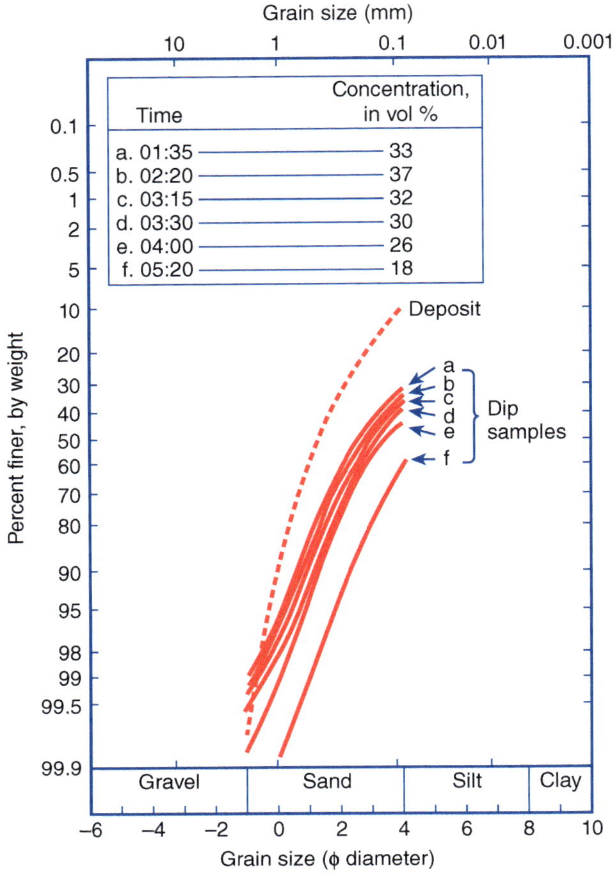

Fig. 4 Cumulative particle-size frequency curves for the March 19, 1982, North Fork Toutle River lahar at Mount St. Helens (Washington, USA) after transformation to volcanic hyperconcentrated flow (VHF), sampled 73 km downstream of the crater source in the early morning of March 20 (modified from Pierson and Scott 1985). Clock time (24-h format) is given in hours and minutes. A sample of the deposit was collected later at the same site. The deposit contained only about 10 wt% fines, while the flow dip samples contained between 30 and 60 wt% fines

because only a fraction of the silt and clay particles in a moving VHF are deposited with the sand and gravel fraction in VHF deposits (Pierson and Scott 1985; Cronin et al. 1999; Wilcox et al. 2014) (Fig. 4). Most of the silt and clay (as well as some fine sand) stays in suspension and is transported farther downstream. Yet enough fines usually are trapped in VHF deposits to inhibit rapid drainage of water during consolidation of the deposits, which allows characteristic features to form in VHF deposits—water-escape structures and soft-sediment deformation (see Chap. "Lahar Deposits—Hyperconcentrated-Flow Type").

3 Lahar Initiation Mechanisms

Lahar initiation involves a two-step sequence of events: mass mobilization and particle-water mixing. Mobilization encompasses the initiation of motion of a mass of solid particles or liquid water or a mixture of both. Particles are typically volcanic rock fragments, but some lahars are composed almost entirely of ice fragments (Cronin et al. 1996; Lube et al. 2009; Waythomas et al. 2013). Physical mixing produces the particle–fluid mixture. Once flow begins, additional rock or ice fragments and liquid water are entrained by scour at the channel bed and sloughing from channel banks or valley sides.

Six basic mechanisms initiate most lahars (Table 3). The first three mechanisms involve initial mobilization of water and solid particles together, and the remaining three involve initial mobilization of liquid water that quickly entrains solid particles. All six mechanisms can generate primary lahars during eruptions, but flank collapses, lake breakouts, glacial outburst floods, and heavy rain can also produce secondary lahars when a volcano is quiescent.

Rare lahar-genesis mechanisms also occur and include (a) expulsion of water or mud slurries directly from fissures on a volcano due to heating and pressurization of groundwater in the edifice during final magma ascent (Minakami et al. 1951; Johnson et al. 2018), (b) liquefaction of moving earthquake-triggered landslides (Scott et al. 2001), and (c) localized seismic compaction, liquefaction, and remobilization of wet static debris-avalanche deposits (Janda et al. 1981; Fairchild 1985; Scott 1988a).

4 Lahar Flow Behavior

Most lahars that have been directly observed and measured in the field have both a VDF component and a VHF component, which have different flow behaviors. This section contrasts the flow behaviors of each flow type with regard to turbulence, flow transformations, and sediment entrainment (bulking)—behaviors that occur in both VDFs and VHFs. These observations are based on results from both field studies and controlled flows in experimental flumes.

4.1 Turbulence

Turbulence, determined by the ratio of inertial forces to viscous forces within a fluid, occurs in both VDFs and VHFs when flow velocities are sufficiently high in proportion to the viscosity of the particle–fluid mixture. Thus, turbulence is dampened with (1) increasing concentration of suspended particles—fine sand, silt, and clay, which increases viscosity (e.g., Wilcox et al. 2014) and (2) decreasing flow velocity. VDFs in field settings typically exhibit flow that appears laminar (smooth flow surfaces) at velocities of only a few meters per second. At higher velocities, turbulence can occur in even highly concentrated lahars. However, turbulence may be present near the bed even while surface flow is laminar (Wang and Larsen 1994), and bed irregularities or channel cross-section changes may trigger turbulent boils in seemingly laminar flow (Fig. 5). Granular agitation also can be observed in shallow coarse-grained VDFs moving at relatively high velocities, presumably the result of interparticle collisions within the flow. In contrast to VDFs, VHFs are observed to be turbulent (in varying degree) in field settings up to sediment concentrations sufficient to dampen turbulence but also to effectively hinder particle settling (Pierson et al. 1996; Cronin et al. 2000; Pierson 2005; Wilcox et al. 2014) (Fig. 4b). Turbulence intensity decreases with increasing fines content (Wan and Wang 1994) and increasing total sediment concentration (Wilcox et al. 2014).

4.2 Flow Transformations

Flow transformations are transitions from one flow type to another caused by changing mass, momentum, or sediment concentration (see Chap. "Deposits Emplaced by Volcanic Wet Flows"). Their occurrence is deduced from observed changes in flow behavior or deposit characteristics. Clay-poor VDFs readily transform to more dilute flow types—to VHFs and eventually to VWFs—by dilution from the admixing of stream water or by selective deposition of sediment (Pierson and Scott 1985; Best 1992; Cronin et al. 1999, 2000; Hodgson and Manville 1999). For transformation to more concentrated flow types to occur, massive quantities of sediment must be entrained in the flow, typically from vigorous scour of channel beds or undercutting and collapse of channel banks (e.g., Best 1992; Pierson 1999; Scott et al. 2005;

a b

Fig. 5 Turbulence in flowing lahars: **a** localized turbulent boil in center of photo (about 1 m high) in an otherwise smooth, laminar-appearing flow surface of a hot (steaming) volcanic debris flow (VDF) on the Sacobia River, east flank of Mount Pinatubo (Philippines), Aug 20, 1991. Flow is left to right. Sediment concentration is 70 vol% solids; flow depth uncertain (~ 1 m?) and flow width is perhaps 15 m. Maximum surface velocity is ~ 5 m/s where surface appears streaked; slurry temperature ~ 50 °C. Apparent slight narrowing and steepening of channel here may have caused flow to transition abruptly from laminar to turbulent flow, with a train of vigorous turbulence continuing downstream of this point 10 m or more. **b** A volcanic hyperconcentrated flow (VHF) in the Pasig-Potrero River, Mount Pinatubo, Sep 1, 1991, that is flowing left to right over a paved highway (left center of photo) about 40 cm deep at center and showing hydraulic jump where it flows down off the right edge of the road. Edge of flow is lapping onto mud-covered road surface in foreground. Suspended-sediment concentration estimated to be about 20 vol%; flow velocity not measured (Pierson 2005). USGS photos by T. Pierson

Wilcox et al. 2014) (Sect. 4.3). Clay-rich VDFs generally do not transform to VHFs (Vallance and Iverson 2015). The high clay content in the matrix prevents both the mixing in of additional water and the segregation and deposition of gravel-size clasts.

4.3 Bulking: Accretion of Mass

Lahars and VWFs commonly add mass (and volume) as they flow downstream—a process termed *bulking* (e.g., Scott 1989; Pierson et al. 1990; Best 1992; Fagents and Baloga 2006). Bulking is common in lahars on steep volcano flanks (typically for at least the first 5 km), but in some cases bulking may continue to occur as far as 50 km from source, depending on channel gradient and flow depth (e.g., Pierson et al. 1990) (Fig. 6). In other cases, reaches dominated by entrainment of sediment may alternate with reaches dominated by deposition (Procter et al. 2010).

Volumetric bulking factors ($V_{final}/V_{initial}$) of at least 3 are generally necessary for transformation of a water flood to VDF, but factors as high as 20 have been documented (Table 4) when additional water is entrained along with additional sediment. Bulking efficiency appears to be related to channel bed and bank erodibility, channel slope, relative channel width (narrow channels vulnerable to undercutting of banks and talus slopes), bank height and steepness (collapse of high banks adds more sediment than low banks), the boundary shear stress applied to the bed and banks by the fluid mixture, and the looseness and degree of saturation of the bed material (Pierson 1999; Hungr et al. 2005; Graettinger et al. 2010; Iverson et al. 2011). Lahar scour depths in unconsolidated sediment can range up to 15 m (Pierson and Scott 1985; Rodolfo 1989; Rodolfo and Arguden 1991; Paguican et al. 2009; Procter et al. 2010). Volumetric bulking rates in natural channels (i.e., flow volume increase per unit of channel length) can be more than 600 m^3/m (Pierson et al. 1990).

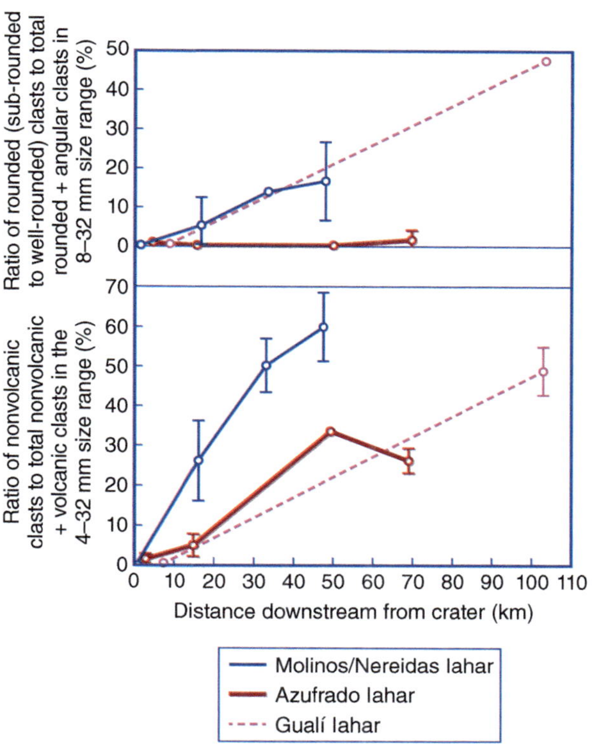

Fig. 6 Rate of volumetric bulking in three lahars from the 1985 eruption of Nevado del Ruiz (Colombia), estimated from rate of increase of rounded stream gravel in lahar deposits (top) and rate of increase of nonvolcanic clast content in deposits (bottom) with distance from source (modified from Pierson et al. 1990). The volcano edifice sits on metamorphic terrain. Error bars indicate one standard deviation; no error bars shown for single samples or where error was too small to show

Sediment typically is added to moving lahars by scour and incorporation of loose particles from channel beds and banks, which can be facilitated in wet valley bottoms by undrained loading and liquefaction of saturated bed material by the weight and shearing action of the overpassing lahar (Iverson et al. 2011; Vallance and Iverson 2015). Incorporation of liquefied bed material adds both sediment and water to a lahar, but it also produces high fluid pressure at the base of the lahar, which reduces bed friction and accelerates flow (Iverson et al. 2011). Sediment can also be added by bank sloughing and landslides from valley sides.

River water can be added to lahars by inflow at stream confluences and by a process we term *water-wave coupling*. Coupling of a wave of river water at a lahar front occurs when the front of a VDF flows faster than the water flowing in the channel and pushes or "bulldozes" the overtaken river water ahead of itself (Cronin et al. 1999; Dinehart 1999) (Fig. 7).[5] Dense lahars should be able to dive under water in a channel if the water depth is greater than the depth of the lahar, as happens when debris flows enter lakes or reservoirs (Weirich 1989), but water depths in steep river channels on volcano flanks are typically shallower than the fronts of large lahars, so lahar submergence is not plausible in such cases. For example, the head of the lahar that descended the Azufrado River at Nevado del Ruiz (Fig. 1) on November 13, 1985, was about 12 m high, coming down a canyon that hosted a

[5] It has been proposed the dilute water "waves" generated at the fronts of some lahars are actually mechanical surface waves moving in tandem with the lahar front (Manville et al. 2000). However, this is not plausible for VDFs because the water-wave celerity would be much slower than the velocity of the lahar front in most channels draining volcanoes, based on the derived celerities using the shallow-water wave equation when realistic water depths are applied.

Table 4 Calculated volumetric bulking factors for lahars flowing dominantly as volcanic debris flows (VDFs)

Source volcano for lahar	River channel, age or date, and initiation mechanism	Approximate bulking factor ($V_{final}/V_{initial}$)	References
Lahars initiating from bulking of volcanic water floods (VWFs)			
Mount St. Helens	North Fork Toutle River, ~ 2.5 ka, lake outbreak flood	3–5	Scott (1988a, b)
Mount St. Helens	North Fork Toutle River, Mar 19–20, 1982, lake outbreak flood	4–5	Pierson (1999)
Nevado de Colima	Naranjo River, ~ 18.5 ka, lake outbreak flood	10	Capra and Macías (2002)
Nevado del Ruiz	Azufrado River, Nov 13, 1985, rapid snowmelt by PDCs	4–5	Pierson et al. (1990)
Nevado del Ruiz	Lagunillas River, Nov 13, 1985, rapid snowmelt by PDCs	3–4	Pierson et al. (1990)
Nevado del Ruiz	Guali River, Nov 13, 1985, rapid snowmelt by PDCs	3	Pierson et al. (1990)
Nevado del Ruiz	Molinas-Nereidas River, Nov 13, 1985; rapid snowmelt by PDCs	4–6	Pierson et al. (1990)
Lahars initiating from liquefaction of landslides (edifice/flank collapse)			
Mount Rainier	White River (Osceola/Greenwater lahar), 5.6 ka, major edifice collapse	1.2–1.6	Vallance and Scott (1997)
Casita volcano	South flank of edifice, Oct 30, 1998, edifice collapse	>2.6	Scott et al. (2005)
Mount Hood	Polallie Creek, Dec 25, 1980, small flank collapse of volcanic sediments	20	Gallino and Pierson (1985)

Initial volumes estimated from approximate water volumes in source lakes producing lake-outbreak lahars, from the area, thickness, and approximate density of snowpack or ice scoured by pyroclastic density currents (PDCs) producing water rapidly from snowmelt, or from the calculated volume of landslides transforming to lahars, based on slide scar volumes. Final volumes calculated from deposit areas and average thicknesses. See Volcano Index for volcano locations and details

shallow stream with a small dam and reservoir. Survivors of this lahar disaster reported a clear-water flood immediately preceding the lahar (Pierson et al. 1990). Water floods preceding lahars also have been reported elsewhere (Best 1992).

Between the wave of clear water and the front of a VDF is a mixing zone of progressively increasing sediment concentration (Rodolfo et al. 1996; Cronin et al. 1999; Dinehart 1999) (Figs. 7 and 8). Mixing zones expand with distance traveled (Pierson and Scott 1985; Cronin et al. 1999), however, the mechanisms involved in mixing and dilution have not been fully resolved (Manville et al. 2000). Rapid fine-grained debris flows can have flow fronts that are turbulent and highly agitated (Scott and Wang 2004), and in experimental submarine debris flows, injection of water beneath debris-flow bodies leads to hydroplaning and breakup of the flow front (Mohrig et al. 1998; Marr et al. 2001; Harbitz et al. 2003; de Blasio et al. 2004)—mechanisms possibly contributing to disruption of VDF fronts and progressive mixing

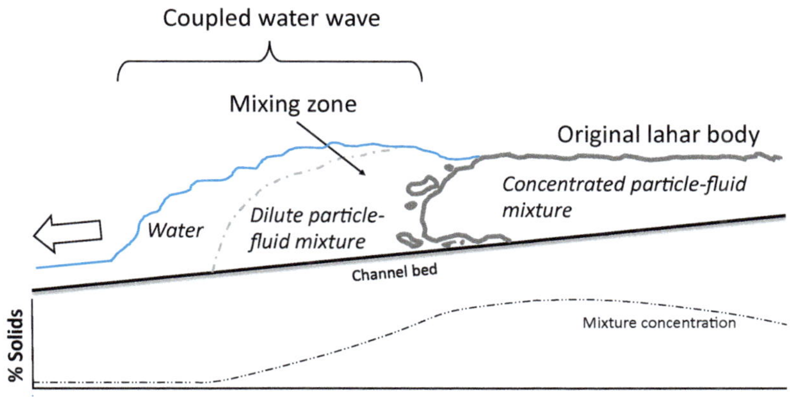

Fig. 7 Schematic longitudinal profile (foreshortened) of the flow front of a volcanic debris flow (VDF) pushing a wave of river water ahead of itself as the water wave couples with the lahar front and becomes progressively more concentrated with time. Mixture concentration begins increasing in the mixing zone, peaks in the main lahar body, and then gradually decreases in the lahar tail. Thus, it is possible for one lahar to include water flow, hyperconcentrated flow, and debris flow in different parts of its flow body

with coupled water waves. Field observations show that with time and distance traveled, coupled water waves become progressively more concentrated with suspended sediment. In some lahars, the more dilute coupled wave produces the highest stage and discharge of the lahar (Cronin et al. 1999; Dinehart 1999).

4.4 Behavior of Volcanic Debris Flows (VDFs)

Lahars (or parts of lahars) moving as debris flow have the consistency of fluid wet concrete. They typically exhibit laminar or nearly laminar flow at low velocities (generally <2 m/s) but become turbulent at relatively high velocities, depending on flow depth, channel slope, and sediment concentration. High-velocity shallow VDFs commonly exhibit supercritical flow, as evidenced by standing waves (antidunes) and hydraulic jumps documented in the field (e.g., Costa and Williams 1984; Rodolfo et al. 1996; Pringle and Cameron 1999).

VDFs flow rapidly, despite their high solids content, due to their high bulk densities (about twice the density of water), somewhat dampened turbulence, and highly liquefied condition. In the partially liquefied bodies of fines-poor experimental debris flows, the pore fluid typically carries 80% or more of the total bed-normal stress (Major and Iverson 1999), meaning that intergranular friction within these water–rock mixtures is drastically reduced. Limited field data suggest that VDF lahars can be up to 60% faster (surface velocities) than water flow of similar depth and slope, and somewhat faster than VHF lahars, when not held back by bouldery flow fronts (Lavigne and Suwa 2004).

Peak flow velocities of large to very large[6] lahars (10^3–10^6 m^3/s) typically exceed 10 m/s on the lower flanks of stratocones but may approach 40 m/s on the upper flanks (Fig. 9). Very large lahars on very steep slopes have been documented >80 m/s (Allstadt 2013). Such lahars typically decelerate to 8–15 m/s by the time they are about 40 km downstream. An important consideration in hazards assessments is *travel time*—the time from lahar initiation at point of origin to lahar arrival at critical locations downstream along the flow route. Travel time is determined by lahar volume, the longitudinal channel profile (average steepness of flow path), and flow mobility (a function

[6] Range of lahar magnitudes are defined by flow discharge, which can be translated to flow volume (Pierson 1998).

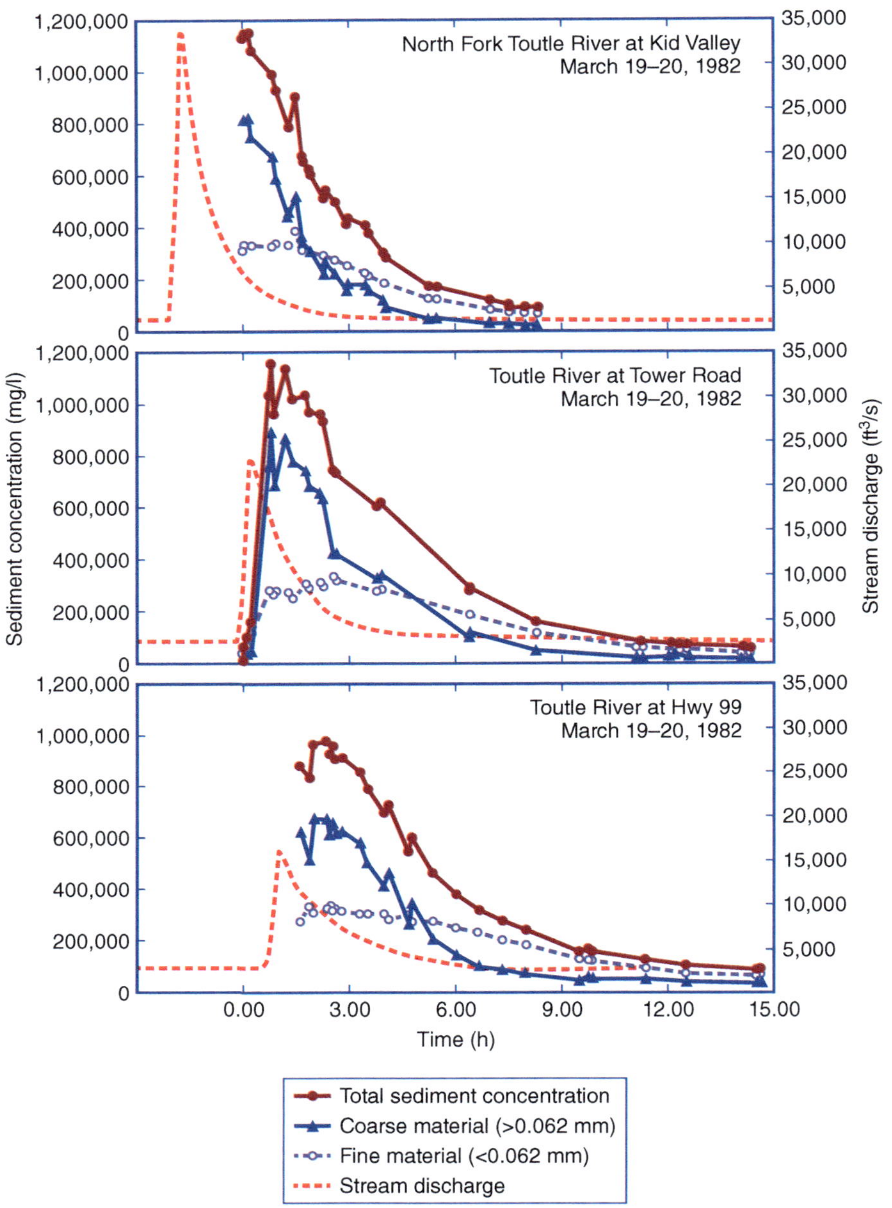

Fig. 8 Discharge hydrographs of a lahar, flowing here mainly as volcanic hyperconcentrated flow, on March 19–20, 1982, at three gaging stations (49 km [Kid Valley], 58 km [Tower Rd], and 73 km [Hwy 99] from source) on the North Fork Toutle River and main stem Toutle River, Mount St. Helens (Washington, USA, modified from Dinehart 1999). The Tower Road and Hwy 99 hydrographs clearly show the time lag of the more concentrated lahar body behind the dilute wave of water at the lahar front (i.e., peak discharge is achieved at the downstream stations before arrival of peak sediment concentration). The peak suspended coarse-material concentration (mostly sand) corresponds to a fines concentration of 300–350 thousand mg/L (equivalent to ~ 11–13 vol% solids), but the data showed that amount of coarse material in suspension began increasing rapidly once the fines concentration exceeded about 8 vol%, although this is not clear in the plots (Dinehart 1999). One $m^3/s = 35.3\ ft^3/s$

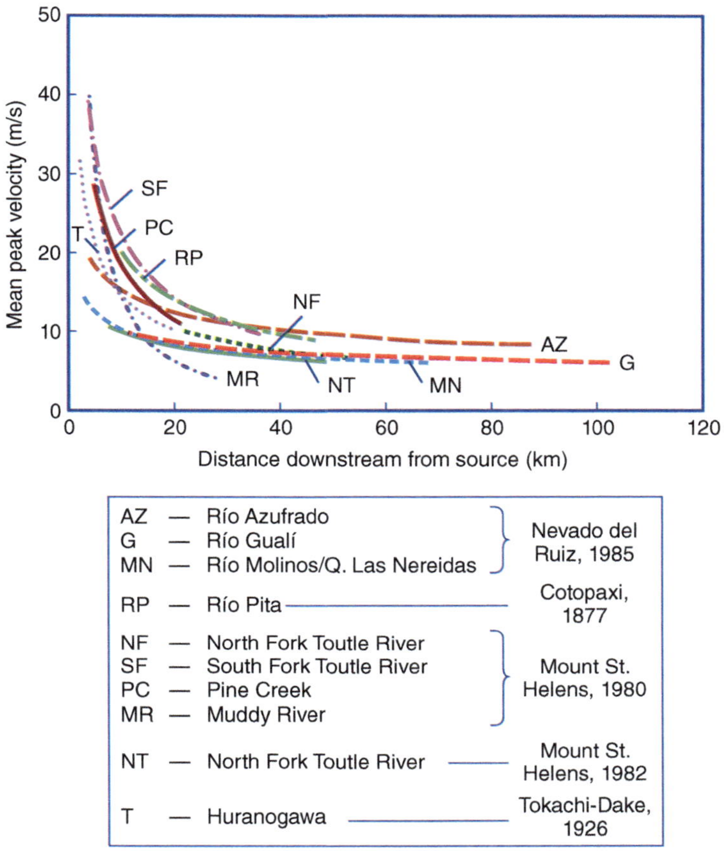

Fig. 9 Change in peak flow velocity with distance downstream for large to very large (10^3–10^6 m^3/s) lahars flowing dominantly as volcanic debris flows (VDFs) (modified from Pierson 1995). Approximate relation of lahar discharge to lahar volume is given in Pierson (1998). All lahars started at zero velocity at zero distance downstream and accelerated to their highest velocities (back calculated) in reaches with steepest slopes and within a few kilometers of their starting points. All were mainly VDFs for their entire flow routes, except for lahar NT, which transformed to hyperconcentrated flow between 27 and 43 km from source. Volcano locations and details are given in the Volcano Index

of composition plus the initial and evolving hydraulic permeability of the flow mixture) (Pierson 1998; George et al. 2022) (Fig. 10).

4.4.1 Flow Front and Levee Formation

Clay-poor VDFs commonly have flow fronts and lateral margins that are enriched in coarse particles due to a three-stage process involving (1) development of inverse grading from the upward migration of coarse particles by the mechanisms of kinetic sieving and squeeze expulsion; (2) conveyance of the cobble- and boulder-size clasts on or near the flow surface to flow fronts by higher velocity at the flow surface; and (3) migration of the coarser clasts in a mixture to flow margins, resulting in large part from pushing aside of frontal coarse clasts (Okuda et al. 1981; Suwa and Okuda 1983; Pierson 1986; Suwa 1988; Suwa et al. 1989; Major and Iverson 1999; Gray and Ancey 2009; Johnson et al. 2012) (Fig. 11). Concentrations of coarse clasts at flow fronts and margins allow more rapid dissipation of pore pressure (Fig. 12), which results in pronounced increases in intergranular friction and the formation of coarse flow levees that laterally confine flows (Major 1997; Major and Iverson 1999). Flow fronts and levees are commonly well developed and boulder rich in relatively small, clay-poor VDFs, but the coarseness of these features depends on the maximum size of clasts being transported. Clay-rich VDFs have minimal clast segregation during flow and relatively few large clasts (weak large clasts tend to disaggregate during transport). Therefore, they are unlikely to develop flow fronts or levees defined by accumulations of boulders.

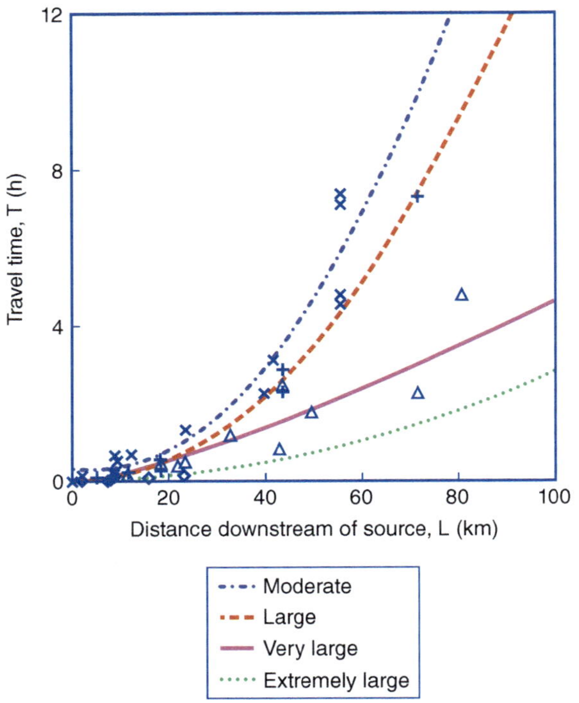

Fig. 10 Travel times for lahars and volcanic debris avalanches to points downstream for flows of moderate size (10^2–10^3 m^3/s), shown by symbol "x"; large size (10^3–10^4 m^3/s), shown by symbol "+"; very large size (10^4–10^6 m^3/s), shown by symbol "Δ"; and extremely large size (> 10^6 m^3/s), shown by symbol "◊" (modified from Pierson 1998). Most of these examples are volcanic debris flows (VDFs), but volcanic and nonvolcanic debris avalanches make up the "extremely large" category. They are included here because some transformed to debris flow en route and because no data were found for pure VDFs of this size. Polynomial best fit curves do not all go through graph origin

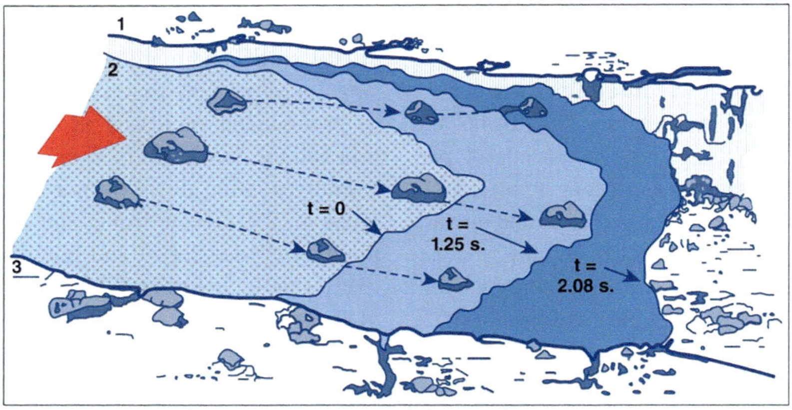

Fig. 11 Stop-action movement of surface boulders and flow-front positions of a small volcanic debris flow, traced from movie frames, illustrating the preferential forward movement of large clasts at the surface during flow (modified from Pierson 1986). This concentrated lahar occurred at Mount St. Helens (Washington, USA) on October 1, 1981. The boulder in the center of the flow was moving twice as fast as the flow front. Largest boulder is about 0.5 m in diameter. Video can be accessed in Costa and Williams (1984)

Fig. 12 Boulder levees emplaced by small clay-poor volcanic debris flows. Note that large clasts do not have matrix material between them. **a** On south flank of Mount Rainier (Washington, USA). **b** On northwest flank of Láscar volcano (Chile). USGS photos by T. Pierson

4.4.2 Hydraulic Damming by Flow Fronts

Boulders and logs accumulated at flow fronts of some VDFs can form a wedge of open-work and highly frictional material that gets pushed ahead of the flow like a deformable, sliding debris dam (Okuda et al. 1981; Pierson 1986; Major and Iverson 1999) (Fig. 13). Liquefied slurry coming behind such coarse flow fronts moves faster than the frictional fronts, so that much of the flow behind the fronts progressively backs up and deepens, although some liquefied material may bypass and get ahead of the fronts. Thus, such VDFs typically have steep flow fronts and peak discharges occurring directly behind the fronts (Fig. 14). Large VDFs may not collect enough coarse debris to form such moving dams at their fronts. Smaller VDFs with highly frictional flow fronts are unlikely to flow fast enough to overtake river water in the channel and develop a coupled water wave.

4.4.3 Surging Behavior

VDFs characteristically exhibit surging or pulsing behavior (Iverson 1997, 2003, 2005; Hungr 2000; Zanuttigh and Lamberti 2007). Resembling downstream-moving waves of debris, surges are a component of unsteady flow, have distinct flow fronts, and move faster than the shallower, slower flow in front of them. Surges commonly develop

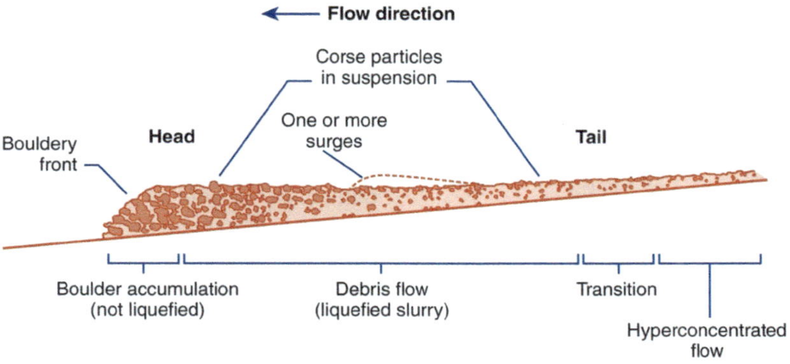

Fig. 13 Schematic longitudinal profile of a relatively small, clay-poor volcanic debris flow with a coarse flow front. It is typical in such flows that sediment concentration decreases steadily from flow front to flow tail, and peak discharge is achieved directly behind the coarse flow front

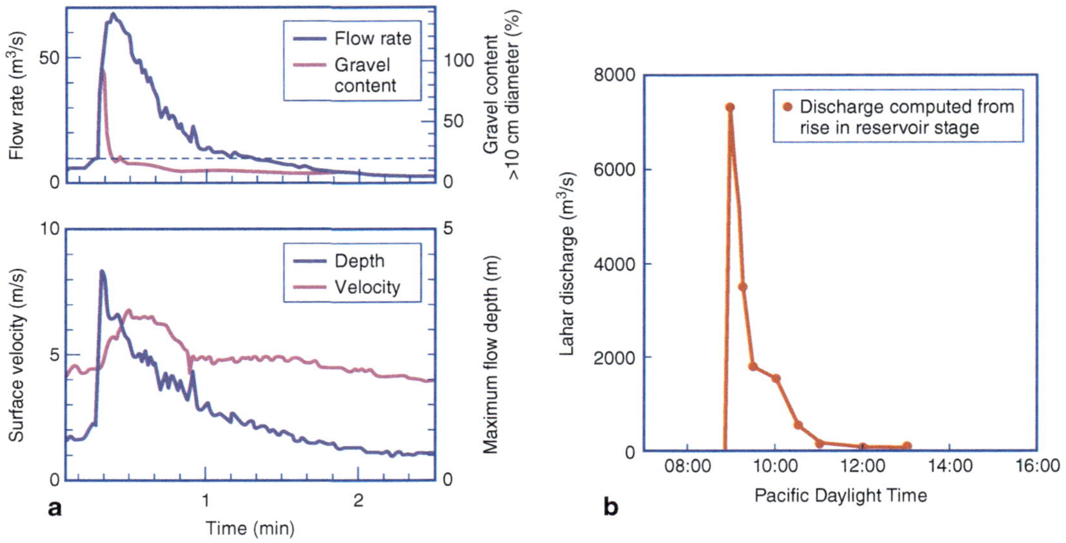

Fig. 14 Lahar hydrographs for flows moving dominantly as volcanic debris flows (VDFs). **a** A surge on a moderate-sized VDF at Mount Yakedake (Japan) showing discharge (flow rate), content of gravel >10 cm, velocity, and depth as flow passes an observation point (modified from Suwa and Yamakoshi 1999). **b** Discharge hydrograph of the very large Pine Creek lahar, obtained from recorder of pool level at Swift Reservoir on the lower southern flank of Mount St. Helens (Washington, USA), into which the lahar flowed on May 18, 1980 (modified from Pierson 1985). The Pine Creek lahar arrived at the reservoir at about 08:54 after traveling 23 km from source; it was triggered at about 08:34

spontaneously and have front velocities in proportion to their size—larger surges move faster than, overtake, and coalesce with smaller surges. Surges within a lahar body can be larger than the initial flow fronts (Iverson 1997) (Fig. 15a, b). In addition, surges may also develop due to transient damming (Pierson 1980) or the staggered confluence of multiple smaller lahars. Surges in relatively small flows commonly concentrate cobbles and boulders at surge fronts, depending on the availability of coarse particles (Iverson 2003, 2005), but some surge fronts have the same size distribution as the material farther back within the lahar body (Fig. 15).

4.5 Behavior of Volcanic Hyperconcentrated Flows (VHFs)

Lahars (or parts of lahars) moving as hyperconcentrated flow are intermediate between debris flow and water flow in sediment concentration, fluid viscosity, and flow behavior (Pierson and Scott 1985; Costa 1988; Cronin et al. 1999; Lavigne and Suwa 2004; Pierson 2005; Wilcox et al. 2014; Vallance and Iverson 2015). They typically have a fluid consistency resembling motor oil or heavy cream and are always observed to be turbulent to some degree. Characteristics of VHFs commonly exhibited in the field include: (1) large quantities of sand and fine gravel being carried in relatively uniform concentrations throughout much of the vertical profile but becoming more concentrated near the bed (Fig. 16); (2) small-scale turbulence being dampened in relatively deep subcritical flows (Fig. 17a) but increasing in shallow supercritical flows, where the flow surface can take on a finely textured "choppy" appearance (Figs. 17b and 18b); (3) particles of fine gravel appearing momentarily at the surface of the flow and then sinking out of sight again; (4) convection cells forming by zonal settling of coarser particles

Fig. 15 Flow surges in debris flows. **a** Turbulent (highly agitated) surge front flowing toward camera. Flow was a rainfall-triggered, nonvolcanic, clay-rich debris flow (8–10 wt% clay) in Jiang Jia Ravine, Yunnan Province, China. Surge front is 3–4 m high and front velocity is 8–10 m/s; note people (circled) on bank in upper right for scale (Scott and Wang 2004). **b** Same flow surge front photographed several seconds later. This surge is larger than the earlier surges and is overriding more slowly moving debris from the previous surge. **c** Same flow surge front photographed a few more seconds later. **d** Surge front in shallow volcanic debris flow in upper North Fork Toutle River valley, northwest of Mount St. Helens (Washington, USA), where snow is partly covering the channel bank. Largest boulders in flow are nearly 1 m in diameter and rolling on the channel bed; front velocity is estimated to be 1–2 m/s. This is same flow that deposited the material in Fig. 1 in Chap. "Lahar Deposits—Debris-Flow Type". USGS photographs: **a**, **b**, and **c** by K. Scott; **d** by T. Pierson

Fig. 15 (continued)

Fig. 16 Vertical distribution of suspended-sand concentration in an experimental hyperconcentrated flow using sand in a bentonite clay suspension (modified from Bradley 1986). It shows that suspended-sand concentration increases markedly near the bed

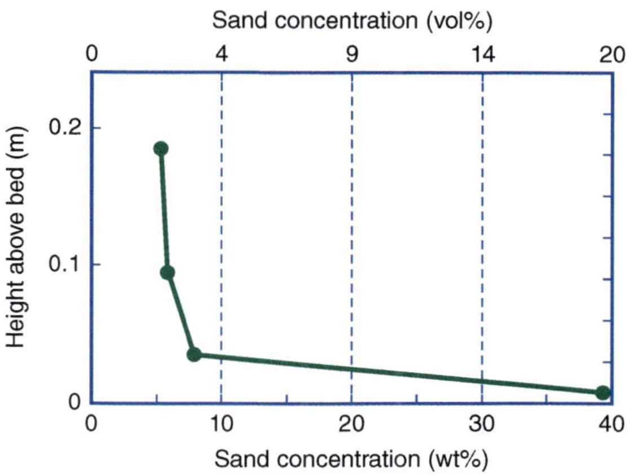

that causes reticulate patterns of upwelling water and fines to appear at the flow surface (Fig. 17a); and (5) flow surges resembling low tidal bores. Hydrographs of VHFs, like VDFs, have steep waxing limbs and gradually decreasing waning limbs (Cronin et al. 1999) (Fig. 8).

VHFs flow at slightly higher velocities than water floods of equivalent depth and channel slope, because hyperconcentrated flows experience less resistance to flow than clear-water flows due to small-eddy suppression in the higher viscosity fluid (Wan and Wang 1994) and to streamlining of dune and antidune bedforms (Simons et al. 1963; Wan 1985). Addition of more bed load to a VHF will increase flow turbulence and decrease velocity, however (Wang and Larsen 1994). At Mount St. Helens, the 1982 VHF (Fig. 8) had a mean velocity that was approximately 10% faster than velocities of several previous VWFs of approximately equivalent depth and discharge, measured at the Tower Road gaging station (Pierson and Scott 1985). At Semeru volcano (Indonesia), VHF surface velocities were 25–30% faster than sediment-laden streamflow surface velocities at equivalent depths ranging from 0.5 to 2 m (Lavigne and Suwa 2004). Where comparisons have been made, surface velocities for turbulent VHFs appear to be slightly slower than those for VDFs at similar channel gradients and flow depths (Lavigne and Thouret 2002; Lavigne and Suwa 2004).

5 Lahar Magnitude and Frequency

Lahars can involve enormous volumes of flowing water and volcaniclastic debris. Total volumes exceeding 10^9 m^3 have been documented (Vallance and Scott 1997; Mothes et al. 1998), but extraordinarily large lahars are not common. Such super-sized lahars are likely to be clay-rich and triggered by large edifice collapses (e.g., Vallance and Scott 1997), but clay-poor lahars of similar size can also occur (e.g., Mothes et al. 1998). Total flow volume typically increases steadily downstream if flow depths and channel gradients are sufficient to induce continued erosion and bulking, most commonly where lahars are confined in steep narrow canyons. In contrast, VDFs tend to be larger and more mobile than nonvolcanic debris flows in mountainous regions around the world (Jakob 2005; Berti and Simoni 2007; Griswold and Iverson 2008), which is likely due to the greater abundance of loose fragmental rock debris, larger volumes of water in unstable impoundments, and more unstable slopes available at volcanic edifices.

Lahars, particularly clay-poor lahars, may have extremely large peak discharges relative to total flow volume, which result from a combination of rapid bulking, high flow velocities, water-wave coupling, and, in some cases, partial hydraulic damming caused by encumbered flow fronts moving slower than average flow velocity

a
b

◀ **Fig. 17** Volcanic hyperconcentrated flow (VHF) surfaces (from Pierson 2005): **a** deep tranquil flow of a VHF showing a relatively smooth, oily-looking surface with a reticulate color pattern in a lahar from Mount Ruapehu (New Zealand). Pattern apparently caused by the formation of convection cells due to locally downward sinking coarse sediment and upwelling of more dilute water and fines (Cronin et al. 1999). Sampled average sediment concentration of this mixture was 46 vol%. Scale bar in lower right estimated to be about 0.5 m. Photo by S. Cronin from Pierson (2005). **b** Shallow supercritical flow in VHF in Pasig-Potrero River, Mount Pinatubo (Philippines), where flow depth is 20–30 cm and estimated velocity (from runup on man's leg) is 2–3 m/s. Flow surface shows a pronounced chop. Concentration was not measured but visually estimated (from mixture consistency) to be 20–30 vol%. Photo by T. Pierson (Pierson 2005)

Fig. 18 Surface waves in a shallow volcanic hyperconcentrated flow (VHF) on the Sacobia River, Mount Pinatubo (Philippines), September 26, 1993, flowing left to right. Suspended sediment concentration was about 23 vol%, flow velocity was about 2.5 m/s, and flow depth was about 50 cm (Pierson 2005). Standing wave (largest wave on left) reflects formation of antidunes on channel bed (supercritical flow). Edge of vertical bank is in foreground. Photo by T. Pierson

of some lahar bodies. Major lahars can achieve peak discharges up to about 100,000 m^3/s on steep slopes within 10 km of their sources (Fig. 19); very limited discharge data for large clay-rich lahars show lower peak discharges (Fairchild 1985, 1987).

Peak discharge of lahars typically attenuates with distance downvalley due to channel widening, decrease or cessation of material entrainment, and sediment deposition as lahars continue flowing down channels of decreasing gradient. Peak discharges are commonly achieved in the steep channels of the upper flanks of volcanoes (Figs. 19 and 20), although flow volume may continue to increase as peak discharge declines (e.g., Pierson et al. 1990). As discharge peaks decrease, lahar bodies stretch out longitudinally, and flow duration increases (Fig. 20).

Lahar frequency is related to (1) type and timing of initiating mechanisms, (2) degree,

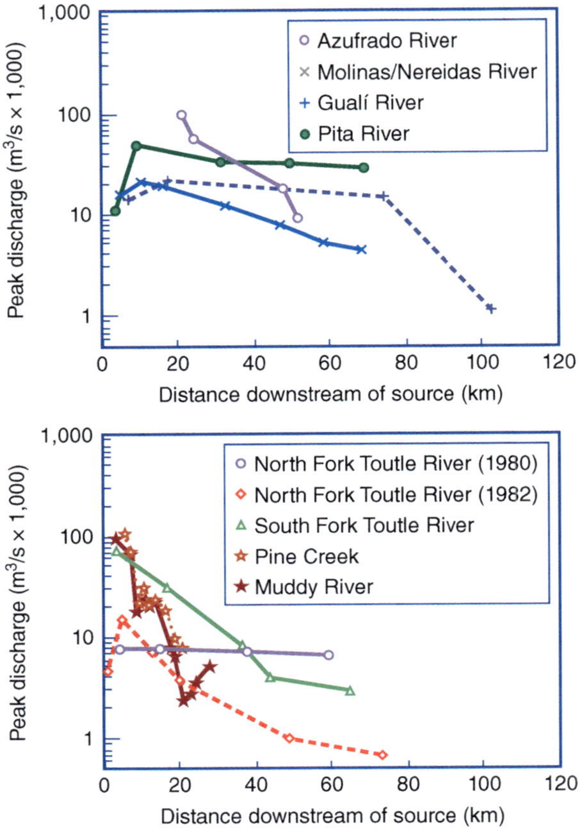

Fig. 19 Variation in peak discharge for very large eruption-triggered lahars (10^7–10^8 m^3 in volume; modified from Pierson 1995), which show discharge decreases with increasing distance downstream from source. Graph shows lahar data from the 1877 Cotopaxi (Ecuador) eruption in the Pita River and three lahars from the 1985 Nevado del Ruiz (Colombia) eruption (channels of the Azufrado, Molinas/Nereidas, and Gualí Rivers). All were clay-poor volcanic debris flows for much of their flow paths, and they show steady downstream decreases in peak discharge—a behavior typical for clay-poor lahars. Estimated error on data points is 20–30%

type, and recency of disturbance to source areas on volcano flanks, (3) degree and extent of hydrothermal alteration within the edifice, (4) steepness and composition of volcano flanks, and (5) lahar magnitude. Some volcanoes, such as Mount Rainier (Washington, USA), show an inverse relationship between frequency and magnitude (Table 5). Rainfall-triggered lahars are common and can be frequent (up to hundreds per year) within the first year or two following major explosive eruptions (Tuñgol and Regalado 1996; Pierson et al. 1996; Umbal 1997). If the disturbance is not ongoing, lahar activity declines back to the predisturbance rate within a few years or a few decades, depending on type and degree of disturbance (Major et al. 2000; Pierson and Major 2014). If disturbance is ongoing, lahar activity can remain high. For example, Sakurajima volcano (Japan) has been continuously active since 1955 and has deposited ash regularly in the upper part of the Nojiri River basin (3.0 km^2 drainage area). As a result of the ash deposition and heavy seasonal rainfall, the Nojiri River produced between about 10 and 20 moderate-size lahars per year in the decades following the onset of this most recent eruptive period (average 18 flows per year or 6 flows/km^2 per year from 1973 to 1977) (Watanabe and Ikeya 1981). For comparison, published frequencies of nonvolcanic debris flows in undisturbed, steep, forested terrain in the western United States and Canada vary from about 0.005 to 0.5 flows/km^2 per year (Osterkamp et al. 1986; Hupp et al. 1987).

Lahar volume is a major determinant of travel distance, though not the only one. Mixture properties (fines content, especially clay content, hydraulic diffusivity, and initial density state), amount of dilution by tributary streams, and flow path steepness and degree of channel

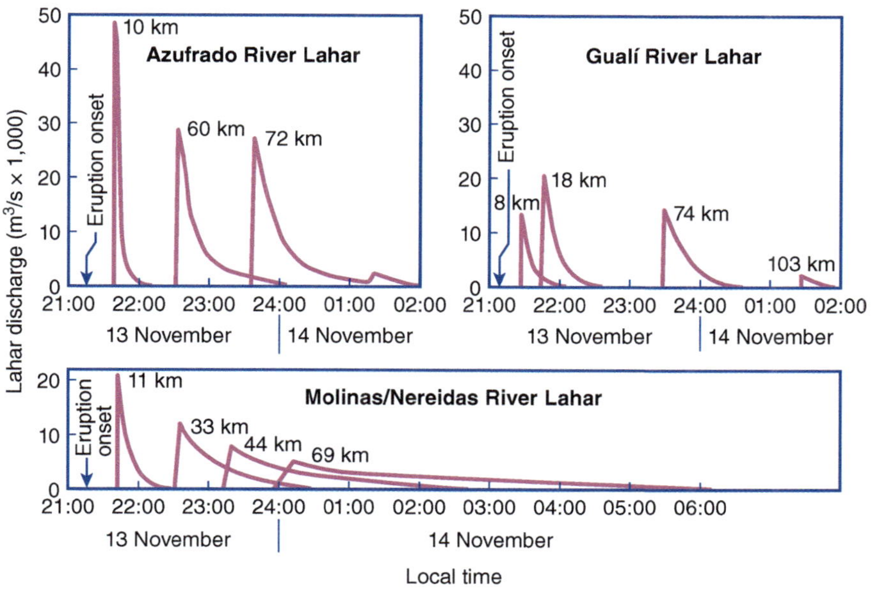

Fig. 20 Downstream changes in magnitude and shape of reconstructed hydrographs for clay-poor lahars (flowing mainly as volcanic debris flows) triggered in three different rivers by rapid snowmelt during the November 13, 1985, eruption of Nevado del Ruiz (Colombia). Modified from Pierson et al. (1990). Peak flow levels obtained from surveyed mudlines; durations obtained from multiple eyewitness accounts

confinement also play important roles in how far lahars will flow. Small lahars (< 10^4 m^3) typically travel less than 10 km and may not even extend beyond the lower flanks of the volcano (Walder and Driedger 1994a, b, 1995). Extremely large lahars (> 10^9 m^3), on the other hand, have been documented to travel over 300 km (Mothes et al. 1998).

6 Lahar Sediment Transport and Deposition

Sediment transported by lahars, including some of the gravel fraction (> 2 mm), is largely carried in suspension above the channel bed. In VDFs, some gravel remains in suspension continuously during flow and even after flow stops, owing to the yield strength of the particle–fluid mixtures. In VHFs, observations suggest that suspension of gravel is intermittent—a function of turbulence acting in concert with lesser degrees of yield strength provided by their more dilute particle–fluid mixtures. Whether VDF or VHF, some coarse particles are too heavy to be suspended in the flow and must be transported solely as bed load—sliding, saltating, or rolling along the channel bed, impelled by fluid drag. Transport of extremely large bed-load clasts, especially by VDFs, is possible (Fig. 21), and diameters of such clasts can be greater than the flow depths of the lahars moving them (Pierson 1985; Scott and Wang 2004). Deposition of such isolated, boulder-size, bed-load clasts occurs largely independently of mechanisms emplacing finer particles moving mainly in suspension, as discussed below. Movement of large "outsized" clasts is halted when the resisting force of bed friction exceeds the impelling force of fluid drag acting on the rolling or sliding clast, and the outsized clasts become stranded (Pierson et al. 1990; Pierson 2005). They may or may not be buried by subsequent deposition by the lahar that emplaced them.

Table 5 Order-of-magnitude frequencies and recurrence intervals for lahars of different size categories at Mount Rainier volcano (Washington, USA), interpreted from documented recent small and moderate lahars (debris flows) at Mount Rainier and from tables summarizing occurrence of prehistoric lahar deposits mapped and studied by Crandell (1971) and Scott et al. (1995)

Magnitude (from Pierson 1998)	Debris volume (m^3)	Frequency (events/1000 calendar years) with explanatory notes	Recurrence interval (years)	References
Small	10^3–10^4	*1000* Documented occurrences in last several decades average 0.9, 2.0, and 2.5 events/year. Maximum is 7 in one year in different drainages. Variation due to frequency of extreme rainstorms and intensity of late-summer glacier melt	1	Walder and Driedger (1995), Scott et al. (1995), Copeland (2009), Legg et al. (2014)
Moderate	10^4–10^6	*100* Six notable events (estimated to be of this magnitude) over 73 years prior to 1980s	10	Walder and Driedger (1995)
Large	10^6–10^7	*10* Ten prehistoric events since 5600 yBP (stratigraphically above Osceola Mudflow)	100	Scott et al. (1995)
Very large	10^7–10^9	*1* Four events since ~ 2500 yBP; seven events since ~ 3600 yBP	1000	Scott et al. (1995)
Extremely large	$>10^9$	*0.1* One prehistoric event (Osceola Mudflow) since 5600 yBP (and probably since ~ 11,000 yBP)	10,000	Scott et al. (1995)

All years given are calendar years (calibrated radiocarbon dates)

Fig. 21 Huge lahar bed-load particles. Boulders were transported unknown distances to these locations downstream of Mount St. Helens (Washington, USA) by an eruption-triggered lahar (mainly volcanic debris flow) on May 18, 1980. **a** Boulder (5–6 m intermediate diameter) emplaced on a low terrace surface (previously forested and logged) of the Muddy River, about 20 km downstream of Mount St. Helens (Washington, USA). Note that mudlines on trees to right of boulder (level of dashed line) indicates that flow depth was less than boulder diameter. Also note that there was minimal deposition on this terrace surface. USGS archive photo. **b** Boulder 6–7 m intermediate diameter also emplaced on a low terrace by the May 18 Muddy River lahar. Dashed line marks flow depth along the far bank. USGS photo by P. J. Carpenter

6.1 Lahar Depositional Mechanisms

Lahars deposit fully or partly suspended sediment by various mechanisms, depending on flow type and depositional location (Table 6), with different depositional mechanisms likely occurring simultaneously in different places within a lahar. Both particle concentration and particle-size distributions determine flow type and play important roles in determining depositional mechanism.

Concentrated lahars (VDFs) mainly deposit sediment by *progressive aggradation*, which is accomplished by incremental accretion of particles on the channel bed or ground surface, usually in response to progressive dilution within the lahar body (Vallance 1994, 2000; Vallance and Scott 1997; Major 1997, 2000; Major and Iverson 1999; Vallance and Iverson 2015) (Fig. 22). Where this mechanism is operative, vertical grading in a deposit reflects sequential emplacement of accreting particles of different grain sizes by the overpassing lahar.

An important but more localized deposition mechanism in VDFs is *edge-friction damming* where VDFs spill out onto broad low-gradient surfaces, where flows thin, lose momentum, and decrease flow depth. At the edges of such spreading and slowing flows, excess pore pressure within the sediment–water mixture decreases due to pore-fluid leakage. As this happens, increased grain-to-grain and grain-to-bed friction at a flow's edge increase and can bring flow to a halt (Major 2000; Major and Iverson 1999). The still-liquefied interior of the flow body is effectively dammed by a highly frictional rim of generally coarser, nonliquefied particles. However, loss of excess pore-fluid pressure by fluid drainage from within the main body of a VDF cannot occur fast enough to frictionally stop flows on steeper slopes or deeper flows (Major 2000), but it can limit lateral spreading of VDFs on broad surfaces through formation of coarse lateral levees (Johnson et al. 2012) that confine flow along its lateral edge and allow a VDF to flow farther than it might otherwise (e.g., Pierson 1980).

Deposition mechanisms in VHFs have not been studied in detail, but physical experiments and field observations suggest that VHFs deposit particles mainly by a two-stage progressive aggradation that involves (a) suspended-load fallout onto a mobile bed, and then (b) accretion

Table 6 Proposed mechanisms by which lahars—volcanic debris flows and hyperconcentrated flows—segregate particles and emplace deposits

Flow type	Mechanism	Description	Deposit characteristics	References
Debris flow	Progressive aggradation at bed through incremental accretion of particles	Suspended particles accreted progressively onto bed surface by the overpassing flow as flow slows and shear forces diminish at lower flow boundary	Massive; poorly sorted and commonly graded; vertical section of flow unit records sequential deposition of head, middle, and tail of lahar body	Vallance (1994, 2000), Vallance and Scott (1997), Major (1997, 2000), Major and Iverson (1999), Vallance and Iverson (2015)
Debris flow	Segregation and emplacement of coarse clasts at flow margins without interstitial matrix	Coarsest clasts in flow migrate to flow surface by kinetic sieving and squeeze expulsion; higher flow velocity at surface selectively moves coarse clasts to front and lateral margins, where they are concentrated	Massive, open-work gravel forming flow levees and lobate margins at flow fronts and margins	Suwa and Okuda (1983), Pierson (1986), Suwa (1988), Major and Iverson (1999), Gray and Ancey (2009), Johnson et al. (2012)
Debris flow	Edge-friction damming	Liquefied slurry impounded behind highly frictional flow fronts and lateral margins acting as dynamic debris dams	Massive; poorly sorted and possibly normally graded deposit interiors with better sorted and commonly coarser margins	Vallance (1994), Major (1997, 2000), Major and Iverson (1999)
Debris flow and hyperconcentrated flow	Stranding of coarse bed load particles	Bed load particles stop rolling or sliding along bed when fluid drag forces unable to keep them moving; clasts may pile up against each other	Within deposits: gravel-size clasts or clast clusters suspended in much finer sediment On deposit surfaces: gravel clasts rest on or protrude out of deposit surface	Dal Cin (1968), Brayshaw (1984), Pierson et al. (1990); Pierson (2005)
Hyperconcentrated flow	Suspension fallout	Individual particles settle out of suspension due to decreased turbulence in low-velocity regions of flow	Nonstratified to faintly stratified; moderately sorted; silty sand flow units that grade into other textures	Cronin et al. (2000), Pierson (2005)
Hyperconcentrated flow	Traction-carpet accretion	Particles and clusters/sheets of particles move along bed by flow drag force; halt when frictional resistance at bed is too great	Weakly stratified layers in relatively uniform sediment	Sohn (1997)
Hyperconcentrated flow	Grounded pumice mats	Floating pumice lowered to bed surface as flow wanes	Layers of pumice sandwiched between beds of denser sand	Vallance and Iverson (2015)

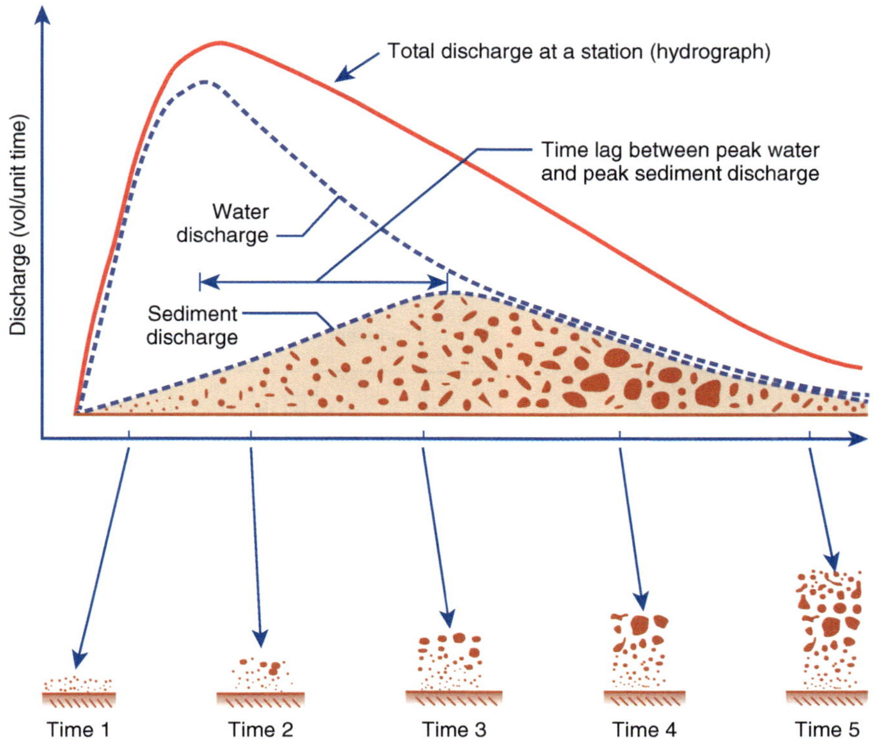

Fig. 22 Schematic hydrograph of a volcanic debris flow (VDF) showing how deposition would occur at one point along the channel by progressive aggradation (also described as incremental particle accretion) during passage of such a lahar (modified from Vallance 1994). This case assumes that the front of the lahar is finer grained than in the main body of the lahar (in other cases the lahar front could contain coarser material). The head of the lahar, depositing sediment at time 1 (indicated by arrow), has low solids concentration and fine particle size. Therefore, the sediment accreted at that time at the base of the eventual deposit is fine grained. At times 2 and 3, both solids concentration and maximum particle size are increasing, so the accreting deposit on the bed becomes progressively coarser in the upward direction. Time 4 occurs when the coarsest part of the lahar is accreting its sediment. Thereafter, sediment concentration and maximum particle size decrease, and the deposit becomes progressively finer grained as the tail of the lahar passes. The result of this depositional scenario is a deposit that is inversely graded in its lower part, coarsest in the middle, and normally graded in its upper part. Different distributions of coarse particles in a lahar result in different grading patterns. A lahar with a relatively uniform size distribution of particles from the lahar's head to its tail emplaces an ungraded deposit

of particles (and sheets of particles) within the mobile bed onto an aggrading static bed surface (Hanes and Bowen 1985; Drake 1990; Sohn 1997; Pierson 2005) (Fig. 23). This general model suggests that particle fallout in VHFs is somewhat random but favors coarser and denser particles, especially when flow slows, concentration decreases, or turbulence dampens. The mobile bed, also termed the *traction carpet*, is a layer of particles in frictional contact with each other, sandwiched between the static bed and the turbulent overpassing fluid; intense bed shear stress is applied to it by the overpassing fluid (Hanes and Bowen 1985; Todd 1989; Sohn 1997; Pierson 2005). Particles are added to the mobile bed by suspension fallout from above and subtracted by accretion to the static bed below. Interfaces between the overpassing fluid and the mobile bed and between the mobile bed and the static bed are gradational. In flows with abundant pumice, the pumice particles commonly float (as mats or rafts) on the flow surface, the fluid being denser than water. Deposition of pumice rafts occurs by grounding

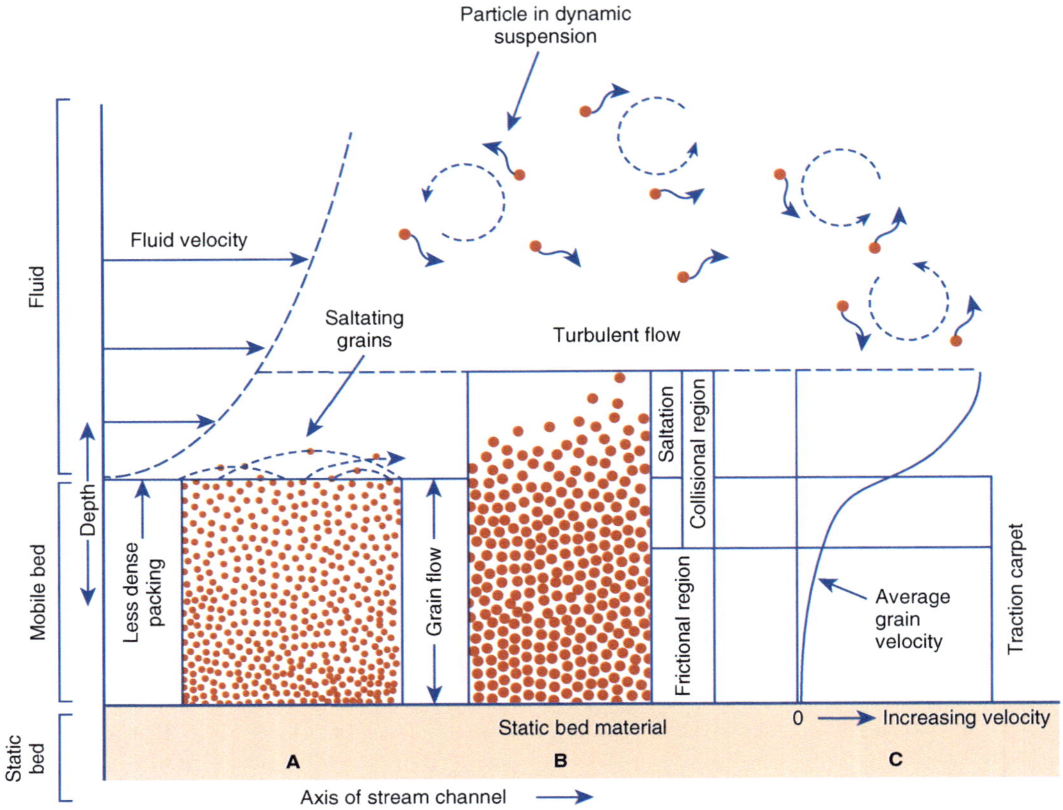

Fig. 23 Schematic longitudinal cross-section view of a concentration-stratified (as in Fig. 16) particle–fluid flow in a flume or channel involving turbulent fluid flow overriding and exerting shear stress on a more densely packed mobile layer of bed material—a *traction carpet* (Hanes and Bowen 1985; Sohn 1997). Field observations of shallow volcanic hyperconcentrated flows (VHFs) support this model for VHF sediment transport, where the boundary between the concentrated rapidly flowing turbulent fluid and the slower movement of the traction carpet is gradual and indistinct (Pierson 2005). The nearly continuous grain contacts and downward densification of a packed bed-material layer can also be driven by gravity alone—dry granular flow (e.g., Drake 1990), or possibly by a combination of gravity and applied shear (Manville et al. 2000). Both models result in similar bed-material layers (models A and B in the diagram). Mobile bed model A (Hanes and Bowen 1985) postulates an internally shearing, high-concentration, collision-dominated particle–fluid flow region that grades downward into a highly frictional zone. Model B (Drake 1990) is based on high-speed photography of dry grain flow experiments, wherein **a** densely packed grains move very little in a quasi-static basal zone (lower part of frictional region), **b** less densely packed grains move together in highly frictional packets of particles in the block-gliding zone (upper part of frictional region), and **c** loosely packed grains in the lower collisional region have significant particle–particle interactions that grade upward to a zone of saltating particles in the upper collisional region. Model C is an interpretation of particle velocities at different levels in the mobile bed (traction carpet)—a synthesis of models A and B (Sohn 1997). Deposition by progressive accretion at the interface between the static bed material and the traction carpet subtracts particles from the traction carpet and adds them to the bed. Fallout of particles from the overpassing fluid adds sediment to the top of the traction carpet

Fig. 24 A fill terrace on far side of the upper Sandy River, about 8 km downstream of the summit of Mount Hood (Oregon, USA). Deposits exposed in cut-bank include those of volcanic debris flows, volcanic hyperconcentrated flows, and volcanic water floods. Terrace height above the stream is 3–4 m. USGS photo by T. Pierson

as discharge wanes and flow retreats off broad surfaces into channels, producing well-developed pumice layers that become sandwiched between beds of lithic sand (Pierson et al. 1996; Vallance and Iverson 2015).

6.2 Sites of Lahar Deposition

Lahar deposits, particularly those emplaced by moderate to very large lahars, are typically found in channels and on debris fans and debris aprons on the lower flanks of volcanic edifices where slopes are less than about 2–3° (Rodolfo 1989; Pierson 1995; Pierson et al. 1996; Vallance and Scott 1997). On the lowest-gradient surfaces and channel beds, deposits typically form broad sheets or bank-to-bank channel fills. Fill-deposit surfaces are relatively smooth and flat if the particles transported were no coarser than fine gravel. Lahars transporting cobbles and boulders have rough surfaces, with those large clasts protruding up out of the finer-grained deposit. However, bank-to-bank channel fills deposited by the main body of a lahar are commonly incised by the more dilute and more turbulent tail of the same lahar. This initial downcutting and bank erosion is quickly augmented by further channel downcutting and widening post-lahar stream erosion. This erosion typically results in terraces of fresh lahar deposits formed along channel banks or valley sides—*fill terraces*. Fill terraces in volcanic terrain are commonly composed of deposits from multiple

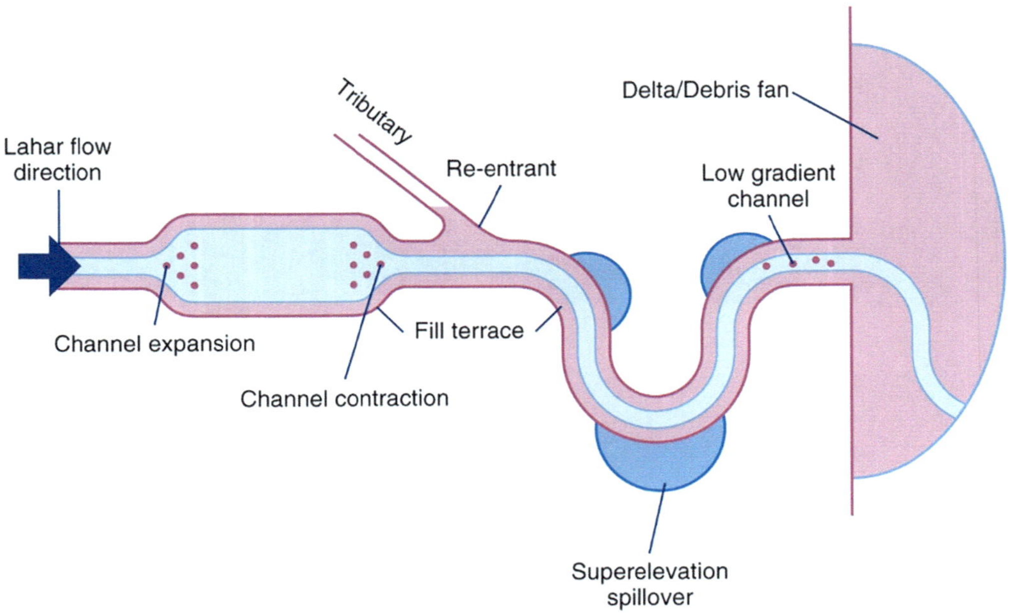

Fig. 25 Locations where lahars flowing down river valleys are most likely to emplace deposits

flow events (both lahars and water floods), formed when new lahars and floods spill onto terrace surfaces (Fig. 24).

Not all deposition by relatively large lahars occurs on slopes less than 2–3°. In steeper channels, deposition occurs mainly along channel banks or valley sides and on terrace surfaces, where flow is shallower (and channel roughness is typically higher) than along the thalweg (the main axis of flow in a channel). In such steeper channels, deposits along the thalweg are patchy and thin. Deposition also occurs in steeper channel reaches and on steeper slopes where localized topography facilitates deposition (e.g., Procter et al. 2010). Small VDFs, particularly those with coarse, highly frictional flow fronts, typically emplace the bulk of their deposits on steeper slopes than do larger flows (e.g., Suwa and Okuda 1983; O'Connor et al. 2001).

Regardless of location on a volcano, channel pattern and morphology influence lahar hydraulics in ways that promote localized deposition by slowing flow velocity, decreasing turbulence, and decreasing flow depth (e.g., Pierson and Scott 1985; Pierson et al. 1990; Procter et al. 2010). Lahar deposition is especially common in certain geomorphic settings in river valleys (Fig. 25), where:

- Channel/fan slope has decreased to a critical gradient (for that particle–fluid mixture).
- Channel sinuosity increases form roughness.
- Bed topography or floodplain vegetation increases bed roughness.
- Flow backs up and stagnates in tributary mouths.
- Eddies cause flow stagnation behind obstructions.
- Channel constrictions or obstructions induce hydraulic damming (velocity decreases).
- Channel expansions cause a critical decrease in flow depth.
- Channelized flow in tight channel bends superelevates and spills out onto terrace surfaces.

Thus, such sites are good places to explore for lahar deposits.

References

Allstadt K (2013) Extracting source characteristics and dynamics of the August 2010 Mount Meager landslide from broadband seismograms. J Geophys Res Earth Surf 118(3):1472–1490. https://doi.org/10.1002/jgrf.20110

Arboleda RA, Martinez MML (1996) 1992 lahars in the Pasig-Potrero River system. In: Newhall CG, Punongbayan RS (eds) Fire and mud: eruptions and lahars of Mount Pinatubo, Philippines. Philippine Institute of Volcanology and Seismology, Quezon City, University of Washington Press, Seattle, pp 1045–1052

Bates RL, Jackson JA (eds) (1987) Glossary of geology. American Geological Institute, Alexandria, Virginia, 788 pp

Berti M, Simoni A (2007) Prediction of debris flow inundation areas using empirical mobility relationships. Geomorphology 90:144–161

Best JL (1992) Sedimentology and event timing of a catastrophic volcaniclastic mass flow, Volcan Hudson, Southern Chile. Bull Volcanol 54:299–318

Beverage JP, Culbertson JK (1964) Hyperconcentrations of suspended sediment. J Hydraulics Div Proc Am Soc Civ Eng 90:117–128

Bradley JB (1986) Hydraulics and bed material transport at high fine suspended sediment concentrations. Dissertation, Colorado State University, Fort Collins, 140 pp

Brayshaw AC (1984) Characteristics and origin of cluster bedforms in coarse-grained alluvial channels. In: Koster EH, Steel RJ (eds) Sedimentology of gravels and conglomerates. Can Soc Pet Geol Mem 10:77–85

Bursik M, Reid J (2004) Lahar in Glass Creek and Owens River during the Inyo eruption, Mono-Inyo Craters, California. J Volcanol Geotherm Res 131:321–331

Cao R, Qian S (1990) Sediment transport characteristics of hyperconcentrated flow with suspended load. In: French RH (ed) Proceedings of the international symposium on hydraulics/hydrology of arid lands, San Diego, California, 30 July–2 August 1990. American Society of Civil Engineers, New York, pp 657–662

Capra L, Macías JL (2002) The cohesive Naranjo debris-flow deposit (10 km^3): a dam breakout flow derived from the Pleistocene debris-avalanche deposit of Nevado de Colima Volcano (México). J Volcanol Geotherm Res 117(1–2):213–235. https://doi.org/10.1016/S0377-0273(02)00245-7

Copeland EA (2009) Recent periglacial debris flows from Mount Rainier, Washington. MS thesis, Oregon State University. https://ir.library.oregonstate.edu/concern/graduate_thesis_or_dissertations/jm214r90j#:~:text=https%3A//ir.library.oregonstate.edu/concern/graduate_thesis_or_dissertations/jm214r90j

Costa JE (1988) Rheologic, geomorphic, and sedimentologic differentiation of water floods, hyperconcentrated flows, and debris flows. In: Baker VR, Kochel RC, Patton PC (eds) Flood geomorphology. Wiley, New York, pp 113–122

Costa JE, Williams GP (1984) Debris-flow dynamics (video). U S Geol Surv Open-File Rep 84-606, 22.5 min

Crandell DR (1971) Postglacial lahars from Mount Rainier Volcano, Washington. U S Geol Surv Prof Pap 677:1–73

Cronin SJ, Neall VE, Lecointre JA, Palmer AS (1996) Unusual "snow slurry" lahars from Ruapehu volcano, New Zealand, September 1995. Geology 24:1107–1110

Cronin SJ, Neall VE, Palmer AS, Lecointre JA (1997) Changes in Whangaehu River lahar characteristics during the 1995 eruption sequence, Ruapehu volcano, New Zealand. J Volcanol Geotherm Res 76:47–61

Cronin SJ, Neall VE, Lecointre JA, Palmer AS (1999) Dynamic interactions between lahars and stream flow: a case study from Ruapehu volcano, New Zealand. Geol Soc Am Bull 111:28–38

Cronin SJ, LeCointre JA, Palmer AS, Neall VE (2000) Transformation, internal stratification, and depositional processes within a channelized, multi-peaked lahar flow. New Zeal J Geol Geophys 43:117–128

Dal Cin R (1968) "Pebble clusters": their origin and utilization in the study of palaeocurrents. Sediment Geol 2(4):233–241

de Bélizal E, Lavigne F, Hadmoko DS, Degeai J-P, Dipayana GA, Mutaqin BW, Marfai MA, Coquet M, Le Mauff B, Robin A-K, Vidal C, Cholik N, Aisyah N (2013) Rain-triggered lahars following the 2010 eruption of Merapi volcano, Indonesia: a major risk. J Volcanol Geotherm Res 261:330–347. https://doi.org/10.1016/j.jvolgeores.2013.01.010

de Blasio, Engvik L, Harbitz CB, Elverhøi A (2004) Hydroplaning and submarine debris flows. J Geophys Res Oceans 109(C1). https://doi.org/10.1029/2002JC001714

Dinehart RL (1999) Sediment transport in the hyperconcentrated phase of the March 19, 1982, lahar. In: Pierson TC (ed) Hydrologic consequences of hot-rock/snowpack interactions at Mount St. Helens volcano, Washington, 1982–84. U S Geol Surv Prof Pap 1586:37–52

Dinehart RL, Ritter JR, Knott JM (1981) Sediment data for streams near Mount St. Helens, Washington, v. 1—1980 water year data. U S Geol Surv Open-File Rep 81-822, 20 pp

Dorava JM, Meyer DF (1994) Hydrologic hazards in the lower Drift River basin associated with the 1989–1990 eruption of Redoubt Volcano, Alaska. In: Miller TP, Chouet BA (eds) The 1989–1990 eruptions of Redoubt Volcano, Alaska. J Volcanol Geotherm Res 62:387–407

Drake TG (1990) Structural features in granular flows. J Geophys Res Solid Earth 95(B6):8681–8696

Druitt TH (1995) Settling behaviour of concentrated dispersions and some volcanological applications. J

Volcanol Geotherm Res 65(1–2):27–39. https://doi.org/10.1016/0377-0273(94)00090-4
Escher BG (1922) On the hot "lahar" (mud flow) of the Valley of Ten Thousand Smokes (Alaska). Proc Sect Sci Koninklijke Akademie Wetenschappen 24(1–7):282–293
Estep J, Dufek J (2013) Discrete element simulations of bed force anomalies due to force chains in dense granular flows. J Volcanol Geotherm Res 254:108–117. https://doi.org/10.1016/j.jvolgeores.2012.12.023
Fagents SA, Baloga SM (2006) Toward a model for the bulking and debulking of lahars. J Geophys Res 111:B10201. https://doi.org/10.1029/2005JB003986
Fairchild LH (1985) Lahars at Mount St. Helens, Washington. Dissertation, University of Washington, 374 pp
Fairchild LH (1987) The importance of lahar initiation process. In: Costa JE, Wieczorek GF (eds) Debris flows/avalanches: process, recognition, and mitigation. Geol Soc Am Rev Eng Geol 7:51–62
Fisher RV, Schmincke H-U (1984) Pyroclastic rocks. Springer-Verlag, Berlin, p 472
Gallino GL, Pierson TC (1985) Polallie Creek debris flow and subsequent dam-break flood of 1980, East Fork Hood River Basin, Oregon. U S Geol Surv Water-Supply Pap 2273, 22 pp
George DL, Iverson RM, Cannon CM (2022) Modeling the dynamics of lahars that originate as landslides on the west side of Mount Rainier, Washington. U S Geol Surv Open-File Rep 2021-1118. https://doi.org/10.3133/ofr20211118
Gerson R (1977) Sediment transport for desert watersheds in erodible materials. Earth Surf Process 2:343–361
Glicken H (1996) Rockslide debris avalanche of May 18, 1980, Mount St. Helens Volcano, Washington. U S Geol Surv Open-File Rep 96-677, 90 pp and 5 plates
Graettinger AH, Manville V, Briggs RM (2010) Depositional record of historic lahars in the upper Whangaehu Valley, Mt. Ruapehu, New Zealand: implications for trigger mechanisms, flow dynamics and lahar hazards. Bull Volcanol 72:279–296. https://doi.org/10.1007/s00445-009-0318-2
Gray JMNT, Ancey C (2009) Segregation, recirculation and deposition of coarse particles near two-dimensional avalanche fronts. J Fluid Mech 629:387–423. https://doi.org/10.1017/S0022112009006466
Griswold JP, Iverson RM (2008) Mobility statistics and automated hazard mapping for debris flows and rock avalanches. U S Geol Surv Sci Invest Rep 2007-5276, 59 pp
Hall ML (1978) Descriptions of volcanic eruptions—Ecuador. Bull Vol Erupt (Volcanol Soc Jpn) 16:64–65
Hampton MA (1975) Competence of fine-grained debris flows. J Sediment Petrol 45:834–844
Hanes DM, Bowen AJ (1985) A granular-fluid model for steady intense bed-load transport. J Geophys Res 90:9149–9158
Harbitz CB, Parker G, Elverhøi A, Marr JG, Mohrig D, Harff PA (2003) Hydroplaning of subaqueous debris flows and glide blocks: analytical solutions and discussion. J Geophys Res Solid Earth 108:B7. https://doi.org/10.1029/2001JB001454
Hodgson KA, Manville VR (1999) Sedimentology and flow behavior of a rain-triggered lahar, Mangatoetoenui Stream, Ruapehu volcano, New Zealand. Geol Soc Am Bull 111:743–754
Hungr O (2000) Analysis of debris flow surges using the theory of uniformly progressive flow. Earth Surf Process Landf 25:483–495
Hungr O, McDougall S, Bovis M (2005) Entrainment of material by debris flows. In: Jakob M, Hungr O (eds) Debris-flow hazards and related phenomena. Praxis-Springer, Berlin, pp 135–158
Hupp CR, Osterkamp WR, Thornton J (1987) Dendrogeomorphic evidence and dating of recent debris flows on Mount Shasta, northern California. U S Geol Surv Prof Pap 1396B
Iverson RM (1997) The physics of debris flows. Rev Geophys 35:245–296
Iverson RM (2003) The debris–flow rheology myth. In: Rickenmann D, Chen CL (eds) Debris-flow hazards mitigation. Mechanics, prediction, and assessment (1). Millpress, Rotterdam, pp 303–314
Iverson RM (2005) Debris-flow mechanics. In: Jakob M, Hungr O (eds) Debris-flow hazards and related phenomena. Praxis-Springer, Berlin, pp 103–134
Iverson RM, LaHusen RG (1993) Friction in debris flows: inferences from large-scale flume experiments. In: Proceedings of American Society of Civil Engineers, hydraulic engineering conference, San Francisco, 25–30 July 1993, pp 1604–1609
Iverson RM, Reid ME, Logan M, LaHusen RG, Godt JW, Griswold JP (2011) Positive feedback and momentum growth during debris-flow entrainment of wet bed sediment. Nat Geosci 4:116–121. https://doi.org/10.1038/ngeo1040
Jakob M (2005) A size classification for debris flows. Eng Geol 79:151–161
Janda RJ, Scott KM, Nolan KM, Martinson HA (1981) Lahar movement, effects, and deposits. In: Lipman PW, Mullineaux DR (eds) The 1980 eruptions of Mount St. Helens, Washington. U S Geol Surv Prof Pap 1250:461–478
Johnson CG, Kokelaar BP, Iverson RM, Logan M, LaHusen RG, Gray JMNT (2012) Grain-size segregation and levee formation in geophysical mass flows. J Geophys Res Earth Surf 117:F01032. https://doi.org/10.1029/2011JF002185
Johnson PJ, Valentine GA, Stauffer PH, Lowry CS, Sonder I, Pulgarín BA, Santacoloma CC, Agudelo A (2018) Groundwater drainage from fissures as a source for lahars. Bull Volcanol 80:39. https://doi.org/10.1007/s00445-018-1214-4
Kilgour G, Manville V, Della Pasqua F, Graettinger A, Hodgson KA, Jolly GE (2010) The 25 September 2007 eruption of Mount Ruapehu, New Zealand:

directed ballistics, surtseyan jets, and ice-slurry lahars. J Volcanol Geotherm Res 191:1–14. https://doi.org/10.1016/j.jvolgeores.2009.10.015

Kjartansson G (1951) Water flood and mudflows. In: Einarsson T, Kjartansson G, Thorarinsson S (eds) The eruption of Hekla 1947–48. Visindafelag Islending (Soc Sci Islandica), Reykjavik II(4):1–51

Komar PD (1988) Sediment transport by floods. In: Baker VR, Kochel RC, Patton PC (eds) Flood geomorphology. Wiley-Interscience, New York, pp 97–111

Lavigne F, Suwa H (2004) Contrasts between debris flows, hyperconcentrated flows and stream flows at a channel of Mount Semeru, East Java, Indonesia. Geomorphology 61:41–58

Lavigne F, Thouret J-C (2002) Sediment transportation and deposition by rain-triggered lahars at Merapi Volcano, Central Java, Indonesia. Geomorphology 49:45–69

Legg NT, Meigs AJ, Grant GE, Kennard P (2014) Debris flow initiation in proglacial gullies on Mount Rainier, Washington. Geomorphology 226:249–260. https://doi.org/10.1016/j.geomorph.2014.08.003

Lowe DR, Williams SN, Leigh H, Connor CB, Gemmell JB, Stoiber RE (1986) Lahars initiated by the 13 November 1985 eruption of Nevado del Ruiz, Colombia. Nature 324:51–53

Lube G, Cronin SJ, Procter JN (2009) Explaining the extreme mobility of volcanic ice-slurry flows, Ruapehu volcano, New Zealand. Geology 37:15–18. https://doi.org/10.1130/G25352A.1

Maizels J (1989) Sedimentology, paleoflow dynamics and flood history of jökulhlaup deposits: paleohydrology of Holocene sediment sequences in southern Iceland sandur deposits. J Sediment Petrol 59:204–223

Major JJ (1997) Depositional processes in large-scale debris-flow experiments. J Geol 105:345–366

Major JJ (2000) Gravity-driven consolidation of granular slurries—implications for debris-flow deposition and deposit characteristics. J Sediment Res 70:64–83

Major JJ (2003) Hindered settling. In: Middleton GV, Church MJ, Coniglio M, Hardie LA, Longstaffe FJ (eds) Encyclopedia of sediments and sedimentary rocks. Kluwer Academic Publishers, Dordrecht, pp 358–360

Major JJ (2023) Subaerial volcaniclastic deposits—influences of initiation mechanisms and transport behaviour on characteristics and distributions. In: Di Capua A, De Rosa R, Kereszturi G, Le Pera E, Rosi M, Watt SFL (eds) Volcanic processes in the sedimentary record: when volcanoes meet the environment. Geol Soc Lond Spec Publ 520:29–100. First published online 12 July 2022. https://doi.org/10.1144/SP520-2021-142

Major JJ, Iverson RM (1999) Debris-flow deposition: effects of pore-fluid pressure and friction concentrated at flow margins. Geol Soc Am Bull 111:1424–1434

Major JJ, Newhall CG (1989) Snow and ice perturbation during historical volcanic eruptions and the formation of lahars and floods—a global review. Bull Volcanol 52:1–27

Major JJ, Janda RJ, Daag AS (1996) Watershed disturbance and lahars on the east side of Mount Pinatubo during the mid-June 1991 eruptions. In: Newhall CG, Punongbayan RS (eds) Fire and mud: eruptions and lahars of Mount Pinatubo, Philippines. Philippine Institute of Volcanology and Seismology, Quezon City, University of Washington Press, Seattle, pp 895–919

Major JJ, Pierson TC, Dinehart RL, Costa JE (2000) Sediment yield following severe volcanic disturbance—a two-decade perspective from Mount St. Helens. Geology 28:819–822

Major JJ, Schilling SP, Pullinger CR, Escobar CD (2004) Debris-flow hazards at San Salvador, San Vicente, and San Miguel volcanoes, El Salvador. In: Rose WI, Bommer JJ, Lopez DL, Carr MJ, Major JJ (eds) Natural hazards in El Salvador. Geol Soc Am Spec Pap 375:89–108

Manville V (2010) An overview of break-out floods from intracaldera lake. Glob Planet Change 70(1–4):14–23

Manville V, Cronin SJ (2007) Breakout lahar from New Zealand's Crater Lake. Eos Trans Am Geophys Union 88(43):441–442

Manville V, White JDL, Hodgson KA (2000) Dynamic interactions between lahars and stream flow: a case study from Ruapehu volcano, New Zealand: discussion and reply discussion. Geol Soc Am Bull 112(7):1149–1152

Marr JG, Harff PA, Shanmugam G, Parker G (2001) Experiments on subaqueous sandy gravity flows: the role of clay and water content in flow dynamics and depositional structures. Geol Soc Am Bull 113(11):1377–1386. https://doi.org/10.1130/0016-7606(2001)113<1377:EOSSGF>2.0.CO;2

Marren PM, Russsel AJ, Rushmer EL (2009) Sedimentology of a sandur formed by multiple jökulhlaups, Kverkfjöll, Iceland. Sediment Geol 213:77–88

Martinez MML, Arboleda RA, Delos Reyes PJ, Gabinete E, Dolan MT (1996) Observations of 1992 lahars along the Sacobia-Bamban River system. In: Newhall CG, Punongbayan RS (eds) Fire and mud: eruptions and lahars of Mount Pinatubo, Philippines. Philippine Institute of Volcanology and Seismology, Quezon City, University of Washington Press, Seattle, pp 1033–1044

Minakami T, Ishikawa T, Yagi K (1951) The 1944 eruption of volcano Usu in Hokkaido, Japan. Bull Volcanol 11:45–157

Mizuyama T, Kobashi S, Ou G (1992) Prediction of debris flow peak discharge. In: Proceedings of interpraevent international symposium, vol 4. Bern, pp 99–108

Mohrig D, Ellis C, Parker G, Whipple KX, Hondzo M (1998) Hydroplaning of subaqueous debris flows. Geol Soc Am Bull 110(3):387–394. https://doi.org/10.1130/0016-7606(1998)110<0387:HOSDF>2.3.CO;2

Mothes PA, Hall ML, Janda RJ (1998) The enormous Chillos Valley lahar: an ash-flow-generated debris flow from Cotopaxi volcano, Ecuador. Bull Volcanol 59:233–244

Nairn IA, Wood CP, Hewson CAY (1979) Phreatic eruptions of Ruapehu: April 1975. New Zeal J Geol Geophys 22:155–173

Naranjo JL, Sigurdsson H, Carey SN, Fritz WJ (1986) Eruption of the Nevado del Ruiz volcano, Colombia, on 13 November 1985: tephra fall and lahars. Science 233:961–963

Neall VE (1976) Lahars: global occurrence, origins, and annotated bibliography. Publ Geol Dep Victoria Univ Wellington 5:1–18

Neuendorf KKE, Mehl JP, Jackson JA (2005) Glossary of geology, 5th edn. American Geological Institute, Alexandria, VA, 779 pp

O'Connor JE, Hardison JH III, Costa JE (2001) Debris flows from failures of Neoglacial-age moraine dams in the Three Sisters and Mt. Jefferson Wilderness Areas, Oregon. U S Geol Surv Prof Pap 1606:1–93

O'Shea BE (1954) Ruapehu and the Tangiwai disaster. New Zeal J Sci Technol B36:174–189

Okuda S, Suwa H, Okunishi K, Yokoyama K (1981) Depositional processes of debris flow at Kamikamihori fan, Northern Japan Alps. Jpn Geomorphol Union Trans 2:353–361

Osterkamp WR, Hupp CR, Blodgett JC (1986) Magnitude and frequency of debris flows and areas of hazard on Mount Shasta, Northern California. U S Geol Surv Prof Pap 1396-C, 21 pp

Oswalt JS, Nichols W, O'Hara JF (1996) Meteorological observations of the 1991 Mount Pinatubo eruption. In: Newhall CG, Punongbayan RS (eds) Fire and mud: eruptions and lahars of Mount Pinatubo, Philippines. Philippine Institute of Volcanology and Seismology, Quezon City, University of Washington Press, Seattle, pp 625–636

Oyagi N (1987) The 1984 Ontake-san landslide and its movement. Trans Jpn Geomorphol Union 8(2):127–144

Paguican EMR, Lagmay AMF, Rodolfo KS, Rodolfo RS, Tengonciang AMP, Lapus MR, Ballatan EG, Obille EC Jr (2009) Extreme rainfall-induced lahars and dike breaching, 30 November 2006, Mayon Volcano, Philippines. Bull Volcanol 71:845–857

Pierson TC (1980) Erosion and deposition by debris flows at Mt Thomas, North Canterbury, New Zealand. Earth Surf Process 5:227–247

Pierson TC (1985) Initiation and flow behavior of the 1980 Pine Creek and Muddy River lahars, Mount St. Helens, Washington. Geol Soc Am Bull 96:1056–1069. https://doi.org/10.1130/0016-7606(1985)96<1056:IAFBOT>2.0.CO;2

Pierson TC (1986) Flow behavior of channelized debris flows, Mount St. Helens, Washington. In: Abrahams AD (ed) Hillslope processes. Allen Unwin, Boston, pp 269–296

Pierson TC (1995) Flow characteristics of large eruption-triggered debris flows at snow-clad volcanoes: constraints for debris-flow models. J Volcanol Geotherm Res 66:283–294

Pierson TC (1998) An empirical method for estimating travel times for wet volcanic mass flows. Bull Volcanol 60:98–109

Pierson TC (1999) Transformation of water flood to debris flow following the eruption-triggered transient-lake breakout from the crater on March 19, 1982. In: Pierson TC (ed) Hydrologic consequences of hot-rock/snowpack interactions at Mount St. Helens volcano, Washington 1982–84. U S Geol Surv Prof Pap 1586:19–36

Pierson TC (2005) Hyperconcentrated flows—transitional process between water flow and debris flow, chap 8. In: Jakob M, Hungr O (eds) Debris-flow hazards and related phenomena. Praxis-Springer, Berlin, pp 159–202

Pierson TC, Major JJ (2014) Hydrogeomorphic effects of explosive volcanic eruptions on drainage basins. Annu Rev Earth Planet Sci 42:469–507. https://doi.org/10.1146/annurev-earth-060313-054913

Pierson TC, Scott KM (1985) Downstream dilution of a lahar: transition from debris flow to hyperconcentrated streamflow. Water Resour Res 21:1511–1524

Pierson TC, Janda RJ, Thouret JC, Borrero CA (1990) Perturbation and melting of snow and ice by the 13 November 1985 eruption of Nevado del Ruiz, Colombia, and consequent mobilization, flow, and deposition of lahars. J Volcanol Geotherm Res 41:17–66

Pierson TC, Daag AS, Delos Reyes PJ, Regalado MTM, Solidum RU, Tubianosa BS (1996) Flow and deposition of posteruption hot lahars on the east side of Mount Pinatubo, July–October 1991. In: Newhall CG, Punongbayan RS (eds) Fire and mud: eruptions and lahars of Mount Pinatubo, Philippines. Philippine Institute of Volcanology and Seismology, Quezon City, University of Washington Press, Seattle, pp 921–950

Pierson TC, Major JJ, Amigo Á, Moreno H (2013) Acute sedimentation response to rainfall following the explosive phase of the 2008–2009 eruption of Chaitén volcano, Chile. Bull Volcanol 75:723. https://doi.org/10.1007/s00445-013-0723-4

Pringle PT, Cameron KA (1999) Eruption-triggered lahar on May 14, 1984. In: Pierson TC (ed) Hydrologic consequences of hot-rock/snowpack interactions at Mount St. Helens volcano, Washington 1982–84. U S Geol Surv Prof Pap 1586:81–103

Procter J, Cronin SJ, Fuller IC, Lube G, Manville V (2010) Quantifying the geomorphic impacts of a lake-breakout lahar, Mount Ruapehu, New Zealand. Geology 38:67–70

Qian Y, Yang W, Zhao W, Cheng X, Zhang L, Xu W (1981) Basic characteristics of flow with hyperconcentration of sediment. In: Proceedings of international symposium on river sedimentation. Chinese Hydraulic Engineering Society, Beijing, pp 175–184

Reid ME, Keith TEC, Kayen RE, Iverson NR, Iverson RM, Brien DL (2010) Volcano collapse promoted by progressive strength reduction: new data from Mount St. Helens. Bull Volcanol 72:761–766. https://doi.org/10.1007/s00445-010-0377-4

Richardson D (1968) Glacier outburst floods in the Pacific Northwest. U S Geol Surv Prof Pap 600-D:79–86

Rickenmann D (1991) Hyperconcentrated flow and sediment transport at steep slopes. J Hydraul Eng 117:1419–1439

Rodolfo KS (1989) Origin and early evolution of lahar channel at Mabinit, Mayon Volcano, Philippines. Geol Soc Am Bull 101:414–426

Rodolfo KS, Arguden AT (1991) Rain-lahar generation and sediment-delivery systems at Mayon volcano, Philippines. In: Fisher RV, Smith GA (eds) Sedimentation in volcanic settings. SEPM (Soc Sediment Geol) Spec Publ 45:71–87

Rodolfo KS, Umbal JV, Alonso RA, Remotigue CT, Paladio-Melosantos ML, Salvador JHG, Evangelista D, Miller Y (1996) Two years of lahars on the western flank of Mount Pinatubo: initiation, flow processes, deposits, and attendant geomorphic and hydraulic changes. In: Newhall CG, Punongbayan RS (eds) Fire and mud: eruptions and lahars of Mount Pinatubo, Philippines. Philippine Institute of Volcanology and Seismology, Quezon City, University of Washington Press, Seattle, pp 989–1013

Russell AJ, Knudsen Ó (1999) Controls on the sedimentology of the November 1996 jökulhlaup deposits, Skeiðarársandur, Iceland. In: Smith ND, Rogers J (eds) Fluvial sedimentology VI. International Association of Sedimentologists, pp 315–329

Schmidt KG (1934) Die Schuttströme am Merapi auf Java nach dem Ausbruch von 1930. De Ingenieur Nederlandsch-Indië 1(7):91–120 (parts 1–7), 1(8):123–134 (parts 8–9), 1(9):143–171 (parts 10–12) (in German)

Scott KM (1988a) Origins, behavior, and sedimentology of lahars and lahar-runout flows in the Toutle-Cowlitz River system. U S Geol Surv Prof Pap 1447-A, 74 pp

Scott KM (1988b) Origin, behavior, and sedimentology of prehistoric catastrophic lahars at Mount St. Helens, Washington. In: Clifton HE (ed) Sedimentologic consequences of convulsive geologic events. Geol Soc Am Spec Pap 229:23–36. https://doi.org/10.1130/SPE229-p23

Scott KM (1989) Magnitude and frequency of lahars and lahar-runout flows in the Toutle-Cowlitz River system. U S Geol Surv Prof Pap 1447-B, 33 pp

Scott KM, Wang Y (2004) Debris flows—geologic process and hazard—illustrated by a surge sequence at Jiangjia Ravine, Yunnan, China. U S Geol Surv Prof Pap 1671, 30 pp, with accompanying video or DVD

Scott KM, Vallance JW, Pringle PP (1995) Sedimentology, behavior, and hazards of debris flows at Mount Rainier, Washington. U S Geol Surv Prof Pap 1547, 56 pp

Scott KM, Macías JL, Naranjo JA, Rodríguez S, McGeehin JP (2001) Catastrophic debris flows transformed from landslides in volcanic terrains: mobility, hazard assessment, and mitigation strategies. U S Geol Surv Prof Pap 1630, 59 pp

Scott KM, Vallance JW, Kerle N, Macías JL, Strauch W, Devoli G (2005) Catastrophic precipitation-triggered lahar at Casita volcano, Nicaragua: occurrence, bulking and transformation. Earth Surf Process Landf 30:59–79. https://doi.org/10.1002/esp.1127

Scrivenor JB (1929) The mudstreams ("lahars") of Gunong Keloet in Java. Geol Mag 66:433–434

Simons DB, Richardson EV, Haushild WL (1963) Studies of flow in alluvial channels, some effects of fine sediment on flow phenomena. U S Geol Surv Water-Supply Pap 1498-G, 46 pp

Smith GA (1986) Coarse-grained nonmarine volcaniclastic sediment: terminology and depositional process. Geol Soc Am Bull 97:1–10

Smith GA, Fritz WJ (1989) Volcanic influences on terrestrial sedimentation. Geology 17:375–376

Smith GA, Lowe DR (1991) Lahars: volcano-hydrologic events and deposition in the debris flow—hyperconcentrated flow continuum. In: Fisher RV, Smith GA (eds) Sedimentation in volcanic settings. SEPM (Soc Sediment Geol) Spec Publ 45:59–70

Sohn YK (1997) On traction-carpet sedimentation. J Sediment Res 67:502–509

Suwa H (1988) Focusing mechanism of large boulders to a debris-flow front. Trans Jpn Geomorphol Union 9:151–178

Suwa H, Okuda S (1983) Deposition of debris flows on a fan surface, Mt. Yakedake, Japan. Z Geomorphol 46:79–101

Suwa H, Yamakoshi T (1999) Sediment discharge by storm runoff at volcanic torrents affected by eruption. Z Geomorphol Suppl Bd 114:63–88

Suwa H, Manaka T, Inaniwa A (1989) Occurrence of debris flows and their scales in the Kamikamihori valley of Mount Yakedake. Annu Disaster Prev Res Inst Koto Univ 32:229–247

Thouret J-C (1990) Effects of the November 1985 eruption on the snowpack and ice cap of Nevado del Ruiz volcano, Colombia. J Volcanol Geotherm Res 41:177–201

Thouret J-C, Ramírez CJ, Gibert-Malengreau B, Vargas CA, Naranjo JL, Vandemeulebrouck J, Valla F, Funk M (2007) Volcano–glacier interactions on composite cones and lahar generation: Nevado del Ruiz, Colombia, case study. Annu Glaciol 45:115–127. https://doi.org/10.3189/172756407782282589

Todd SP (1989) Stream-driven, high-density gravelly traction carpets: possible deposits in the Trabeg Conglomerate Formation, SW Ireland and some theoretical considerations of their origin. Sedimentology 36(4):513–530. https://doi.org/10.1111/j.1365-3091.1989.tb02083.x

Tuñgol NM, Regalado MTM (1996) Rainfall, acoustic flow monitor records, and observed lahars of the Sacobia River in 1992. In: Newhall CG, Punongbayan RS (eds) Fire and mud: eruptions and lahars of Mount Pinatubo, Philippines. Philippine Institute of Volcanology and Seismology, Quezon City, University of Washington Press, Seattle, pp 1023–1032

Umbal J (1997) Five years of lahars at Pinatubo volcano: declining but still potentially lethal hazards. J Geol Soc Philipp 52:1–19

Umbal JV, Rodolfo KS (1996) The 1991 lahars of southwestern Mount Pinatubo and evolution of the lahar-dammed Mapanuepe Lake. In: Newhall CG, Punongbayan RS (eds) Fire and mud: eruptions and lahars of Mount Pinatubo, Philippines. Philippine Institute of Volcanology and Seismology, Quezon City, University of Washington Press, Seattle, pp 951–970

Vallance JW (1994) Experimental and field studies related to the behavior of granular mass flows and the characteristics of their deposits. Dissertation, Michigan Technological University, 197 pp

Vallance JW (2000) Lahars. In: Sigurdsson H, Houghton BF, McNutt SR, Rymer H, Stix J (eds) Encyclopedia of volcanoes. Academic Press, San Diego, pp 601–616

Vallance JW, Iverson RM (2015) Lahars and their deposits. In: Sigurdsson H, Houghton B, McNutt SR, Rymer H, Stix J (eds) The encyclopedia of volcanoes, 2nd edn. Elsevier, Amsterdam, pp 649–664. https://doi.org/10.1016/B978-0-12-385938-9.00037-7

Vallance JW, Scott KM (1997) The Osceola Mudflow from Mount Rainier: sedimentology and hazard implications of a huge clay-rich debris flow. Geol Soc Am Bull 109:143–163. https://doi.org/10.1130/0016-7606(1997)109<0143:TOMFMR>2.3.CO;2

Vallance JW, Scott KM, Pringle PT (1995) Sedimentology, behavior, and hazards of debris flows at Mount Rainier, Washington. U S Geol Surv Prof Pap 1547, 56 pp

Van Bemmelen RW (1949) The geology of Indonesia, 1A. Government Printing Office, The Hague

Voight B (1990) The 1985 Nevado del Ruiz volcano catastrophe: anatomy and retrospection. J Volcanol Geotherm Res 42(1–2):151–188

Voight B, Sousa J (1994) Lessons from Ontakesan: a comparative analysis of debris avalanche dynamics. Eng Geol 38:261–297. https://doi.org/10.1016/0013-7952(94)90042-6

Voight B, Glicken H, Janda RJ, Douglass PM (1981) Catastrophic rockslide avalanche of May 18. In: Lipman PW, Mullineaux DR (eds) The 1980 eruptions of Mount St. Helens, Washington. U S Geol Surv Prof Pap 1250:347–377

Voight B, Janda RJ, Glicken H, Douglass PM (1983) Nature and mechanics of the Mount St. Helens rockslide-avalanche of 18 May 1980. Geotechnique 33:243–273. https://doi.org/10.1680/geot.1983.33.3.243

Walder JS, Driedger CL (1994a) Geomorphic change caused by outburst floods and debris flows at Mount Rainier, Washington, with emphasis on Tahoma Creek valley. U S Geol Surv Water-Resour Invest Rep 93-4093:1–93

Walder JS, Driedger CL (1994b) Rapid geomorphic change caused by glacial outburst floods and debris flows along Tahoma Creek, Mount Rainier, Washington, USA. Arctic Alpine Res 26:319–327

Walder JS, Driedger CL (1995) Frequent outburst floods from South Tahoma glacier, Mount Rainier, U.S.A.; relation to debris flows, meteorological origin and implications for subglacial hydrology. J Glaciol 41(137):1–9. https://doi.org/10.3189/S0022143000017718

Waldron HH (1967) Debris flow and erosion control problems caused by the ash eruptions of Irazu volcano, Costa Rica. U S Geol Surv Bull 1241-I, 37 pp

Wan Z (1985) Bed material movement in hyperconcentrated flow. J Hydraul Eng 111:987–1002

Wan Z, Song T (1987) The effect of fine particles on vertical concentration distribution and transport rate of coarse particles. In: White WR (ed) Topics in fluvial hydraulics. Proceedings of the 22nd congress international association for hydraulic research. Ecole Polytechnique Federale, Lausanne, Switzerland, pp 80–85

Wan Z, Wang Z (1994) Hyperconcentrated flow. Balkema, Rotterdam, p 290

Wang Z, Larsen P (1994) Turbulent structure of water and clay suspensions with bed load. J Hydraul Eng 120:577–599

Watanabe M, Ikeya H (1981) Investigation and analysis of volcanic mud flows on Mt. Sakurajima, Japan. In: Proceedings of symposium on erosion and sediment transport measurement, Florence, Italy. Int Assoc Hydrol Sci Publ 133:245–256

Waythomas CF (2014) Water, ice, and mud: lahars and lahar hazards at ice- and snow-clad volcanoes. Geol Today 30(1):34–39

Waythomas CF, Pierson TC, Major JJ, Scott WE (2013) Voluminous ice-rich lahars generated during the 2009 eruption of Redoubt Volcano, Alaska. J Volcanol Geotherm Res 259:389–413

Weirich FH (1989) The generation of turbidity currents by subaerial debris flows, California. Geol Soc Am Bull 101(2):278–291. https://doi.org/10.1130/0016-7606(1989)101<0278:TGOTCB>2.3.CO;2

Wilcox AC, O'Connor JE, Major JJ (2014) Rapid reservoir erosion, hyperconcentrated flow, and

downstream deposition triggered by breaching of 38 m tall Condit Dam, White Salmon River, Washington. J Geophys Res Earth Surf 119:1376–1394. https://doi.org/10.1002/2013JF003073
Wilson CJN (1985) The Taupo eruption, New Zealand. II. The Taupo ignimbrite. Phil Trans R Soc Lond Ser A Math Phys Sci 314(1529):229–310. https://doi.org/10.1098/rsta.1985.0020
Xu J (2002) Implication of relationships among suspended sediment size, water discharge, and suspended sediment concentration, the Yellow River basin, China. CATENA 49:289–307
Xu J (2003) Hyperconcentrated flows in the lower Yellow River as influenced by drainage basin factors. Z Geomorphol 47:393–410
Yang H, Wei F, Hu K, Wang C (2014) Determination of the suspension competence of debris flows based on particle size analysis. Int J Sediment Res 29:73–81
Zanuttigh B, Lamberti A (2007) Instability and surge development in debris flows. Rev Geophys 45:RG3006

Lahar Deposits—Debris-Flow Type

Abstract

Concentrated lahars moving as debris flow are termed volcanic debris flows (VDFs). The unstratified deposits of individual VDFs commonly contain particles ranging from clay size to boulders and may contain blocks 10 m or more in diameter. Deposit volumes can exceed 1 km^3. Relatively small amounts of clay-size particles in VDF slurries play important roles in the mechanics of sediment transport and deposition, thus two categories of VDF deposits are recognized—those that are clay rich ($\geq$ 5 weight percent clay or about 2 volume percent) and those that are clay poor (< 5 weight percent). VDF deposits are common in volcanic sedimentary sequences, primarily at the bases of intermediate-composition volcanic edifices and in downstream river valleys. Clay-rich VDF deposits may be emplaced more than 100 km downstream of source. Their key characteristics include uniform color and sedimentary texture, extremely poor sorting, notably firm/hard consistencies in outcrop, bubble vesicles in matrix material, mainly normal grading, and intergranular voids found only at deposit margins. Clay-poor deposits are similar but more commonly found relatively close to source and exhibit very poor sorting, sandier matrix textures, variable grading of all types, slightly less firm/hard consistencies, and crude rounding and pockmarking of boulders.

1 Overview

Deposits emplaced by volcanic debris flows (VDFs) are typically coarse, massive, relatively thin, and poorly sorted mixtures of rock fragments and water (Fig. 1). It may take hours to days for water to drain from freshly emplaced mixtures that are rich in fines. Rare deposits of highly mobile flows composed largely of ice particles appear to have similar characteristics (e.g., Cronin et al. 1996; Kilgour et al. 2010; Pierson and Janda 1994; Waitt et al. 1994; Lube et al. 2009; Waythomas et al. 2013; Waythomas 2014) but only until the ice particles melt. Ice-slurry lahars have no chance of being preserved in their original form in the geologic record, so deposit characterization of VDFs in this chapter and volcanic hyperconcentrated flows (VHFs) in Chap. "Lahar Deposits—Hyperconcentrated-Flow Type" apply only to deposits emplaced by flows composed dominantly of rock particles.

There are two basic categories of VDF deposits—clay poor and clay rich. The importance of this distinction is explained in Sect. 2.1 in

T. C. Pierson et al., *Deposits of Volcanic Wet Flows*, Advances in Volcanology,
https://doi.org/10.1007/978-3-031-66574-5_5

Fig. 1 Lobe of freshly emplaced volcanic debris flow (VDF) deposit from Mount St. Helens (Washington, USA), having an estimated solids concentration of 65–70% by volume. This deposit was emplaced about 15 min before the photograph was taken on a terrace surface about a meter above the channel bed; the terrace surface slopes from upper left to lower right in photograph. White patches are remnants of a thin snow layer on the terrace surface. Note that pore water had not yet begun to drain out of the granular deposit matrix. The mixture was still partially liquefied and could be induced to flow farther by pushing on it when the photo was taken. Tape measure shows "1F" mark, which is 30.5 cm. USGS photograph by T. Pierson

Chap. "Lahars—Process Overview". Clay-poor deposits contain less than 5 wt% (1.9 vol%) clay-size particles in their matrix fraction (particles < 2 mm) (Vallance and Iverson 2015); the amount of clay plus silt is usually less than 30 wt% (11 vol%) (Pierson 1986; Scott 1988b). Clay-rich VDF deposits contain between about 5 and 40 wt% clay-size particles (commonly including clay minerals), and the fines fraction of the matrix can be as much as 65 wt% (Vallance and Scott 1997; Vallance and Iverson 2015).

VDF deposits are encountered on debris aprons on the lower flanks of volcanoes and in incised river valleys extending well beyond volcanic edifices. Particle sizes typically range from clay-size to boulders but may contain blocks 10 m in diameter or more (Fig. 2). Some deposits can have volumes more than 1 km^3 (Carrasco-Núñez et al. 1993; Vallance and Scott 1997; Mothes et al. 1998; Capra and Macías 2000, 2002).

VDF deposits are a principal component of volcanic sedimentary sequences typically found at the bases of intermediate-composition volcanic edifices and in valleys farther downstream (Smith 1986; Zernack et al. 2009, 2011). Although common in volcanic landscapes, they received only limited stratigraphic and sedimentologic analysis prior to 1980 (e.g., Schmidt 1934; Crandell and Waldron 1956; Mullineaux and Crandell 1962; Schmincke 1967; Crandell

Fig. 2 Block (approximately 10 m in diameter) deposited on a broad debris fan, transported about 15 km by a lahar (dominantly volcanic debris flow) triggered by rapid snowmelt during the 1877 eruption of Cotopaxi volcano (Ecuador), in background. USGS photograph by T. Pierson

1971). In proximal zones (see Chap. "Scaling and Timing of Volcanic Wet Flows in Response to Eruptions"), VDF deposits are commonly interlayered with lava flows and deposits of pyroclastic density currents (PDCs), tephra falls, alpine glaciers, and rare volcanic hyperconcentrated flows (VHFs). In distal zones tens to hundreds of kilometers downstream of source volcanoes, they are typically interbedded with abundant VHF deposits, volcanic water-flood (VWF) deposits, tephra beds, and occasionally eolian deposits (Vessell and Davies 1981; Mathisen and Vondra 1983; Smith 1987a, b; Waresback and Turbeville 1990; Nakayama and Yoshikawa 1997; Donoghue and Neall 2001; Kataoka and Nakajo 2002; Zernack et al. 2009).

2 Landscape-Scale Characteristics

2.1 Deposit Architecture and Morphology

VDFs emplace sedimentary bodies that have three-dimensional shapes, structures, and bounding surfaces (collectively known as deposit architecture) that can resemble those of debris-flow deposits in nonvolcanic terrain. However, VDF deposits are commonly much larger and travel farther downstream than their nonvolcanic counterparts. Architectural features depend on the size of the VDF, the proportion of gravel-size clasts, and the amount of clay within the matrix.

a
b

◀ **Fig. 3** Depositional environments affecting sediment-body shape of volcanic debris-flow (VDF) deposits; dashed lines show edges of sediment bodies. **a** Broad debris fan/apron—a volcaniclastic debris fan on north flank of Mount St. Helens (Washington, USA) known as the Pumice Plain. Deposits on surface are dominantly pumiceous pyroclastic density–current deposits, interbedded with minor lahar deposits, and underlain by the 1980 volcanic debris-avalanche deposit. USGS photo by T. Pierson, 2005. **b** Narrow valley fill—lahar sediment body partly filling incised river valley (South Fork Toutle River) downstream of Mount St. Helens (Washington, USA); volcano edifice is just out of photo at upper right. Some of valley fill was deposited on May 18, 1980. Subsequent incision (~ 10 m) created the vegetated fill terraces. USGS photo by K. Spicer, 2007

2.1.1 Sediment-Body Shape

The plan-view shape of VDF deposits is determined by the topography of the depositional zone. Deposition on the lower flanks of volcanoes typically produces broad, thick, fan-shaped, and downslope-thinning wedges of sediment termed *debris aprons*, which are made up of coalescing fans of volcaniclastic debris (Fig. 3a). Debris aprons are termed *ring plains* where they partially or entirely surround the volcanic edifice (e.g., Davidson and De Silva 2000; Donoghue and Neall 2001; Zernack et al. 2011). Debris aprons commonly extend 20–40 km out from the volcanic vent, bury preexisting topography to varying degrees, can cover hundreds of square kilometers and can be stratigraphically and laterally complex due to alternating episodes of erosion and deposition, as well as overlapping of deposits with other flow types (Smith 1986; Procter et al. 2009; Zernack 2021). Growth of debris aprons is typically driven by combined lahar, PDC, and tephra-fall deposition and is most vigorous during syneruptive and early posteruptive periods, when large volumes of primary volcaniclastic sediments remain available for erosion and redistribution (Smith 1987a, b; Palmer et al. 1993; Donoghue and Neall 2001; Zernack 2021). Incised river valleys extending beyond debris aprons at the base of a volcano may contain sinuous valley-fill sedimentary bodies that are perhaps only hundreds of meters wide but tens to hundreds of kilometers long (Fig. 3b). Syneruptive and posteruptive erosion of valley fills typically leave raised fill terraces along valley sides. In broad valleys or on plains, large clay-rich VDF deposits may spread out farther over broad areas than clay-poor VDFs of equal volume, due to the ability of clay-rich flows to remain liquefied longer and continue to flow when unconfined (Vallance and Scott 1997).

Thicknesses of VDF deposits are variable, reflecting both the gradient of depositional zones and VDF clay content (Fig. 4). Individual depositional units of clay-poor VDFs most commonly range from about 0.5 to 2 m in thickness, although thicker accumulations (up to 20 m thick) form over topographic depressions or in areas where damming or flow constrictions occur (Vallance and Iverson 2015). Stacks of multiple flow units on debris aprons can be tens to hundreds of meters thick and may be interbedded with VHF, VWF, volcanic debris-avalanche (VDA), and PDC deposits (Smith 1987a; Zernack et al. 2011; Zernack 2021). Individual flow units of large clay-rich VDFs can be up to 100 m thick (e.g., Capra and Macías 2002; Capra et al. 2002). Notably, individual depositional VDF units can be difficult to differentiate in a sedimentary sequence unless there are contrasting textural differences or weathering horizons between depositional units (Suwa and Okuda 1983; Major 1996, 1997). This blurring of flow-unit boundaries may give the impression that a thick deposit was emplaced as a single flow unit, when instead it was a sequence of several thinner flow units of similar material that were emplaced over a relatively short time interval. This is also a problem when interpreting some PDC deposits (e.g., Brand et al. 2014).

Large clay-poor VDFs typically produce sediment deposits characterized by relatively shallow channel incision that is caused by erosion during the waning, more dilute and more turbulent part of the flow that emplaced the deposit. Streamflow that resumes following lahar passage typically continues that incision during periods of high flow, which produces flights of

a

b

c

◀ **Fig. 4** Variation in flow depths and deposit thicknesses of lahars moving mainly as volcanic debris flows (VDFs). **a** Lahar-impacted channel (Lagunilla River) on the lower east flank of Nevado del Ruiz volcano (Colombia) just after the November 13, 1985, eruption and passage of a clay-poor VDF down this channel. Flow was right to left. A prominent fill terrace composed of lahar deposit is visible on the far right bank several meters below the peak mudline; the two people (right side of photo) are standing on the same terrace surface on the near left bank. USGS photo by T. Pierson. **b** Freshly emplaced lahar deposit (2–3 m thick) on terrace in foreground and thin skim of deposit on the snow-covered far valley wall to mark peak-flow mudline on the lower north flank of Redoubt Volcano (Alaska, USA), shortly after passage of an eruption-triggered ice-rich lahar, upper Drift River valley (Dorava and Meyer 1994). USGS photo by T. Miller, 1990. **c** Thick deposit of clay-rich VDF burying a bridge structure, which was carried to this wide reach in the main stem of Toutle River, 50–60 km downstream of Mount St. Helens (Washington, USA), on May 18, 1980. Deposit is at least 5 m thick at this location. Relatively smooth deposit surface reflects paucity of bouldery clasts at the surface of the deposit here. Note the limited incision by the river into the clay-rich deposit a few weeks after deposit emplacement. USGS photo by R. Schuster, date unknown

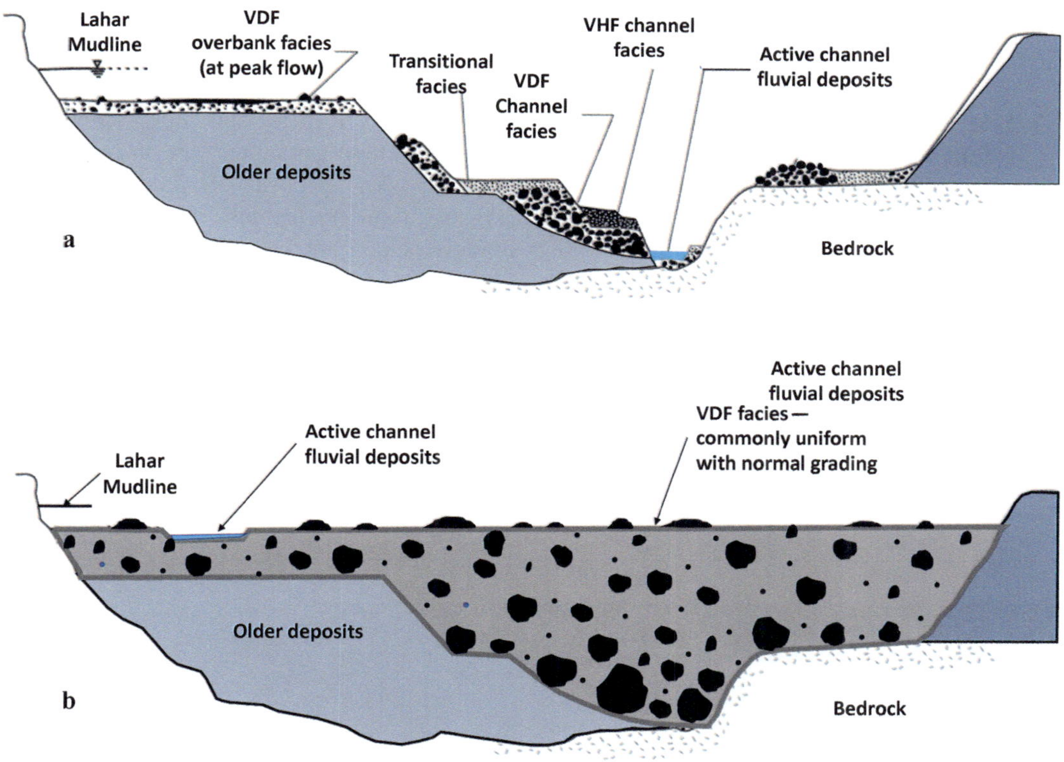

Fig. 5 Schematic channel cross-sections filled with remnants of lahar deposits, emplaced by lahars flowing mainly as volcanic debris flows (VDFs). Mudlines indicate levels of peak flow in each case. **a** Clay-poor VDF deposit forming a flight of inset fill terraces that are typically composed of different lithofacies sequentially emplaced by the same lahar and by subsequent streamflow. Deposit types include those emplaced by VDF, VHF, and VWF. **b** Clay-rich VDF deposit forming a thick fill of lahar sediment having a similar lithofacies throughout

inset terraces (Figs. 4a and 5a). Thin veneers of VWF deposits typically cap inset terraces. Unless there are older higher terraces preserved in a valley, the highest valley-fill terraces are the remnants of the original lahar deposit, and lower inset terraces typically host younger streamflow deposits on their surfaces. In some cases, mudlines on valley sides above the highest terraces represent the level of maximum flow depth during lahar passage. In other cases, the level of the highest terraces mark peak-flow surface elevation. This depositional sequence leaves clay-poor VDF deposits with different lithofacies at different levels (see Sects. 2.3 in Chap. "Scaling and Timing of Volcanic Wet Flows in Response to Eruptions" and 2.2).

Large clay-rich VDFs, on the other hand, typically emplace thick sediment bodies having one basic lithofacies. In most cases there is only limited fluvial incision into fresh clay-rich deposits, at least in the short term, due to the resistance of clay-rich deposits to erosion or to the fact that subsequent streamflow may be shunted to the valley side to form a new channel there (e.g., Figs. 4c and 5b).

2.1.2 Structures and Bounding Surfaces

Landscape-scale structures in clay-poor VDF deposits commonly include bars, boulder clusters, and isolated large clasts surrounded by finer-grained material. Deposit margins commonly exhibit lateral flow levees and lobate flow-unit fronts having a high concentration of gravel-size clasts. Upper surfaces of VDF sediment bodies may show evidence of post-depositional dewatering and may be slightly convex in narrow valleys. Lower bounding surfaces may, in some cases, show disconformities or paraconformities with underlying units—evidence of little to no basal erosion in deposition zones; in other cases these surfaces may show evidence of significant erosion. Clay-rich VDF deposits do not exhibit structures formed by segregation of gravel-size clasts, but small hummocks inherited from transformed debris avalanches are occasionally found in some large clay-rich deposits (e.g., Scott et al. 1995; see Sect. 2.1.2.2).

Bars, Boulder Clusters, and Isolated Large Clasts

Clay-poor VDFs selectively deposit boulder bars, boulder clusters, and large isolated boulders that become stranded on the channel bed when flow velocities and flow competence decrease. These structures are uncommon in clay-rich VDF deposits. Boulder point bars (Fig. 6a) form on the insides of bends where flow velocities are less than on the outsides. They are morphologically comparable to point bars in fluvial deposits. Mid-channel longitudinal bars (Fig. 6b) typically form downstream of major flow obstacles or obstructions, where boulders moving as bed load settle out in the eddy wake of the obstacle. Boulder clusters form when one boulder, moving as bed load, is stranded by decreasing flow competence or

a

b

Fig. 6 Boulder bars formed by clay-poor volcanic debris flows (VDFs): **a** bouldery point-bar deposit on inside of a left-sweeping river bend, looking downstream (1985 Chinchiná River VDF, Nevado del Ruiz, Colombia); largest boulder is about 1.5 m in diameter. **b** Longitudinal boulder bar developed behind a flow obstruction (bedrock knob) on valley side (Chinchiná River VDF); view is looking upstream to the bedrock knob. USGS photographs by T. Pierson stopped upstream of a flow obstruction, such as a tree (Fig. 7). Once one boulder is grounded, additional boulders can be added (upstream or downstream of the first), forming a growing flow obstruction. In other cases, large boulders may be left grounded on the channel bed as isolated single clasts (Fig. 2). The particles making up boulder bars and clusters sometimes show AB-plane imbrication, which is indicative of particle-by-particle deposition.

A type of longitudinal bar unique to clay-poor VDF deposits, a *whaleback bar*, was first recognized in 1980 lahar deposits at Mount St. Helens (Washington, USA) (Scott 1988a). Such accretionary bars have a smooth, highly streamlined appearance (Fig. 8) and comprise a core of unstratified coarse gravel that is covered by a smooth accreted coating of finer particles. These coatings tend to be normally graded. Six such features left along the flow path of the May 18, 1980, South Fork Toutle River lahar at Mount St. Helens averaged 79 m in length, 23 m in width, and 2.5 m in height. Whaleback bars superficially resemble glacial drumlins, but whaleback bars typically have their highest and widest points near the downstream end of the bars whereas drumlins are usually tallest and widest at the upstream end (Scott 1988a).

Hummocks

Deposits of clay-rich VDFs evolving from flank- or edifice-collapse debris avalanches may contain mounds—hummocks—that are composed of coherent but highly fractured pieces of the failed edifice or country rock that were carried as large blocks in the lahars until becoming

Fig. 7 Boulder cluster formed by clay-poor volcanic debris flow (VDF) in 1985 (Chinchiná River, Nevado del Ruiz, Colombia). Structure is on a low terrace, mostly downstream of tree that acted as flow obstruction; flow direction from left to right (looking across terrace with channel axis behind camera). Note mudline on tree marking peak flow depth at this site and debris piled against upstream side of tree, as well. Person is 1.75 m tall. The amount of run-up on the upstream side of the tree trunk suggests a peak-stage velocity of about 3 m/s at this location. USGS photograph by T. Pierson

Fig. 8 Streamlined longitudinal bar (*whaleback bar*) formed by 1980 South Fork Toutle River lahar flowing mainly as volcanic debris flow at this location (Mount St. Helens, Washington, USA) along the channel axis. Note that peak-stage mudline on the far bank is slightly higher than bar crest. Bar crest is approximately 2.5 m above its base. Cutbank exposure reveals contact between fresh VDF deposit and underlying fluvial gravel. USGS photo by K. Scott

grounded (Scott et al. 1995; Capra and Macías 2002; Capra et al. 2002; Zernack 2021). These small hills protrude from the otherwise flat surfaces of clay-rich VDF deposits (Vallance and Iverson 2015) (Fig. 9). Smaller semi-coherent pieces of the edifice may also be present as deformed fragile clasts seen in outcrop faces. If small hummocks are present, a volcanic debris-avalanche origin is indicated.

Deposit Margins

Frontal deposit margins of relatively small clay-poor VDF deposits are commonly coarse grained, steep fronted and lobate in plan view (Fig. 10). Both frontal and lateral margins characteristically exhibit accumulations of segregated coarse clasts (mainly cobbles and boulders) that may or may not have intergranular voids. Segregation of coarse clasts at deposit margins is uncommon in large clay-poor deposits emplaced by rapidly moving flows and rare or absent in clay-rich VDF deposits (Vallance and Iverson 2015).

Flow levees can develop along flow margins of clay-poor VDF deposits, particularly where segregation and deposition of coarse gravel-size clasts is sufficiently pronounced. Levees form relatively continuous ridges of undifferentiated bulk material, segregated gravel-size clasts, or tangles of coarse woody debris (commonly anchored by standing trees), and they tend to be strong enough to provide some lateral constraint on flow width during lahar passage (Fig. 11). Levees are rare in clay-rich VDF deposits, but if they do occur they are typically ridges of undifferentiated flow material. Levees of undifferentiated debris also occur in clay-poor VDF deposits near their point of origin, where there was insufficient time for size segregation of coarse clasts to have occurred (Fig. 11a, b). Multiple lahar

Fig. 9 Flat surface of a prehistoric clay-rich lahar deposit at Mount Rainier (Washington, USA) exhibiting a few relatively small hummocks. This is the "Greenwater Lahar" deposit in the upper White River valley (Crandell 1971), reinterpreted by Scott et al. (1995) as the Osceola Mudflow (lahar) deposit. The hummocks on this otherwise flat terrace surface (a recently planted tree farm) indicate its sector-collapse origin. Largest hummock on left is about 12 m in diameter and 3 m high. USGS photo by K. Scott

surges in a channel can leave nested levees (Fig. 11c), and levee formation can be facilitated by the additional frictional resistance of trees growing along flow margins (Fig. 11d).

Bounding Surfaces

Top surfaces of VDF deposits are flat to slightly convex. Surfaces of bouldery, clay-poor VDFs are commonly studded with cobbles and boulders protruding out of the surface plane to varying degrees (Fig. 12a, b). In deposits of fine-grained VDFs and clay-rich flows, top surfaces tend to be smooth or have only sparse large clasts protruding (Fig. 12c, d), depending on coarse clast availability. VDF deposit surfaces may also exhibit "tree wells" near deposit margins where flow velocities were low (Fig. 13). These features are vertically oriented cylindrical cavities (tree wells or tree molds) left behind after trees encased in VDF deposits died and their wood rotted away. Tree wells can form in VHF and VWF deposits as well but are seldom preserved—the looser deposits tend to collapse into the cavities and fill them.

VDF (and VHF) deposit surfaces may show evidence of deposit dewatering during bulk consolidation (see Sect. 3.6 in Chap. "Deposit Identification and Interpretation"), particularly in clay-poor deposits. Dewatering expels muddy pore water upward, which carries suspended clay, silt, and fine sand. Surface wash from expelled pore water can produce fine-scale erosional features on the deposit surface and deposit thin layers of fine sand, silt, and clay (*cap layers*) in shallow depressions on the deposit surface (Cronin et al. 2000). Cap layers are typically only a few millimeters thick but can accumulate to several to tens of centimeters in deeper depressions that collect surface drainage from a large area of the slightly undulating

Fig. 10 Lobate frontal debris-flow margins in an active flow that was still slowly moving when photo was taken. It was a slow, multi-lobed; nonvolcanic debris flow that was being deposited by edge-friction damming on an active debris fan at Mount Thomas, Canterbury (New Zealand). USGS photographs by T. Pierson

deposit surface. When cap layers are later buried by other deposits, they form laminae on contact boundaries between individual depositional units in vertical sections.

Lower bounding surfaces of VDF deposits may or may not show evidence of erosion, depending on the basal shear stress imposed by the flow at the depositional site. Well preserved VDF and VHF deposits are commonly found on terraces, where spillover onto those surfaces occurred at low velocities, and organic-rich buried soils and even leaf litter are preserved. Other depositional sites, particularly on steeper slopes, may show evidence of basal erosion by lahars.

2.2 Lithofacies

Lithofacies in VDF deposits typically are limited to diamicts—very poorly to extremely poorly sorted, unstratified, muddy sandy gravels (msG) or muddy gravelly sands (mgS). They can be clast supported (framework supported) or matrix supported. Clast-supported structures are more common in clay-poor VDF deposits, and matrix supported structures are more common in clay-rich VDF deposits. Typical lithofacies codes include Dmm, Dmg, Dcm, Dcg, Gmm, and Gcm (see Table 2 in Chap. "Deposit Identification and Interpretation"). In both clay-poor and clay-rich deposits, spaces between gravel-size clasts are filled with tightly packed matrix material (e.g., sand and mud), leaving no gaps or voids except in two instances: (1) voids that occur between gravel clasts at deposit margins and in flow levees in clay-poor VDF deposits (giving the deposit an open-work structure, code Gmo) and (2) small (mm-scale) air bubble vesicles that typically form in VDF matrix material.

All VDF depositional units form single beds and are never internally stratified—this is a hallmark characteristic of all debris-flow deposits. However, both inverse and normal grading are common within single depositional units, which

Fig. 11 Flow levees formed by small to moderate-size, clay-poor volcanic debris flows (VDFs): **a** sharp-crested levee of undifferentiated flow material with minimal segregation of coarse clasts in crater at Mount St. Helens (Washington, USA); note ice axe for scale at sharp crest of levee. Photo taken a few days after emplacement. **b** Broad-crested levee of mostly undifferentiated flow material at margin of a months-old clay-poor, rain-triggered VDF deposit in Muddy Fork of Sandy River, Mount Hood (Oregon, USA). **c** Openwork, size-segregated, nested boulder levees on edge of VDF-eroded gully, west flank of Láscar volcano (Chile); note person for scale. **d** Levee formed of woody debris by a rain-triggered VDF, Eliot Branch of Middle Fork Hood River on north flank of Mount Hood (Oregon, USA). Photo taken 9 months after emplacement. Smooth area on right is VDF deposit surface. USGS photographs by T. Pierson

can sometimes give the impression of stratification. Even stratigraphic sequences including multiple VDF depositional units may or may not be clearly stratified (see Sect. 3.1).

Clay-poor VDF depositional units from a single lahar can have a variety of lithofacies due to segregation of clasts by size and density and to flow transformations. Deposits emplaced along channel axes tend to be clast supported and have the coarsest particles—the VDF *channel facies* (Fig. 14a). This is because clasts too large or heavy to be suspended move as bed load in the deepest and fastest part of the current. Typical facies codes include Dcm, Dcg, and Gcm. In contrast, deposits emplaced near peak-flow levels, high on valley sides and on surfaces of high terraces, tend to be matrix-supported, finer grained overall, and ungraded—the VDF *overbank facies* (Fig. 14b) (Scott 1988a, b; Vallance and Iverson 2015). Facies codes Dmm, Dmg, Gmm are commonly applicable. These differences are best developed in deposits of large lahars where gravity segregation of large gravel-size clasts occurred during flow. Small clay-poor

Fig. 12 Surfaces of volcanic debris flow (VDF) deposits: **a** heavily boulder-studded surface of clay-poor VDF deposit in South Tahoma Creek at Mount Rainier (Washington, USA). **b** Moderately boulder-studded surface of a VDF deposit in the upper Eliot Branch valley at Mount Hood (Oregon, USA), which occurred in November 2006. **c** Lightly boulder-studded surface of the distal deposit of the 1985 Azufrado River VDF that destroyed the town of Armero, downstream of Nevado del Ruiz volcano (Colombia). **d** Smooth surface of the 1980 North Fork Toutle River VDF deposit (clay-rich and free of any coarse clasts on the surface) about 60 km downstream of Mount St. Helens (Washington, USA), photographed 3 months after deposition. USGS photos by T. Pierson

VDF deposits also commonly have a *marginal facies* that comprises clast-supported cobbles and boulders that may have voids between clasts. A *transitional facies* is sometimes also described, which has characteristics transitional between VDF and VHF deposits.

Clay-rich VDFs that have transformed from VDAs are commonly much larger than clay-poor VDFs and commonly fill entire valleys as they flow. Their deposits typically exhibit a relatively uniform *axial facies* (roughly equivalent to the channel facies in clay-poor flows), which is normally graded and shows relatively little textural change in vertical sections (Vallance and Scott 1997; Vallance and Iverson 2015). However, longitudinal textural changes may be observed in these sediment bodies that characterize proximal, medial, and distal axial facies (Vallance and Scott 1997; Vallance and Iverson 2015). Thin (0.1–1 m) veneers of deposit plastered on valley side slopes contain finer gravel-size clasts and comprise the *valley-side facies*. Some clay-rich VDF deposits may also include a *hummocky facies*—zones containing abundant large debris blocks (inherited from an initiating debris avalanche) (Vallance and Scott 1997; Vallance and Iverson 2015). Otherwise, the facies in clay-rich VDF sediment bodies commonly show only relatively minor differences.

Fig. 13 Tree well, 0.4–0.5 m in diameter, is preserved in a moss-covered, ~240-year-old, clay-poor volcanic debris-flow deposit in Zigzag River valley downstream of Mount Hood (Oregon, USA). Living trees are growing on the deposit surface. Photo by Lynd Wieman, used with permission

2.3 Erosion and Effects on Vegetation

Clay-poor VDFs commonly scour to bedrock in steep confined channels, eroding much or all of the soil and loose sediment present in a channel or valley (Fig. 15a). Bedrock or tightly packed alluvium surfaces can be markedly gouged, striated, or planed smooth by abrasion from the flow (Fig. 15b, c). High-velocity clay-poor VDFs on erodible beds tend to cut vertically instead of laterally (Fig. 16), and the eroded channels typically have a somewhat U-shaped cross-section. However, slow VDFs that spread out on relatively gentle slopes (such as on alluvial fans) can flow without eroding or disturbing even delicate underlying materials such as leaf litter. Experimental studies indicate that erosion (and entrainment) of unconsolidated bed sediment by debris flows depends in part on the degree to which the bed sediment is liquefiable when overridden by a flow, which is determined by sufficient water content and looseness of compaction (Iverson et al. 2011).

Although clay-rich VDFs can achieve high velocities and can be erosive, they tend to be less erosive than clay-poor VDFs relative to their volume—clay-rich flows have lower bulking factors. For example, the bulking factor for the enormous Osceola Mudflow from Mount Rainier (Washington, USA) was about 2, with most of the erosion of material occurring in the steeper upper part of the flow path (Vallance and Scott 1997). Bulking factors for clay-poor VDFs are commonly 3–6 (Table 4 in Chap. "Lahars—Process Overview").

Damage to trees and other vegetation by VDFs can be severe (Fig. 17; Table 3 in Chap. "Deposit Identification and Interpretation"), but type and degree of damage depends on flow magnitude and on the position of the affected vegetation relative to the main axis of flow. VDFs in lower gradient valleys typically show a damage gradient—a progression of damage that is most severe in the axial flow region (where flow was deep and fast) to least severe at the lateral margins of the flow path (where flow was slow and shallow). Large lahars in steep terrain break off and carry away large trees near the axis of flow, leaving behind splintered stumps (Fig. 17a). Small trees knocked down but still anchored are subject to intense abrasion along their downstream-pointing stems, often sharpening them to bayonet-like points (Fig. 17b). Trees strong enough to remain standing in high-velocity flow typically suffer intense damage up to and often slightly above submersion depth—bark removal and intense abrasion of underlying wood on the upstream side, but with little damage on the downstream side (Fig. 17c). Gravel may be embedded in the wood facing the oncoming flow (Fig. 17d). Farther out at the flow margin where flow velocities are low, tree trunks typically acquire mud coatings (Fig. 17e), the thickness of which depends on the clay and silt content of the slurry.

a b

Fig. 14 Channel and overbank facies in the deposit of a large, 7- to 8-m-deep, clay-poor volcanic debris flow (VDF) that occurred in 1985 in the Chinchiná River valley, 47 km from the source volcano, Nevado del Ruiz (Colombia). **a** Channel facies of the lahar deposited along main axis of river channel near the channel bed; largest clasts are boulder-size and note pencil for scale. View is upstream on the inside of a channel bend. **b** Relatively fine-grained overbank facies, deposited on sloping bank at peak stage of the same lahar in the same river cross-section but about 7 m above the channel bed. Largest clasts are 1–2 cm in diameter; note pencil for scale leaning against vertical face. View is downstream on the true right bank of the channel. This thin deposit overlies pasture grass that was not eroded. USGS photos by T. Pierson, taken about 2 weeks after emplacement

3 Outcrop-Scale Characteristics of Deposits

3.1 Stratification and Bed Forms

VDF depositional units form single beds without any internal stratification, and therefore no bed forms can be preserved within single depositional units. Single flow units form massive (nonstratified) beds that, for clay-poor VDFs, typically range from about 0.5 to 2 m in thickness (Fig. 18). Clay-rich VDFs can produce depositional units ("beds") many meters in thickness (e.g., Vallance and Scott 1997).

True stratification does not occur in VDF deposits, but layering of multiple flow units can occur, giving the appearance of stratification, particularly where flow units are thin. Stacks of successive VDF depositional units are commonly found on raised terrace surfaces, but they are only apparent if there are sharp contacts between flow units (Fig. 18). Such sharp contacts can be provided by sole layers (see below), weathering horizons, buried soils, and thin cap layers (Fig. 12c). Without such features along contacts, particularly when flow units are deposited in quick succession, multiple debris-flow deposits in a vertical sequence can be difficult to differentiate and can resemble a single thick flow unit (Suwa and Okuda 1983; Major 1996, 1997). Thus, spillover from multiple surge peaks on a terrace during passage of a single large lahar is unlikely to produce discernable beds.

Contacts between VDF deposits and other underlying sediments are typically sharp and may be conformable, unconformable, or disconformable (Fig. 1 in Chap. "Deposit Identification and Interpretation"). Underlying units may be eroded (Fig. 19a), faulted (Fig. 19b), or deformed (Fig. 19c) by the weight and shear forces imposed by the overriding lahar. Contacts are typically conformable where flow velocities were low. Load structures and sediment

Fig. 15 Field evidence of erosion by clay-poor volcanic debris flows (VDFs) triggered by the 1985 eruption of Nevado del Ruiz (Colombia); flow was toward camera in all three examples: **a** near total removal of soil and vegetation on canyon walls by the Azufrado River lahar, about 50 km downstream from the volcano. **b** Bedrock striated by passing lahar at same site as photo **a** (note pencil for scale in bottom center of frame). **c** Erosion of clay-rich soil in lee of partially buried boulder by Chinchiná River lahar. USGS photographs by T. Pierson taken about 2 weeks after lahar passage

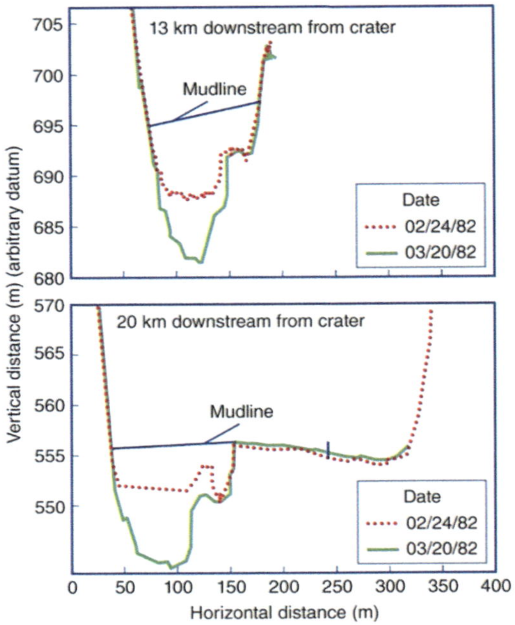

Fig. 16 Cross-sections of channels (10× vertical exaggeration) incised by a clay-poor volcanic debris flow in unconsolidated sediment on March 19, 1982, on the North Fork Toutle River at Mount St. Helens (Washington, USA) (modified from Pierson and Scott 1985). Tilted "mudlines" show elevation differences of peak flow due to superelevation during flow. Channels were cut into the relatively erodible 1980 debris-avalanche deposit

dikes may disrupt basal contacts where deposits are superimposed on still-liquefied underlying deposits.

Sole layers (Fig. 20) are texturally distinct basal subunits that are associated with some VDF deposits (Schmincke 1967; Crandell 1971; Scott 1988a, b; Hodgson and Manville 1999). Identified types of sole layers include (1) slightly compacted sandy layers, sometimes showing faint stratification—the Type 1 sole layer of Scott (1988a); (2) highly compacted layers of matrix-supported muddy sand and fine gravel, sometimes showing sub-horizontal foliation not related to bedding—Type 2 sole layer (Scott 1988a); and (3) compacted, clast-supported pebbly layers of crushed gravel clasts—Type 3 sole layer (Scott 1988a). Contacts between sole layers and overlying diamicts forming the main part of the deposit can be either relatively sharp or gradational over multiple centimeters. Color is similar for sole layers and the overlying part of the flow unit.

Type 1 sandy sole layers (Fig. 20a), found in clay-poor VDF deposits, grade progressively upward into coarser sediment mixtures (no sharp contact), and they likely form by progressive aggradation and incremental accretion of sand beneath the highly dilute flow fronts of lahars pushing a coupled wave of dilute river water (see Fig. 7 in Chap. "Lahars—Process Overview"). In long runout clay-poor lahars, the sandy sole layer may gradually increase in thickness with distance traveled to progressively become the whole flow-unit deposit, reflecting transformation of a VDF to a VHF (Pierson and Scott 1985; Scott 1988a; Cronin et al. 1999).

A highly compacted Type 2 sole layer (Fig. 20b) formed at the base of the clay-rich North Fork Toutle River lahar (Mount St. Helens) in 1980 from basal accretion of lahar sediment at the lower flow boundary. This layer ranged from 8 to 52 cm in thickness (less than 10–15% of total deposit thickness). High compaction in this channel facies sole layer is probably achieved by the combined effect of the clay content of the matrix and the high normal stress imposed by the deep (5–8 m) overpassing lahar (Janda et al. 1981; Scott 1988a). The layer locally exhibits a crude layering, as well as a more pervasive foliation parallel to the flow boundary. The contact with the overlying deposit is sharp in some places and diffuse in others. Interestingly, this sole layer was observed to extend from the base of the deposit up onto the upstream sides of tree trunks that were standing in the flow path (Janda et al. 1981; Scott 1988a).

Type 3 clast-supported gravelly sole layers (Fig. 20c) likely form by accretionary layering at the boundary shear zone in extremely large, deep, high-velocity VDFs, where prolonged normal and shear stresses imposed by the flow are sufficient to crush some clasts (Scott 1989). These sole layers are highly compacted, have sharp upper contacts, and have been observed to range in thickness from 45 to 110 cm in channel facies deposits emplaced by a lake breakout

a b c d e

◀ **Fig. 17** Types and degrees of damage to trees by volcanic debris flows (VDFs). **a** Tree stump with splintered top, from which a live tree was broken off and removed by the 2006 Eliot Branch VDF downstream of Mount Hood (Oregon, USA). Field notebook (20 × 15 cm) for scale. Flow was left to right. **b** Abrasion and sharpening of downstream-pointing tree branches and stems to dagger-like points ("bayonet trees") by the 1980 North Fork Toutle River lahar downstream of Mount St. Helens (Washington, USA). **c** Abrasion of bark and wood on upstream side of tree trunk (Douglas fir) by the 2006 Eliot Branch VDF at Mount Hood. Trunk diameter is about 40 cm. **d** Pebble-size rock fragment embedded in wood of a live tree by the force of a lahar flowing into it downstream of Calbuco volcano (Chile) in 2015. **e** Mud coatings (2–3 cm thick) on tree trunks preserved along the low-velocity flow margins of a clay-rich VDF (tree trunks are 50–80 cm in diameter). USGS photos **a**, **c**, **d**, **e** by T. Pierson: **b** from USGS archive

Fig. 18 Vertical sequence of three prehistoric volcanic debris-flow deposits in a riverbank near the confluence of the North Fork Toutle and South Fork Toutle Rivers, about 58 km downstream of Mount St. Helens (Washington, USA). Note handle of folding shovel in lower left (~ 50 cm long) for scale. Textural differences and distinctive contacts differentiate the units. Uppermost massive fine-grained unit is essentially ungraded. Middle unit has a slightly inversely graded base and a normally graded top. Coarse lowermost unit above shovel is mainly normally graded, with a narrow inversely graded zone at the base. USGS photo by T. Pierson

lahar that was up to 30 m deep (Scott 1989). Intact clasts in these layers have rough, hackly surfaces (Scott 1989).

3.2 Gravel-Size Clast Framework

Channel facies of clay-poor VDF deposits tend to be clast supported. Overbank facies of VDF deposits and clay-rich VDF deposits, on the other hand, tend to be matrix supported. Flow fronts, lateral margins, and flow levees of small clay-poor VDF deposits commonly have open-work structures. However, deposits especially rich or poor is gravel-size clasts may not reflect these general observations.

3.3 Fabric

Many VDF deposits do not exhibit discernable fabrics in outcrop, because gravel-size clasts commonly have equant shapes. In units where large clasts are more elongate or tabular, clast orientations may be random, show a flow-parallel orientation (Vallance and Scott 1997), or show an imbricate fabric with long axes (a-axes) of clasts oriented parallel to flow and a-b planes dipping upstream (Lindsay 1968; Mills 1984, 1991; Major and Voight 1986; Major 1996, 1998; Vallance and Scott 1997). However, where highly frictional flow fronts or flow margins induce flow compression, long axes tend to be transverse to the flow direction and subparallel

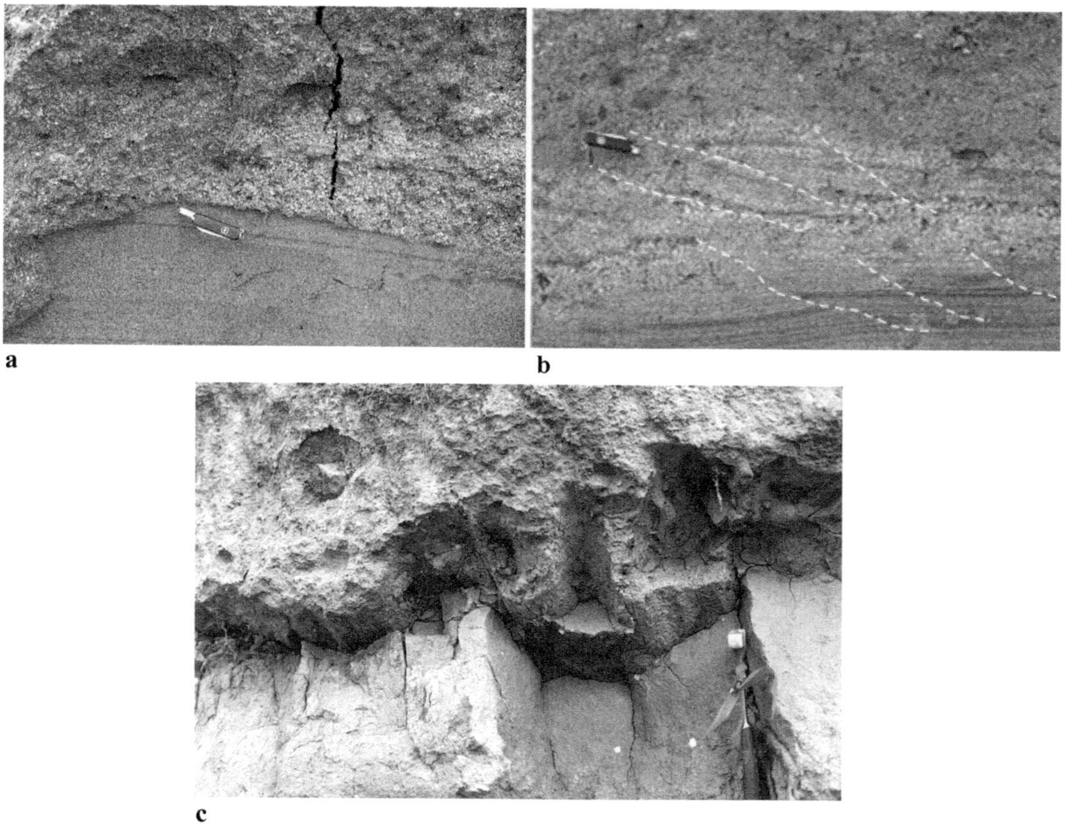

Fig. 19 Contacts between volcanic debris flow (VDF) deposits and underlying sediment exposed in vertical riverbanks, which show evidence of erosion, bed shear, and loading by overriding flows. **a** Truncated stratified sand beds eroded by a VDF (Bamban River, Mount Pinatubo, Philippines). Flow was right to left. **b** Small thrust faults in stratified sand beneath a VDF deposit (Bamban River, Mount Pinatubo). Flow was right to left. **c** Flutes (grooves) in soil overridden by a large VDF from Cotopaxi volcano (Ecuador). View is upward toward the overhanging lahar deposit in a vertical outcrop face. Flow was toward camera. Rock hammer for scale; square end of hammer head is 2.5 × 2.5 cm. USGS photos by T. Pierson

to flow margins (Mills 1984; Major and Voight 1986; Major 1998). Imbrication of large clasts has also been observed in boulder bars and clast clusters (e.g., Pierson et al. 1990).

3.4 Grading

Grading is common in single VDF depositional units (beds) but is not always present. Thick clay-rich VDF units are generally normally graded. Single clay-poor VDF units can be inversely graded throughout, normally graded throughout, inversely graded at the base and normally graded at the top of the deposit, or ungraded altogether (Figs. 18 and 21).

Grading is mainly produced by the progressive incremental accretion of particles from the base of the flow onto the bed surface as the size of coarsest clasts varies in the overpassing lahar flow body (Vallance 1994; Vallance and Iverson 2015) (see Fig. 22 in Chap. "Lahars—Process Overview"). Small particles laid down before large particles produce inverse grading, and vice versa. However, other mechanisms are likely to be at work, as well. Large clasts suspended, in part, by collisions with the bed and other particles in a matrix-rich mixture while

Fig. 20 Sole layers in volcanic debris flow (VDF) deposits. **a** Sandy Type 1 sole layer at base of upper lahar deposit (above thin lamination) from Osorno volcano (Chile). It is the pebble-free zone above the cap layer on top of the underlying lahar deposit; note that the upper boundary of the sole layer is not sharp. **b** Compact matrix-supported Type 2 sole layer in the May 18, 1980, North Fork Toutle River VDF deposit at Mount St. Helens (Washington, USA). It is the layer on which the hammer is resting. This sole layer, sitting on fluvial river gravel, has been partially eroded and exposed by later high flow in the river. Note that the sole layer is more resistant to erosion than the overlying main body of the fine-grained lahar deposit and that it contains rounded river pebbles that were entrained by the lahar. **c** Fines-depleted, clast-supported, Type 3 sole layer ("ball-bearing" bed) in the channel facies of the prehistoric PC1 lahar deposit at Mount St. Helens (Scott 1988a, b). It is the slightly eroded clast-supported gravel unit behind shovel (handle is 55 cm long) on the North Fork Toutle River about 2 km upstream of the South Fork Toutle River confluence. High river flows have washed out some of the less cohesive sole layer material, giving it the appearance of a separate depositional unit, but digging into the deposit revealed that the contact between the sole layer and the material above was gradational. USGS photos: **a** by T. Pierson; **b**, **c** by K. Scott

the lahar is flowing may settle downward once the sediment–water slurry comes to rest, producing normal grading. This is plausible where the concentration of large clasts is low enough so as not to hinder settling, and also where dewatering of the deposit is slow and the mixture remains liquefied for sufficient time, such as in clay-rich VDFs (Vallance and Scott 1997; Vallance and Iverson 2015). Another mechanism is fostered in clay-poor VDFs where clast-rich mixtures become vertically size segregated during flow (Sect. 4.4.1 in Chap. "Lahars—Process Overview"), with largest clasts being transported near the flow surface (Pierson 1986; Suwa 1988;

Fig. 21 Well-developed inverse grading in deposit of a single clay-poor volcanic debris flow that is overlying loose fluvial sand on the surface of a terrace. It was emplaced by the March 19, 1982, lahar in the North Fork Toutle River valley at Mount St. Helens (Washington, USA). The lahar overflowed onto this high terrace during peak flow. Note how the semi-consolidated lahar deposit overhangs the looser and raveling fluvial sand unit below, casting a shadow. Shovel is 75 cm long. USGS photograph by T. Pierson

Iverson 1997; Gray and Ancey 2009; Johnson et al. 2012).

3.5 Fluid-Escape Structures

Deposit consolidation and drainage of low-permeability deposits of clay-rich VDFs may be too slow for dewatering structures to form—they are not reported in such deposits, but in the deposits of some clay-poor VDFs, particularly from flows that were close to the transition point to hyperconcentrated flow, pore-water drainage is sufficiently rapid to form structures (see Sect. 3.6 in Chap. "Deposit Identification and Interpretation"). Dewatering pipes produced by upward flow of pore water during deposit compaction have been observed in some VDF deposits (Fig. 22a). Cap layers also form on the top surfaces of some VDF deposits (Fig. 22b, c).

3.6 Shear Deformation

Vertical velocity profiles and shear-rate distributions within natural debris flows are highly variable, changing between different flows within a given channel and between the fronts and tails of individual flows (Nagl et al. 2020). Maximum shear rates can be concentrated near the base of the flow, near the surface, or at different levels within the vertical velocity profile. Shear strength of rock clasts transported by VDFs typically exceeds the shear forces generated in those flows, although collisions between clasts can cause clast fracturing or breakage.

Weak fragile clasts (pieces of highly fractured rock debris, semiconsolidated sediment, or soil clods up to meters in diameter) may or may not be deformed by shear, depending on the amount of shear stress they are exposed to within VDFs or within debris blocks in VDAs that later transform to lahars (Fig. 23a, b).

Fig. 22 Fluid-escape structures in clay-poor volcanic debris flow (VDF) deposits. **a** Downstream-slanting water-escape pipes in vertical face of a VDF deposit overlying stratified sand in the Pasig-Potrero River valley (loose, fines-depleted sand raveled out of the pipes) downstream of Mount Pinatubo (Philippines). Flow direction was right to left. **b** Localized cap layer formed in a surface depression on a fill terrace composed of a clay-poor VDF deposit on the Pasig-Potrero River downstream of Mount Pinatubo. Shovel blade is about 16 cm wide. **c** Thin cap layer on upper surface of an eruption-triggered VDF deposit that was emplaced on the surface of a paleoterrace (and underlying subsequent overbank flood deposits) in the Sandy River about 70 km downstream of Mount Hood (Oregon, USA) (Pierson et al. 2011). Scraper shaft is 5 mm in diameter. USGS photos by T. Pierson

a

b

Fig. 23 Deformed and undeformed fragile clasts in volcanic debris flow (VDF) deposits: **a** highly deformed fragile clast suspended within a debris block in the clay-rich Naranjo lahar deposit downstream of Nevado de Colima (Mexico). The lahar transformed from a volcanic debris avalanche from the volcano (see Capra and Macías 2002), and the intense deformation in this block likely occurred during transport by the debris avalanche. Note that some of the dark clast material was injected into the light-colored surrounding matrix material. Blue pen above deformed clast gives scale. USGS photo by T. Pierson. **b** Rounded but undeformed soft sediment clast (a semi-consolidated mudstone) in an apparent debris-flow facies layer sandwiched between hyperconcentrated-flow facies within the same lahar deposit about 80 km downstream of Mount Hood (Oregon, USA). Scraper tool is about 38 cm long. Photo by L. Siebert

3.7 Material Consistency

VDF deposits are, in general, highly consolidated due to efficient packing of poorly sorted granular mixtures and the presence of small amounts of clay. Depending on clay content, they may show some degree of stickiness and plasticity when moist.

Clay-poor VDF deposits are typically firm to very firm when moist and slightly hard to hard when dry (see Table 4 in Chap. "Deposit Identification and Interpretation" for criteria). Wet hand samples tend to be nonsticky and nonplastic to slightly plastic. At outcrops, these deposits usually can be dug with a shovel with moderate to high difficulty. Matrix-rich areas of an outcrop face break off in chunks when struck by a pick or shovel.

Clay-rich VDF deposits tend to be very firm to extremely firm when moist, hard to very hard when dry, and slightly sticky to sticky and slightly plastic to plastic when wet. Deposits generally cannot be dug with a shovel. Some hot VDF deposits can be even harder, indurated nearly as hard as rock (Arguden and Rodolfo 1990). Heat may accelerate chemical reactions leading to diagenetic recrystallization and cementation within some fresh hot deposits, but the mechanism is uncertain. More commonly, indurated zones within older unconsolidated VDF deposits are formed by cementation from groundwater precipitates such as silica, calcium carbonate, or iron oxides over hundreds to thousands of years. Such cementation commonly occurs above or between fine-grained low-permeability zones within deposits.

3.8 Evidence of Heat

Deposits of hot VDFs may or may not show evidence of heat. A "hot" VDF cannot come close to having the temperatures of PDCs, however, because debris flows, by definition, are saturated with liquid water. Therefore, the average temperature of a VDF must be less than 100 °C. Some uniformly hot flows may develop hard, concrete-like surfaces (e.g., Arguden and Rodolfo 1990), while others known to be significantly hotter than ambient temperatures may not show any evidence of heat (e.g., Pierson et al. 1996). In some cases, however, relatively cool lahar can transport large gravel-size clasts much hotter than 100 °C, and such clasts and can create localized blind fumaroles and locally char adjacent vegetation. For example, steaming dacite boulders (lava dome fragments) were stranded on low terraces a few hours after being transported in the 1982 North Fork Toutle River lahar at Mount St. Helens. One boulder between 2 and 3 m in diameter had a surface still hot enough to melt rubber-coated work gloves hours after emplacement and an interior that was weakly incandescent after a piece was broken off with a hammer at night (Dinehart 1999; M. Uhrich, U.S. Geological Survey, written communication, 2019). Yet gaging station records show that this and other nearby hot boulders were transported by a clay-poor VDF that was cooler than 20 °C on average (Dinehart 1999). The lahar had traveled 58 km and was transforming to a VHF when the boulders were deposited. In another example, a deposit believed to have been emplaced by a hot lahar (but possibly by a PDC) at Kelut volcano (Indonesia) in 1919 measured 360 °C in fumaroles on the deposit surface a few days after emplacement (Kemmerling 1921, as cited in Crandell 1971).

The presence of charcoal or prismatically jointed gravel-size clasts in a deposit—strong evidence of initial high heat—can be misleading in lahar deposits. This is because cold lahars can entrain such materials from eroded pyroclastic deposits (Fig. 24). In such cases these indicators of high heat would be found side by side with evidence of low temperature (e.g., uncharred vegetation), supporting the interpretation that the emplacing flow was a lahar.

Fig. 24 Radial fractures in a prismatically jointed block of high-silica andesite in a volcanic debris-flow deposit downstream of Mount Hood (Oregon, USA), which was sourced from block-and-ash flow deposits originating from collapse of a hot lava dome. Shovel handle shaft is about 55 cm long. USGS photo by T. Pierson

4 Particle-Scale Characteristics of Deposits

4.1 Particle Composition

VDF deposits typically are polymictic and contain all three types of volcanic particles—primary volcaniclastic, secondary volcaniclastic, and epiclastic (Major and Scott 1988; Scott 1988a, b; Pierson et al. 1990; Major et al. 2005) (see Sects. 2 in Chap. "Deposits Emplaced by Volcanic Wet Flows" and 4 in Chap. "Deposit Identification and Interpretation"). This compositional variety is due to the propensity of VDFs to erode and entrain large quantities of sediment along their flow paths (e.g., Scott 1988a). However, lahar deposits encountered close to source may be oligomictic or monomictic (e.g., Pierson 1985; Pierson et al. 1996; see Sect. 4.1 in Chap. "Deposit Identification and Interpretation" for definitions of terms). Accidental exotic clasts in VDF deposits (even very thick ones) tend to be well mixed throughout the entire vertical profile and not just confined to the zone just above the basal contact (Vallance and Scott 1997), which contrasts with basal concentrations in debris-avalanche and some PDC deposits. Some large exotic clasts can be composed of fragile assemblages of other sediments or debris blocks, originating from debris blockages by outbreaks of debris-dammed lakes or channel bank collapse (Scott 1988a; Major et al. 2005).

4.2 Particle Roundness, Shape, Fracturing, and Surface Texture

Particle roundness is a function of particle size, as well as clast lithology, degree of weathering or alteration, and distance traveled (Pierson 1985; Pierson et al. 1990). Dense, hard sand grains and fine gravel clasts remain angular in most clay-poor VDF deposits, showing little or no evidence of rounding during transport, even in long-runout lahars. Rounding of cobble- and boulder-size clasts of fresh, hard rock in such VDFs starts to become evident after transport of a few kilometers. Degree of rounding further increases with boulder-size particles due to the higher impact forces sustained in collisions between large clasts (Okuda et al. 1980). Angular boulders can become subrounded or even rounded after transport distances of only a few tens of kilometers (Pierson 1985; Pierson et al. 1990) (Fig. 25). Softer hydrothermally altered rock clasts found in clay-rich VDF deposits may show greater degrees of rounding over a range of clast sizes; some rounding of source-rock clasts does occur in these more viscous debris mixtures (Vallance and Scott 1997). Importantly, the average degree of rounding of gravel-size clasts in VDF deposits can also increase due to entrainment of rounded stream gravel from riverbeds and channel banks (Scott 1988a). However, the surfaces of stream-rounded clasts are distinctly smoother than those rounded by interparticle collisions during

Fig. 25 Rounded andesite boulder showing a roughly pitted (pockmarked) surface texture, which was abraded and fractured during transport of 15 km or less by one or more lahars (and possibly watery debris floods) in a steep ravine on the lower flank of Mount Hood (Oregon, USA). It was emplaced at this site by the Eliot Branch VDF in November 2006. Surface pitting is uncharacteristic of rounding in streamflow. Mobile phone (11 cm long) for scale. USGS photograph by T. Pierson

VDF transport. Limited-distance rounding of boulders also can occur in more dilute flows—VHFs and water floods—if discharge is high enough to induce intense inter-particle collisions during bed-load transport (e.g., Scott and Gravlee 1968). VDF deposits containing high percentages of low-density clasts such as pumice or scoria, or other soft materials, typically show rounding in all gravel clast sizes, although such particles may already be well-rounded (for example, from abrasion by PDC transport) before being entrained by VDFs.

Roughly equant shapes are typical for most gravel-size volcaniclastic particles, though fragments from highly fractured lavas and tuffs, as well as sand-size ash particles, can be platy or elongate in shape. Particle shapes in VDF deposits are largely inherited from fragmentation of lava during eruption or from breakage during rapid cooling and emplacement as primary volcaniclastic source deposits.

Fracturing and breakage of gravel-size clasts occurs during lahar transport, although it is rare to find large clasts that are fractured but not broken apart (Fig. 25); most fractured clasts likely break into smaller clasts. Impact forces in collisions between boulders and stationary objects in VDF channels can produce impact forces of 10–100 kN on contact areas measuring just 15 cm^2 (Okuda et al. 1980). Such force is sufficient to fracture boulders in VDFs, but large intact clasts showing pervasive fine-scale fracturing (as occurs in VDA deposits) do not occur. High-impact particle–particle collisions also cause many boulder surfaces to be rough and pitted (Fig. 25), although high-pressure abrasive grinding must also within VDFs. Highly pitted boulder surfaces are more likely in VDF deposits than in other wet-flow deposits.

4.3 Particle-Size Distribution and Sedimentary Texture

VDF deposits exhibit a wide distribution of particle sizes, from clay-size particles to boulders or even to blocks up to 10 m in diameter, and sedimentary textures characteristically are muddy sandy gravels and gravelly muddy sands (Fig. 26). Clay-rich VDF deposits characteristically have a wider (more poorly sorted) distribution of particle sizes than granular VDF deposits (Scott 1988a, b; Vallance and Scott 1997) (Fig. 23 in Chap. "Deposit Identification and Interpretation"). Except for voids in openwork deposits sometimes found at marginal facies and vesicles in deposit matrices, there are no open spaces between gravel-size clasts in VDF deposits; all spaces are tightly packed with matrix material.

VDF deposits characteristically have bimodal or polymodal size distributions, and sorting is typically very poor to extremely poor—sorting values are generally greater than 3.0φ when calculated as Inclusive Graphic Standard Deviation (Folk and Ward 1957; Scott 1988a, b; Arguden and Rodolfo 1990) (see Sect. 4.3 in Chap. "Deposit Identification and Interpretation"). However, some fine-grained deposits can have sorting values in the range 2.0–3.0φ (e.g., Sharp and Nobles 1953; Scott 1971, 1988a; Pierson 1986; Pierson et al. 1990; Scott et al. 1995; Vallance et al. 1995; Vallance and Scott 1997). Sorting generally improves and bimodal distributions evolve toward unimodal distributions with distance away from source and from bottom to top in vertical sections of deposits in clay-poor VDFs (Scott 1988a). Clay-rich VDF deposits characteristically have a wider (more poorly sorted) distributions of particle sizes (sorting values $>4.0\varphi$) and show less downstream fining than clay-poor deposits. Skewness is usually either neutral or positive in VDF deposits. Positive skewness means the fine tail of the distribution is larger than the coarse tail (e.g., Scott 1971, 1988a; Pierson et al. 1990; Scott et al. 1995; Vallance and Scott 1997). Kurtosis values are most commonly slightly less than or close to 1.00 (platykurtic to mesokurtic) (Pierson 1985; Scott 1988a; Vallance and Scott 1997), indicating that the log-transformed distributions of VDF sediment size are either close to a normal distribution or slightly flat-peaked.

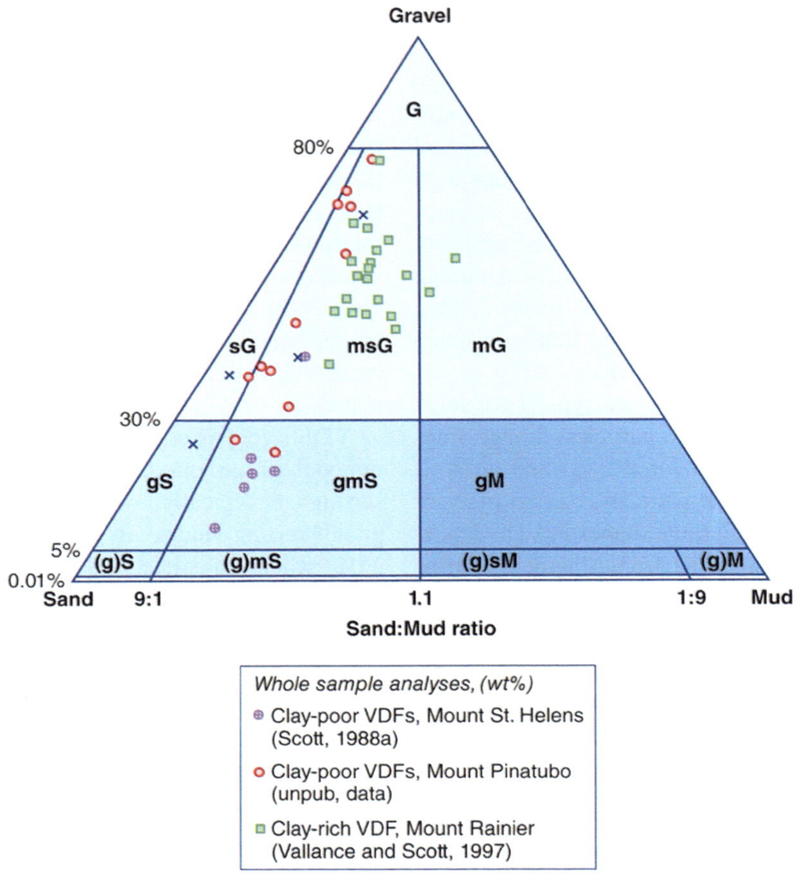

Fig. 26 Textural classification of sedimentary deposits containing gravel (G), sand (S), and mud (silt + clay) (M) (modified from Folk 1954), with plotted values for clay-rich and clay-poor volcanic debris flow (VDF) deposits from Mount St. Helens (Washington, USA), Mount Pinatubo (Philippines), and Mount Rainier (Washington, USA). Where texture label letters are capitalized, this size is dominant in the deposit (e.g., G, S, or M). Where those letters are lower case, that size is a minor component (e.g., g, s, or m). Where the letter is in parentheses, it is a very minor component. Variations in gravel and mud content greatly affect the appearance and properties of flows and their deposits

Published differentiation methods for identifying debris-flow deposits, such as mean grain size versus sorting plots (Buller and McManus 1972; Pe and Piper 1975), CM plots (coarsest particles versus median grain size) (Passega 1957; Bull 1962), and maximum particle size versus bed thickness (Nemec and Steel 1984) may have some usefulness in localized studies but do not appear to be universally valid. Similar overlap of distribution fields based on particle-size properties has been observed for PDC and tephra-fall deposits (Walker 1971).

4.4 Vesicles and Intergranular Voids

Vesicles formed by gas (mainly air) bubbles in VDF debris mixtures are typically about 1 mm in diameter and are common in both VDF and nonvolcanic debris-flow deposits (Sharp and Nobles 1953; Bull 1962; Johnson 1965; Crandell 1971; Fisher and Schmincke 1984; Costa 1984; Major and Voight 1986; Arguden and Rodolfo 1990; Smith and Lowe 1991) (Fig. 27 in Chap. "Deposit Identification and

Interpretation"). Vesicles result from air ingested by the flow and possibly from air escaping from overridden substrates or from steam created adjacent to hot entrained clasts. Tight packing of particles and the presence of fines in VDF mixtures create the low permeability in water-saturated mixtures to prevent entrapped air bubbles from readily migrating or escaping. Vesicles can compose up to 5% of the deposit volume (Major and Voight 1986). Those formed from ingested air appear to be randomly distributed within high-concentration deposits (viscous intergranular fluid) but can migrate and be concentrated beneath large gravel-size clasts or fine-grained cap layers in dilute mixtures. Vesicles originating from a substrate source can be concentrated at the base of the deposit. Other types of interstitial voids, such as irregular gaps or spaces between particles do not occur in VDF deposits except in marginal facies. In deposit interiors, all such spaces are filled with matrix material.

5 Key Characteristics of Volcanic Debris Flow (VDF) Deposits

VDF deposits can resemble other types of volcaniclastic sediments, particularly VHF deposits, coarse VWF deposits, PDC deposits, and some volcano-glacial deposits. A summary table of all relevant characteristics of VDF deposits is found in Chap. "Deposit Identification: Key and Summary Tables", and a gallery of outcrop photographs is in Chap. "Annotated Outcrop Photographs: Volcanic Wet-Flow Deposits and Other Unconsolidated Volcanic Deposits". Some of the most useful identification criteria are summarized below. Note that no single deposit characteristic is unique to a particular flow type, and thus no single characteristic is entirely sufficient for emplacement process identification. Field observations and physical characteristics of sediment must be evaluated together for an accurate assessment of a deposit's origin.

The characteristics listed below and in Chap. "Deposit Identification: Key and Summary Tables" are based on evaluation of fresh deposits. Fresh deposits and freshly impacted flow paths provide the best opportunities for identifying emplacement process, prior to the complicating effects of weathering, mineral alteration, chemical staining, and cementation of the sediment particles themselves; erosion of landforms; and decay of affected vegetation.

- **Landscape-scale criteria**
  (1) Flat (to slightly convex), boulder-studded deposit surfaces (for clay-poor VDF deposits) or mainly smooth, slightly undular surfaces (for clay-rich VDF deposits).
  (2) Intense abrasion damage to standing trees, extending from channel bed level to maximum flow height.
  (3) Abrasively sharpened, downstream-pointing "bayonet" tree tips and branches are common.
  (4) Mud coatings on tree trunks are characteristic—thin (up to a few mm) for clay-poor VDFs and thicker (a few mm to a few cm) for clay-rich VDFs.
  (5) Lateral flow levees are common in clay-poor VDF deposits but not always present.
  (6) Lobate flow fronts of open-work concentrations of coarsest clasts are common, mainly in small, clay-poor VDF deposits.
- **Outcrop-scale criteria**
  (1) Massive internal structure (no internal stratification) within a single depositional unit/bed is always observed.
  (2) Deposits are uniform and homogeneous in color and sedimentary texture.
  (3) Spatial clustering of gravel-size clasts is absent except at some deposit margins within individual depositional units.
  (4) Grading is common, both inverse and normal types.
  (5) Fabric is typically weak.
  (6) Firm/hard consistency (high degree of consolidation) is found in all VDF deposits but particularly well developed in clay-rich deposits; samples generally

break off in chunks rather than as individual particles when dug or struck with tools.

(7) Dewatering structures are possible but rare.

(8) Evidence of high heat (> 100 °C) is only found in highly localized zones around large gravel clasts that were hot when transported; some dispersed charcoal or prismatically jointed clasts may be present if eroded PDC deposits containing these materials were source material for the lahar.

- **Particle-scale criteria**

(1) Polymictic particle compositions are common unless deposit is sampled close to a monomictic or oligomictic sediment source.

(2) Crude (rough-textured) rounding of cobbles and boulders is typical in first-cycle volcaniclastic sediment; fine pebbles and sand are not rounded. Smoothly rounded gravel-size clasts can be present if mature river gravel was entrained in the lahar.

(3) Visible fractures are rare in large gravel-size clasts; fracturing during transport commonly results in breakage and creation of two or more unfractured clasts.

(4) Samples are very poorly to extremely poorly sorted in most cases.

(5) Small air bubble vesicles are commonly preserved in muddy matrix material.

(6) No large (> 1 cm) intergranular voids are found within deposits, except in the marginal facies of some deposits (at flow fronts or lateral margins or in flow levees) or where decayed wood fragments have left behind empty molds of the original fragment shapes. Spaces between gravel-size clasts are normally packed tight with smaller clasts and muddy matrix material.

References

Arguden AT, Rodolfo KS (1990) Sedimentologic and dynamic differences between hot and cold laharic debris flows of Mayon volcano, Philippines. Geol Soc Am Bull 102:865–876

Brand BD, Mackaman-Lofland C, Pollock NM, Bendaña S, Dawson B, Wichgers P (2014) Dynamics of pyroclastic density currents: conditions that promote substrate erosion and self-channelization—Mount St Helens, Washington (USA). J Volcanol Geotherm Res 276:189–214. https://doi.org/10.1016/j.jvolgeores.2014.01.007

Bull WB (1962) Relation of textural (CM) patterns to depositional environment of alluvial-fan deposits. J Sediment Petrol 32:211–216

Buller AT, McManus J (1972) Simple metric statistics used to recognise different environments. Sedimentology 18:1–21

Capra L, Macías JL (2000) Pleistocene cohesive debris flows at Nevado de Toluca volcano, central Mexico. J Volcanol Geotherm Res 102:149–168. https://doi.org/10.1016/S0377-0273(00)00186-4

Capra L, Macías JL (2002) The cohesive Naranjo debris-flow deposit (10 km^3): a dam breakout flow derived from the Pleistocene debris-avalanche deposit of Nevado de Colima Volcano (México). J Volcanol Geotherm Res 117(1–2):213–235. https://doi.org/10.1016/S0377-0273(02)00245-7

Capra L, Macías JL, Scott KM, Abrams M, Garduño-Monroy VH (2002) Debris avalanches and debris flows transformed from collapses in the Trans-Mexican Volcanic Belt, Mexico—behavior and implications for hazard assessment. J Volcanol Geotherm Res 113:81–110. https://doi.org/10.1016/S0377-0273(01)00252-9

Carrasco-Núñez G, Vallance JW, Rose WI (1993) A voluminous avalanche-induced lahar from Citlaltepetl volcano, Mexico: implications for hazard assessment. J Volcanol Geotherm Res 59:35–46. https://doi.org/10.1016/0377-0273(93)90076-4

Costa JE (1984) The physical geomorphology of debris flows. In: Costa JE, Fleishe PJ (eds) Developments and applications of geomorphology. Springer, Berlin and New York, pp 268–317

Crandell DR (1971) Postglacial lahars from Mount Rainier Volcano, Washington. U S Geol Surv Prof Pap 677:1–73

Crandell DR, Waldron HH (1956) A recent volcanic mudflow of exceptional dimensions from Mt. Rainier, Washington. Am J Sci 254(6):349–362

Cronin SJ, Neall VE, Lecointre JA, Palmer AS (1996) Unusual "snow slurry" lahars from Ruapehu volcano, New Zealand, September 1995. Geology 24:1107–1110

Cronin SJ, Neall VE, Lecointre JA, Palmer AS (1999) Dynamic interactions between lahars and stream flow: a case study from Ruapehu volcano, New Zealand. Geol Soc Am Bull 111:28–38

Cronin SJ, LeCointre JA, Palmer AS, Neall VE (2000) Transformation, internal stratification, and depositional processes within a channelized, multi-peaked lahar flow. N Z J Geol Geophys 43:117–128

Davidson JP, De Silva S (2000) Composite volcanoes. In: Sigurdsson H, Houghton B, McNutt SR, Rymer H, Stix J (eds) Encyclopedia of volcanoes. Academic Press, San Diego, pp 663–681

Dinehart RL (1999) Sediment transport in the hyperconcentrated phase of the March 19, 1982, lahar. In: Pierson TC (ed) Hydrologic consequences of hot-rock/snowpack interactions at Mount St. Helens volcano, Washington, 1982–84. U S Geol Surv Prof Pap 1586:37–52

Donoghue SL, Neall VE (2001) Late Quaternary constructional history of the southeastern Ruapehu ring plain, New Zealand. N Z J Geol Geophys 44(3):439–466. https://doi.org/10.1080/00288306.2001.9514949

Dorava JM, Meyer DF (1994) Hydrologic hazards in the lower Drift River basin associated with the 1989–1990 eruption of Redoubt Volcano, Alaska. In: Miller TP, Chouet BA (eds) The 1989–1990 eruptions of Redoubt Volcano, Alaska. J Volcanol Geotherm Res 62:387–407

Fisher RV, Schmincke H-U (1984) Pyroclastic rocks. Springer-Verlag, Berlin, p 472

Folk RL (1954) The distinction between grain size and mineral composition in sedimentary rock nomenclature. J Geol 62:344–359

Folk RL, Ward WC (1957) Brazos River bar: a study in the significance of grain size parameters. J Sediment Petrol 27:3–26

Gray JMNT, Ancey C (2009) Segregation, recirculation and deposition of coarse particles near two-dimensional avalanche fronts. J Fluid Mech 629:387–423. https://doi.org/10.1017/S0022112009006466

Hodgson KA, Manville VR (1999) Sedimentology and flow behavior of a rain-triggered lahar, Mangatoetoenui Stream, Ruapehu volcano, New Zealand. Geol Soc Am Bull 111:743–754

Iverson RM (1997) The physics of debris flows. Rev Geophys 35:245–296

Iverson RM, Reid ME, Logan M, LaHusen RG, Godt JW, Griswold JP (2011) Positive feedback and momentum growth during debris-flow entrainment of wet bed sediment. Nat Geosci 4:116–121. https://doi.org/10.1038/ngeo1040

Janda RJ, Scott KM, Nolan KM, Martinson HA (1981) Lahar movement, effects, and deposits. In: Lipman PW, Mullineaux DR (eds) The 1980 eruptions of Mount St. Helens, Washington. U S Geol Surv Prof Pap 1250:461–478

Johnson AM (1965) A model for debris flow. Dissertation, Pennsylvania State University

Johnson CG, Kokelaar BP, Iverson RM, Logan M, LaHusen RG, Gray JMNT (2012) Grain-size segregation and levee formation in geophysical mass flows. J Geophys Res Earth Surf 117:F01032. https://doi.org/10.1029/2011JF002185

Kataoka K, Nakajo T (2002) Volcaniclastic resedimentation in distal fluvial basins induced by large-volume explosive volcanism. The Ebisutoge—Fukuda tephra, Plio-Pleistocene boundary, central Japan. Sedimentology 49:319–334

Kemmerling GLL (1921) De geologie en geomorfologie van den Idjen. Het Idjen Hoogland. Kon Natuurk Ver Monografie II. G Koff & Co, Weltevreden-Batavia

Kilgour G, Manville V, Della Pasqua F, Graettinger A, Hodgson KA, Jolly GE (2010) The 25 September 2007 eruption of Mount Ruapehu, New Zealand: directed ballistics, surtseyan jets, and ice-slurry lahars. J Volcanol Geotherm Res 191:1–14. https://doi.org/10.1016/j.jvolgeores.2009.10.015

Lindsay JF (1968) The development of clast fabric in mudflows. J Sediment Petrol 38:1242–1253

Lube G, Cronin SJ, Procter JN (2009) Explaining the extreme mobility of volcanic ice-slurry flows, Ruapehu volcano, New Zealand. Geology 37:15–18. https://doi.org/10.1130/G25352A.1

Major JJ (1996) Experimental studies of deposition by debris flows: process, characteristics of deposits and effects of pore-fluid pressure. Dissertation, University of Washington, 341 pp

Major JJ (1997) Depositional processes in large-scale debris-flow experiments. J Geol 105:345–366

Major JJ (1998) Pebble orientation on large, experimental debris-flow deposits. Sediment Geol 117:151–164

Major JJ, Scott KM (1988) Volcaniclastic sedimentation in the Lewis River valley, Mount St. Helens, Washington—processes, extent, and hazards. U S Geol Surv Bull 1383-D, 38 pp

Major JJ, Voight B (1986) Sedimentology and clast orientations of the 18 May 1980 southwest-flank lahars, Mount St. Helens, Washington. J Sediment Petrol 56:691–705

Major JJ, Pierson TC, Scott KM (2005) Debris flows at Mount St. Helens, Washington, USA. In: Jakob M, Hungr O (eds) Debris-flow hazards and related phenomena. Praxis-Springer, Berlin, pp 685–731

Mathisen ME, Vondra CF (1983) The fluvial and pyroclastic deposits of the Cagayan Basin, Northern Luzon, Philippines—an example of non-marine volcaniclastic sedimentation in an interarc basin. Sedimentology 30:369–392

Mills HH (1984) Clast orientation in Mount St. Helens debris-flow deposits, North Fork Toutle River, Washington. J Sediment Petrol 54:626–634

Mills HH (1991) Three-dimensional clast orientation in glacial and mass-movement sediments: compilation and preliminary analysis. U S Geol Surv Open-File Rep 90–128, 71 pp

Mothes PA, Hall ML, Janda RJ (1998) The enormous Chillos Valley lahar: an ash-flow-generated debris flow from Cotopaxi volcano, Ecuador. Bull Volcanol 59:233–244

Mullineaux DR, Crandell DR (1962) Recent lahars from Mount St. Helens, Washington. Geol Soc Am Bull 73(7):855–870. https://doi.org/10.1130/0016-7606(1962)73[855:RLFMSH]2.0.CO;2

Nagl G, Hübl J, Kaitna R (2020) Velocity profiles and basal stresses in natural debris flows. Earth Surf Process Landf 45(8):1764–1776. https://doi.org/10.1002/esp.4844

Nakayama K, Yoshikawa S (1997) Depositional processes of primary to reworked volcaniclastics on an alluvial plain; an example from the Lower Pliocene Ohta tephra bed of the Tokai Group, central Japan. Sediment Geol 107:211–229. https://doi.org/10.1016/S0037-0738(96)00022-X

Nemec W, Steel RJ (1984) Alluvial and coastal conglomerates: their significant features and some comments on gravelly mass-flow deposits. In: Koster EH, Steel RJ (eds) Sedimentology of gravels and conglomerates. Can Soc Pet Geol Mem 10:1–31

Okuda S, Suwa H, Okunishi K, Yokoyama K, Nakano M (1980) Observations on the motion of a debris flow and its geomorphological effects. Z Geomorphol Suppl Bd 35:142–163

Palmer BA, Purves AM, Donoghue SL (1993) Controls on accumulation of a volcaniclastic fan, Ruapehu composite volcano, New Zealand. Bull Volcanol 55:176–189. https://doi.org/10.1007/BF00301515

Passega R (1957) Texture as characteristic of clastic deposition. Am Assoc Pet Geol Bull 41:1952–1984

Pe GG, Piper DJW (1975) Textural recognition of mudflow deposits. Sedimentology 13:303–306

Pierson TC (1985) Initiation and flow behavior of the 1980 Pine Creek and Muddy River lahars, Mount St. Helens, Washington. Geol Soc Am Bull 96:1056–1069. https://doi.org/10.1130/0016-7606(1985)96<1056:IAFBOT>2.0.CO;2

Pierson TC (1986) Flow behavior of channelized debris flows, Mount St. Helens, Washington. In: Abrahams AD (ed) Hillslope processes. Allen Unwin, Boston, pp 269–296

Pierson TC, Janda RJ (1994) Volcanic mixed avalanches: a distinct eruption-triggered mass-flow process at snow-clad volcanoes. Geol Soc Am Bull 106:1351–1358

Pierson TC, Scott KM (1985) Downstream dilution of a lahar: transition from debris flow to hyperconcentrated streamflow. Water Resour Res 21:1511–1524

Pierson TC, Janda RJ, Thouret JC, Borrero CA (1990) Perturbation and melting of snow and ice by the 13 November 1985 eruption of Nevado del Ruiz, Colombia, and consequent mobilization, flow, and deposition of lahars. J Volcanol Geotherm Res 41:17–66

Pierson TC, Daag AS, Delos Reyes PJ, Regalado MTM, Solidum RU, Tubianosa BS (1996) Flow and deposition of posteruption hot lahars on the east side of Mount Pinatubo, July–October 1991. In: Newhall CG, Punongbayan RS (eds) Fire and mud: eruptions and lahars of Mount Pinatubo, Philippines. Philippine Institute of Volcanology and Seismology, Quezon City; University of Washington Press, Seattle, pp 921–950

Pierson TC, Pringle PT, Cameron KC (2011) Magnitude and timing of downstream channel aggradation in response to a dome-building eruption at Mount Hood, Oregon. Geol Soc Am Bull 123:3–20

Procter JN, Cronin SJ, Zernack AV (2009) Landscape and sedimentary response to catastrophic debris avalanches, western Taranaki, New Zealand. Sediment Geol 220(3–4):271–287. https://doi.org/10.1016/j.sedgeo.2009.04.027

Schmidt KG (1934) Die Schuttströme am Merapi auf Java nach dem Ausbruch von 1930. De Ingenieur Nederlandsch-Indië 1(7):91–120 (parts 1–7), 1(8):123–134 (parts 8–9), 1(9):143–171 (parts 10–12) (in German)

Schmincke H-U (1967) Graded lahars in the type sections of the Ellensburg Formation, south-central Washington. J Sediment Petrol 37:438–448

Scott KM (1971) Origin and sedimentology of 1969 debris flows near Glendora, California. U S Geol Surv Prof Pap 750-C:242–247

Scott KM (1988a) Origins, behavior, and sedimentology of lahars and lahar-runout flows in the Toutle-Cowlitz River system. U S Geol Surv Prof Pap 1447-A:74 pp

Scott KM (1988b) Origin, behavior, and sedimentology of prehistoric catastrophic lahars at Mount St. Helens, Washington. In: Clifton HE (ed) Sedimentologic consequences of convulsive geologic events. Geol Soc Am Spec Pap 229:23–36. https://doi.org/10.1130/SPE229-p23

Scott KM (1989) Magnitude and frequency of lahars and lahar-runout flows in the Toutle-Cowlitz River system. U S Geol Surv Prof Pap 1447-B:33 pp

Scott KM, Gravlee GC Jr (1968) Flood surge on the Rubicon River, California—hydrology, hydraulics and boulder transport. U S Geol Surv Prof Pap 422-M:40 pp

Scott KM, Vallance JW, Pringle PP (1995) Sedimentology, behavior, and hazards of debris flows at Mount Rainier, Washington. U S Geol Surv Prof Pap 1547:56 pp

Sharp RP, Nobles LH (1953) Mudflow of 1941 at Wrightwood, southern California. Geol Soc Am Bull 64:547–560

Smith GA (1986) Coarse-grained nonmarine volcaniclastic sediment: terminology and depositional process. Geol Soc Am Bull 97:1–10

Smith GA (1987a) The influence of explosive volcanism on fluvial sedimentation: the Deschutes Formation (Neogene) in central Oregon. J Sediment Petrol 57:613–629

Smith GA (1987b) Sedimentology of volcanism-induced aggradation in fluvial basins: examples from the

Pacific Northwest, U.S.A. In: Ethridge FG, Flores RM, Harvey MD (eds) Recent developments in fluvial sedimentology. Soc Econ Paleontol Mineral Spec Publ 39:217–228

Smith GA, Lowe DR (1991) Lahars: volcano-hydrologic events and deposition in the debris flow—hyperconcentrated flow continuum. In: Fisher RV, Smith GA (eds) Sedimentation in volcanic settings. SEPM (Soc Sediment Geol) Spec Publ 45:59–70

Suwa H (1988) Focusing mechanism of large boulders to a debris-flow front. Trans Jpn Geomorphol Union 9:151–178

Suwa H, Okuda S (1983) Deposition of debris flows on a fan surface, Mt. Yakedake, Japan. Z Geomorphol 46:79–101

Vallance JW (1994) Experimental and field studies related to the behavior of granular mass flows and the characteristics of their deposits. Dissertation, Michigan Technological University, 197 pp

Vallance JW, Iverson RM (2015) Lahars and their deposits. In: Sigurdsson H, Houghton B, McNutt SR, Rymer H, Stix J (eds) The encyclopedia of volcanoes, 2nd edn. Elsevier, Amsterdam, pp 649–664. https://doi.org/10.1016/B978-0-12-385938-9.00037-7

Vallance JW, Scott KM (1997) The Osceola Mudflow from Mount Rainier: sedimentology and hazard implications of a huge clay-rich debris flow. Geol Soc Am Bull 109:143–163. https://doi.org/10.1130/0016-7606(1997)109<0143:TOMFMR>2.3.CO;2

Vallance JW, Scott KM, Pringle PT (1995) Sedimentology, behavior, and hazards of debris flows at Mount Rainier, Washington. U S Geol Surv Prof Pap 1547:56 pp

Vessell RD, Davies KK (1981) Nonmarine sedimentation in an active fore arc basin. In: Ethridge FG, Flores RM (eds) Nonmarine depositional environments: models for exploration. Soc Econ Paleontol Mineral Spec Publ 31:34–45

Waitt RB, Gardner CA, Pierson TC, Major JJ, Neal CA (1994) Unusual ice diamicts emplaced during 15 December 1989 eruption of Redoubt Volcano, Alaska. J Volcanol Geotherm Res 62:409–428

Walker GPL (1971) Grain-size characteristics of pyroclastic deposits. J Geol 79:696–714

Waresback DB, Turbeville BN (1990) Evolution of a Plio-Pleistocene volcanogenic-alluvial fan: the Puye Formation, Jemez Mountains, New Mexico. Geol Soc Am Bull 102:298–314

Waythomas CF (2014) Water, ice, and mud: lahars and lahar hazards at ice- and snow-clad volcanoes. Geol Today 30(1):34–39

Waythomas CF, Pierson TC, Major JJ, Scott WE (2013) Voluminous ice-rich lahars generated during the 2009 eruption of Redoubt Volcano, Alaska. J Volcanol Geotherm Res 259:389–413

Zernack AV (2021) Volcanic debris-avalanche deposits in the context of volcaniclastic ring plain successions—a case study from Mt. Taranaki. In: Roverato M, Dufresne A, Procter J (eds) Volcanic debris avalanches. Advances in volcanology, vol 6. Springer, Cham, pp 211–254. https://doi.org/10.1007/978-3-030-57411-6_9

Zernack AV, Procter JN, Cronin SJ (2009) Sedimentary signatures of cyclic growth and destruction of stratovolcanoes: a case study from Mt. Taranaki, New Zealand. Sed Geol 220(3–4):288–305. https://doi.org/10.1016/j.sedgeo.2009.04.024

Zernack AV, Cronin SJ, Neall VE, Procter JN (2011) A medial to distal volcaniclastic record of an andesite stratovolcano: detailed stratigraphy of the ring-plain succession of south-west Taranaki, New Zealand. Int J Earth Sci (Geol Rundsch) 100:1937–1966. https://doi.org/10.1007/s00531-010-0610-6

Lahar Deposits—Hyperconcentrated-Flow Type

Abstract

Dilute lahars moving as hyperconcentrated flow are termed volcanic hyperconcentrated flows (VHFs). Their deposits commonly comprise faintly stratified sandy beds with minor gravel lenses and isolated large gravel clasts, but deposit characteristics can vary with solids concentration of the flow. Some dilute lahars flow as VHFs from start to finish, but flow transformations in long-runout lahars typically result in both VDF-type and VHF-type lithofacies in a single lahar deposit. Key features of VHF deposits include emplacement in wide, shallow, vertically banked, incised channels; plane-parallel bedding contacts that are faint and diffuse (sharp contacts may be present but are uncommon); dominantly low-angle cross-bedding; outsized clasts on deposit surfaces; slightly firm/hard consistencies; fine-grained cap layers, consolidation laminations, and other dewatering structures that disrupt bedding; and moderate to poor sorting.

1 Overview

Volcanic hyperconcentrated flow (VHF) deposits typically comprise sandy, faintly stratified beds with minor gravel lenses and isolated large gravel clasts (Fig. 1), but deposit characteristics can occupy the full spectrum of features between volcanic debris flow (VDF) and volcanic water flood (VWF) deposits. Many lahar deposits exhibit both VDF-type and VHF-type deposits in the same depositional flow unit, and if clay-poor VDFs flow far enough, such lahars typically evolve into pure VHFs, sometimes called *lahar-runout flows* (e.g., Janda et al. 1981; Scott 1988a, b, 1989). Some dilute lahars flow as VHFs from start to finish.

Where VDF is the dominant component of a lahar, VHF deposition generally occurs at only two locations: (1) on the channel bed beneath the flow front at the mixing interface between the coupled wave of river water being pushed ahead of the highly concentrated main body of the lahar (see Sect. 4.3 in Chap. "Lahars—Process Overview") and (2) on the channel bed beneath the tail of the lahar (where progressive dilution causes the transformation from VDF back to VHF). This kind of VDF-dominant lahar leaves VHF deposits at the base of the main lahar deposit and in inset terraces within the channel cut into the main body deposit toward the end of the lahar's passage; these inset deposits are not always preserved (Vallance and Iverson 2015). Where VHF is the dominant component of a lahar, aggradational VHF deposition typically sandwiches a thin zone of VDF sediment between a thin basal VHF zone and a thicker zone emplaced by the VHF main body of the lahar (e.g., Pierson et al. 2013).

T. C. Pierson et al., *Deposits of Volcanic Wet Flows*, Advances in Volcanology,
https://doi.org/10.1007/978-3-031-66574-5_6

a

b

Fig. 1 Volcanic hyperconcentrated flow (VHF) deposit emplaced by an eruption-triggered lahar on April 4, 2009, in the Drift River valley about 30 km from source, downstream of Redoubt Volcano (Alaska, USA). **a** Top surface of VHF deposit; pencil points downstream. Lighter-colored smooth areas are locally preserved cap layers. **b** Vertical section of same deposit with several separate surge-deposited subunits. Each subunit is faintly laminated, subhorizontally bedded fine to medium sand at the base with gravelly sand at the top. Deposit sits on coarse fluvial gravel. Top of section has a continuous thick (~2 cm) cap layer of thinly laminated beds of fine sand and mud emplaced during the waning stages of lahar flow. Flow was right to left. Scraper shaft is 5 mm thick. USGS photos by T. Pierson, 6 weeks after deposition

Descriptions of deposits documented to have been emplaced by volcanic or nonvolcanic hyperconcentrated flows are rare (Pierson and Scott 1985; Scott 1988a; Pierson et al. 1996; Dinehart 1999; Pringle and Cameron 1999; Cronin et al. 1997, 1999, 2000; Thouret et al. 1998; Wilcox et al. 2014), although deposition by VHF has been accurately inferred in many cases (e.g., Smith 1986; Smith 1987a, b; Wells and Harvey 1987; Hackett and Houghton 1989; Runkel 1990; Smith 1991a, b; Smith and Lowe 1991; Rodolfo and Arguden 1991; Major et al. 1996; Manville et al. 2005; Procter et al. 2009; Zernack et al. 2009, 2011).

2 Landscape-Scale Characteristics of Deposits

2.1 Deposit Architecture and Morphology

VHF sediment bodies are hard to visualize in three dimensions, because they commonly grade vertically and laterally into VDF and VWF deposits. They are typically preserved as sediment layers accreted onto surfaces of low terraces, as flow units intercalated between VDF flow units, or as channel-fill remnants within channels incised into the fresh VDF deposits emplaced by the same lahar. Hyperconcentrated-flow deposits are not common in nonvolcanic settings and have not been well described there, although one exception is (Wilcox et al. 2014), who documented the connections between flow concentration, changing particle-size distribution, and flow behavior in detail.

2.1.1 Sediment-Body Shape

VHF sediment bodies occupy three-dimensional space much as water-flood deposits do—as flat-surfaced depositional sheets hundreds of meters wide on fans, debris aprons or ring plains, or narrowly confined when filling incised channels. Thin units on fans may be referred to as sheetflood deposits (Smith 1987a, Smith 1991a, b). Individual VHF depositional units have been observed to range in thickness from several tens of centimeters up to 2.5 m (Scott 1988a; Pierson et al. 1996; Cronin et al. 2000), although deposits that could appear to be single depositional units (no obvious internal erosion surfaces) may reach thicknesses of 30 m (Smith 1986). Shallow, vertically sided, incised channels are common on VHF deposit surfaces. It is common for VHF deposits to be interlayered with VDF deposits, particularly when the bulk sediment concentration of the lahar is near the transition between flow types.

Thicker sediment bodies comprising multiple flow units form where channel gradients decrease to threshold values of 0.01 or less (e.g., (Pierson et al. 1996), locations that are usually at least 20–30 km from volcanic vents. VHF sediment bodies (as with VDF sediment bodies) taper in thickness with distance from source. At some volcanoes, VHF deposits may be largely confined within the vertically walled channels that are incised into the surfaces of larger VDF sediment bodies. This scenario occurs when the dilute lahar tail initiates incision due to a pronounced increase in flow turbulence and to the vulnerability to erosion of sediments emplaced earlier by the same lahar. Such incision results in the formation of low, vertically sided, inset terraces (Fig. 2). Such inset VHF deposits form narrow but long tabular sediment bodies.

2.1.2 Structures and Bounding Surfaces

Bars, Boulder Clusters, Isolated Large Clasts, and Berms

Bar deposits are rare or absent in deposits emplaced by broad shallow VHFs (e.g., Smith 1986), although point bar deposits

Fig. 2 Volcanic hyperconcentrated flow (VHF) deposit exposed in vertically walled bank of shallow channel that was incised by the tail of the same lahar that emplaced the deposit—the March 19–20, 1982, North Fork Toutle River lahar that was triggered by an outbreak flood of meltwater from the crater of Mount St. Helens (Washington, USA) (Pierson and Scott 1985; Scott 1988a; Dinehart 1999; Pierson 1999). The lahar transformed from a volcanic debris flow (VDF) to a VHF farther upstream over a channel reach of several tens of kilometers. Note that boulders transported by the lahar are mainly on or near the surface of the deposit. View is looking upstream. USGS photo by T. Pierson

could be expected in highly channelized flows. Longitudinal boulder bars have been observed along some VHF flow paths but were likely emplaced by more watery lahar tails (e.g., Pringle and Cameron 1999). The general lack of bars (accumulations of gravel-size bed-load clasts) is probably due to the high tractive forces applied to large clasts on the channel bed by VHFs, largely preventing deposition during high-discharge flow and reserving deposition for the brief final waning stages of flow (Fig. 2). Flat-topped berms of deposited suspended sediment (devoid of gravel-size clasts) are the more common large structures to form in VHF channels. These berms are emplaced along the insides of channels bends and at other points of decreased flow velocity and turbulence (Fig. 3). Such berms can be tens of meters wide and meters thick (Dinehart 1999). Upper surfaces of such berms can form within centimeters of peak stage level (Dinehart 1999; Cronin et al. 1999).

Isolated outsized clasts and small clusters of gravel clasts of normal-density rock types (up to boulder size) are a common feature on the surfaces of VHF deposits that were emplaced near flow axes where flow velocities were high (Figs. 1a, and 2). Waning stages of VHFs are probably too brief for the accumulation of bars. Outsized clasts are seldom found buried within single VHF depositional units, presumably because the smooth sandy beds of aggrading VHF channels and the large drag forces applied to large clasts promote continued transport by rolling or sliding along the bed until flow discharge wanes, depth decreases, and large clasts moving as bed load become abruptly stranded. In sequences of multiple VHF deposits reflecting aggradation under unsteady flow condition, outsized clasts are commonly buried within sandy beds, and the

a b

Fig. 3 Volcanic hyperconcentrated flow (VHF) berms (up to 30-m wide) emplaced along right bank of the Toutle River—a lower energy channel-margin setting, 83 km downstream of Mount St. Helens (Washington, USA) crater (the lahar source), exposed only hours before being photographed when flow stage dropped during passage of the March 19–20, 1982, lahar (from (Dinehart 1999). **a** Broad berm surface, view looking upstream, with active channel visible in upper right of photo. **b** Berm surface (view toward bridge pier on true right bank); deeper channel flow visible in upper right corner. Glistening surfaces of berms is due to vertically expelled pore water with fine sediment, creating thin sheetflow on berm surfaces that is forming micro-erosion features and cap layers. Pore-water expulsion is due to rapid consolidation of the water-saturated sediments. At both sites, berms formed on insides of bends. USGS photos by R. Dinehart plane-parallel bedding of the sand commonly extends uninterrupted around outsized clasts. Scour structures around such clasts are rare.

Deposit Margins

VHFs do not form marginal flow levees composed of coarse clasts as VDFs do, although marginal accumulations of woody debris are possible. Neither do VHF deposits have distinct lateral margins, because they commonly grade laterally to fluvial-type deposits—low flow velocities and low turbulence at VHF margins cause fallout of suspended sediment and localized transformation to water flow.

Bounding Surfaces

Many VHF surfaces are studded with isolated gravel-size clasts or clast clusters, as noted above, but they are also commonly capped with layers of pumice (or scoria) when abundant amounts of these low-density particles are available for transport. Clasts of pumice of varying size may be transported by VHFs as floating rafts of particles if particle density is less than that of the transporting fluid. When flow discharge drops, stranded rafts of particles are lowered to the bed surface and form pumice mats that cap beds of higher density particles (Fig. 4). Floating organic debris can be mixed in with the pumice mats. Pumice clasts denser than the transporting fluid may also move as bed-load sheets and be deposited as pumice layers interbedded with normal-density sand grains.

Top surfaces of VHF deposits are more commonly draped by thin (1–30 mm), sharply bounded cap layers of finely laminated very fine sand, silt, and clay, commonly localized in shallow swales on the deposit surface (see Sect. 3.6.1 in Chap. "Deposit Identification and Interpretation") (Cronin et al. 2000) (Figs. 1, 3, and 5). Cap layers also form on some VDF deposits.

Fig. 4 Grounded mat or "raft" of pumice and wood particles left on the floodplain of the upper Chaitén River, downstream of Chaitén volcano (Chile) by a waning, dilute, volcanic hyperconcentrated flow (VHF) or a muddy water flood. The pumice layer caps sand and gravel beds beneath. View is looking upstream, with channel to the right. Shovel blade is 16 cm wide. USGS photos by T. Pierson

a b c

Fig. 5 Surface of the volcanic hyperconcentrated flow (VHF) deposit emplaced April 4, 2009, in the Drift River valley downstream of Redoubt Volcano (Alaska, USA) (Waythomas et al. 2013). Deposit remnants are preserved as fill terraces following reincision of the river following lahar passage. **a** Broad flat deposit surface with small, isolated clasts and clast clusters on the deposit surface. View is looking upstream. **b** Cap layer closeup showing flaking erosion by wind following desiccation, exposing the darker coarser deposit beneath. Pencil points downstream. **c** Flat, smooth VHF deposit surface with a cap layer is on the right (shovel blade is 16 cm wide). Edge of deposit exhibits rivulet channels from sheetflow drainage of expelled pore water into the depression on the left (created by melting of ice block). USGS photos by T. Pierson, 6 weeks after deposit emplacement

2.2 Lithofacies

Lithofacies of VHF deposits overlap with those of VDF and VWF, and all three types of deposits can be interbedded. Facies of typical sediment types include (1) massive to faintly sub-horizontal beds of poorly sorted, gravel and gravel-sand mixtures ranging from sandy gravel to gravelly sand; (2) massive to faintly planar-bedded, poorly sorted, coarse sand with gravel lenses and stringers; (3) distinctly planar-bedded, poorly sorted, coarse sand and gravelly sand with mixed pumice and lithic-grain compositions; and (4) massive to faintly planar-bedded fine to medium sand with sub-horizontal silty partings (Pierson and Scott 1985; Smith 1987a; Scott 1988a; Cronin et al. 1999, 2000; Pierson 2005b; Pierson et al. 2013; Waythomas et al. 2013). Sediment types emplaced by high-energy flow along main channel axes encompass a lithofacies association termed *channel facies*, which may include interbedded VDF deposits. VHF sediment emplaced at lower velocities on marginal valley sides, floodplains, or terraces comprise the *overbank facies* (Pierson and Scott 1985; Scott 1988a; Cronin et al. 2000) (Figs. 3, and 6).

Channel-facies deposits are commonly massive and contain the coarsest average grain size and massive textures (Cronin et al. 2000). Emplacement of such coarse unstratified deposits in the channel thalweg (zone of deep, fast flow) may be (1) by true hyperconcentrated flow where intense turbulence and high sediment concentration permit the random accumulation of particles by incremental accretion of particles on the bed or (2) by flow stratification with a coarse underflow moving and depositing as debris flow or as traction-induced granular flow (Cronin et al. 2000; Manville and White 2003). Channel-facies deposits generally comprise either (1) unstratified, cobble- or boulder-rich, massive to thickly bedded sediment if the VHF was highly concentrated or (2) sharply to faintly stratified coarse sand and gravel if deposits were emplaced by more-dilute flows. Channel-facies deposits tend to become progressively finer grained and more distinctly bedded with distance downstream. Thus, mean grain size decreases and sorting values improve with distance away from the volcano (Pierson and Scott 1985; Cronin et al. 1997, 2000).

Overbank-facies deposits comprise finer-grained, predominantly massive to faintly bedded, slightly muddy sand, but both the overbank and channel facies may have out-sized clasts that are found mainly on deposit surfaces in both facies types (Pierson and Scott 1985; Smith 1987a). Overall, average grain size tends to become finer with lateral distance from the channel axis and with distance downstream (Scott 1988a). VHF lithofacies are commonly traceable over hundreds of meters along and across VHF sediment bodies (Smith 1987a); (Cronin et al. 2000), whereas beds in

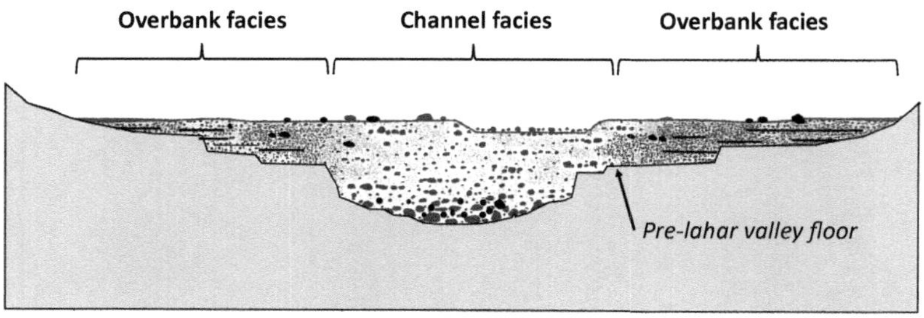

Fig. 6 Schematic cross-section of a typical lahar deposit emplaced mainly as volcanic hyperconcentrated flow (VHF). Coarse material in lower part of buried channel could be a volcanic debris flow (VDF) depositional unit or a particularly coarse VHF unit, which grades upward into finer-grained VHF units. Deposits in the shallow new channel on the fill surface likely contain volcanic water-flood sediment. Overall, coarsest sediment is concentrated in the channel facies

water-flood deposits are much more likely to suffer disruption by thalweg shifts and localized scour/fill.

2.3 Erosion and Effects on Vegetation

Direct observations of rapid turbulent VHFs indicate that this flow type can be highly erosive in channels of unconsolidated material. At Mount Pinatubo following its 1991 eruption, vigorous erosion involving both vertical scour and infill (bed elevation changes up to 10 m/day) and lateral shifting and widening of channels (channel position changes up to 10 m/day) were observed on the lower flanks of the edifice in response to monsoon rainfall (Pierson et al. 1996).

VHFs typically create wide, relatively shallow, vertically walled channels in valleys with mobile bed material (Pierson et al. 1996; Pierson 2005b) (Figs. 2, 5, and 7). The higher the sediment concentration in hyperconcentrated flows, the deeper the eroded channels are relative to their width (Xu 1999). The hyperconcentrated tails of lahars having VDF main bodies commonly erode such rectangular channels into the VDF deposits just emplaced by the main body of the lahar (Pierson 1986; Rodolfo and Arguden 1991). Such incision can emanate from knickpoints ("steps" in the longitudinal profile), and the upstream migration of knickpoints has been observed in active flows (Pringle and Cameron 1999). Such knickpoints are occasionally preserved (Fig. 8).

VHF flows can cause abrasion to standing tree trunks, fences, walls, etc., but these dilute lahars typically do less damage to vegetation than VDFs. Some tree trunks may be debarked on their upstream sides, but serious damage is commonly confined to two zones—near the bed (up to about 50 cm above) where the impact and abrasion by saltating bed load and intermittently suspended gravel clasts is strongest, and near the level of peak stage, where tree trunks can be struck by floating debris (Pierson 2005a).

Fig. 7 Wide rectangular channel with smooth sandy bed, cut by a volcanic hyperconcentrated flow (VHF) in 1993 in the Sacobia River, downstream of Mount Pinatubo (Philippines). View is looking upstream. USGS photo by T. Pierson

Fig. 8 Small knickpoint in lahar channel (looking upstream), preserved after passage of volcanic hyperconcentrated flow (VHF) in Sacobia River, Mount Pinatubo (Philippines) (Pierson 2005b). It was observed being eroded by the tail of the lahar. The VHF shown in Fig. 18 (Chap. "Lahars—Process Overview) was photographed at this location several days before. USGS photo by T. Pierson

Branches of trees and bushes are commonly bent but not broken or significantly abraded. Although crenulations in tree bark may collect particles of sand and silt or a thin skim of silt and fine sand may coat some surfaces, coats muddy matrix material does not adhere to vegetation (Pierson 2005a).

3 Outcrop-Scale Characteristics of Deposits

3.1 Stratification and Bed Forms

Bedding characteristics in VHF deposits appear to depend largely on position within a channel (high-energy vs. low-energy depositional sites), as well as on concentration of the flow mixture and caliber of sediment being transported. Types and degree of stratification in VHF deposits merge with those of VDF and VWF deposits at either end of the concentration spectrum.

Bedding tends to be mostly or entirely absent in deposits of high-concentration VHFs (those approaching VDF concentrations) (Fig. 9). Such deposits closely resemble VDF deposits, except for subtle visual hints of stratification and lower matrix mud contents than in VDF sediment. Such massive gravelly VHF deposits probably result mainly from incremental deposition of gravel bed load being transported in traction along the bed along with some coarse particle fallout from intermittent suspension (Smith 1986).

In sand-dominated VHF deposits, stratification is usually present but typically faint and difficult to see in outcrop. Where present, strata appear as sub-horizontal planar beds of poorly sorted and medium to very coarse sand with few internal erosion surfaces (Pierson and Scott 1985; Scott 1988a; Smith 1986; Cronin et al.

Fig. 9 Volcanic hyperconcentrated flow (VHF) deposit 82 km downstream of Mount Hood (Oregon, USA) along the Sandy River. The VHF deposit (in upper two thirds of exposure) is transitioning upward from faintly bedded sand with consolidation laminations to an unstratified gravelly facies that was emplaced by a more concentrated part of the lahar, as the deposit was being continuously accreted about 1500 years ago. Flow was from left to right. The lower third of the exposure is a fluvial deposit of fine sand. USGS photo by T. Pierson

2000; Wilcox et al. 2014). Where sandy VHF deposits were produced by flow-front mixing in a concentrated lahar, they typically underlie the coarser deposits emplaced by the main body of the flow (Fig. 10). In overbank facies (low-energy depositional environment), sandy deposits may contain more fines than channel facies deposits and may be more weakly stratified or even apparently unstratified (Fig. 11). In channel-facies deposits where heterogeneous sediment is being transported (high-energy depositional environment), bedding contacts are more sharply delineated with abrupt changes in mean grain size, which can be accentuated when mixtures of hydraulically equivalent pumice granules and lithic sand grains are transported together (Fig. 12). Discontinuous gravel lenses (often composed of pumice clasts) and single outsized gravel clasts surrounded by sand (pumice or lithic clasts) are occasional bedding features in channel facies deposits (Fig. 13).

Pseudostratification in VHF deposits is provided by two important types of dewatering structures that mainly form postdepositionally—thin, muddy, sub-horizontal layers sandwiched between layers of coarser sand: consolidation laminations, which form internally within deposits, and cap layers, which form on the surfaces of deposits (Fig. 14) (Sect. 3.4.1 in Chap. "Deposit Identification and Interpretation"). Both can help identify VHF deposits. Consolidation laminations form by translocation of fines during sediment consolidation following emplacement, and cap layers

a b

Fig. 10 Vertical channel-bank outcrops of the March 19–20, 1982, North Fork Toutle River lahar downstream of Mount St. Helens (Washington, USA) (Pierson and Scott 1985; Pierson 1999; Dinehart 1999). The lahar transformed from mainly volcanic debris flow (VDF) to volcanic hyperconcentrated flow (VHF) by the time it had flowed 40 to 60 km. **a** Two distinct facies are present in this outcrop—faintly plane-parallel-bedded sand emplaced by VHF (below) and gravelly diamict emplaced by VDF (above); the transformation was ongoing. **b** At this location farther downstream, only the plane-parallel-bedded sand is present (transformation was complete in this part of the channel). Outsized clasts were stranded on the deposit surface. Lahar flow was right to left, and the entire deposit was emplaced over a period of 12 h. The lahar was triggered by a snowmelt outbreak flood caused by an explosion from a growing lava dome in the snow-filled crater. Shovel blade (in **b**) is 16 cm wide. USGS photo by T. Pierson

Fig. 11 Sand-dominated volcanic hyperconcentrated flow (VHF) deposit with subhorizontal plane-parallel stratification so faint that it is revealed only by slight etching of the outcrop face by wind and rain. Deposit is overbank facies emplaced by same lahar that emplaced deposit shown in Fig. 10. Flow was right to left. Pencil stuck in deposit surface for scale. USGS photo by T. Pierson

a
CM
IN
b

◄ **Fig. 12** Distinctly stratified sandy volcanic hyperconcentrated flow (VHF) deposits (channel facies) emplaced in high-energy, main-channel depositional environments. **a** Subhorizontal, plane-parallel- bedded, medium lithic sand and coarse pumiceous sand to fine gravel deposited by a small lahar (23 vol% sediment concentration) in the Sacobia River, Mount Pinatubo, Philippines (Pierson 2005b). This deposit was photographed immediately after the lahar passed, and the vertical outcrop face was exposed by erosion from the lahar tail. Stratification is so prominent due to the contrasting colors and sizes of the lithic and pumice sands. Flow was left to right. **b** Subhorizontal plane-parallel bedding in a VHF deposit composed of silty medium to coarse lithic sand and pumice granules/small pebbles in Chaitén River, 10 km downstream of Chaitén volcano (Chile). Note small scour channels with coarser fills (arrows) in what had been a shallow, actively aggrading, smooth, sand channel bed during deposition. Flow was toward the camera. Folding shovel blade is 16 cm wide. USGS photos by T. Pierson

Fig. 13 Outsized clasts (normal density dacite) within volcanic hyperconcentrated flow (VHF) deposit in upper Sandy River, about 20 km downstream of Mount Hood (Oregon, USA). Flow direction appears to have been approximately normal to the outcrop face, with channel facies material on the left, grading into overbank facies on the right. Note pinching out of gravelly lenses and internal cross-cutting scour surfaces in the transition zone. Handle of trenching tool (about 4 cm in diameter) for scale. USGS photo by T. Pierson

can form both by direct suspension fallout of fines in slowly moving waning flow in channel embayments and by postdeposition expulsion of fines-rich pore water to deposit surfaces during consolidation (Lowe 1975; Cronin et al. 2000; Pierson 2005b; see Chap. "Deposit Identification and Interpretation"). Both are created by the concentration of clay, silt, and very fine sand carried by pore water being expelled from deposits compacting under their own weight. Consolidation laminations are presumed to form at irregular intervals by the sub-horizontally

a
CM
IN
b

◀ **Fig. 14** Consolidation laminations and cap layers in sandy volcanic hyperconcentrated flow (VHF) deposits. **a** Massive VHF deposit separated by consolidation laminations in 1500-yr-old lahar deposit, Sandy River, 82 km downstream from source on Mount Hood (Oregon, USA). Note that laminations have limited length, are locally nonplanar, and do not appear to be entirely bedding-conformable. Flow was left to right. **b** A straight, light-colored, bedding-conformable cap layer in a VHF deposit (just above center of photo) and a non-bedding-conformable, light-colored, wavy consolidation lamination (just below center of photo) in a deposit emplaced by the April 4, 2009, lahar in Drift River valley, about 38 km downstream of Redoubt Volcano (Alaska, USA) (Waythomas et al. 2013). The lower lamination is crenulated by soft-sediment deformation. The tape measure in lower left of photo is 17 mm wide. Flow was left to right. USGS photos by T. Pierson

directed injection of pressurized muddy pore fluid beneath less permeable layers; they are typically only 1–3 mm thick. Cap layers form on the surfaces of freshly emplaced deposits. They can be thicker than consolidation laminations on flat surfaces (up to ~ 5 mm) and even thicker where depressions on the deposit surface allow deep ponding and suspension fallout of muddy waning flow. Cap layers do not cross-cut surrounding beds; consolidation laminations commonly do. Although not uniquely diagnostic, these layers are much more common in VHF deposits than in any other type of deposit, and they are often associated with the development of dish structures.

A common bed-form type in VHF deposits, other than the commonly encountered subhorizontal planar bedding, is low-angle (<15°) cross stratification (Smith 1986). Low-angle cross-bedding results from the development of antidunes in upper-flow-regime (supercritical) flow. Localized low-angle cross-beds also form from shallow scour and fill structures, but these are relatively uncommon in low-gradient channels, due to the high rates of channel-bed aggradation associated with VHFs. Development of chute-and-pool structures has been reported in active VHFs (Pringle and Cameron 1999), but it has not yet been recognized in deposits, to our knowledge. Low-angle cross-bedding is also found in water-flood and in dilute pyroclastic density–current deposits. However, high-angle (>15°) cross-bedding has not been found in sandy VHF deposits, and its absence is one of the key criteria for identification of these deposits (Pierson and Scott 1985; Smith 1986; Cronin et al. 2000; Pierson 2005b).

3.2 Gravel-Size Clast Framework

High concentrations of gravel-size clasts are uncommon in VHF deposits. Sand-dominated sedimentary textures are the norm, so gravel clasts in VHF deposits are typically matrix supported. However, gravel can be concentrated locally in lenses or stringers having clast-supported structures. Openwork structures are rare if present at all.

3.3 Fabric

Long-axis orientations of gravel-size clasts in VHF deposits in the Cascade Range (USA) were found to be bimodal according to clast size (Smith 1986). In that study, long axes of clasts > 15 cm were oriented perpendicular to flow direction, owing to rolling traction transport along the bed. Long axes of smaller clasts, inferred to have been carried principally by intermittent suspension, were mainly oriented parallel to flow direction. Imbrication of platy clasts (AB planes dipping upstream) was present but weakly developed.

3.4 Grading

Grading is variable in VHF deposits although not always present within beds emplaced by VHFs (Smith 1986; Thouret et al. 1998) (e.g., Figs. 11, and 12b). Grading is found, however, in deposits emplaced by VHFs that are changing their solids concentration as they flow. Inverse grading typically occurs during deposition by a

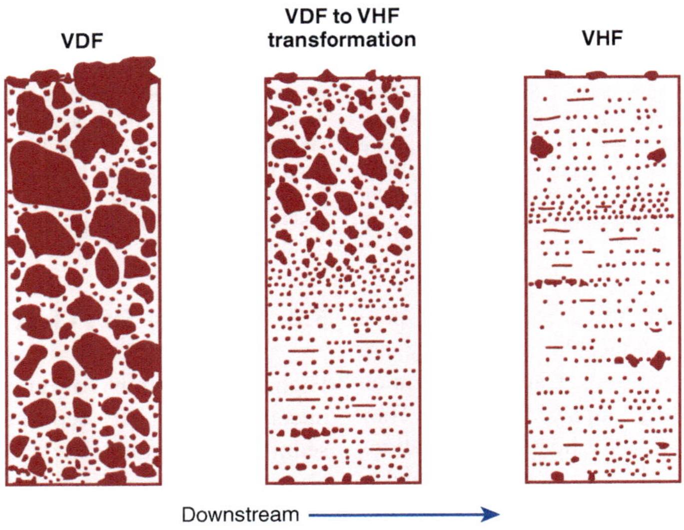

Fig. 15 Schematic representation of longitudinal changes in deposits caused by transformation from volcanic debris flow (VDF) to volcanic hyperconcentrated flow (VHF) in a lahar as it progresses downstream. This transformation is commonly seen as a progressive thickening of a faintly stratified sandy sole layer. Full transformation from VDF to VHF may occur over a few to many kilometers, depending on rate of in-stream dilution or dilution by tributary stream water addition. Transformation can also go from VHF to VDF if a lahar is initiated by a water flood that erodes channel sediment. Modified from Pierson and Scott (1985)

VHF that is becoming more concentrated (suspending coarser clasts) and depositing progressively coarser sediment on the channel bed as it flows (Vallance and Iverson 2015). Mixing of a lahar with river water at the flow front (see Fig. 7 of Chap. "Lahars—Process Overview") can cause such inverse grading (e.g., Figs. 14a, and 15 [middle column]) (Pierson and Scott 1985; Scott 1988a; Cronin et al. 2000). Progressive dilution and fining of particles in a lahar as it flows downstream results in a normally graded deposit (Vallance and Iverson 2015).

Full flow transformation of a VDF to VHF is common in long-runout lahars and can be inferred from progressive changes in grading, bedding, and sediment textures in longitudinal sequences of vertical sections of deposits over many kilometers of channel (e.g., Pierson and Scott 1985) (Fig. 15). Where this occurs, a sandy sole layer progressively thickens (Fig. 15, middle column) with distance downstream, and the overlying gravelly deposit progressively thins (Pierson and Scott 1985; Scott 1988a; Cronin et al. 2000). However, full transformation from VDF to VHF or from VHF to VDF is not a requisite condition for grading in a lahar deposit.

3.5 Fluid-Escape Structures and Soft-Sediment Deformation

Dish structures, dewatering pipes, and soft-sediment deformation (crenulated and convoluted bedding) have been observed in deposits of witnessed VHFs (Cronin et al. 2000) and in volcaniclastic deposits inferred to have been deposited by VHFs (Lowe and LoPiccolo 1974; Lowe 1975; Hodgson 1993). VHFs have properties that favor the formation of such structures in their deposits—rapid deposition in an aggradational environment, loose packing with fluid-filled pore spaces, and silt and clay suspended in the pore fluid to decrease permeability. Rapid upward expulsion of pore water and liquefaction due to undrained loading can result.

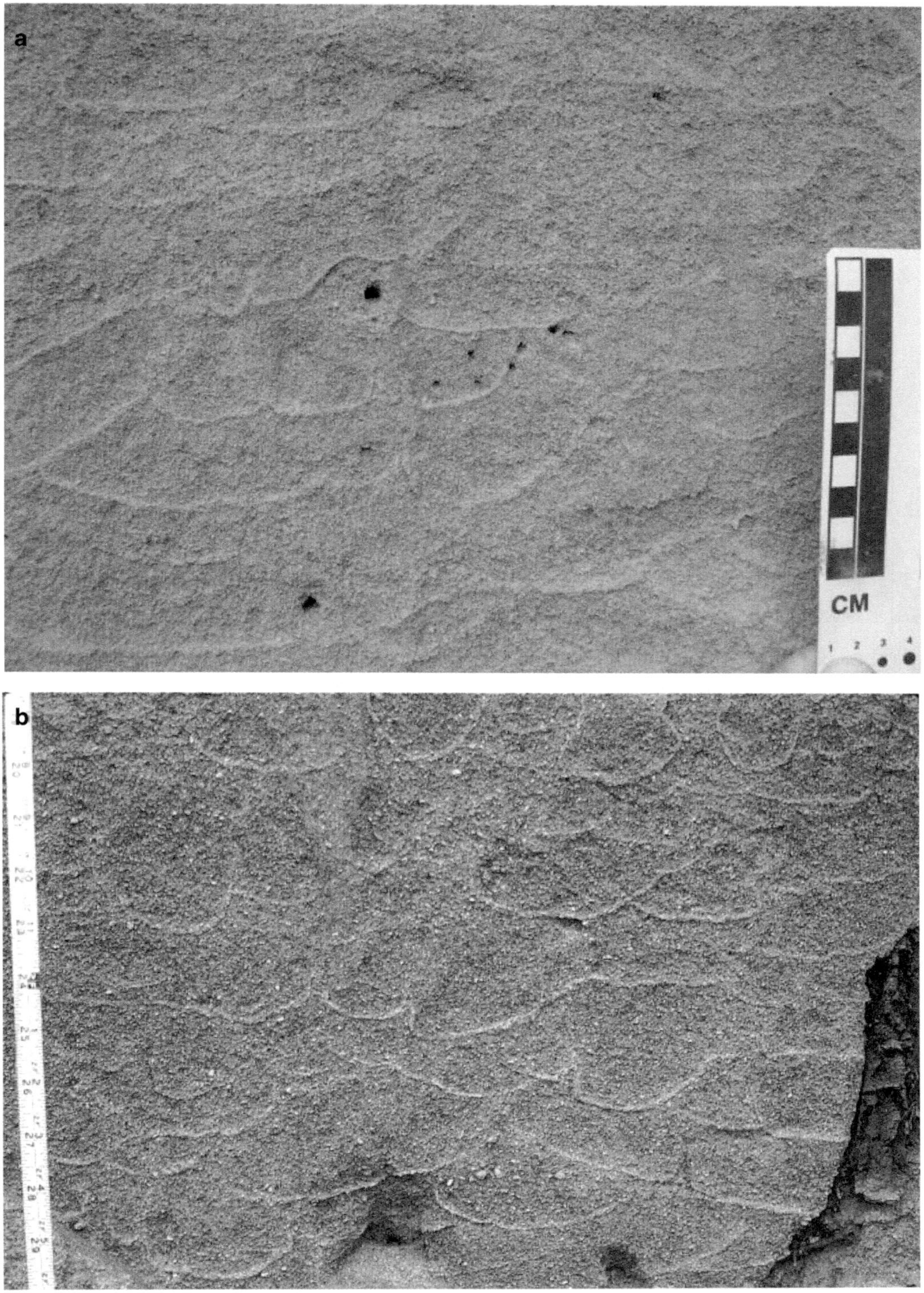

Fig. 16 Dish structures in volcanic hyperconcentrated flow (VHF) deposits. **a** Dish structures in lower part of a VHF deposit on the Sandy River about 81 km downstream from Mount Hood (Oregon, USA). Dark small holes are insect burrows. USGS photo by T. Pierson. **b** Dish structures in the VHF phase of the National Lahar deposit, Mount Rainier (Washington, USA) (Scott et al. 1995); tape scale on left is in inches. USGS photo by K. Scott

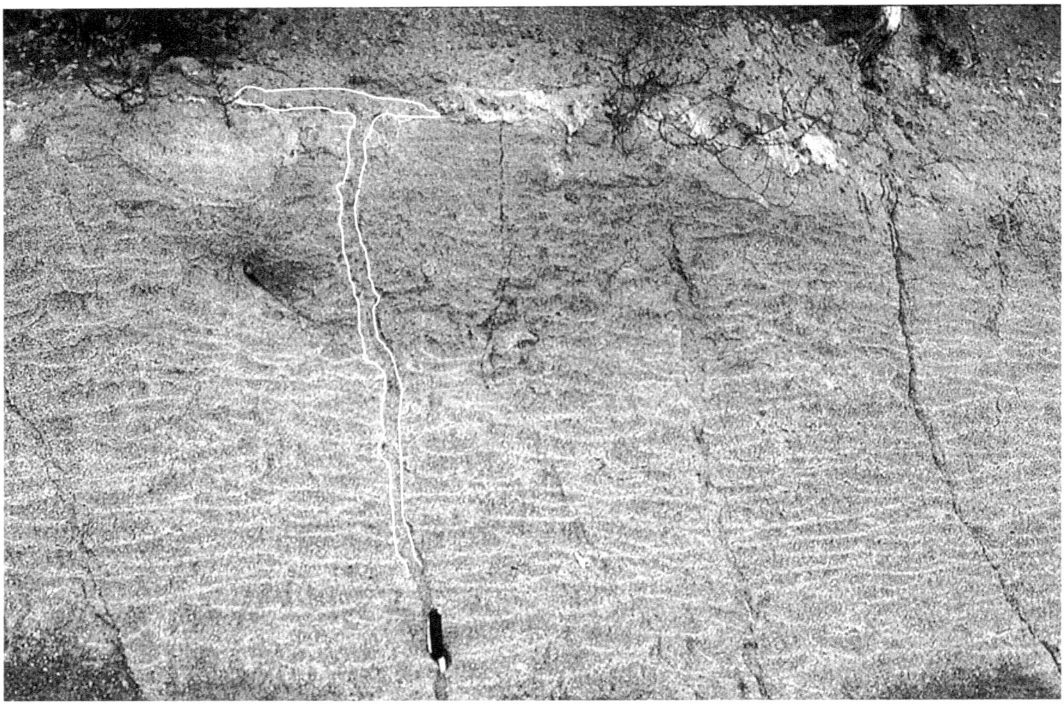

Fig. 17 Dewatering pipes (and dish structures) in massive sand of a volcanic hyperconcentrated flow (VHF) deposit at Ngamoki quarry along Whangaehu River, Mount Ruapehu (New Zealand). Cross-sections of small sand mounds ("sand blows") are preserved at the tops of the pipes on the paleosurface of the deposit. White lines outline a pipe and the sand mound above it. Pocketknife case beneath the outlined pipe is about 5 cm long. USGS photo by T. Pierson

Dish structures are common in VHF deposits but not observed in other volcanic wet-flow deposits or in pyroclastic density–current (PDF), tephra-fall (TF), or glacial till deposits. In vertical cross-section these structures are formed from concave-upward segments of disrupted consolidation laminations or cap layers, resembling small bowls or plates in cross-section (Fig. 16). Individual dishes are on the order of 10 cm across, with upward-turned edges of adjacent dishes meeting to form upward-pointing 'V's. They commonly occur in the same outcrops as dewatering pipes, but dish structures appear to be formed by a more diffuse upward migration of water during rapid deposit consolidation, as opposed to the focused upward flow that forms pipes. Laminations can be deformed into dish structures within hours to days of deposit emplacement. (Cronin et al. 2000).

Dewatering pipes originate in the lower third or half of a deposit and extend upward as narrow irregular conduits filled with well-sorted sand from which fines have been removed (Fig. 17) (Sect. 3.5.1 in Chap. "Deposit Identification and Interpretation"). The pipes tend to flare out within 10 to 20 cm of the top of the deposit. Examples are well developed in dilute lahar deposits downstream of Mount Ruapehu (New Zealand) at Ngamoki quarry. Original bedding within the flare-out zone is disrupted, and coherent fragments of the original deposit float in well-sorted sand. The presence of rounded mounds on the top of the flare-out zones suggests that these were "sand blows" (sand boils) and that the pore water had escaped under considerable pressure. The presence of sand mound structures at the top ends of the pipes rebuts the argument that the pipes are simply rain-wash erosional rills on the outcrop face.

Soft-sediment deformation, made visible by the disruption and contortion of originally flat-lying strata, is formed by the liquefaction and convolution of beds of water-saturated sediment resulting from rapid loading by additional sediment. Upward injections of liquefied soft sediment commonly form flame structures (Cronin et al. 2000), which penetrate overlying strata (Fig. 18).

3.6 Material Consistency

Consistency of fresh VHF deposits is intermediate between that of VDF and VWF deposits—typically softer and more friable than in VDF deposits but firmer and harder than in water-flow deposits, owing to the intermediate amounts of fines preserved in VHF deposits. They are moderately consolidated, with a slightly firm consistency when moist and a slightly hard consistency when dry; outcrops break into fragile chunks when sampled. Outcrops of sandy VHF deposits characteristically form vertical banks that are relatively stable, but outcrops of coarse massive VHF deposits are characteristically unstable (Pierson et al. 1996; Cronin et al. 2000). When digging into them with a shovel, resistance to digging is noticeable, but the sand usually disaggregates completely. Older, chemically weathered deposits can be significantly harder and more cemented, however).

3.7 Evidence of Heat

Evidence of heat (i.e., temperatures > 100 °C) are not expected in VHF deposits because the flows emplacing them are composed dominantly of liquid water. As with VDF deposits, however, it is possible that charcoal fragments or prismatically jointed gravel clasts derived from source PDC deposits could be entrained and transported by VHFs.

4 Particle-Scale Characteristics

4.1 Particle Composition

VHF deposits include monomictic, oligomictic, or polymictic particle mixtures, depending on sediment source, and they commonly contain primary volcaniclastic, secondary volcaniclastic, and epiclastic particles, as do VDF deposits (see Sect. 4.1 in Chap. "Deposit Identification and Interpretation" for definitions of particle composition). VHFs are generally more erosive than VDFs, owing to their generally higher levels of turbulence. Sediment entrainment is particularly effective at lower channel gradients through bank erosion and channel widening, so sediment can be incorporated from channel banks well beyond the flanks of the source volcano (see Chap. "Lahars—Process Overview"). Low-density pumice particles, if present in VHF deposits, tend to be segregated from lithic particles in beds and lenses (Fig. 19), unlike in VDF deposits where grains of contrasting particle densities are better mixed, even though some density segregation may occur. Segregated pumice layers on tops of VHF deposits resemble tephra-fall deposits, and correlation with pumice deposits on higher surfaces may be necessary to verify origin. The pumice layer in Fig. 19a was emplaced at a time when no tephra falls were occurring.

4.2 Particle Roundness, Shape, Fracturing, and Surface Texture

In near-source channels on the flanks of volcanoes, primary gravel-size clasts composed of normal-density volcanic rock are mainly angular, but gravel entrained from bed and bank sediment may be rounded to varying degrees. Pumice clasts and fragile clasts are typically rounded following transport (Fig. 20). The percentage of rounded gravel clasts increases with distance traveled, mainly due to entrainment of more well-rounded fluvial gravel.

a

b

◀ **Fig. 18** Convolute bedding (soft-sediment deformation) in a complex, sandy volcanic hyperconcentrated flow (VHF) deposit that was emplaced in multiple surge units by a rainfall-triggered lahar on the Middle Fork Hood River downstream of Mount Hood (Oregon, USA) in 2006 (Poole 2015). **a** Protrusions of sandy VHF deposit that were injected upward into the overlying volcanic debris flow (VDF) unit that had been emplaced minutes after the VHF unit. Both units were still liquefied when deformation occurred. Shaft of scraper is 5 mm in diameter. **b** A bulbous protrusion of sandy VHF deposit is injected into the overlying VDF deposit at a location a few tens of meters away from the outcrop in photo **a**. This photo also shows pathway of pore water, fines, and sand having been squirted upward (arrow) into the overlying deposit. Overhanging unit contact between the two arrows reflects the high fines content and cohesion of the upper unit. The outcrop face is about one meter tall. USGS photos by T. Pierson

a b

Fig. 19 Segregation of pumice particles (coarse sand and fine gravel) from denser lithic sand particles in volcanic hyperconcentrated flow (VHF) deposits. **a** Pumice gravel layer on top of lithic VHF deposits along the Sacobia River on the lower east flank of Mount Pinatubo (Philippines). **b** Pumice gravel layers interbedded with lithic (rhyolite) sand in outcrop of VHF deposit along the Chaitén (Blanco) River about 10 km downstream of Chaitén volcano (Chile). The thickest pumice bed is about 5 cm thick. USGS photos by T. Pierson

Normal-density rock particles finer than 2 mm typically remain very angular during transport, as is the case in VDF and VWF deposits, as well. Particles in VHF deposits are generally equant in shape, due to the predominantly equant shape of source volcaniclastic rock fragments. Gravel-size-clast surfaces may be rough and pitted due to earlier transport by VDF or smooth on clasts entrained from riverbeds.

4.3 Particle-Size Distribution and Sedimentary Texture

VHF deposits vary from poorly sorted, massive, coarse, boulder-bearing muddy sandy gravel to well-sorted, well-laminated medium sand (Pierson and Scott 1985; Cronin et al. 2000). Particle-size distributions generally are unimodal or bimodal (Fig. 21), although they may be polymodal if samples include both the fine gravel that had been in suspension during flow and coarser clasts moved as bed load (Scott 1988a; Cronin et al. 1999). Total fines content preserved in VHF deposits usually ranges from about 3 to 10 wt%—generally less than for VDF lahar deposits but more than in volcanic water-flood deposits (Pierson and Scott 1985; Scott 1988a; Cronin et al. 1997; Thouret et al. 1998). However, the fines content in flowing VHFs is much greater than that in deposits—20–60 wt% silt and clay in flowing VHFs

Fig. 20 Large rounded fragile clast of semi-consolidated Pliocene Sandy River Mudstone (about 1.5 × 2 m), probably entrained by bank collapse from farther upstream and carried down the Sandy River by a volcanic hyperconcentrated flow from Mount Hood (Oregon, USA) about 1500 years ago. Visible increments on scale are 10 cm. Clast is resting on the paleochannel bed. Photo by L. Siebert

(Pierson and Scott 1985; Pierson et al. 1990; Cronin et al. 1999) (Fig. 4 of Chap. "Lahars—Process Overview"). Most of the silt and clay in VHFs continues to stay in suspension after deposits containing mainly sand and gravel deposits are emplaced.

Sorting values for VHF deposits are typically intermediate between those of VDF deposits and water-flood deposits, namely in the range of about 0.9 to 2.0φ for dominantly sandy deposits, and most commonly in the range of 1.1 to 1.6φ in samples from flowing VHFs (Scott 1988a; Cronin et al. 2000). Sorting values for gravelly VHF deposits can range up to 3.0φ, however (Pierson and Scott 1985). Skewness is generally positive (excess of fines) (Scott 1988a; Cronin et al. 2000). Mean particle size decreases both with distance downstream from source and with time during passage of a VHF past a point (Fig. 21).

4.4 Vesicles and Intergranular Voids

We have not observed gas- or air-bubble vesicles uniformly dispersed within VHF deposits, nor have we seen them reported in the literature. We have seen a few cases, however, where bubble vesicles have been merged and concentrated beneath impermeable cap layers at the tops of deposits (Fig. 22). This is likely due to the ease with which air bubbles might move upward through the lower viscosity pore fluid in freshly emplaced VHF deposits. Without a thick and

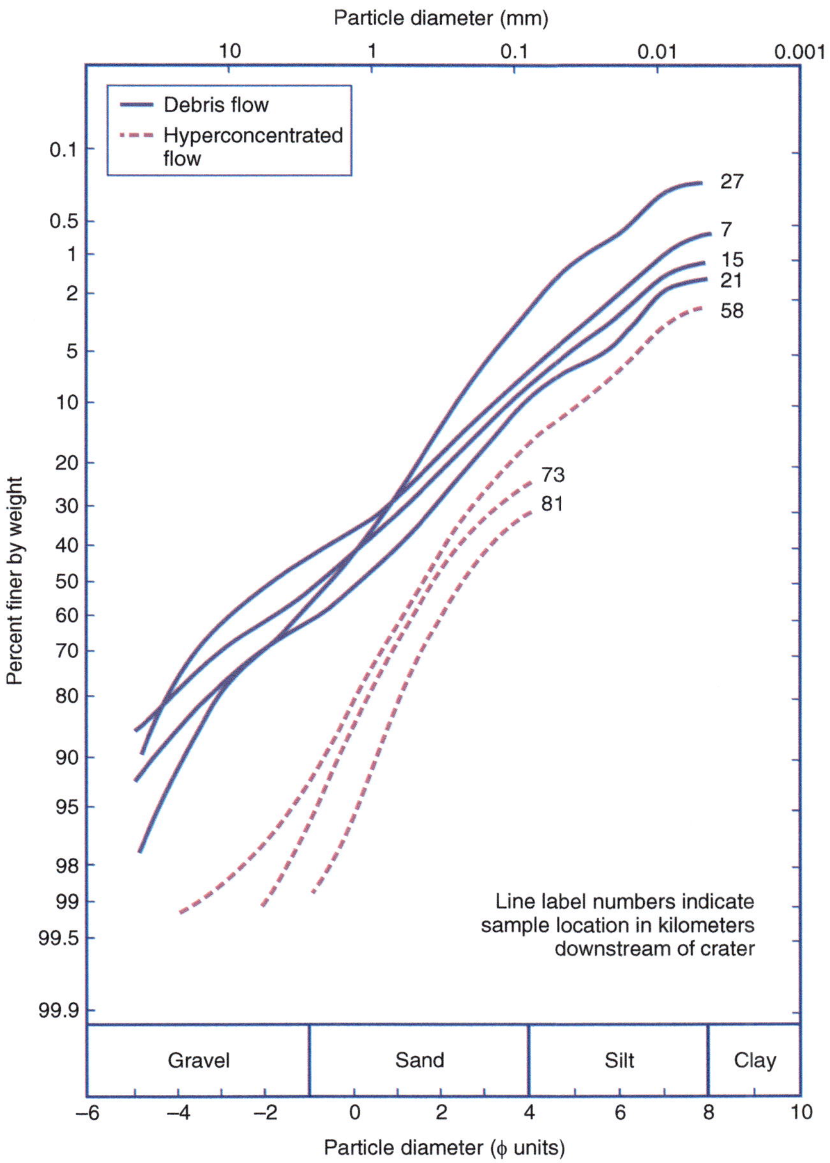

Fig. 21 Cumulative particle-size distributions of deposit samples from the March 19–20, 1982, lahar in the North Fork Toutle and Toutle Rivers, collected at different distances downstream of the crater of Mount St. Helens (Washington, USA), which was the lahar source. The lahar initiated as a lake-breakout water flood, quickly transformed to volcanic debris flow (VDF)—those deposits shown with solid lines. The lahar gradually transformed with distance downstream to volcanic hyperconcentrated flow (VHF)—those deposits shown with dashed lines (Pierson and Scott 1985; Dinehart 1999). Line numbers show distance downstream in kilometers from the Mount St. Helens crater. Modified from Pierson and Scott (1985)

Fig. 22 Large, merged, and flattened air bubble vesicles concentrated beneath an impermeable silty cap layer in a volcanic hyperconcentrated flow (VHF) deposit, downstream of Calbuco volcano (Chile). **a** Extracted piece of the cohesive deposit being held with cap layer on top, viewed in cross-section. Numerous large vesicles are visible (as dark spots) in the sand matrix. **b** Underside of cap layer turned upside down, with a small field of large, flattened, irregularly shaped vesicles localized to the right of the knife blade and indicated by the bracket. USGS photos by T. Pierson

continuous layer of fines capping the deposit surface, air bubbles could escape to the atmosphere before the deposit solidified. Larger intergranular voids are absent in VHF deposits.

5 Key Characteristics of Volcanic Hyperconcentrated Flow (VHF) Deposits

VHF deposits can resemble other types of volcaniclastic sediments, particularly some VDF, VWF, pyroclastic density–current (PDC), and tephra-fall (TF) deposits. A summary table of all relevant characteristics of VHF deposits, together with an identification key, is found in Chap. "Deposit Identification: Key and Summary Tables", and a gallery of outcrop photographs is in Chap. "Annotated Outcrop Photographs: Volcanic Wet-Flow Deposits and Other Unconsolidated Volcanic Deposits". Some of the most useful identification criteria are listed below and are described in detail in the sections above. However, any one characteristic can be found in at least one other type of deposit. Therefore, no single deposit characteristic is entirely sufficient for emplacement process identification. Field observations and physical characteristics of sediment must be evaluated together for an accurate assessment of a deposit's origin.

The characteristics listed below and in Chap. "Deposit Identification: Key and Summary Tables" are based on evaluation of fresh deposits. Fresh deposits and freshly impacted flow paths provide the best opportunities for identifying emplacement process, prior to the complicating effects of weathering, mineral alteration, chemical staining, and cementation of the sediment particles themselves; erosion of landforms; and decay of affected vegetation.

- **Landscape-scale criteria**
  (1) Flat deposit surfaces are typical, commonly with cobbles and boulders sitting on top surfaces.
  (2) Low to moderate abrasion commonly damages standing trees, mainly just above the bed and near peak-flow height.
  (3) Thin skim coats of silty sand commonly found on tree trunks that were submerged; no thick coatings of mud occur.
  (4) Channels eroded by VHF are commonly wide and relatively shallow, with vertical banks; channel beds typically show evidence of aggradation.
  (5) No lateral flow levees are present.

- **Outcrop-scale criteria**
  (1) Both massive and stratified deposits are possible; both may be found in different parts of a single sediment body.
  (2) Sub-horizontal plane-parallel is dominant type; it is commonly faint and diffuse. Sharp bedding contacts are found in channel facies with contrasting types of particles.
  (3) Low-angle (upper flow regime) cross-bedding is common; high-angle cross-bedding (ripple/dune type) may be possible.
  (4) Outsized clasts (pebbles, cobbles, small boulders) are common on deposit surfaces of dominantly sandy deposits but much less commonly buried within single depositional units.
  (5) Imbrication of elongate or platy clasts may be present.
  (6) Deposits have a slightly firm consistency when moist and a slightly hard consistency when dry; outcrops break into fragile chunks.
  (7) Cap layers are common on deposit surfaces.
  (8) Thin, sub-horizontal laminations of clayey silt to silty fine sand are common within deposits; these can be buried cap layers or consolidation laminations. Consolidation laminations are commonly disrupted and deformed into dish structures.
  (9) Dewatering structures in VHF deposits—dish structures, fluid-escape pipes, soft-sediment deformation—are common where sudden burial of fresh deposits causes undrained loading. Soft-sediment deformation is common only in VHF deposits.

- **Particle-scale criteria**
  (1) Some fines are preserved in sandy matrix of deposits—more than in VWF deposits.
  (2) Minimal rounding of first-cycle lithic gravel-size clasts can occur; entrained fluvial gravel from streambeds may be well rounded, and incorporated pumice clasts are typically well rounded.
  (3) Sorting is moderate to very poor (0.9 to 3.0φ); most commonly in the range of 1.1 to 1.6φ for sandy deposits.
  (4) Gas/air bubble vesicles have been reported only under impermeable cap layers.

References

Cronin SJ, Neall VE, Palmer AS, Lecointre JA (1997) Changes in Whangaehu River lahar characteristics during the 1995 eruption sequence, Ruapehu volcano, New Zealand. J Volcanol Geotherm Res 76:47–61

Cronin SJ, Neall VE, Lecointre JA, Palmer AS (1999) Dynamic interactions between lahars and stream flow: a case study from Ruapehu volcano, New Zealand. Geol Soc Am Bull 111:28–38

Cronin SJ, LeCointre JA, Palmer AS, Neall VE (2000) Transformation, internal stratification, and depositional processes within a channelized, multi-peaked lahar flow. New Zeal J Geol Geophys 43:117–128

Dinehart RL (1999) Sediment transport in the hyperconcentrated phase of the March 19, 1982, lahar. In: Pierson TC (ed) Hydrologic consequences of hot-rock/snowpack interactions at Mount St. Helens volcano, Washington, 1982–84, vol 1586. US Geology Survey Professional Paper, pp 37–52

Hackett WR, Houghton BF (1989) A facies model for a Quaternary andesitic composite volcano: Ruapehu, New Zealand. Bull Volcanol 51:51–68

Hodgson KA (1993) Late Quaternary lahars from Mt. Ruapehu in the Whangaehu River valley, North Island, New Zealand. Dissertation, Massey University, 242 pp

Janda RJ, Scott KM, Nolan KM, Martinson HA (1981) Lahar movement, effects, and deposits. In: Lipman PW, Mullineaux DR (eds) The 1980 eruptions of Mount St. Helens, Washington, vol 1250. US Geology Survey Professional Paper, pp. 461–478

Lowe DR (1975) Water escape structures in coarse-grained sediments. Sedimentology 22:157–204

Lowe DR, LoPiccolo RD (1974) The characteristics and origins of dish and pillar structures. J Sediment Petrol 44:484–501

Major JJ, Janda RJ, Daag AS (1996) Watershed disturbance and lahars on the east side of Mount Pinatubo during the mid-June 1991 eruptions. In: Newhall CG, Punongbayan RS (eds) Fire and mud: eruptions and lahars of Mount Pinatubo, Philippines. Philipp Inst Volcanol Seismol, Quezon City, and University of Washington Press, Seattle, pp 895–919

Manville VR, White JDL (2003) Granular mass flows at the base of sediment-laden floods, and the roles of flow competence and flow capacity in the deposition of stratified boulder sands. Sediment Geol 155:157–173

Manville V, Newton EH, White JDL (2005) Fluvial responses to volcanism: resedimentation of the 1800a Taupo ignimbrite in the Rangitaiki River catchment, North Island, New Zealand. Geomorphology 65:49–70

Pierson TC (1986) Flow behavior of channelized debris flows, Mount St. Helens, Washington. In: Abrahams AD (ed) Hillslope processes. Allen Unwin, Boston, pp 269–296

Pierson TC (1999) Transformation of water flood to debris flow following the eruption-triggered transient-lake breakout from the crater on March 19, 1982. In: Pierson TC (ed) Hydrologic consequences of hot-rock/snowpack interactions at Mount St. Helens volcano, Washington 1982–84, vol 1586. US Geology Survey Professional Paper, pp 19–36

Pierson TC (2005a) Distinguishing between debris flows and floods from field evidence in small watersheds. US Geology Survey Fact Sheet 2004–3142, 4 pp

Pierson TC (2005b) Hyperconcentrated flows—transitional process between water flow and debris flow (Chapter 8). In: Jakob M, Hungr O (eds) Debris-flow hazards and related phenomena. Praxis-Springer, Berlin, pp 159–202

Pierson TC, Scott KM (1985) Downstream dilution of a lahar: transition from debris flow to hyperconcentrated streamflow. Water Resour Res 21:1511–1524

Pierson TC, Janda RJ, Thouret JC, Borrero CA (1990) Perturbation and melting of snow and ice by the 13 November 1985 eruption of Nevado del Ruiz, Colombia, and consequent mobilization, flow, and deposition of lahars. J Volcanol Geotherm Res 41:17–66

Pierson TC, Major JJ, Amigo Á, Moreno H (2013) Acute sedimentation response to rainfall following the explosive phase of the 2008–2009 eruption of Chaitén volcano, Chile. Bull Volcanol 75:723. https://doi.org/10.1007/s00445-013-0723-4

Pierson TC, Daag AS, Delos Reyes PJ, Regalado MTM, Solidum RU, Tubianosa BS (1996) Flow and deposition of posteruption hot lahars on the east side of Mount Pinatubo, July-October 1991. In: Newhall CG, Punongbayan RS (eds), Fire and mud: eruptions and lahars of Mount Pinatubo, Philippines: Philippines Institute Volcanology Seismological, Quezon City, and University of Washington Press, Seattle, pp 921–950

Poole M (2015) Sedimentologic changes in the deposits of an evolving lahar-flood, Hood River Basin, Mount Hood, Oregon. MSci Thesis, Portland State University, 130 pp

Pringle PT, Cameron KA (1999) Eruption-triggered lahar on May 14, 1984 In: Pierson TC (ed) Hydrologic consequences of hot-rock/snowpack interactions at Mount St. Helens volcano, Washington 1982–84, vol 1586. US Geology Survey Professional Paper, pp 81–103

Procter JN, Cronin SJ, Zernack AV (2009) Landscape and sedimentary response to catastrophic debris avalanches, western Taranaki, New Zealand. Sediment Geol 220(3–4):271–287. https://doi.org/10.1016/j.sedgeo.2009.04.027

Rodolfo KS, Arguden AT (1991) Rain-lahar generation and sediment-delivery systems at Mayon volcano, Philippines. In: Fisher RV, Smith GA (eds) Sedimentation in volcanic settings, vol 45. SEPM (Social Sediment Geology) Special Publication, pp 71–87

Runkel AC (1990) Lateral and temporal changes in volcanogenic sedimentation: analysis of two Eocene sedimentary aprons, Big Bend region, Texas. J Sediment Petrol 60:747–760

Scott KM (1988a) Origins, behavior, and sedimentology of lahars and lahar-runout flows in the Toutle-Cowlitz River system. US Geology Survey Professional Paper 1447-A, 74 pp

Scott KM (1988b) Origin, behavior, and sedimentology of prehistoric catastrophic lahars at Mount St. Helens, Washington. In: Clifton HE (ed) Sedimentologic consequences of convulsive geologic events, vol 229. Geology Survey Professional Paper, pp 23–36. https://doi.org/10.1130/SPE229-p23

Scott KM (1989) Magnitude and frequency of lahars and lahar-runout flows in the Toutle-Cowlitz River system. US Geology Survey Professional Paper 1447-B, 33 pp

Scott KM, Vallance JW, Pringle PP (1995) Sedimentology, behavior, and hazards of debris flows at Mount Rainier, Washington, vol 1547. US Geology Survey Professional Paper, 56 pp

Smith GA (1986) Coarse-grained nonmarine volcaniclastic sediment: terminology and depositional process. Geol Soc Am Bull 97:1–10

Smith GA (1987a) The influence of explosive volcanism on fluvial sedimentation: the Deschutes Formation (Neogene) in central Oregon. J Sediment Petrol 57:613–629

Smith GA (1987b) Sedimentology of volcanism-induced aggradation in fluvial basins: examples from the Pacific Northwest, U.S.A. In: Ethridge FG, Flores RM, Harvey MD (eds) Recent developments in fluvial sedimentology, vol 39. Society Economic Paleontology Mineral Special Publication pp 217–228

Smith GA (1991a) Facies sequences and geometries in continental volcaniclastic sediments. In: Fisher RV, Smith GA (eds) Sedimentation in volcanic settings, vol 45. SEPM (Society Sediment Geology) Special Publication, pp 109–121

Smith RCM (1991b) Post-eruption sedimentation on the margin of a caldera lake, Taupo Volcanic Centre, New Zealand. Sediment Geol 74:89–138

Smith GA, Lowe DR (1991) Lahars: volcano-hydrologic events and deposition in the debris flow—hyperconcentrated flow continuum. In: Fisher RV, Smith GA (eds) Sedimentation in volcanic settings, vol 45. SEPM (Social Sediment Geology) Special Publication, pp 59–70

Thouret J-C, Abdurachman KE, Bourdier J-L, Bronto S (1998) Origin, characteristics, and behavior of lahars following the 1990 eruption of Kelud volcano, eastern Java (Indonesia). Bull Volcanol 59:460–480

Vallance JW, Iverson RM (2015) Lahars and their deposits. In: Sigurdsson H, Houghton B, McNutt SR, Rymer H, Stix J (eds) The encyclopedia of volcanoes, 2nd edn. Elsevier, Amsterdam, pp 649–664. https://doi.org/10.1016/B978-0-12-385938-9.00037-7

Waythomas CF, Pierson TC, Major JJ, Scott WE (2013) Voluminous ice-rich lahars generated during the 2009 eruption of Redoubt Volcano, Alaska. J Volcanol Geotherm Res 259:389–413

Wells SG, Harvey AM (1987) Sedimentologic and geomorphic variations in storm-generated alluvial fans, Howgill Fells, northwest England. Geol Soc Am Bull 98:182–198

Wilcox AC, O'Connor JE, Major JJ (2014) Rapid reservoir erosion, hyperconcentrated flow, and downstream deposition triggered by breaching of 38 m tall Condit Dam, White Salmon River, Washington. J Geophys Res Earth Surf 119:1376–1394. https://doi.org/10.1002/2013JF003073

Xu J (1999) Erosion caused by hyperconcentrated flow on the Loess Plateau of China. CATENA 36:1–19

Zernack AV, Cronin SJ, Neall VE, Procter JN (2011) A medial to distal volcaniclastic record of an andesite stratovolcano: detailed stratigraphy of the ring-plain succession of south-west Taranaki, New Zealand. Int J Earth Sci (Geol Rundsch) 100:1937–1966. https://doi.org/10.1007/s00531-010-0610-6

Zernack AV, Procter JN, Cronin SJ (2009) Sedimentary signatures of cyclic growth and destruction of stratovolcanoes: a case study from Mt. Taranaki, New Zealand. Sed Geol 220(3–4):288–305. https://doi.org/10.1016/j.sedgeo.2009.04.024

Volcanic Debris Avalanches—Process Overview

Abstract

Volcanic debris avalanches (VDAs) start as rapid gravitational collapses (landslides) of weakened volcanic edifices, which can vary from limited-volume flank failures that leave relatively shallow scars on volcano sides to enormous edifice-collapse failures involving the volcano summit and core, leaving deep-seated, bowl-shaped, open-sided depressions. VDA movement begins as sliding but quickly progresses to thorough fracturing of edifice rock, crushing disaggregation of some of that rock, and subsequent highly inertial, particle–fluid flow of rock fragment mixtures and other debris that are typically wet but unsaturated. Intergranular spaces are occupied by air and volcanic gases as well as minor amounts of liquid water. VDAs differ from nonvolcanic rock avalanches; they tend to involve larger volumes of material, produce higher proportions of fine-grained material, and have proportionally longer runout distances. Moving VDAs typically transport numerous debris-avalanche blocks—fractured but still-intact pieces of the source edifice, which can vary up to hundreds of meters in diameter and which form prominent positive-relief features in VDA deposits (hummocks). VDAs either directly deposit their debris masses, typically within a few tens of kilometers of their source, or transform to volcanic debris flows (VDFs) that can flow much farther downstream.

1 Definitions and Context

Volcanic debris avalanches (VDAs) are rapid gravitational mass movements that begin as sliding failures of parts of volcanic edifices and then rapidly transform to flows of non-coherent, unsorted, unsaturated particle–fluid mixtures of rock fragments, soil, and commonly ice and wood fragments, plus other debris entrained during transport, mixed with varying amounts of air, volcanic gases, and commonly liquid water in the dilated interstices between particles (e.g., Ui 1983, 1989; Voight et al. 1983; Schuster and Crandell 1984; Siebert 1984, 1996; Ui et al. 2000; van Wyk de Vries and Davies 2015; Roverato et al. 2021). VDAs are not water-saturated overall, although localized pockets of saturated material may occur in some avalanches. Particle–fluid interactions occur in primarily dry or damp material, in contrast to the water-saturated mixtures in lahars. These complex mass movements typically transport enormous volumes of material and emplace extended aprons of relatively thick, poorly sorted deposits of fragmental rock debris at the bases of their source volcanoes, both on land and under the sea. Although submarine debris-avalanche deposits have textural and large-scale morphological similarities to their terrestrial counterparts, submarine deposit outcrop-scale textures generally are not accessible and discussion of VDA deposits in this volume focuses on subaerial examples.

T. C. Pierson et al., *Deposits of Volcanic Wet Flows*, Advances in Volcanology,
https://doi.org/10.1007/978-3-031-66574-5_7

Edifice collapse during the 1980 eruption of Mount St. Helens (Washington, USA) was a seminal event that changed the way volcanologists think about how volcanoes behave. VDA deposits were not widely recognized or understood prior to then although they had been identified elsewhere (Siebert and Roverato 2021). Landslide origins for VDA deposits had been noted in Indonesia (Neumann van Padang 1939), Alaska (Griggs 1920), and for collapse of oceanic shield volcanoes in Hawaii and the Canary Islands (Moore 1964; von Buch 1825). Volcanologists in Japan began to see a distinction between deposits of lahars and VDAs prior to 1980 (Murai 1961; Mizuno 1964; Ando and Yamagishi 1975; Nakamura 1978; Moriya 1980). The edifice collapse at the onset of the May 18, 1980, Mount St. Helens eruption (Figs. 1 and 2) demonstrated that VDA deposits (Fig. 3) were unusually thick and morphologically and texturally distinctive in comparison with deposits of other types of volcanic particle–fluid flows (see Chap. "Volcanic Debris-Avalanche Deposits").

After the 1980 edifice collapse, the first such event witnessed and documented at the time of occurrence, geologists began to identify VDA deposits at many other volcanoes (e.g., Fig. 4) and compile regional and global inventories. In the four decades since the 1980 Mount St. Helens eruption, more than a thousand collapse events from nearly 600 volcanoes have been documented from deposits or inferred from source areas, along with an average of more than one regional or global review annually, almost equally divided between subaerial and submarine events (Dufresne et al. 2021a; Siebert and Roverato 2021; Siebert and Reid 2023).

VDAs differ from their nonvolcanic counterparts (e.g., Hsü 1975; McSaveney 1978; Plafker and Ericksen 1978; Scott et al. 2001; Strom et al. 2019) in that they tend to involve larger volumes of material, produce higher proportions of fine-grained material, and have proportionally longer runout distances (Dufresne et al. 2021a). Both have similar spreading behaviors that correlate well with deposit area, although VDAs typically have longer runouts than nonvolcanic rock

Fig. 1 Initial May 18, 1980 events at Mount St. Helens (Washington, USA), viewed from the west. Rapidly descending slide block I (lighter-colored area below left arrow) was estimated to be moving at about 100 m/s (Voight et al. 1983). Dark clouds upslope of the slide block are the beginnings of the lateral blast—exploding rock debris and gas escaping from the suddenly depressurized magma body that slide block I had been sitting over. The right arrow marks the approximate starting point for the head of the initial slide block; a second slide block (not visible in this image) is moving at the right of slide block I. Photograph by Robert Rogers, used with permission

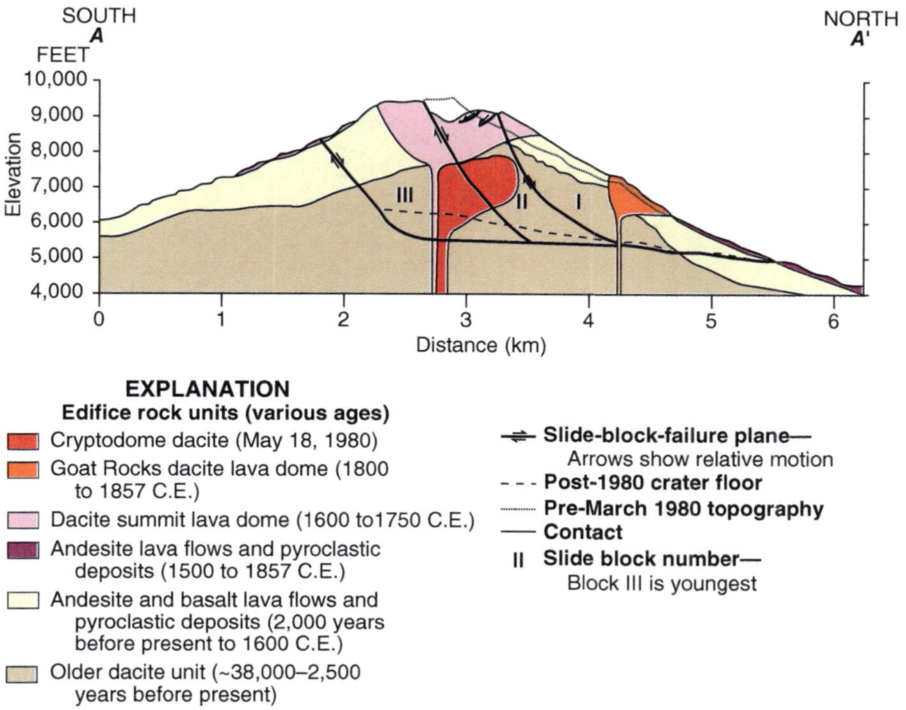

Fig. 2 Mount St. Helens (Washington, USA) volcanic edifice units prior to 1980 CE edifice collapse. Cross-sectional pre-collapse view of retrogressive slide blocks I, II, and III, as viewed from the east. Pierson et al. (2018), modified from Glicken (1996)

avalanches at comparable volumes (Dufresne et al. 2021a) (Fig. 5a). Lucas et al. (2014) noted a correlation between volume and sliding velocity with reduced friction coefficients promoting long runouts for landslides on Earth and planetary bodies. Volcanic failures produce greater volumes with scars that are typically deeper-seated and more bowl-shaped than debris-avalanche scars in nonvolcanic terrain. This is displayed in volume versus drop height plots (Fig. 5b) and shows that nonvolcanic avalanches typically have smaller volumes for similar drop heights than VDAs (Dufresne et al. 2021a).

VDA hazards include not only the direct impacts from the avalanches themselves but also indirect impacts from (1) laterally directed explosions produced when collapses depressurize newly intruded magma, (2) tsunamis generated when avalanches flow into the sea or the failure plane extends below sea level, and (3) hazards from any tephra fall (TF), pyroclastic density currents (PDCs), or lahars triggered by edifice collapses. Casualties from tsunamis strongly dominate the known fatalities caused by debris avalanches worldwide (Siebert 1984; Begét 2000; Silver et al. 2009; Day 2015; Siebert and Roverato 2021).

2 Initiation and Mobilization

Volcanic debris avalanches have various underlying causes (Fig. 6), all of which can lead to structural failure and gravitational collapse of parts of active and inactive volcanic edifices. Compared to nonvolcanic mountain slopes underlain only by fractured solid rock, volcanic edifices are typically constructed of alternating layers and lenses of steeply sloping fractured lava rock (flows, dikes, and domes) plus loose fragmental debris (PDC, TF, and lahar deposits)—all held in place mainly by internal

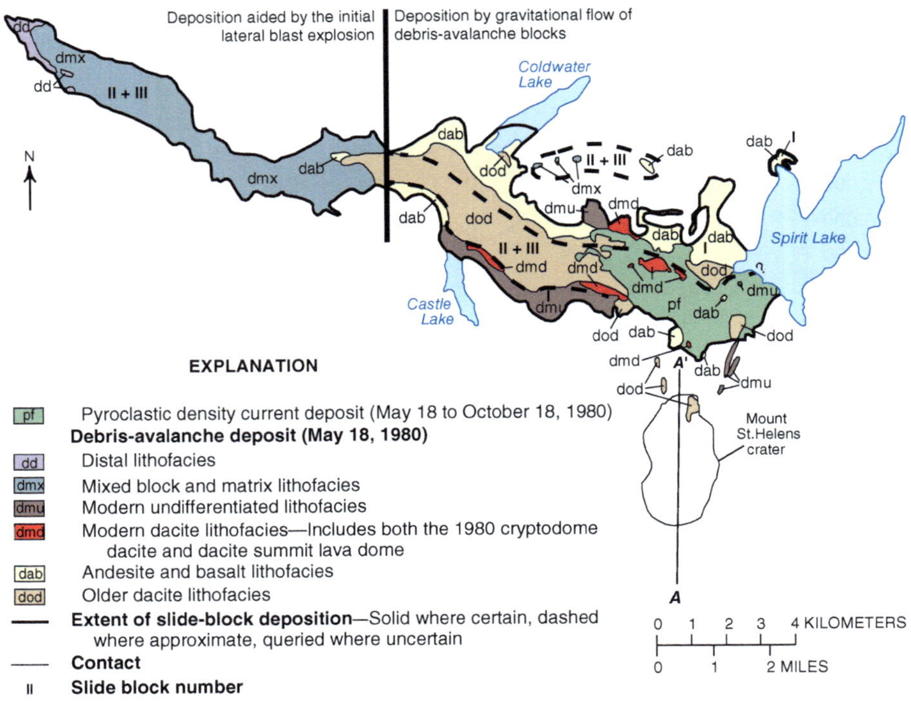

Fig. 3 Distribution of edifice units in the 1980 Mount St. Helens volcanic debris avalanche (VDA) deposit (Washington, USA). Roman numerals indicate inferred depositional areas of the individual VDA slide blocks shown in Fig. 2. Pierson et al. (2018), modified from Glicken (1996)

friction. Edifices can be further weakened by hydrothermal alteration—chemical reactions that convert initially hard aluminosilicate minerals in volcanic rock to softer, weaker mineral species (López and Williams 1993; Watters and Delahaut 1995; Frank 1995; Watters et al. 2000; Reid et al. 2001; Ward and Day 2006; John et al. 2008). These chemical reactions are accelerated in edifices by groundwater that is acidified by dissolved volcanic volatiles and driven upward by magmatic heat through permeable units or along fracture systems into edifices (Hurwitz et al. 2003; Ball et al. 2018). Reid et al. (2010) noted that although hydrothermal alteration could influence failure at other stratovolcanoes it was not a dominant factor at the 1980 Mount St. Helens collapse.

Volcanoes particularly susceptible to gravitational failure include steep-sided intermediate-composition to silicic stratocones and lava dome complexes, but flatter mafic oceanic shield volcanoes also can be vulnerable, as commonly seen in island groups such as the Hawaiian and Canary Islands. In a few cases, gravitational failure of whole subaerial volcanic fields has been documented, such as at Heart Mountain in Wyoming, USA (Malone et al. 2017) and in megaslides in the Margakunt Plateau of Utah, USA—mass movements rivaling volumes of submarine VDAs (Hacker et al. 2014; Biek et al. 2019). In a few cases, enormous gravitational failure involving large parts of subaerial volcanic fields has been documented, such as the Heart Mountain landslide (Absaroka volcanic field) in Wyoming, USA (Malone et al. 2017) and Margakunt Plateau megaslides in Utah, USA—subaerial mass movements rivaling the volumes of submarine VDAs (Hacker et al. 2014; Biek et al. 2019).

Lava domes and stratocones situated on or near potential caldera ring faults may produce debris avalanches that flow inward onto caldera

Fig. 4 Hummocky topography of massive >45 km^3 Pleistocene volcanic debris avalanche (VDA) deposit from Mount Shasta (California, USA) (Crandell 1989). USGS photo by S. Brantley

floors rather than producing outward-spreading VDAs. This is common where syneruptive caldera collapse destabilizes the foundation beneath such edifices (Belousov et al. 2005). This type of failure occurred at Rinjani volcano (Indonesia) in 1257 CE, when a large portion of the stratovolcano collapsed into the newly forming Samalas caldera (Lavigne et al. 2013). Large syneruptive and posteruptive collapses of caldera walls themselves can also produce significant VDAs. At Fernandina volcano (Galápagos Islands) in 1988, nearly 1 km^3 of the caldera wall collapsed inward during an effusive eruption (Chadwick et al. 1991). Caldera wall collapses at Crater Lake (Oregon, USA) that deposited large debris avalanches on the caldera floor (Bacon et al. 2002) and a caldera rim collapse at Piton de la Fournaise (Réunion Island) (Merle et al. 2008; Michon et al. 2023) occurred long after caldera formation.

The stability of a volcanic edifice greatly depends on the highly variable strength properties of the rock composing it (Watters et al. 2000; Zimbelman et al. 2004; Heap and Violay 2021). Strength loss in rock is attributed to both the degree of rock fracturing and hydrothermal alteration of the rock, both of which increase with time in active volcanic edifices (Kilburn and Voight 1998; John et al. 2008; Heap et al. 2021). Edifice stability has been analyzed by many authors who have shown that stability is controlled by parameters including those shown in Fig. 6 and listed below (e.g., Day 1996; McGuire 1996; Russo et al. 1996; van Wyk de Vries and Borgia 1996; Elsworth and Voight 1996; van Wyk de Vries and Francis 1997; Voight and Elsworth 1997; Keating and McGuire 2000; Reid et al. 2000, 2001, 2010; Voight 2000; Lagmay et al. 2000; Hurwitz et al. 2003; Reid 2004; Thomas et al. 2004; Wooller et al. 2004; Norini and Lagmay 2005; Capra 2006; Tibaldi et al. 2006; Reid and Brien 2006; Watt et al. 2009; Wooller et al. 2009; Andrade and van Wyk de Vries 2010; Tormey 2010; Paguican et al. 2012, 2014; Capra et al. 2013; Schaefer et al. 2015; van Wyk de Vries and

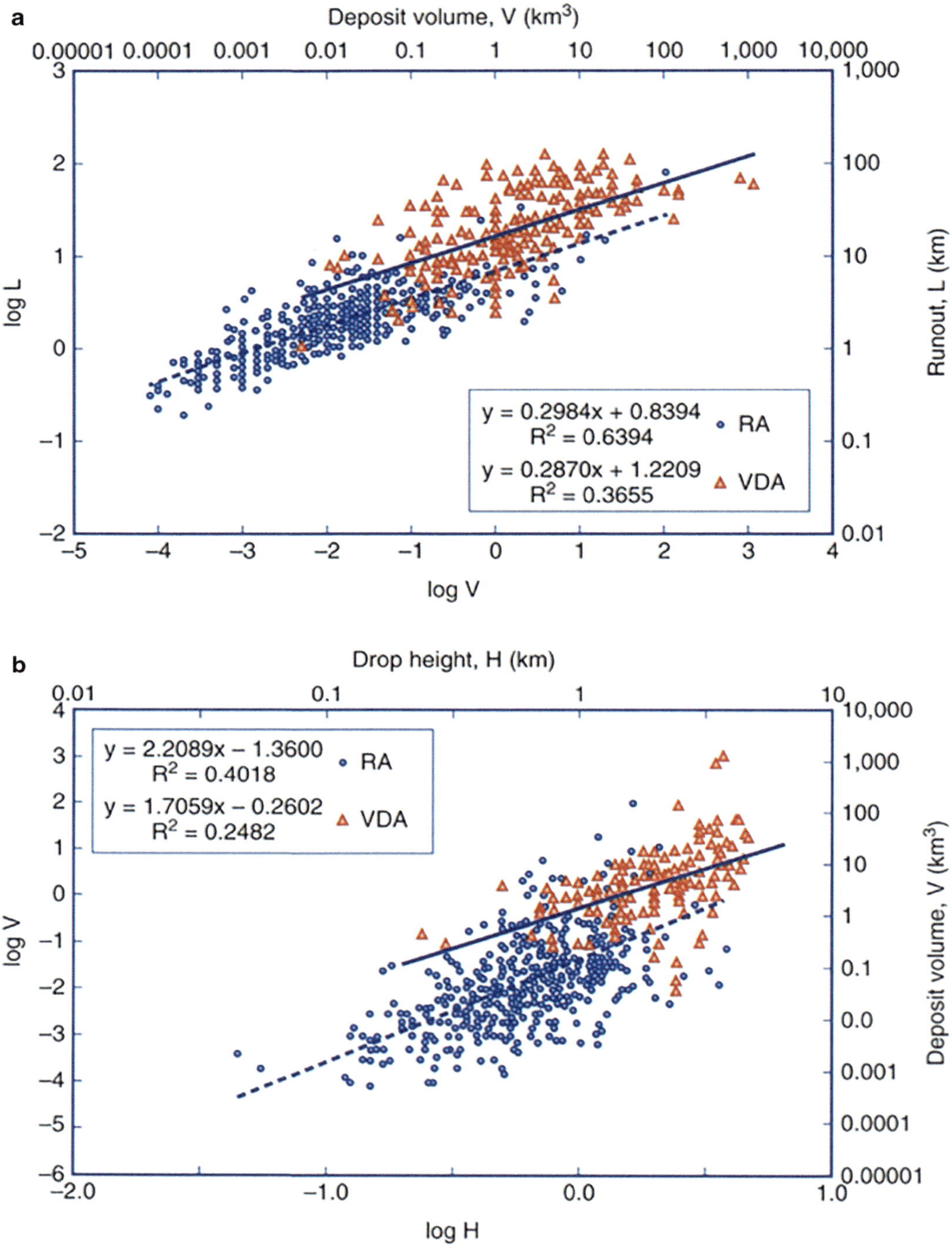

Fig. 5 Comparison of runout distance (L), deposit bulk volume (V), and drop height (H) for volcanic debris avalanches (VDAs) and nonvolcanic debris avalanches (rock avalanches) (RAs) (modified from Dufresne et al. 2021a). Orange triangles = VDAs; blue circles = RAs. Solid trend lines are for VDAs, dashed lines for nonvolcanic avalanches. **a** Bulk avalanche volume versus runout distance. **b** Drop height versus bulk volume

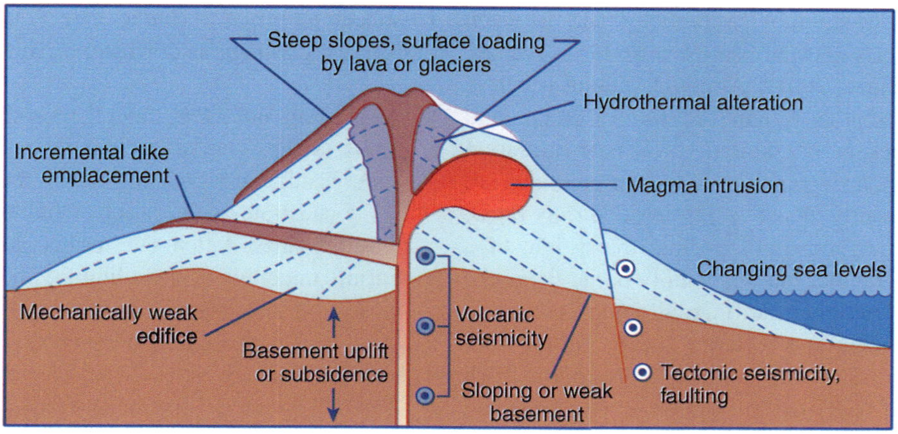

Fig. 6 Schematic representation of a few of the many volcanic and nonvolcanic factors contributing to edifice instability. Modified from Siebert et al. (2019), which was modified from McGuire (1996). Voight and Elsworth (1997) noted as many as 48 destabilizing factors from magmatic-intrusion-related collapses alone

Davies 2015; Norini et al. 2019; Lavallée and Kendrick 2021; Roverato et al. 2021; Siebert and Reid 2023):

- Edifice geometry (height, slope steepness, shape)
- Strength of the underlying basement foundation and the slope of contact with basement
- Types, densities, and orientations of fabric elements and structural discontinuities within the edifice (joints, fractures, faults, dikes, beds of fragmented rock)
- Regional tectonic stresses, strain, faulting
- Groundwater distribution and hydraulic conductivity within the edifice
- Distribution of thermally and mechanically elevated fluid pressures within the edifice
- Thermal and mechanical stress from magma intrusion (microfracturing and strength reduction)
- Gravitational spreading of the volcanic edifice or its foundation
- Climate and sea level changes, debuttressing of steep slopes by glacier shrinkage.

Most factors acting to reduce edifice stability operate over long time periods (10^2–10^5 years). Shorter-term dynamic triggers of failure operate on the scale of seconds to months; these include (1) rapid pore-pressure increase from heavy rainfall (Kerle et al. 2003; Devoli et al. 2009), (2) mechanical strain-weakening and deformation of rock and rapid mechanical and hydrothermal pressurization of pore fluids from intrusion of magma to high levels in the edifice (Elsworth and Voight 1995, 1996; Voight and Elsworth 1997; Donnadieu and Merle 1998; Reid 2004; Famin and Michon 2010; Reid et al. 2010; del Potro et al. 2013), and (3) stress regime modification by earthquakes or volcanic explosions (magmatic or phreatic). An important point, however, is that even in the absence of rapidly acting dynamic triggers, failures can still occur (and without warning) when the more slowly acting, destabilizing processes gradually reduce stability to a critical threshold.

3 Flow Behavior

Movement of VDAs is highly complex, in part controlled by variable average volumetric water contents ranging from near 0 to about 30% and distributions of liquid water within avalanche masses. Initial movement typically involves rigid-body frictional sliding at the outset before disaggregation and deformation of the failed mass occurs. Mass flowage takes over following disaggregation, and then flow mechanics are principally controlled by inertial and frictional forces between moving particles and between the particles and the bed (Savage 1984; Iverson and Vallance 2001). Although liquid water is often present in moving debris avalanches and local pockets of saturation or near-saturation may exist, intergranular pore-water pressures are predominantly negative (at least during rapid, highly energetic flow), so that the weight of a flowing granular mass is mainly borne by grain-to-grain frictional and inertial contacts (Savage 1984; Iverson and Vallance 2001), and the bulk of the mass is unsaturated. Despite the high degree of frictional interaction among hard and rigid VDA components, flow velocities can be high. The velocity of slide block I of the 1980 VDA at Mount St. Helens was about 50 m/s (calculated from photograph sequences) within about 26 s after initial failure, and over the next 10 s it accelerated to 70–80 m/s before slowing (Voight 1981). The avalanche front traveled approximately 29 km in about 10 min—a mean frontal velocity of about 48 m/s (Voight et al. 1981; Glicken 1996). Velocities greater than 50 m/s have also been measured or back calculated for other large VDAs in early stages of movement, including those at Lastarria volcano, Chile/Argentina (~ 80 m/s) (Naranjo and Francis 1987), and Mount Meager, Canada (64 m/s) (Guthrie et al. 2012).

VDAs typically move through a progression from initial rock sliding, commonly involving multiple retrogressing slide blocks (Fig. 2), progressive disaggregation of those slide blocks into smaller blocks, and eventual transformation to largely unsaturated, highly inertial particle–fluid flow of fragmented rock debris (e.g., Voight et al. 1981, 1983; Ui et al. 1986; Glicken 1996; Siebert 1996; Takarada et al. 1999; Andrade and van Wyk de Vries 2010; Thompson et al. 2010; van Wyk de Vries and Davies 2015). Slide blocks and particles move directly in response to gravity, i.e., they are not transported by a separate flowing fluid. The degree to which particles move differentially with respect to each other varies greatly in time and space and with the degree of bulk fragmentation and disaggregation within avalanche bodies. Several models have been proposed to characterize VDA motion at different points in its downslope evolution:

1. A *translational slide* model, whereby listric faults (steep, curved normal faults at landslide heads) are created within the edifice or its substrate at the onset of failure. These failure planes flatten out farther downslope during unconfined spreading of the sliding mass, with lower-angle faults transitioning to basal shear zones (décollements) (Andrade and van Wyk de Vries 2010; Paguican et al. 2014). This model has been used to explain the features of Socompa (Chile/Argentina), Parinacota (Chile/Bolivia), and Antuco (Chile) VDAs (Wadge et al. 1995; van Wyk de Vries et al. 2000, 2001; Clavero et al. 2002; Andrade and van Wyk de Vries 2010; Paguican et al. 2012, 2014; Romero et al. 2022), Mombacho volcano (Nicaragua) (Shea et al. 2008) and Iriga volcano (Philippines) (Paguican et al. 2012).
2. A *multiple shear zone* model, which assumes pervasive rock fragmentation at the onset of failure due to sudden decompression and dilation of the failing rock mass. This mechanism, which would account for a greater degree of bulk fragmentation and shear distribution than would occur in a simple rockslide, has been proposed at Mount St. Helens (Glicken 1991, 1996) and at Taranaki (Egmont) volcano (New Zealand) where multiple discrete plug masses are bounded by discrete shear zones (Roverato et al. 2015). This model—a special case of the first model—also can account for smaller secondary hummocks above secondary shear planes shallower than the basal décollement (Ui et al. 1986; Glicken 1996; Paguican et al. 2014, 2021). However, no zones of concentrated shear were observed in the more highly disaggregated Abona VDA deposit of Tenerife volcano (Canary Islands), where Makris et al. (2023) noted randomly distributed networks of shear displacement throughout the deposit (and in the upper substrate) that are interpreted to reflect chaotic

ephemeral changes in stress fields and shear accommodation during flow.

3. A *plug-flow* model for translational avalanche movement, where the bulk of the avalanche material (particularly parts with interlocking debris-avalanche blocks) moves as an undeforming plug bounded on the sides and bottom by relatively thin shear zones (Glicken 1996; Takarada et al. 1999). This model has been attributed to VDAs at Mount St. Helens, and Tashirodake and Ontake volcanoes (Japan) (Voight et al. 1983; Glicken 1996; Takarada et al. 1999). Entrainment of snow, ice, or wet sediment may lead to liquefaction within the basal shear zone and greater avalanche mobility (Clavero et al. 2002; Caballero and Capra 2011).

Regardless of the early-stage mechanism of motion, shear within an avalanching mass becomes more broadly distributed throughout the vertical profile of an avalanche body as it progresses down valley (Voight et al. 1981; Glicken 1996; Takarada et al. 1999; Roverato et al. 2015). Debris-avalanche blocks and large gravel-size clasts consequently become progressively smaller (from abrasion, breaking apart, or sorting), slightly rounder, and more deformed (faulted or plastically deformed). Both the relative proportion of matrix and the degree of mixing within the avalanche body also increase with distance from source (e.g., Ui et al. 1986; Glicken 1996; Takarada et al. 1999; Morelli et al. 2010; Perinotto et al. 2015; Dufresne et al. 2021b).

Most lateral collapses of volcanic edifices result initially in highly inertial debris-avalanche movement (Vallance and Iverson 2015), unaffected by water saturation of the material (Weidinger et al. 2014; Makris et al. 2020). Localize pockets of water may be carried by some VDAs but sedimentologic features of deposits mainly reflect unsaturated flow and minimal debris mixing, at least along much of their flow paths (e.g., uneven distribution of matrix material and localized clusters of gravel-size clasts (Siebert et al. 2004; Yoshida and Sugai 2010; Roverato et al. 2015)). A significant change in flow behavior occurs, however, if VDAs transform to volcanic debris flows (VDFs), i.e., concentrated lahars, as they flow downslope. Partial or complete transformation from VDA to VDF can occur either shortly after failure or after the avalanche has traveled some distance (e.g., Palmer et al. 1991; Carrasco-Núñez et al. 1993; Voight and Sousa 1994; Vallance and Scott 1997; Capra and Macías 2000, 2002; Evans et al. 2001; Scott et al. 2001, 2005; Capra et al. 2002; Bernard et al. 2009, 2019; Roberti et al. 2017; Dufresne et al. 2021b; Paguican et al. 2021). We hypothesize that transformation to VDF depends on these conditions:

(1) Sufficient disaggregation of the avalanche mass occurs, such that necessary small pore spaces are created to confine and pressurize enough intergranular pore water to allow undrained loading and partial liquefaction.
(2) Sufficient liquid water is already present in the avalanche mass or is entrained while moving downslope to allow full water saturation and liquefaction of a sufficient volume of avalanche material. A volume concentration of at least 15–20 vol% water is required for a debris mixture to flow with low intergranular friction (see Sect. 3.3 in Chap. "Deposits Emplaced by Volcanic Wet Flows").
(3) A sufficient proportion of very fine particles must be present in the disaggregating avalanche mass or be entrained while moving downslope to facilitate entrapment and pressurization of porewater. Transformations to VDFs are most likely where failed edifices are hydrothermally altered and rich in clay (Vallance and Scott 1997; Scott et al. 2001).

As highly agitated avalanche masses decelerate, the fragmental debris masses expel air and water while the spaces between particles diminish in size. With continued slowing and consolidation, fines-rich debris masses within an avalanche body that meet the enumerated conditions above can attain full water saturation and begin creating excess positive pore pressures during the final late stage of consolidation (Pierson 1981; George et al. 2022). If the excess pore pressures are high enough, the saturated parts of the avalanche body can liquefy and continue to flow after the unsaturated parts have halted. Importantly, such transformations of VDAs to VDFs extend VDA travel distances many tens of kilometers farther downstream.

Parts of some VDAs can remobilize as liquefied slurries hours after initial emplacement. This was observed for the first time on May 18, 1980, at Mount St. Helens. Aerial observations revealed that liquefied slurries of avalanche debris began flowing directly from the new VDA deposit at several locations about 5 h after initial deposition of the avalanche mass (Janda et al. 1981; Fairchild 1987; Scott 1988). These liquefied streams of muddy slurry coalesced to form the massive North Fork Toutle River lahar (Fig. 7), a clay-rich VDF (see Chaps. "Lahars—Process Overview" and "Lahar Deposits—Debris-Flow Type") that eventually traveled about 100 km to the Columbia River. Generation of the muddy slurry coincided with a period of continuous low-amplitude seismicity in that area related to the ongoing eruption, which may have triggered localized but extensive liquefaction of the deposit (Fairchild 1985, 1987). Other factors such as slumping of saturated portions of the VDA deposit and melting of ice fragments from disaggregated glaciers and snowpack that were entrained in the avalanche mass (Janda et al. 1981) may have been in play, so the definitive triggering mechanism is uncertain.

VDA material contains little water and remains unsaturated during flow in relatively arid settings and in edifices containing little groundwater. High internal friction within such flows can abruptly slow and halt initially rapid avalanches, leaving deposits with thick, steep fronts. Examples of such "dry" flow behavior are inferred from avalanches at Jocotitlán (Mexico) (Siebe et al. 1992), Socompa (van Wyk de Vries et al. 2001), Santa Ana (El Salvador) (Siebert et al. 2004), and Asama (Japan) volcanoes (Yoshida and Sugai 2010). Unsaturated VDAs moving as highly frictional particle–fluid flows commonly produce

Fig. 7 Volcanic debris flow (VDF) deposit (surface in foreground) that formed by liquefaction of parts of the static 1980 Mount St. Helens (Washington, USA) volcanic debris avalanche (VDA) deposit. This concentrated lahar began flowing within hours of emplacement of the avalanche deposit. The relatively flat, light-colored VDF deposit covers most of the VDA deposit surface at this location, about 13 km from the source crater, although large hummocks are visible near the valley walls. Downed trees on valley walls were swept by passage of the lateral blast. USGS photo by K. Scott, July 1980

longitudinal and transverse ridges such as those observed in the VDA deposits at Shiveluch (Kamchatka) (Belousov et al. 1999), Lastarria volcano (Naranjo and Francis 1987; Rodríguez et al. 2020), and Tutupaca volcano (Peru) (Valderrama et al. 2016, 2018).

A wide range of other models have been invoked for VDA motion to explain their high mobility. These mechanisms include, but are not restricted to, acoustic fluidization (Melosh 1987, 2015), dynamic basal fragmentation (Davies et al. 2010), basal friction reduction due to incorporation of fine-grained substrate material (Bernard et al. 2008), and intense crushing or frictional melting of rock along basal slide planes where decreasing kinetic energy and runout velocity are thought to also lead to more intense and prolonged frictional contacts with the bed (Voight and Sousa 1994; Schneider and Fisher 1998; Takarada et al. 1999; Legros et al. 2000; Ui et al. 2000; Clavero et al. 2002; Weidinger et al. 2014; Dufresne et al. 2021b). Detailed discussion of these processes is beyond the focus of this book, and we refer the reader to summaries in Paguican et al. (2021).

4 Material Properties and Sediment Transport

Major components of unsaturated VDAs include (1) *unconsolidated material*—disaggregated and relatively loose solid rock fragments ranging from clay-size grains to >10-m boulders, and (2) *debris-avalanche blocks*—highly fractured, unconsolidated to semi-consolidated pieces of the failed edifice up to hundreds of meters in diameter that have been transported relatively intact to their point of deposition (Glicken 1991, 1996). Unconsolidated material can be subdivided by size into matrix material (particles ≤2 mm diameter) and gravel-size clasts (particles >2 mm diameter), and it can range in composition from localized concentrations of monomictic fragments derived from comminution of single lithology debris-avalanche blocks to *mixed material*—thoroughly mixed aggregations of polymictic fragments derived from different lithologies, including substrate materials. The proportion of matrix material within an avalanche mass typically increases distally and toward the flow margins due to progressive disaggregation of debris-avalanche blocks and fractured gravel-size clasts (Ui 1983; Palmer et al. 1991; Alloway et al. 2005; Zernack et al. 2009; Dufresne et al. 2021b). VDA deposits may include variable and sometimes significant amounts of deeply eroded and entrained substrate rock, such as occurred at Socompa (Wadge et al. 1995) or Mombacho volcanoes (Shea et al. 2008) or through incorporation of surficial soil and sediment scoured from the ground surface (Palmer et al. 1991; Bernard et al. 2008).

The amount, type, and localization of groundwater within volcanic edifices have important implications for both edifice stability and post-failure avalanche behavior (e.g., Reid 2004; Delcamp et al. 2016; Ball et al. 2018). Infiltrating meteoric groundwater, originating as rainfall, snowmelt, or glacial melt, percolates downward through unsaturated material within the edifice in response to gravity. It may encounter low-permeability barriers (e.g., dikes, sills, or zones of hydrothermally altered rock) and be trapped as saturated zones (perched aquifers) in the upper part of the edifice, or it may continue to move downward through more isotropically permeable material until it encounters the regional water table. Pore pressures developed within these saturated zones are from gravity flow and hydrostatic head. In contrast, pressurized hydrothermal groundwater that is heated by hot rock in shallow magma bodies beneath active volcanoes or by ascending magma may move hot groundwater upward to produce destabilizing pore pressures within the edifice (Reid 2004; Ball et al. 2018). Direct measurements of saturation zones within edifices are valuable but rare (Keller et al. 1979; Finn et al. 2007), and most of the thinking about groundwater in volcanoes comes from numerical modeling (e.g., Hurwitz et al. 2003; Delcamp et al. 2016; Ball et al. 2018). It is likely that localized perched

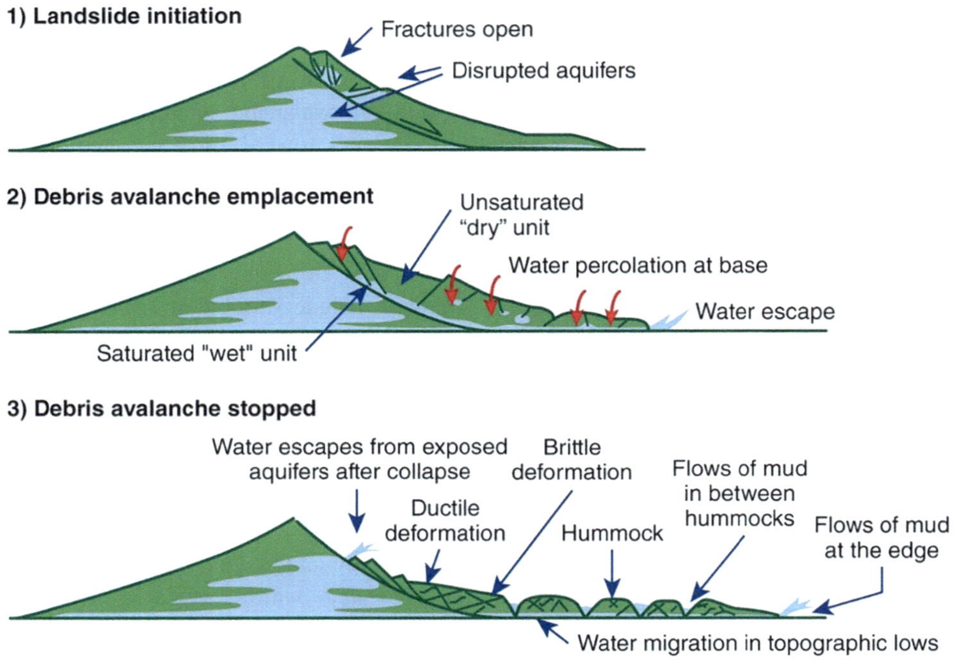

Fig. 8 Model of water migration and influence during landslide initiation and mass transport (modified from Delcamp et al. 2016)

aquifers within pre-failure edifices can produce more localized water saturation and explain why some parts of VDAs liquefy while others do not, such as was documented in the 1980 VDA deposit at Mount St. Helens (Janda et al. 1981) (Fig. 7). Delcamp et al. (2016) developed the implications of rapid water release and lahar generation due to truncation of perched aquifers within volcanic edifices during and after collapse (Fig. 8).

There is growing evidence to suggest that volcanic edifices may contain both unsaturated and saturated zones (e.g., Delcamp et al. 2016; Dufresne et al. 2021b; Paguican et al. 2021). Failure of an edifice containing localized perched aquifers could result in an avalanche mass having a distribution of different physical properties and localized partial liquefaction. Limited saturated zones at the level of the main failure plane could result in undrained loading and high pore pressures (friction reduction) concentrated at the slide base, which could enhance the mobility of the slide mass without overall liquefaction of the mass. Concentration of some water at the avalanche base could result in a highly mobile avalanche without transformation to VDF (Clavero et al. 2002; Caballero and Capra 2011; Iverson et al. 2011; Delcamp et al. 2016). Thus, even in relatively "dry" unsaturated avalanche masses (including extraterrestrial examples (Legros 2002)), water may play a role in reducing internal friction and promoting mobility (Johnson 1978; Voight and Sousa 1994). In such cases where water distribution within VDA bodies varies widely, diverse structures and sediment textures may be found both between and within VDA deposits. Contingent on the degree to which VDAs truncate edifice aquifers, deposits can display both saturated regions with ductile deformation and non-saturated regions with brittle deformation, often in close contact (Delcamp et al. 2016).

In summary, most lateral collapses of volcanic edifices produce debris avalanches that are mainly unsaturated during flow (Vallance and Iverson 2015), and thus most VDA deposits retain sedimentologic features characteristic of unsaturated flow (Siebert et al. 2004; Yoshida and Sugai 2010; Roverato et al. 2015). However, some edifice collapses produce rock debris containing enough liquid to become fully saturated after some amount of jostling and undrained loading during early stages of flow. In these instances, transformation to debris flow occurs and liquefaction of the water-saturated mass becomes the essential factor affecting debris mobility.

5 Magnitude and Frequency

VDAs resulting from lateral collapse of volcanic edifices, once thought to be extremely rare, are now known to be common at many volcanoes. The sustained interest in lateral-collapse phenomena in the decades since the 1980 eruption of Mount St. Helens has led to a proliferation of studies of edifice-failure deposits in varied geological settings. More than a thousand lateral-collapse events ≥ 0.1 km^3 in volume have been identified from more than 600 volcanoes from deposits and inferred source areas (Dufresne et al. 2021a; Siebert and Roverato 2021; Siebert and Reid 2023). The vast majority of these are Quaternary, but VDA deposits have been identified even in Precambrian terranes (Trofimovs et al. 2004; Roverato 2016).

Subaerial VDAs from single volcano sources can huge—reaching tens of cubic kilometers in bulk volume, covering >1000 km^2, and commonly producing the largest volcaniclastic deposits from a single volcanic flow on or near a volcano (Ui et al. 2000; Siebert 2002). Even larger low-angle gravitational failures involving large volcanic fields and nonvolcanic basement rock have also been noted in the geologic record—the well-known >2000 km^3 Eocene Heart Mountain detachment in northwestern Wyoming (Mitchell et al. 2015) and two comparably catastrophic Oligocene and Miocene collapses from the Marysvale volcanic field in Utah, one with a volume of ~ 3000 km^3 (Hacker et al. 2014; Biek et al. 2019). Collapses of large shield volcanoes at oceanic volcanic islands can generate equally enormous submarine avalanches involving greater debris volumes and impacting larger areas than most subaerial VDAs—with volumes to about 3000 km^3, runouts as far as 230 km, and covering as much as 23,000 km^2 of sea floor (Moore et al. 1989; Satake et al. 2002).

Frequency of VDA occurrences at individual volcanoes is variable. Average frequency at any one volcano depends on rates of magma supply and edifice reconstruction, as well as on the operative destabilizing factors and triggers. Volcanoes experiencing rapid reconstruction with high magma extrusion rates and formation of steep slopes are more likely to have repeated failures, and volcanoes with repeated failures are found worldwide. Ten VDAs >0.01 km^3 (7 greater than 0.1 km^3) occurred at Las Derrumbadas volcano (Mexico) during an eruptive episode about 2 ka ago and lasting perhaps decades to centuries (Guilbaud et al. 2023). Within the past 2500 years, at least a dozen avalanches have taken place at Augustine Volcano, Alaska (average return period of about 200 years) (Begét and Kienle 1992; Waitt and Begét 2009). Many collapses have occurred since the late Pleistocene at Bandai volcano (Japan) (Mimura and Endo 1997; Chiba and Kimura 2001) and at Taranaki (Egmont) volcano (New Zealand) (Zernack et al. 2012; Zernack 2021), with almost all the Taranaki VDAs well over a cubic kilometer in volume. Most major stratovolcanoes in Kamchatka (Ponomareva et al. 2006) and the Cascade Range (Siebert and Vallance 2017) have undergone collapse at least once during their life spans, although in other arcs such as the Philippines, the frequency is lower. For the entire volcanic arcs of Japan and Kamchatka, collapse frequencies average about one per century (Ui et al. 2000; Ponomareva et al. 2006). Edifice-collapse frequency on a global scale since 1500 CE for VDAs ≥ 0.1 km^3 in volume is more than five times per century and about seven times per century since 1800 CE (Siebert and Roverato 2021). Collapse frequencies that include smaller volume events (0.01–0.1 km^3) reach 9 per century during this interval (Siebert and Reid 2023). However, fewer of these smaller VDAs were reported in earlier centuries, so the nearly three dozen 0.01–0.1 km^3 failures reported worldwide since 1800 CE may be more representative of the average rate for these events.

Lateral blasts are rarely associated with edifice collapses coincident with magmatic eruptions, despite the well-known coincident occurrences at Mount St. Helens in 1980 and Bezymianny (Kamchatka) in 1956 (Gorshkov 1959; Hoblitt et al. 1981; Kieffer 1981a, b; Moore and Sisson 1981; Waitt 1981). Generation of collapse-related lateral blasts seem to require the coincidence of decompression-generated explosions from a magma body high in an edifice with collapsing slide blocks precisely positioned to deflect explosion forces laterally (Belousov et al. 1999, 2007; Siebert 2002). A vertical blast was reported at Lamington volcano (Papua New Guinea) when the basal failure plane truncated only the upper tip of the magma body (Belousov et al. 2020). When magma is farther below the summit, collapse can be completed before the onset of magmatic eruptions, leaving an open vent that results in more typical vertical explosions with pumiceous PDCs, as occurred at Shiveluch volcano in 1964 (Gorshkov and Dubik 1970; Belousov 1995). When magma is even deeper, phreatic eruptions can occur where hydrothermal systems decompress but no juvenile magma reaches the surface, as at Bandai volcano in 1888 (Nakamura 1978).

Importantly, VDAs can occur in the absence of eruptive activity, even though it is more common for VDAs (particularly the largest avalanches) to coincide with eruptions. Nevertheless, about a fifth of documented post-1500 CE VDAs ≥ 0.1 km^3 worldwide have occurred during periods of volcanic quiescence (Siebert and Roverato 2021), thus possibly occurring without warning. This proportion increases to almost a third when VDAs with volumes 0.01–0.1 km^3 are added to the events sampled (Siebert and Reid 2023). Only relatively young VDAs are considered in this analysis because the preservation potential of geologic evidence for small eruptions being coincident with avalanches (typically thin tephra deposits) decreases with time (Dufresne et al. 2021a; Siebert and Roverato 2021). Yet the fact that VDAs can occur with or without eruptive triggers adds unpredictability to when a VDA might occur, particularly at edifices known to be structurally weakened (e.g., Reid et al. 2001; Finn et al. 2007; John et al. 2008). This unpredictability increases the risk to downstream communities situated in potential hazard zones, because travel times to populated areas could be short (an hour or less) (Pierson 1998; George et al. 2022).

6 Sediment Deposition

As with lahars discussed earlier, different depositional mechanisms are likely to be operative in different parts of VDA bodies. Early movement of collapsing volcanic edifices is likely controlled by basal friction along one (or a few) slide planes or narrow basal shear zones that separate accelerating debris-avalanche slide blocks from the underlying edifice structure, as was inferred for the early motion of the Mount St. Helens VDA in 1980 (Voight et al. 1981, 1983). In avalanches rich in large interlocking debris-avalanche blocks (particularly in proximal zones), it is likely that plug flow dominates material transport and that deposition occurs largely through *en masse* frictional freezing (McPhie et al. 1993; Takarada et al. 1999). In the distal part of an avalanche where shear is distributed throughout the flowing mass (and debris-avalanche blocks and gravel-size clasts are relatively small and not tightly packed), deposition by progressive aggradation at the base of the flow (Schneider and Fisher 1998; Dufresne et al. 2021b) may occur, as has been demonstrated for lahars and PDCs (e.g. Vallance 1994, 2000; Major 1997; Branney and Kokelaar 2002).

References

Alloway B, McComb P, Neall V, Vucetich C, Gibb J, Sherburn S, Stirling M (2005) Stratigraphy, age, and correlation of voluminous debris-avalanche events from an ancestral Egmont volcano: implications for coastal plain construction and regional hazard assessment. J R Soc New Zeal 35:229–267

Ando S, Yamagishi H (1975) Hill topography on the nuée ardente deposits of Shikaribetsu volcano, Hokkaido. Bull Volcanol Soc Jpn 20:31–36 (in Jpn with Engl abstr)

Andrade SD, van Wyk de Vries B (2010) Structural analysis of the early stages of catastrophic stratovolcano flank-collapse using analogue models. Bull Volcanol 72:771–789. https://doi.org/10.1007/s00445-010-0363-x

Bacon CR, Gardner JV, Mayer LA, Buktenica MW, Dartnell P, Ramsey DW, Robinson JE (2002) Morphology, volcanism, and mass wasting in Crater Lake, Oregon. Geol Soc Am Bull 114(6):675–692. https://doi.org/10.1130/0016-7606(2002)114<0675:MVAMWI>2.0.CO;2

Ball JL, Taron J, Reid ME, Hurwitz S, Finn C, Bedrosian P (2018) Combining multiphase groundwater flow and slope stability models to assess stratovolcano flank collapse in the Cascade Range. J Geophys Res Solid Earth 123(4):2787–2805. https://doi.org/10.1002/2017JB015156

Begét JE (2000) Volcanic tsunamis. In: Sigurdsson H, Houghton BF, McNutt SR, Rymer H, Stix J (eds) Encyclopedia of volcanoes. Academic Press, San Diego, pp 1005–1013

Begét JE, Kienle J (1992) Cyclic formation of debris avalanches at Mount St. Augustine volcano. Nature 356:701–704

Belousov AB (1995) The Shiveluch volcanic eruption of 12 November 1964—explosive eruption provoked by failure of the edifice. J Volcanol Geotherm Res 66:357–365. https://doi.org/10.1016/0377-0273(94)00072-O

Belousov A, Belousova M, Voight B (1999) Multiple edifice failures, debris avalanches and associated eruptions in the Holocene history of Shiveluch volcano, Kamchatka, Russia. Bull Volcanol 61:324–342. https://doi.org/10.1007/s004450050300

Belousov A, Walter TR, Troll VR (2005) Large-scale failures on domes and stratocones situated on caldera ring faults: sand-box modeling of natural examples from Kamchatka, Russia. Bull Volcanol 67(5):457–468. https://doi.org/10.1007/s00445-004-0387-1

Belousov A, Voight B, Belosouva M (2007) Directed blasts and blast-generated pyroclastic density currents: a comparison of the Bezymianny 1956, Mount St Helens 1980, and Soufrière Hills, Montserrat 1997 eruptions and deposits. Bull Volcanol 69:701–740. https://doi.org/10.1007/s00445-006-0109-y

Belousov A, Belousova M, Hoblitt R, Patia H (2020) The 1951 eruption of Mount Lamington, Papua New Guinea: devastating directed blast triggered by small-scale edifice failure. J Volcanol Geotherm Res 401:106947. https://doi.org/10.1016/j.jvolgeores.2020.106947

Bernard B, van Wyk de Vries B, Barba D, Leyrit H, Robin C, Alcaraz S, Samaniego P (2008) The Chimborazo sector collapse and debris avalanche: deposit characteristics as evidence of emplacement mechanisms. J Volcanol Geotherm Res 176(1):36–43. https://doi.org/10.1016/j.jvolgeores.2008.03.012

Bernard B, van Wyk de Vries B, Leyrit H (2009) Distinguishing volcanic debris avalanche deposits from their reworked products: the Perrier sequence (French Massif Central). Bull Volcanol 71:1041–1056. https://doi.org/10.1007/s00445-009-0285-7

Bernard K, van Wyk de Vries B, Thouret J-C (2019) Fault textures in volcanic debris-avalanche deposits and transformations into lahars: the Pichu Pichu thrust lobes in south Peru compared to worldwide avalanche deposits. J Volcanol Geotherm Res 371:116–136

Biek RF, Rowley PD, Hacker DB (2019) The gigantic Markagunt and Sevier gravity slides resulting from mid-Cenozoic catastrophic mega-scale failure of the Marysvale volcanic field, Utah, USA. Geol Soc Am Field Guide 56:1–121

Branney MJ, Kokelaar P (2002) Pyroclastic density currents and the sedimentation of ignimbrites. Geol Soc Lond Mem 27:1–143

Caballero L, Capra L (2011) Textural analysis of particles from El Zaguán debris avalanche deposit, Nevado de Toluca volcano, Mexico: evidence of flow behavior during emplacement. J Volcanol Geotherm Res 200:75–82

Capra L (2006) Abrupt climatic changes as triggering mechanisms of massive volcanic collapses. J Volcanol Geotherm Res 155:329–333

Capra L, Macías JL (2000) Pleistocene cohesive debris flows at Nevado de Toluca volcano, central Mexico. J Volcanol Geotherm Res 102:149–168. https://doi.org/10.1016/S0377-0273(00)00186-4

Capra L, Macías JL (2002) The cohesive Naranjo debris-flow deposit (10 km^3): a dam breakout flow derived from the Pleistocene debris-avalanche deposit of Nevado de Colima volcano (México). J Volcanol Geotherm Res 117(1–2):213–235. https://doi.org/10.1016/S0377-0273(02)00245-7

Capra L, Macías JL, Scott KM, Abrams M, Garduño-Monroy VH (2002) Debris avalanches and debris flows transformed from collapses in the Trans-Mexican Volcanic Belt, Mexico—behavior and implications for hazard assessment. J Volcanol Geotherm Res 113:81–110. https://doi.org/10.1016/S0377-0273(01)00252-9

Capra L, Bernal JP, Carrasco-Núñez G, Roverato M (2013) Climatic fluctuations as a significant contributing factor for volcanic collapses. Evidence from Mexico during the Late Pleistocene. Glob Planet Change 100:194–203

Carrasco-Núñez G, Vallance JW, Rose WI (1993) A voluminous avalanche-induced lahar from Citlaltepetl volcano, Mexico: implications for hazard assessment. J Volcanol Geotherm Res 59:35–46. https://doi.org/10.1016/0377-0273(93)90076-4

Chadwick WW Jr, De Roy T, Carrasco A (1991) The September 1988 intracaldera avalanche and eruption at Fernandina volcano, Galapagos Islands. Bull Volcanol 53(4):276–286

Chiba S, Kimura J-I (2001) Geology and growth history of Bandai volcano, Tohoku-Honshu arc, Japan—analysis of volcanic activity by tephrochronology. Jpn Mag Mineral Petrol Sci 30:126–156 (in Jpn with Engl abstr)

Clavero JE, Sparks RSJ, Huppert HE, Dade WB (2002) Geological constraints on the emplacement mechanism of the Parinacota debris avalanche, northern Chile. Bull Volcanol 64:40–54

Crandell DR (1989) Gigantic debris avalanche of Pleistocene age from ancestral Mount Shasta volcano, California, and debris-avalanche hazard zonation. U S Geol Surv Bull 1861:32 pp

Davies T, McSaveney M, Kelfoun K (2010) Runout of the Socompa volcanic debris avalanche, Chile: a mechanical explanation for low basal shear resistance. Bull Volcanol 72:933–944

Day SJ (1996) Hydrothermal pore fluid pressure and the stability of porous, permeable volcanoes. In: McGuire WJ, Jones AP, Neuberg J (eds) Volcano instability on the earth and other planets. Geol Soc Lond Spec Publ 110:77–93

Day SJ (2015) Volcanic tsunamis. In: Sigurdsson H, Houghton B, McNutt SR, Rymer H, Stix J (eds) The encyclopedia of volcanoes, 2nd edn. Elsevier, Amsterdam, pp 993–1009

del Potro R, Hürlimann M, Pinkerton H (2013) Modelling flank instabilities on stratovolcanoes: parameter sensitivity and stability analyses of Teide, Tenerife. J Volcanol Geotherm Res 256:50–60. https://doi.org/10.1016/j.jvolgeores.2013.02.003

Delcamp A, Roberti G, van Wyk de Vries B (2016) Water in volcanoes: evolution, storage and rapid release during landslides. Bull Volcanol 78:87. https://doi.org/10.1007/s00445-016-1082-8

Devoli G, Cepeda J, Kerle N (2009) The 1998 Casita volcano flank failure revisited—new insights into geological setting and failure mechanisms. Eng Geol 105:65–83

Donnadieu F, Merle O (1998) Experiments on the indentation process during cryptodome intrusions. New insights into Mount St. Helens deformation. Geology 26:79–82

Dufresne A, Siebert L, Bernard B (2021a) Distribution and geometric parameters of volcanic debris avalanche deposits. In: Roverato M, Dufresne A, Procter J (eds) Volcanic debris avalanches. Advances in volcanology, vol 6. Springer, Cham, pp 75–90. https://doi.org/10.1007/978-3-030-57411-6_4

Dufresne A, Zernack A, Bernard K, Thouret JC, Roverato M (2021b) Sedimentology of volcanic debris avalanche deposits. In: Roverato M, Dufresne A, Procter J (eds) Volcanic debris avalanches. Advances in volcanology, vol 6. Springer, Cham, pp 175–210. https://doi.org/10.1007/978-3-030-57411-6_8

Elsworth D, Voight B (1995) Dike intrusion as a trigger for large earthquakes and the failure of volcano flanks. J Geophys Res Solid Earth 100(B4):6005–6024

Elsworth D, Voight B (1996) Evaluation of volcano flank instability triggered by dyke intrusion. In: McGuire WJ, Jones AP, Neuberg J (eds) Volcano instability on the earth and other planets. Geol Soc Lond Spec Publ 110:45–53. https://doi.org/10.1144/GSL.SP.1996.110.01.03

Evans SG, Hungr O, Clague JJ (2001) Dynamics of the 1984 rock avalanche and associated distal debris flow on Mount Cayley, British Columbia, Canada; implications for landslide hazard on dissected volcanoes. Eng Geol 61:29–51. https://doi.org/10.1016/S0013-7952(00)00118-6

Fairchild LH (1985) Lahars at Mount St. Helens, Washington. Dissertation, University of Washington, 374 pp

Fairchild LH (1987) The importance of lahar initiation process. In: Costa JE, Wieczorek GF (eds) Debris flows/avalanches: process, recognition, and mitigation. Geol Soc Am Rev Eng Geol 7:51–62

Famin V, Michon L (2010) Volcano destabilization by magma injections in a detachment. Geology 38:219–222

Finn CA, Deszcz-Pan M, Anderson ED, John DA (2007) Three-dimensional geophysical mapping of rock alteration and water content at Mount Adams, Washington: implications for lahar hazards. J Geophys Res Solid Earth 112(B10). https://doi.org/10.1029/2006JB004783

Frank D (1995) Surficial extent and conceptual model of hydrothermal system at Mount Rainier, Washington. J Volcanol Geotherm Res 65:51–80

George DL, Iverson RM, Cannon CM (2022) Modeling the dynamics of lahars that originate as landslides on the west side of Mount Rainier, Washington. U S Geol Surv Open-File Rep 2021-1118. https://doi.org/10.3133/ofr20211118

Glicken H (1991) Sedimentary architecture of large volcanic debris avalanches. In: Fisher RV, Smith GA (eds) Sedimentation in volcanic settings. Soc Sediment Geol Spec Publ 45:99–106

Glicken H (1996) Rockslide debris avalanche of May 18, 1980, Mount St. Helens volcano, Washington. U S Geol Surv Open-File Rep 96-677, 90 pp and 5 plates

Gorshkov GS (1959) Gigantic eruption of the volcano Bezymianny. Bull Volcanol 20:77–109. https://doi.org/10.1007/BF02596572

Gorshkov GS, Dubik YM (1970) Gigantic directed blast at Shiveluch volcano (Kamchatka). Bull Volcanol 34:261–288. https://doi.org/10.1007/BF02597790

Griggs RF (1920) The great Mageik landslide. Ohio J Sci 20:325–354

Guilbaud M, Chédeville C, Molina-Guadarrama ÁN, Pineda-Serrano JC, Siebe C (2023) Volcano-sedimentary processes at Las Derrumbadas rhyolitic twin domes, Serdán-Oriental basin, Eastern Trans-Mexican Volcanic Belt. Geol Soc Lond Spec Publ 520:165–189. https://doi.org/10.1144/SP520-2021-144

Guthrie RH, Friele P, Allstadt K, Roberts N, Evans SG, Delaney KB, Roche D, Clague JJ, Jakob M (2012) The 6 August 2010 Mount Meager rock slide-debris flow, Coast Mountains, British Columbia: characteristics, dynamics, and implications for hazard and risk assessment. Nat Hazards Earth Syst Sci 12:1277–1294. https://doi.org/10.5194/nhess-12-1277-2012

Hacker DB, Biek RF, Rowley PD (2014) Catastrophic emplacement of the gigantic Markagunt gravity slide, southwest Utah—implications for hazards associated with sector collapse of volcanic fields. Geology 42(11):943–946. https://doi.org/10.1130/G35896.1

Heap MJ, Violay ME (2021) The mechanical behaviour and failure modes of volcanic rocks: a review. Bull Volcanol 83(5):1–47. https://doi.org/10.1007/s00445-021-01447-2

Heap MJ, Baumann T, Gilg HA, Kolzenburg S, Ryan AG, Villeneuve M, Russell JK, Kennedy LA, Rosas-Carbajal M, Clynne MA (2021) Hydrothermal alteration can result in pore pressurization and volcano instability. Geology 49(11):1348–1352. https://doi.org/10.1130/G49063.1

Hoblitt RP, Miller CD, Vallance JW (1981) Origin and stratigraphy of the deposit produced by the May 18 directed blast. In: Lipman PW, Mullineaux DR (eds) The 1980 eruptions of Mount St. Helens, Washington. U S Geol Surv Prof Pap 1250:401–419

Hsü KJ (1975) Catastrophic debris streams (Sturzstroms) generated by rockfalls. Geol Soc Am Bull 86:129–140. https://doi.org/10.1130/0016-7606(1975)86<129:CDSSGB>2.0.CO;2

Hurwitz S, Kipp KL, Ingebritsen SE, Reid ME (2003) Groundwater flow, heat transport, and water table position within volcanic edifices: implications for volcanic processes in the Cascade Range. J Geophys Res 108(B12). https://doi.org/10.1029/2003JB002565

Iverson RM, Vallance JW (2001) New views of granular mass flows. Geology 29:115–118

Iverson RM, Reid ME, Logan M, LaHusen RG, Godt JW, Griswold JP (2011) Positive feedback and momentum growth during debris-flow entrainment of wet bed sediment. Nat Geosci 4:116–121. https://doi.org/10.1038/ngeo1040

Janda RJ, Scott KM, Nolan KM, Martinson HA (1981) Lahar movement, effects, and deposits. In: Lipman PW, Mullineaux DR (eds) The 1980 eruptions of Mount St. Helens, Washington. U S Geol Surv Prof Pap 1250:461–478

John DA, Sisson TW, Breit GN, Rye RO, Vallance JW (2008) Characteristics, extent, and origin of hydrothermal alteration at Mount Rainier volcano, Cascades Arc, USA: implications for debris-flow hazards and mineral deposits. J Volcanol Geotherm Res 175:289–314

Johnson B (1978) Blackhawk landslide, California, U.S.A. Dev Geotech Eng 14(A):481–504

Keating BH, McGuire WJ (2000) Island edifice failures and associated tsunami hazards. Pure Appl Geophys 157:899–955

Keller GV, Grose LT, Murray JC, Skokan CK (1979) Results of an experimental drill hole at the summit of Kilauea volcano, Hawaii. J Volcanol Geotherm Res 5:345–385

Kerle N, van Wyk de Vries B, Oppenheimer C (2003) New insight into the factors leading to the 1998 flank collapse and lahar disaster at Casita volcano, Nicaragua. Bull Volcanol 65(5):331–345. https://doi.org/10.1007/s00445-002-0263-9

Kieffer SW (1981a) Blast dynamics at Mount St. Helens on 18 May 1980. Nature 291:568–570. https://doi.org/10.1038/291568a0

Kieffer SW (1981b) Fluid dynamics of the May 18 blast at Mount St. Helens. In: Lipman PW, Mullineaux DR (eds) The 1980 eruptions of Mount St. Helens, Washington. U S Geol Surv Prof Pap 1250:379–400

Kilburn CRJ, Voight B (1998) Slow rock fracture as eruption precursor at Soufriere Hills volcano, Montserrat. Geophys Res Lett 25(19):3665–3668. https://doi.org/10.1029/98GL01609

Lagmay AMF, van Wyk de Vries B, Kerle N, Pyle DM (2000) Volcano instability induced by strike-slip faulting. Bull Volcanol 62:331–346. https://doi.org/10.1007/s004450000103

Lavallée Y, Kendrick JE (2021) A review of the physical and mechanical properties of volcanic rocks and magmas in the brittle and ductile regimes. In: Papale P (ed) Forecasting and planning for volcanic hazards, risks, and disasters. Elsevier, Amsterdam, pp 153–238. https://doi.org/10.1016/B978-0-12-818082-2.00005-6

Lavigne F, Degeai J-P, Komorowski J-C, Guillet S, Robert V, Lahitte P, Oppenheimer C, Stoffel M, Vidal CM, Surono, Pratomo I, Wassmer P, Hajdas I, Danang Sri Hadmoko DS, de Belizal E (2013) Source of the great A.D. 1257 mystery eruption unveiled, Samalas volcano, Rinjani volcanic complex, Indonesia. Proc Nat Acad Sci 110(42):16742–16747. https://doi.org/10.1073/pnas.1307520110

Legros F (2002) The mobility of long-runout landslides. Eng Geol 63:301–331. https://doi.org/10.1016/S0013-7952(01)00090-4

Legros F, Cantagrel J-M, Devouard B (2000) Pseudotachylyte (frictionite) at the base of the Arequipa volcanic landslide deposit (Peru):

implications for emplacement mechanisms. J Geol 108:601–611. https://doi.org/10.1086/314421
López DL, Williams SN (1993) Catastrophic volcanic collapse; relation to hydrothermal processes. Science 260:1794–1796. https://doi.org/10.1126/science.260.5115.1794
Lucas A, Mangeney A, Ampuero JP (2014) Frictional velocity-weakening in landslides on earth and on other planetary bodies. Nat Commun 5(1):1–9. https://doi.org/10.1038/ncomms4417
Major JJ (1997) Depositional processes in large-scale debris-flow experiments. J Geol 105:345–366
Makris S, Manzella I, Cole P, Roverato M (2020) Grain size distribution and sedimentology in volcanic mass-wasting flows: implications for propagation and mobility. Int J Earth Sci 109:2679. https://doi.org/10.1007/s00531-020-01907-8
Makris S, Roverato M, Davila-Harris P, Cole P, Manzella I (2023) Distributed stress fluidisation: insights into the propagation mechanisms of the Abona volcanic debris avalanche (Tenerife) through a novel method for indurated deposit sedimentological analysis. Front Earth Sci 11:1177507. https://doi.org/10.3389/feart.2023.1177507
Malone DH, Craddock JP, Schmitz MD, Kenderes S, Kraushaar B, Murphey CJ, Nielsen S, Mitchell TM (2017) Volcanic initiation of the Eocene Heart Mountain slide, Wyoming, USA. J Geol 125(4):439–457
McGuire WJ (1996) Volcano instability: a review of contemporary themes. In: McGuire WJ, Jones AP, Neuberg J (eds) Volcano instability on the earth and other planets. Geol Soc Lond Spec Publ 110:1–23
McPhie J, Doyle M, Allen R (1993) Volcanic textures—a guide to the interpretation of textures in volcanic rocks. Centre for Ore Deposit and Exploration Studies, University of Tasmania, Hobart, p 196
McSaveney MJ (1978) Sherman glacier rock avalanche, Alaska, USA. In: Voight B (ed) Natural phenomena. Rockslides and avalanches, vol 1. Elsevier, New York, pp 197–258. https://doi.org/10.1016/B978-0-444-41507-3.50014-3
Melosh J (1987) The mechanics of large rock avalanches. Rev Eng Geol VII:41–50
Melosh J (2015) Acoustic fluidization: what it is and is not. In: Bridging the gap III: impact cratering in nature, experiments, and modeling, vol 1861, p 1004
Merle O, Michon L, Bachèlery P (2008) Caldera rim collapse: a hidden volcanic hazard. J Volcanol Geotherm Res 177(2):25–530. https://doi.org/10.1016/j.jvolgeores.2008.06.011
Michon L, Gayer E, Lucas A, Bellin F, Gougeon M (2023) The 1965 Mahavel landslide (Réunion Island, Indian Ocean): morphology, volumes, flow dynamics, and causes of a rock avalanche in tropical setting. J Geophys Res Earth Surf 128(6):e2022JF006944. https://doi.org/10.1029/2022JF006944
Mimura K, Endo H (1997) Repeated collapse and reconstruction of Bandai volcano as revealed in the large outcrop of debris deposits on the Southwest foot. J Volcanol Soc Jpn 42:321–330 (in Jpn with Engl abstr)
Mitchell TM, Smith SAF, Anders MH, Di Toro G, Nielsen S, Cavallo A, Beard AD (2015) Catastrophic emplacement of giant landslides aided by thermal decomposition: Heart Mountain, Wyoming. Earth Planet Sci Lett 411:199–207. https://doi.org/10.1016/j.epsl.2014.10.051
Mizuno Y (1964) Landforms associated with volcanic debris flows at the foot of Zao volcano. Hirosaki Univ Fac Educ Bull 13:23–32 (in Jpn)
Moore JG (1964) Giant submarine landslides on the Hawaiian Ridge. U S Geol Surv Prof Pap 501-D:D95–D98
Moore JG, Sisson TW (1981) Deposits and effects of the May 18 pyroclastic surge. In: Lipman PW, Mullineaux DR (eds) The 1980 eruptions of Mount St. Helens, Washington. U S Geol Surv Prof Pap 1250:421–438
Moore JG, Clague DA, Holcomb RT, Lipman PW, Normark WR, Torresan ME (1989) Prodigious submarine landslides on the Hawaiian Ridge. J Geophys Res 94:17465–17484
Morelli S, Garduño Monroy VH, Gigli G, Falorni G, Arreygue Rocha E, Casagli N (2010) The Tancitaro debris avalanche: characterization, propagation, and modeling. J Volcanol Geotherm Res 193:93–105
Moriya I (1980) "Bandaian eruption" and landforms associated with it. Collection of articles in memory of retirement of Prof. K. Nishimura from Tohoku University, pp 214–219 (in Jpn with Engl abstr)
Murai I (1961) A study of the textural characteristics of pyroclastic flow deposits in Japan. Bull Earthq Res Inst Tokyo Univ 39:133–248
Nakamura Y (1978) Geology and petrology of Bandai and Nekoma volcanoes. Tohoku Univ Sci Rep Ser 3(14):67–119
Naranjo JA, Francis P (1987) High velocity debris avalanche at Lastarria volcano in the north Chilean Andes. Bull Volcanol 49:509–514
Neumann van Padang M (1939) Über die vielen tausend Hügel in westlichen Vorlande des Raoeng-Vulkans (Ostjava). Ing Nederlandsch-Indie 6(4):35–41
Norini G, Lagmay A (2005) Deformed symmetrical volcanoes. Geology 33(7):605–608. https://doi.org/10.1130/G21565.1
Norini G, Agliardi F, Crosta G, Groppelli G, Zuluaga MC (2019) Structure of the Colima volcanic complex: origin and behaviour of active fault systems in the edifice. In: Varley N, Connor CB, Komorowski J-C (eds) Volcán de Colima, active volcanoes of the world. Springer, Berlin
Paguican EMR, van Wyk de Vries B, Lagmay A (2012) Volcano-tectonic controls and emplacement kinematics of the Iriga debris avalanches (Philippines). Bull Volcanol 74:2067–2081. https://doi.org/10.1007/s00445-012-0652-7

Paguican EMR, van Wyk de Vries B, Lagmay A (2014) Hummocks: how they form and how they evolve in rockslide-debris avalanches. Landslides 11:67–80. https://doi.org/10.1007/s10346-012-0368-y

Paguican EMR, Roverato M, Yoshida H (2021) Volcanic debris avalanche transport and emplacement mechanisms. In: Roverato M, Dufresne A, Procter J (eds) Volcanic debris avalanches. Advances in volcanology, vol 6. Springer, Cham, pp 143–173. https://doi.org/10.1007/978-3-030-57411-6_7

Palmer BA, Alloway BV, Neall VE (1991) Volcanic debris-avalanche deposits in New Zealand—lithofacies organization in unconfined, wet-avalanche flows. In: Fisher RV, Smith GA (eds) Sedimentation in volcanic settings. SEPM Spec Publ 45:89–98

Perinotto H, Schneider J-L, Bachèlery P, Le Bourdonnec FX, Famin V, Michon L (2015) The extreme mobility of debris avalanches: a new model of transport mechanism. J Geophys Res Solid Earth 120:8110–8119. https://doi.org/10.1002/2015JB011994

Pierson TC (1981) Dominant particle support mechanisms in debris flows at Mt. Thomas, New Zealand, and implication for flow mobility. Sedimentology 28:49–60

Pierson TC (1998) An empirical method for estimating travel times for wet volcanic mass flows. Bull Volcanol 60:98–109

Pierson TC, Siebert L, Harpel CJ, Scott KM (2018) Geologic field-trip guide of volcaniclastic sediments from snow- and ice-capped volcanoes—Mount St. Helens, Washington, and Mount Hood, Oregon. U S Geol Surv Sci Invest Rep 2017-5022-F:1–97. https://doi.org/10.3133/sir20175022F

Plafker G, Ericksen GE (1978) Nevados Huascarán avalanches, Peru. In: Voight B (ed) Rockslides and avalanches, 1. Elsevier, Amsterdam, pp 277–314

Ponomareva VV, Melekestsev IV, Dirksen OV (2006) Sector collapses and large landslides on Late Pleistocene-Holocene volcanoes in Kamchatka, Russia. J Volcanol Geotherm Res 158:117–138. https://doi.org/10.1016/j.jvolgeores.2006.04.016

Reid ME (2004) Massive collapse of volcano edifices triggered by hydrothermal pressurization. Geology 32:373–376. https://doi.org/10.1130/G20300.1

Reid ME, Brien D (2006) Assessing massive flank collapse at stratovolcanoes using 3-D slope stability analysis. In: Evans SG, Mugnozza GS, Strom A, Hermanns RL (eds) Landslides from massive rock slope failure. NATO Sci Ser 49:445–458. Springer, Dordrecht, pp 445–458. https://doi.org/10.1007/978-1-4020-4037-5_24

Reid ME, Christian SB, Brien DL (2000) Gravitational stability of three-dimensional stratovolcano edifices. J Geophys Res 105(B3):6043–6056. https://doi.org/10.1029/1999JB900310

Reid ME, Sisson TW, Brien DL (2001) Volcano collapse promoted by hydrothermal alteration and edifice shape, Mount Rainier, Washington. Geology 29:779–782. https://doi.org/10.1130/0091-7613(2001)029<0779:VCPBHA>2.0.CO;2

Reid ME, Keith TEC, Kayen RE, Iverson NR, Iverson RM, Brien DL (2010) Volcano collapse promoted by progressive strength reduction: new data from Mount St. Helens. Bull Volcanol 72:761–766. https://doi.org/10.1007/s00445-010-0377-4

Roberti G, Friele P, van Wyk de Vries B, Ward B, Clague JJ, Perotti L, Giardino M (2017) Rheological evolution of the Mount Meager 2010 debris avalanche, southwestern British Columbia. Geosphere 13(2):369–390. https://doi.org/10.1130/GES01389.1

Rodríguez I, Páez J, Maxmillian S, van Wyk de Vries B, Godoy B (2020) Dynamics and physical parameters of the Lastarria debris avalanche, Central Andes. J Volcanol Geotherm Res 402:106990. https://doi.org/10.1016/j.jvolgeores.2020.106990

Romero JE, Moreno H, Polacci M, Burton M, Guzmán D (2022) Mid-Holocene lateral collapse of Antuco volcano (Chile): debris avalanche deposit features, emplacement dynamics, and impacts. Landslides 10:1–8. https://doi.org/10.1007/s10346-022-01865-z

Roverato M (2016) The Montesbelos mass-flow (southern Amazonian craton, Brazil): a Paleoproterozoic volcanic debris avalanche deposit? Bull Volcanol 78:49. https://doi.org/10.1007/s00445-016-1043-2

Roverato M, Cronin S, Procter J, Capra L (2015) Textural features as indicators of debris avalanche transport and emplacement, Taranaki volcano. Geol Soc Am Bull 127:3–18. https://doi.org/10.1130/B30946.1

Roverato M, Dufresne A, Procter J (eds) (2021) Volcanic debris avalanches. In: Advances in volcanology. Springer, Cham, 358 pp. https://doi.org/10.1007/978-3-030-57411-6

Russo G, Giberti G, Sartoris G (1996) The influence of regional stresses on the mechanical stability of volcanoes, Stromboli (Italy). Geol Soc Lond Spec Publ 110:65–75

Satake S, Smith JR, Shinozaki K (2002) Three-dimensional reconstruction and tsunami model of the Nuuanu and Wailau giant landslides, Hawaii. In: Takahashi E, Lipman PW, Garcia MO, Naka J, Aramaki S (eds) Hawaiian volcanoes: deep underwater perspectives. Am Geophys Union Geophys Monogr 128:333–348. https://doi.org/10.1029/GM128p0333

Savage SB (1984) The mechanics of rapid granular flows. Adv Appl Mech 24:289–376

Schaefer LN, Kendrick JE, Oommen T, Lavallée Y, Chigna G (2015) Geomechanical rock properties of a basaltic volcano. Front Earth Sci 3(29). https://doi.org/10.3389/feart.2015.00029

Schneider JL, Fisher RV (1998) Transport and emplacement mechanisms of large volcanic debris avalanches: evidence from the northwest sector of Cantal volcano (France). J Volcanol Geotherm 83:141–165

Schuster RL, Crandell DR (1984) Catastrophic debris avalanches from volcanoes: In: Proceedings of IVth

international conference on landslides, Toronto, pp 567–572
Scott KM (1988) Origins, behavior, and sedimentology of lahars and lahar-runout flows in the Toutle-Cowlitz River system. U S Geol Surv Prof Pap 1447-A:74 pp
Scott KM, Macías JL, Naranjo JA, Rodríguez S, McGeehin JP (2001) Catastrophic debris flows transformed from landslides in volcanic terrains: mobility, hazard assessment, and mitigation strategies. U S Geol Surv Prof Pap 1630:59 pp
Scott KM, Vallance JW, Kerle N, Macías JL, Strauch W, Devoli G (2005) Catastrophic precipitation-triggered lahar at Casita volcano, Nicaragua: occurrence, bulking and transformation. Earth Surf Process Landf 30:59–79. https://doi.org/10.1002/esp.1127
Shea T, van Wyk de Vries B, Pilato M (2008) Emplacement mechanisms of contrasting debris avalanches at Volcán Mombacho (Nicaragua), provided by structural and facies analysis. Bull Volcanol 70:899. https://doi.org/10.1007/s00445-007-0177-7
Siebe C, Komorowski J-C, Sheridan MF (1992) Morphology and emplacement of an unusual debris-avalanche deposit at Jocotitlán volcano, central Mexico. Bull Volcanol 54:573–589. https://doi.org/10.1007/BF00569941
Siebert L (1984) Large volcanic debris avalanches: characteristics of source areas, deposits, and associated eruptions. J Volcanol Geotherm Res 22:163–197. https://doi.org/10.1016/0377-0273(84)90002-7
Siebert L (1996) Hazards of large volcanic debris avalanches and associated eruptive phenomena. In: Scarpa R, Tilling RI (eds) Monitoring and mitigation of volcano hazards. Springer-Verlag, Berlin, pp 541–572
Siebert L (2002) Landslides resulting from structural failure of volcanoes. In: Evans SG, DeGraff JV (eds) Catastrophic landslides: effects, occurrence, and mechanisms. Geol Soc Am Rev Eng Geol 15:209–235. https://doi.org/10.1130/REG15-p209
Siebert L, Reid ME (2023) Lateral edifice collapse and volcanic debris avalanches: a post-1980 Mount St. Helens perspective. Bull Volcanol 85(61). https://doi.org/10.1007/s00445-023-01662-z
Siebert L, Roverato M (2021) A historical perspective on lateral collapse and volcanic debris avalanches. In: Roverato M, Dufresne A, Procter J (eds) Volcanic debris avalanches. Advances in volcanology, vol 6. Springer, Cham, pp 11–50. https://doi.org/10.1007/978-3-030-57411-6_2
Siebert L, Vallance JW (2017) Large volume edifice failures in the Cascade Range of southern British Columbia to northern California. In: International association of volcanology and chemistry of the earth's interior 2017 scientific assembly, Portland OR, 14–18 Aug 2017, abstr VH23A-181
Siebert L, Kimberly P, Pullinger CR (2004) The voluminous Acajutla debris avalanche from Santa Ana volcano, western El Salvador, and comparison with other Central American edifice-failure events. In: Rose WI, Bommer JJ, López DL, Carr MJ, Major JJ (eds) Natural hazards in El Salvador. Geol Soc Am Spec Pap 375:5–23. https://doi.org/10.1130/0-8137-2375-2.5
Siebert L, Reid ME, Vallance JW, Pierson TC (2019) When volcanoes fall down—catastrophic collapse and debris avalanches (ver. 1.2, August 2019). U S Geol Surv Fact Sheet 2019-3023, 6 pp. https://doi.org/10.3133/fs20193023
Silver E, Day S, Ward S, Hoffman G, Llanes P, Driscoll N, Appelgate B, Saunders S (2009) Volcano collapse and tsunami generation in the Bismarck volcanic arc, Papua New Guinea. J Volcanol Geotherm Res 186:210–222. https://doi.org/10.1016/j.jvolgeores.2009.06.013
Strom A, Li L, Lan H (2019) Rock avalanche mobility: optimal characterization and the effects of confinement. Landslides 16(8):1437–1452. https://doi.org/10.1007/s10346-019-01181-z
Takarada S, Ui T, Yamamoto Y (1999) Depositional features and transportation mechanism of valley-filling Iwasegawa and Kaida debris avalanches, Japan. Bull Volcanol 60:508–522. https://doi.org/10.1007/s004450050248
Thomas ME, Petford N, Bromhead EN (2004) The effect of internal gas pressurization on volcanic edifice stability: evolution towards a critical state. Terra Nova 16:312–317
Thompson N, Bennett MR, Petford N (2010) Development of characteristic volcanic debris avalanche deposit structures: new insight from distinct element simulations. J Volcanol Geotherm Res 192:191–200
Tibaldi A, Bistacchi A, Pasquarè FA, Vezzoli L (2006) Extensional tectonics and volcano lateral collapses: insights from Ollagüe volcano (Chile-Bolivia) and analogue modelling. Terra Nova 18:282–289. https://doi.org/10.1111/j.1365-3121.2006.00691.x
Tormey D (2010) Managing the effects of accelerated glacial melting on volcanic collapse and debris flows: Planchon-Peteroa volcano, Southern Andes. Glob Planet Change 74:82–90
Trofimovs J, Cas RAF, Davis BK (2004) An archean submarine volcanic debris avalanche deposit, Yilgarn craton, western Australia, with komatiite, basalt and dacite megablocks: the product of dome collapse. J Volcanol Geotherm Res 138:111–126. https://doi.org/10.1016/j.jvolgeores.2004.06.008
Ui T (1983) Volcanic dry avalanche deposits-identification and comparison with nonvolcanic debris streams deposits. J Volcanol Geotherm Res 18:135–150. https://doi.org/10.1016/0377-0273(83)90006-9
Ui T (1989) Discrimination between debris avalanche and other volcaniclastic deposits. In: Latter JH (ed) Volcanic hazards. International association of volcanology and chemistry of the earth's interior proceedings in volcanology, vol 1, pp 201–209
Ui T, Kawachi S, Neall VE (1986) Fragmentation of debris avalanche material during flowage—evidence

from the Pungarehu Formation, Mount Egmont, New Zealand. J Volcanol Geotherm Res 27:255–264

Ui T, Takarada S, Yoshimoto M (2000) Debris avalanches. In: Sigurdsson H, Houghton B, McNutt SR, Rymer H, Stix J (eds) Encyclopedia of volcanoes. Academic Press, San Diego, pp 617–626

Valderrama P, Roche O, Samaniego P, van Wyk de Vries B, Bernard K, Marino J (2016) Dynamic implications of ridges on a debris avalanche deposit at Tutupaca volcano (Southern Peru). Bull Volcanol 78:14. https://doi.org/10.1007/s00445-016-1011-x

Valderrama P, Roche O, Samaniego P, van Wyk de Vries B, Araujo G (2018) Granular fingering as a mechanism for ridge formation in debris avalanche deposits: laboratory experiments and implications for Tutupaca volcano, Peru. J Volcanol Geotherm Res 349:409–418. https://doi.org/10.1016/j.jvolgeores.2017.12.004

Vallance JW (1994) Experimental and field studies related to the behavior of granular mass flows and the characteristics of their deposits. Dissertation, Michigan Technological University, 197 pp

Vallance JW (2000) Lahars. In: Sigurdsson H, Houghton BF, McNutt SR, Rymer H, Stix J (eds) Encyclopedia of volcanoes. Academic Press, San Diego, pp 601–616

Vallance JW, Iverson RM (2015) Lahars and their deposits. In: Sigurdsson H, Houghton B, McNutt SR, Rymer H, Stix J (eds) The encyclopedia of volcanoes, 2nd edn. Elsevier, Amsterdam, pp 649–664. https://doi.org/10.1016/B978-0-12-385938-9.00037-7

Vallance JW, Scott KM (1997) The Osceola Mudflow from Mount Rainier: sedimentology and hazard implications of a huge clay-rich debris flow. Geol Soc Am Bull 109:143–163. https://doi.org/10.1130/0016-7606(1997)109<0143:TOMFMR>2.3.CO;2

van Wyk de Vries B, Borgia A (1996) The role of basement in volcano deformation. Geol Soc Lond Spec Publ 110(1):95–110. https://doi.org/10.1144/GSL.SP.1996.110.01.07

van Wyk de Vries B, Davies TR (2015) Landslides, debris avalanches, and volcanic gravitational deformation. In: Sigurdsson H (ed) The encyclopedia of volcanoes, 2nd edn. Elsevier, Amsterdam, pp 665–685. https://doi.org/10.1016/B978-0-12-385938-9.00038-9

van Wyk de Vries B, Francis P (1997) Catastrophic collapse at stratovolcanoes induced by gradual volcano spreading. Nature 387(6631):387–390. https://doi.org/10.1038/387387a0

van Wyk de Vries B, Kerle N, Petley D (2000) Sector collapse forming at Casita volcano, Nicaragua. Geology 28:167–170

van Wyk de Vries B, Self S, Francis PW, Keszthelyi L (2001) A gravitational spreading origin for the Socompa debris avalanche. J Volcanol Geotherm Res 105:225–247. https://doi.org/10.1016/S0377-0273(00)00252-3

Voight B (1981) Time scale for the first moments of the May 18 eruption. In: Lipman PW, Mullineaux DR (eds) The 1980 eruptions of Mount St. Helens, Washington. U S Geol Surv Prof Pap 1250:69–86

Voight B (2000) Structural stability of andesitic volcanoes and lava domes. Phil Trans R Soc Lond A 358:1663–1703. https://doi.org/10.1098/rsta.2000.0609

Voight B, Elsworth D (1997) Failure of volcano slopes. Geotechnique 47:1–31. https://doi.org/10.1680/geot.1997.47.1.1

Voight B, Sousa J (1994) Lessons from Ontakesan: a comparative analysis of debris avalanche dynamics. Eng Geol 38:261–297. https://doi.org/10.1016/0013-7952(94)90042-6

Voight B, Glicken H, Janda RJ, Douglass PM (1981) Catastrophic rockslide avalanche of May 18. In: Lipman PW, Mullineaux DR (eds) The 1980 eruptions of Mount St. Helens, Washington. U S Geol Surv Prof Pap 1250:347–377

Voight B, Janda RJ, Glicken H, Douglass PM (1983) Nature and mechanics of the Mount St. Helens rockslide-avalanche of 18 May 1980. Geotechnique 33:243–273. https://doi.org/10.1680/geot.1983.33.3.243

von Buch L (1825) Physikalische Beschreibung der Kanarischen Inseln, vol 2. Hofdruckerei von Königlichen Akademie, Berlin

Wadge G, Francis PW, Ramirez CF (1995) The Socompa collapse and avalanche event. J Volcanol Geotherm Res 66:309–336. https://doi.org/10.1016/0377-0273(94)00083-S

Waitt RB (1981) Devastating pyroclastic density flow and attendant air fall of May 18—stratigraphy and sedimentology of deposits. In: Lipman PW, Mullineaux DR (eds) The 1980 eruptions of Mount St. Helens, Washington. U S Geol Surv Prof Pap 1250:439–458

Waitt RB, Begét JE (2009) Volcanic processes and geology of Augustine volcano, Alaska. U S Geol Surv Prof Pap 1762:1–78, 2 plates, scale 1:25,000

Ward SN, Day S (2006) Particulate kinematic simulations of debris avalanches: interpretation of deposits and landslide seismic signals of Mount Saint Helens, 1980 May 18. Geophys J Int 167(2):991–1004. https://doi.org/10.1111/j.1365-246X.2006.03118.x

Watt SFL, Pyle DM, Naranjo JA, Mather TA (2009) Landslide and tsunami hazard at Yate volcano, Chile as an example of edifice destruction on strike-slip fault zones. Bull Volcanol 71:559–574

Watters RJ, Delahaut WD (1995) Effect of argillic alteration on rock mass stability. In: Haneburg WC, Anderson SA (eds) Clay and shale slope instability. Geol Soc Am Rev Eng Geol 10:139–150

Watters RJ, Zimbelman DR, Bowman SD, Crowley JK (2000) Rock mass strength assessment and significance to edifice stability, Mount Rainier and Mount Hood, Cascade Range volcanoes. Pure Appl Geophys 157:957–976. https://doi.org/10.1007/s000240050012

Weidinger JT, Korup O, Munack H, Altenberger U, Dunning SA, Tippelt G, Lottermoser W (2014) Giant rockslides from the inside. Earth Planet Sci Lett 389:62–73. https://doi.org/10.1016/j.epsl.2013.12.017

Wooller L, van Wyk de Vries B, Murray JB, Rymer H, Meyer S (2004) Volcano spreading controlled by dipping substrata. Geology 32(7):573–576. https://doi.org/10.1130/G20472.1

Wooller L, van Wyk de Vries B, Cecchi E, Rymer H (2009) Analogue models of the effect of long-term basement fault movement on volcanic edifices. Bull Volcanol 71(10):1111–1131. https://doi.org/10.1007/s00445-009-0289-3

Yoshida H, Sugai T (2010) Grain size distribution of the sediments from the 24 ka sector collapse of Asama volcano, Japan. Trans Jpn Geomorph Union 31(2):193–201. https://doi.org/10.4116/jaqua.45.123

Zernack AV (2021) Volcanic debris-avalanche deposits in the context of volcaniclastic ring plain successions—a case study from Mt. Taranaki. In: Roverato M, Dufresne A, Procter J (eds) Volcanic debris avalanches. Advances in volcanology, vol 6. Springer, Cham, pp 211–254. https://doi.org/10.1007/978-3-030-57411-6_9

Zernack AV, Procter JN, Cronin SJ (2009) Sedimentary signatures of cyclic growth and destruction of stratovolcanoes: a case study from Mt. Taranaki, New Zealand. Sed Geol 220(3–4):288–305. https://doi.org/10.1016/j.sedgeo.2009.04.024

Zernack AV, Cronin SJ, Bebbington M, Price R, Smith IEM, Steward RB, Procter JN (2012) Forecasting catastrophic stratovolcano collapse: a model based on Mt Taranaki, New Zealand. Geology 40:983–986

Zimbelman DR, Watters RJ, Firth IR, Breit GN, Carrasco-Núñez G (2004) Stratovolcano stability assessment methods and results from Citlaltépetl, Mexico. Bull Volcanol 66(1):66–79. https://doi.org/10.1007/s00445-003-0296-8

Volcanic Debris-Avalanche Deposits

Abstract

Volcanic debris avalanches (VDAs) emplace characteristically multicolored, hummocky deposits that are typically coarse, unusually thick, poorly sorted, and contain a wider range of components and sedimentary textures than the deposits of other volcanic wet flows. Debris-avalanche blocks, a component unique to VDA deposits (and to some deposits of lahars transformed from VDAs), commonly preserve pieces of recognizable rock units from the source volcano. Rock colors within debris-avalanche blocks can be both highly varied and intense. This is due to the presence of originally multi-hued and interbedded lavas and volcaniclastic units that formed the failed edifice structure, as well as to any clay-rich, hydrothermally altered rocks from the chemically altered edifice core that are included. The somewhat coherent debris-avalanche blocks in VDA deposits are surrounded by more thoroughly homogenized and neutrally colored mixed material, which is matrix material that also contains gravel-size clasts up to meters in diameter. Key characteristics of VDA deposits include areas of highly irregular (hummocky) topography, juxtaposition of rock units having sharply contrasting textures and colors, unstratified deposits of unusual thickness (may exceed 100 m), pervasive fracturing of hard clasts with pieces fitting together like jigsaw puzzles, large clasts that are dominantly angular, polymictic particle mixtures, and the absence of bubble vesicles in matrix material.

1 Overview

Deposits emplaced by volcanic debris avalanches (VDAs) are typically coarse, unusually thick, poorly sorted, and contain a wider range of components and sedimentary textures than the deposits of other volcanic wet flows. These components include rock fragment grains <2 mm in diameter (*matrix material*), larger fragments of solid rock up to meters in diameter (*gravel-size clasts*), zones of sheared and deformed mixtures of gravel-size clasts and matrix material (*mixed material*), and *debris-avalanche blocks* of varying size (centimeters to hundreds of meters in diameter). Debris-avalanche blocks are relatively intact portions of the failed volcanic edifice that typically contain some hard gravel-size clasts, masses of pervasively fractured lava rock (disrupted lava domes, cryptodomes, dikes, and flows), and interbedded layers of unconsolidated volcaniclastic deposits. Such potentially deformable rock masses are typically transported *en masse* relatively intact and relatively undeformed by VDAs, commonly preserving some original stratigraphy from the flanks of the source volcano.

T. C. Pierson et al., *Deposits of Volcanic Wet Flows*, Advances in Volcanology,
https://doi.org/10.1007/978-3-031-66574-5_8

Debris-avalanche blocks are not necessarily "block size," as defined in particle-size classifications (see Chap. "Deposit Identification and Interpretation"), and they are not the same as slide blocks (defined in Chap. "Volcanic Debris Avalanches—Process Overview"), although slide blocks may disintegrate to form debris-avalanche blocks. Gravel-size clasts and debris-avalanche blocks can be of similar size, but these clasts differ in being hard, coherent, mechanically separable, and durable fragments of solid rock. In general, VDA deposits are unique among other types of fragmental deposits at volcanoes in having abrupt and sharp juxtapositions of intense colors between adjacent fragments of unaltered and hydrothermally altered volcanic rock (see Sect. 2.2 in Chap. "Volcanic Debris-Avalanche Deposits"). Color contrasts occur between and within different debris-avalanche blocks, as well as between blocks and the more neutral grayish browns and brownish grays that are typical of mixed material.

Large debris-avalanche blocks have also been termed *megablocks*, *breccia blocks*, and *megaclasts* (e.g., Ui 1983; Siebert 1984, 1991); Dufrense et al. 2021b). While some debris-avalanche blocks can be > 500 m across (Ui 1983) and encompass volumes up to 10^7 m^3 (Glicken 1986, 1996), many are < 1 m in diameter, which renders the prefix 'mega' inappropriate for all debris-avalanche blocks (Bernard et al. 2021). VDAs also can incorporate trees and other vegetation as they move downslope, as well as significant amounts of rock debris and soil from the substrate. This varied componentry leads to morphological and textural features that are distinctive. Furthermore, partial or total transformation from VDA to volcanic debris flow (VDF) during or following emplacement (e.g.. Vallance and Scott 1997; Scott et al. 2001; Dufresne et al. 2021b; see Chap. "Volcanic Debris Avalanches—Process Overview") provides even greater diversity to textural and sedimentological characteristics within single VDA deposits.

VDA deposits can be confined to the lower flanks of a volcano or extend tens of kilometers beyond the source edifice. Depending on edifice and failure sizes, volumes of VDA deposits from subaerial stratovolcanoes or lava domes often vary from 0.1 km^3 to > 10 km^3 and can blanket > 1000 km^2 of terrain. Deposits from failures of oceanic shield volcanoes can be several thousand km^3 in volume (Moore et al. 1989). The occurrence of VDA deposits worldwide indicates large-scale edifice failure is a relatively common event in the life cycle of volcanoes. Surprisingly, this recognition came only in the late twentieth century. Although VDAs had been identified as originating from landslides prior to 1980 (Siebert and Roverato 2021), they were not widely investigated, recognized, or understood as a volcanic hazard until after the 1980 eruption of Mount St. Helens produced a VDA that was well documented (Voight et al. 1981, 1983; Glicken 1996).

2 Landscape-Scale Characteristics of Deposits

Volcanic debris-avalanche deposits differ in morphology and in complexity from other types of volcanic fragmental deposits (Ui 1983, 1985, 1989; Siebert 1984, 1996; Ui et al. 1986a, b, 2000; Palmer et al. 1991; Glicken 1996; Takarada et al. 1999; Clavero et al. 2002; Salinas and López-Blanco 2010; van Wyk de Vries and Davies 2015; Roverato et al. 2021). Spatially, topographically, and compositionally complex VDA deposits are commonly characterized by:

(1) Evidence for a combination of ductile and brittle mechanical behavior during emplacement.
(2) Widely distributed *hummocks* (hills) on the surfaces of deposits.
(3) Closed depressions (often forming lakes or ponds) on surfaces between hummocks.
(4) Sharply defined and topographically prominent lateral and frontal margins (sometimes with steep marginal levees).
(5) Complex patterns of faults, folds, transverse to longitudinal ridges, and flow levees.

(6) Very thick (to > 10^2 m) sediment bodies composed of heterogeneous, variably stratified, extremely poorly sorted sediment.

Such variety in physical characteristics in any one VDA deposit indicates both the range of physical properties and the range of applied stresses encountered in a moving avalanche body. The varying mechanical properties and applied forces within VDAs leads to important contrasting emplacement behaviors and deposit morphologies (Valderrama et al. 2016; Roberti et al. 2017; Rodríguez et al. 2020).

2.1 Deposit Architecture and Morphology

2.1.1 Sediment-Body Shape

VDA deposit bodies (Fig. 1) form thick (tens to hundreds of meters), down-slope-thinning wedges, which can be narrowly bounded in valleys or spread broadly on unconfined slopes. Deposit surfaces can display a complex arrangement of faulting and folding (Fig. 2), with normal, strike-slip, and thrust faults extending into the interior of deposits that demonstrate large-scale brittle deformation during emplacement (e.g., Shea and van Wyk de Vries 2008; Paguican et al. 2012; Makris et al. 2023b).

Fresh deposit surfaces typically have extremely rugged and chaotic topographies and very irregular drainage patterns, whereas older weathered and vegetated deposits may have more regular and subdued surface features that may lead geologists to assume other origins. The lateral and frontal margins of VDA deposits typically are tens of meters high and steep (Siebe et al. 1992; Wadge et al. 1995), although post-avalanche sedimentation, erosion, or transformation to volcanic debris flows (VDFs) may subdue or obscure margins. Interior parts of avalanche deposits can be over 100 m thick (Glicken 1996). Such thicknesses far exceed those of any other deposits of volcanic wet flows.

2.1.2 Hummocks

Hummocks are mounds or hills of varying size and shape, commonly accompanied by intervening closed depressions, on the surfaces of VDA deposits. They are dramatic and distinctive features of such deposits (Fig. 3). Freshly deposited hummocks have angular shapes and appear planimetrically as sharply defined equant, elongated, or irregular hills or elongate ridges; they become more rounded and subdued as they erode and weather. Large avalanche deposits can contain hundreds to thousands of hummocks (Yoshida et al. 2012; Herrick et al. 2013) that range from a meter up to about 200 m in height and more than a kilometer in length. They can form from (1) multiple upward-protruding points on a single, very large debris-avalanche block; (2) single debris-avalanche blocks or several adjacent debris-avalanche blocks protruding above the level of the surrounding mixed material in the deposit; or (3) dense concentrations of gravel-size clasts within the mixed material (Glicken 1996). Hummocks represent spreading-induced brittle deformation of tilted and rotated debris-avalanche blocks in the original failure mass and result from the interplay of normal, thrust, and strike-slip faulting during emplacement (Paguican et al. 2012, 2014). Hummocks most commonly form extensional horst and graben morphology or boudinage structures, where individual debris-avalanche blocks are pulled apart (Voight 1981; van Wyk de Vries and Davies 2015). Their spatial distribution, size, and general appearance depend on origin, distance from source, and age (degree of erosion and vegetation). Small VDA deposits tend to have bouldery irregular surface topographies, as well, but pronounced hills and depressions typically are not well developed (Crandell and Fahnestock 1965).

Orientation of hummocks varies widely but is influenced by topography. Long axes of elongate hummocks are generally aligned with the direction of flow in extensional environments, and dominantly perpendicular to flow direction in more distal compressional environments

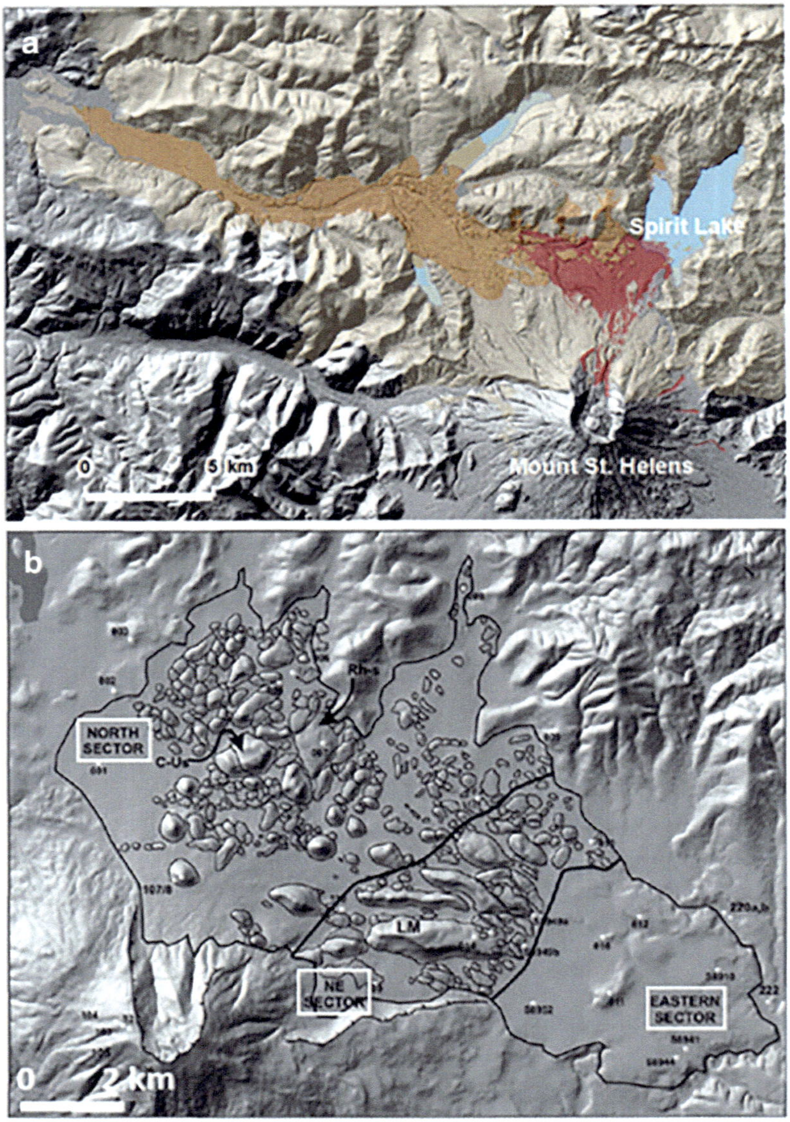

Fig. 1 Depositional terrain of volcanic debris avalanche (VDA) deposits. **a** The 2.5 km^3 1980 Mount St. Helens VDA deposit, Washington, USA (in orange-brown color) is confined within the North Fork Toutle River valley (associated lateral-blast deposit shown in light brown). Deposit is overlain by post-May 18 red-colored pumiceous pyroclastic-flow deposits. Modified from Pierson et al. (2018) and Lipman and Mullineaux (1981). **b** The 2.8 km^3 VDA deposit at Jocotitlán volcano (Mexico) in laterally unconfined terrain. Black lines delineate margins of north, northeastern and eastern sectors (white boxes) of Jocotitlán deposits (modified from Salinas and López-Blanco (2010)

(e.g., Paguican et al. 2012; Yoshida 2014; Roberti et al. 2017). Alternating compressional and extensional environments may occur as slope angles vary or topographic barriers are encountered. Compressional regimes may also result from sudden deceleration when avalanches slow to a stop (Roberti et al. 2017) or enter shallow submarine terrain. Flow-parallel hummocks in subaerial VDAs commonly transition to elongated ridges perpendicular to flow direction after entering submarine environments (e.g., at the Unzen volcano

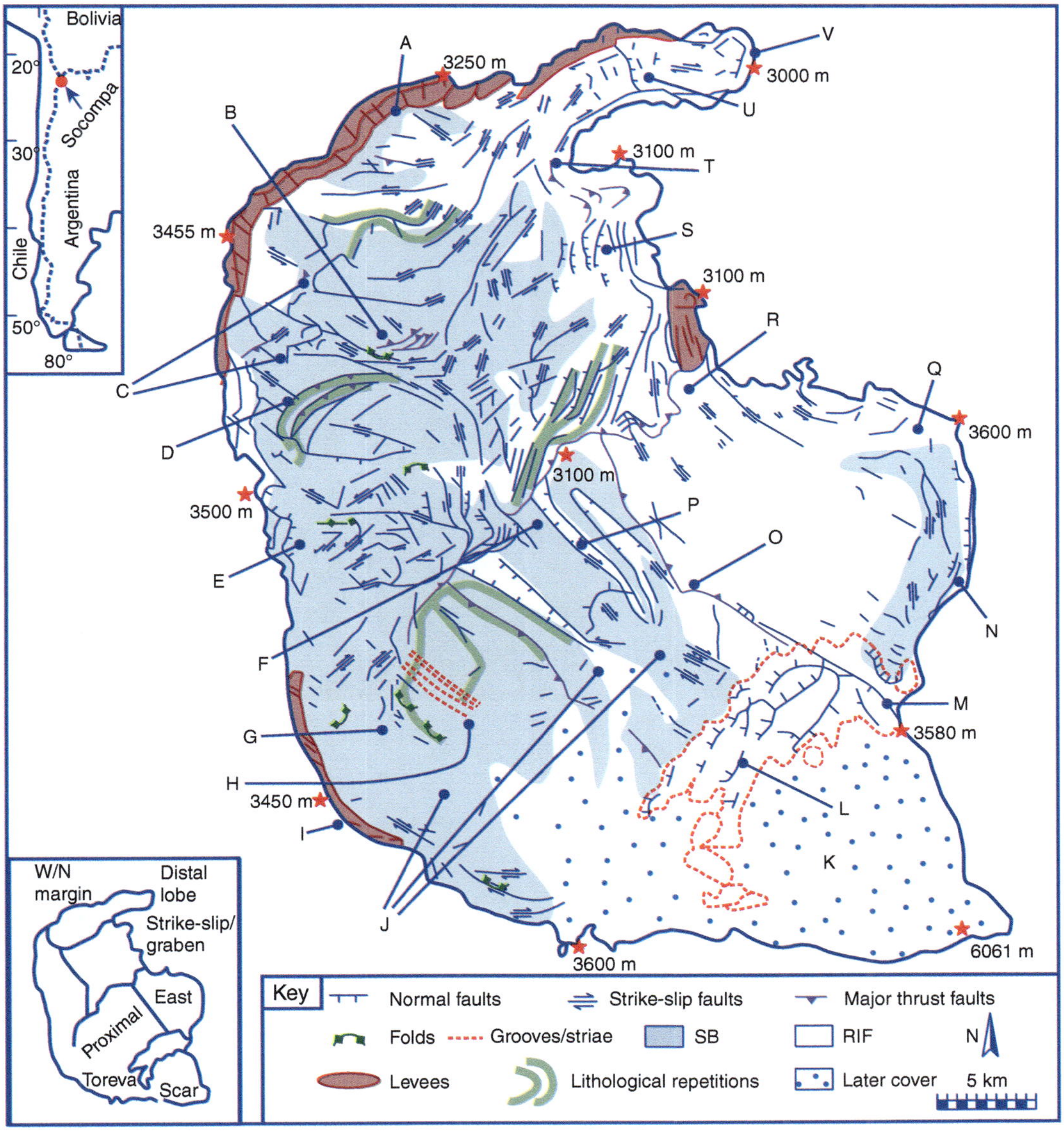

Fig. 2 Structural map of Socompa debris-avalanche deposit (Chile/Argentina) (modified from Shea and van Wyk de Vries 2008). Failure scar (covered) is in lower right side of map. Highlighted features include A. Faulted contact with levee. B. Isolated back thrust area. C. Large-throw normal faults. D. Early thrusts cut by conjugate strike-slips. E. Area of backflow sliding off marginal slope. F. Debris units continue through median scarp but are deflected. G. Median scarp starts as en-echelon folds, thrusts, and strike-slips. H. Longitudinal grooves. I. Levee and marginal strike slips. J. Areas of dense normal faulting. K. Area covered by later deposits and lava flows. L. Toreva faults with up to 300 m throw. M. Buried scar margin. N. 200-m-high fault scarp. O. Longitudinal thrusts. P. Spreading debris units. Q. Abrupt change in flow direction of northeast unit. R. Northeast unit lineations continue through median scarp. S. Eastward spreading indicated by grabens. T. Strike-slip and graben structures meet distal structures here. U. Transverse faults in middle and longitudinal strike-slips at margins. V. Back-facing normal faults at terminus. Unit RIF is the basal reconstituted ignimbrite facies—dominantly fine materials from the remobilized ignimbrite substratum. Unit SB is the Socompa Breccia facies—blocks and matrix. The toreva block zone (large proximal blocks that slid without disaggregating) is outlined by black dashed lines

Fig. 3 Volcanic debris avalanche (VDA) hummocks at different scales. **a** Hummocks of West Island VDA deposit at Augustine Volcano (Alaska), typically < 10 m tall (Siebert et al. 1989, 1995; Waitt and Begét 2009). **b** Small conical hummock of 1980 Mount St. Helens (Washington, USA) VDA deposit (Glicken 1996) formed from disintegration of a debris-avalanche block. Horizontal trim line on hillside beyond upper part of hummock in this 1984 photo marks upper limit of VDA during emplacement; below this line trees blown down by the lateral blast were removed by the avalanche. **c** Large 120-m-tall hummock in VDA deposit on north side of Jocotitlán volcano (Mexico) (Dufresne et al. 2010). **d** Mount St. Helens avalanche deposit near Coldwater Lake (upper left). Note the numerous ponds in the closed depressions on deposit surface. **e** Surface of distal 1980 Mount St. Helens avalanche deposit about 20 km downstream of the volcano, near toe of deposit. Hummocks here are relatively small and subdued. Field of view is roughly one kilometer across. **f** Hummocks from Shasta Valley VDA deposit (California, USA) (Crandell 1989). The flat-topped hummock in center distance is about 140 m tall and a nearby hummock out of view is about 3 km long. Closed depressions between hummocks at Shasta often contain ponds or small lakes, although the one seen here is an artificial reservoir. Photos **a, b, c, f** by L. Siebert, Smithsonian; **d, e** USGS (unknown)

complex (Japan) in 1792, at Augustine Volcano (Alaska) in 1883, at Santa Ana (El Salvador) volcano (Siebert et al. 1987, 1995, 2004), and at Kharimkotan (Harimkotan) volcano (Russia) in 1933 (Belousov and Belousova 1996).

The spatial distribution of debris-avalanche blocks, and therefore the distribution of hummocks within a deposit, is usually not uniform because debris-avalanche blocks experience a type of selective sorting: larger blocks become grounded closer to the volcano than smaller blocks, and breakage during transport reduces the size of debris-avalanche blocks and increases the amount of mixed material in proportion to distance traveled (Ui et al. 1986b, 1986, 1996; Siebert et al. 1996; Takarada et al. 1999). Where large rigid debris-avalanche blocks are surrounded by finer grained matrix material, such blocks may come to a stop when they become grounded on the valley floor. More deformable material may flow around them, thinning and leaving these debris-avalanche blocks to protrude high out of the surrounding deposit (Palmer et al. 1991; Glicken 1996). Concentrations of hummocks are found where moving debris-avalanche blocks meet topographic obstructions—bedrock hills or ridges on the valley floors, such as at Raung volcano (Indonesia) (Siebert 2002) and at Santa Ana volcano (Siebert et al. 2004). (Yoshida et al. 2012) noted that the rate of decrease in hummock size was inversely related to the mobility of the avalanche and that hummock morphology parameters correlate with avalanche volume and mobility. If a debris avalanche transforms directly to a VDF, some relatively small debris-avalanche blocks may continue to be transported and thus form small hummocks on much smoother lahar-deposit surfaces (Vallance and Scott 1997; Scott et al. 1995). Hummock densities vary widely, and larger deposits can display broad flat, hummock-free areas.

2.1.3 Closed Depressions

Closed depressions on VDA deposit surfaces form as (1) irregularly shaped inter-hummock areas caused by gaps between juxtaposed debris-avalanche blocks, (2) grabens formed during emplacement, and (3) depressions formed by collapse of surface material into void space below (Glicken 1996) (Fig. 4). Collapse depressions as large as 240 m across and 50 m deep can form over primary voids (created during flow) or over voids created by melting of entrained glacier ice (Glicken 1986, 1996; Branney and Gilbert 1995). The arrangement of hummocks and intervening closed depressions on deposit surfaces leads to chaotic topographies and very irregular drainage patterns.

2.1.4 Other Morphologic Features

Failure Scars

Failure terminology for VDAs is not standardized. In this volume a lateral edifice failure is termed an *edifice collapse* if a significant proportion of the edifice is removed and transported laterally (in contrast to vertical caldera collapse), including part of the volcanic core and the summit (Fig. 5a–c). A relatively small failure from an edifice side is a *flank collapse* if it does not involve the volcano summit (Fig. 5d). Failure scars from lateral edifice collapses, also termed *sector collapses* (although Bernard et al. (2021) note ambiguities in use of the term), are characteristically deep (to ~ 10^3 m depth, normal to original ground surface). They are breached depressions that are open on the downslope side, with relatively flat floors and steep head walls and sub-parallel side walls (Siebert 1984). Scar geometries vary widely, with some scars having irregular side walls or a triangular shape with a wider breach (Bernard et al. 2021). In contrast to edifice-collapse scars, which can attain caldera-sized dimensions (Siebert 1984), flank-collapse scars (Fig. 5d) are smaller, relatively shallow depressions carved into the sides of edifices (Inokuchi 1985; Voight and Sousa 1994). Post-avalanche eruption deposits often partially or completely fill VDA failure scars.

Volcanic edifice collapse scars differ significantly from those producing nonvolcanic landslides (Fig. 6). Volcanic examples are typically larger and commonly excavate more deeply into a mountain side. Prior to 1980, geomorphic interpretations of VDA source areas varied

Fig. 4 Closed depressions on surfaces of volcanic debris avalanche (VDA) deposits. **a** Irregular elongate depression, containing a pond ~ 10 m wide, between large debris-avalanche blocks at Mount St. Helens (Washington, USA). **b** Small circular collapse depression on the 1980 Mount St. Helens VDA deposit. **c** Large vegetated closed depression on avalanche deposit from Parinacota volcano (Chile/Bolivia) (right) (Clavero et al. 2002). Pomerape volcano is behind, at left. Photos **a, b** by T. Pierson, USGS, **c** by L. Siebert, Smithsonian

widely. Interpretations ranged from erosional or fissuring origin to products of major explosive eruptions creating explosion craters or calderas, or eruptive or noneruptive subsidence forming grabens or sector grabens by down-dropping of a segment of the edifice (Siebert and Roverato 2021). Source area terminology after 1980 has also varied due to the ambiguity of many terms. Source areas have often been referred to as calderas, craters, or depressions, sometimes with morphological modifiers such as *horseshoe-shaped*, or *breached*. Most VDA scars, however, are U-shaped in plan view, with breaches somewhat broader than failure side walls; they are distinct from the narrow breaches typical of erosional crater-draining channels (Siebert 1984). However, actual morphology can vary considerably (Bernard et al. 2021). The 1980 collapse

Fig. 5 Gravitational collapse scars on volcanoes. **a** Edifice-collapse scar at Mount St. Helens (Washington, USA), resulting from a magma-induced edifice failure on May 18, 1980 is 2×3.5 km in size. USGS photo by L. Topinka, 1981. **b** Collapse scar at Galunggung volcano (Indonesia) at upper left with hummocky debris-avalanche terrain (the so-called "Ten Thousand Hills of Tasikmalaya") extending beyond lower right margin of image (Bronto 1989). Arrow shows avalanche travel direction. Map data ©2022 Google. Image modified from Siebert and Roverato (2021). **c** Collapse scar (right) of Pacaya volcano (Guatemala) with avalanche hummock in foreground (Vallance et al. 1995). Very small topographic bump midway down left-hand horizon is left side of scar, which is largely infilled by the post-collapse MacKenney cone forming summit in this 1988 view from the south. Lateral margin of collapse scar on right side of photo is about 4 km long. Smithsonian photo by L. Siebert. **d** Shallow 0.6×1.3-km flank-collapse scar on side of Ontake volcano (Japan), resulting from an earthquake-induced failure in 1984 (Endo et al. 1989). USGS photo by T. Pierson

scar at Mount St. Helens has mostly been referred to as a *crater*, although it and many other such scars attain caldera-sized dimensions. The term *caldera* has been used by some with both genetic and morphological implications, and genetic modifiers such as *avalanche* or *erosion* have been used to distinguish these features from true calderas. The term *amphitheater* has been used to avoid caldera comparisons, but architectural amphitheaters are fully enclosed and not open at one end. The nonvolcanic landslide community uses the term *scarp* when referring to two-dimensional headwall or sidewall cliffs of an avalanche failure site, whereas *scar* refers to the entire three-dimensional structure of a failure site. Bernard et al. (2021) preferred the use of the term *volcanic landslide scar* for these features.

Proximal Features

Massive intact sections of failed volcanic edifices, *toreva blocks* (Reiche 1937), are found at some volcanoes, commonly within the collapse scar or up to a few kilometers downslope. Toreva blocks are remnants of large, frictionally hindered primary or retrogressive slide blocks that have avoided disaggregation and incorporation into the avalanche body (Paguican et al. 2014; Wadge et al. 1995). These enormous blocks commonly display backward rotation

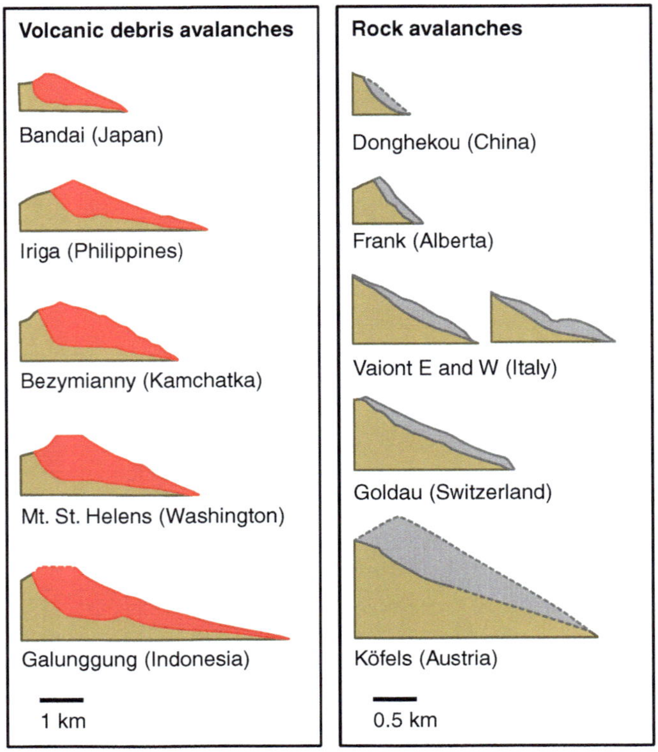

Fig. 6 Differences in cross-sectional failure-scar geometry between volcanic and nonvolcanic rock slope failures (with no vertical exaggeration). The tendency for volcanic failure scars to be deep explains the typically greater volumes for volcanic debris avalanches (VDAs) in comparison to nonvolcanic rock avalanches. VDA profiles and Frank and Goldau rock avalanche profiles are after Siebert (1984), Donghekou rock avalanche profile is from Qi et al. (2011), Vaiont profile is from Dykes and Bromhead (2018), and Köfels profile is from Prager et al. (2009). **Note difference in scale bars between the two slope failure types**. Volcano name font colors show failures associated with magmatic eruptions (red), phreatic eruptions (blue) and noneruptive events (black). Figure modified from Dufresne et al. (2021a)

(inferred from dips of bedding in the blocks), appear to have moved along the same failure planes (listric faults) as their coincident VDAs, and in some cases transition morphologically to large downslope hummocks (e.g., Wadge et al. 1995; Belousov et al. 1999; Andrade and van Wyk de Vries 2010; Paguican et al. 2014, 2021; Dufresne and Geertsema 2020). Toreva blocks reach 2.5 km in length at Socompa volcano (Chile/Argentina) (Francis et al. 1985) (Fig. 2) and double that at Barú volcano (Panama), where those toreva blocks have a combined volume of about 5 km^3 (Siebert et al. 2006; Herrick et al. 2013) (Fig. 7).

Modeling of edifice collapse processes commonly shows the formation of a prominent central graben high on the edifice at the onset of failure (as observed beginning to form at Mount St. Helens weeks before catastrophic failure on May 18, 1980). The graben migrates downslope during failure, and at the end of the failure forms a depression transverse to the failure direction at the base of the failure scar between the edifice and the downslope hummocky terrain of the VDA deposit (Andrade and van Wyk de Vries 2010; Paguican et al. 2014). Deposition of post-collapse lavas or volcaniclastic deposits often obscures the graben.

Medial to Distal Features

Morphologic features commonly observable on or adjacent to VDA deposits include lateral levees, transverse ridges, longitudinal ridges, longitudinal flow banding, and ground deformation of

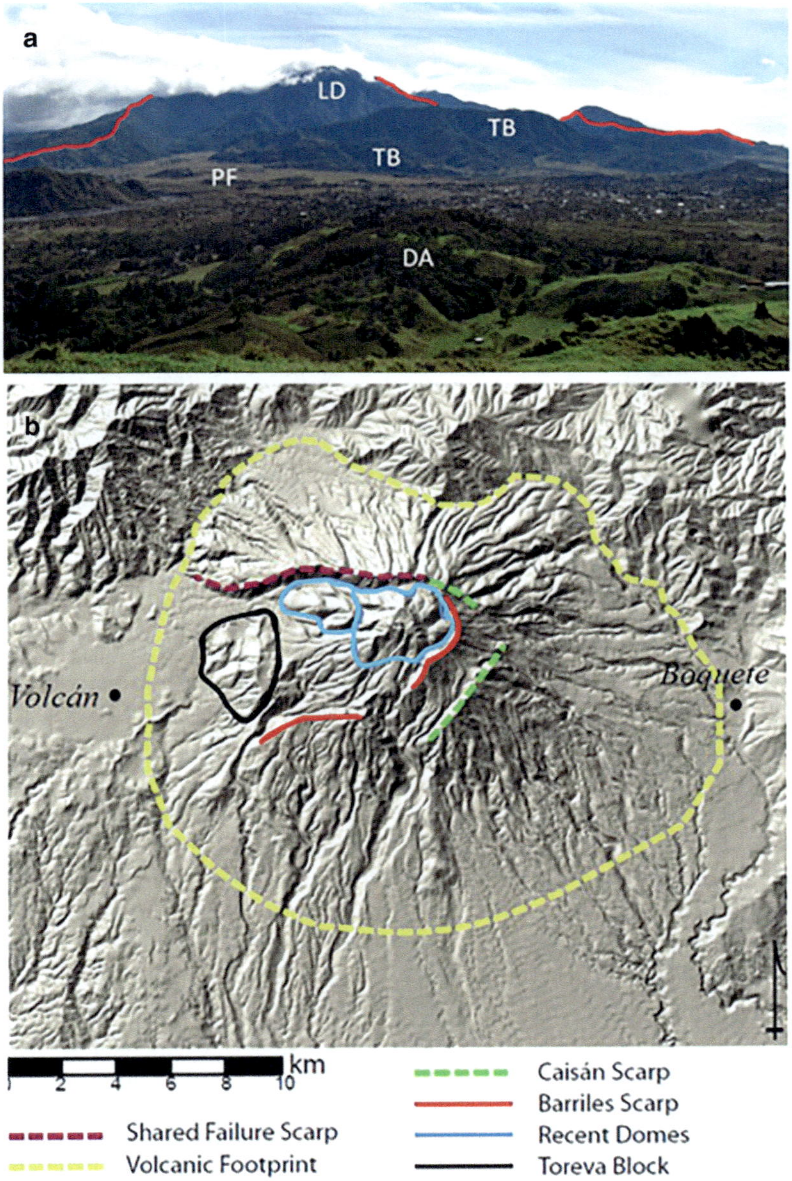

Fig. 7 Toreva blocks at Barú volcano (Panama). **a** Solid red lines: margins of 6 × 10 km collapse scarp, viewed from the west; LD: Post-collapse lava dome complex; TB: Backward-tilted toreva blocks within collapse scarp, PF: post-collapse pyroclastic-flow deposits; DA: Barilles VDA deposit. Smithsonian photo by L. Siebert. **b** Toreva blocks (black outline) and failure scarps at Barú volcano. Image modified from Herrick et al. (2013)

terrain at VDA deposit margins. Some of these features can also be found in VDF and pyroclastic density–current (PDC) deposits but typically occur at a smaller scale in those deposits.

Lateral levees (Fig. 8), which can be as high as 50 m above the surface of a VDA deposit and as wide as 1 km are significantly higher than the those of VDFs and PDCs, which typically do not exceed a few meters in height. Levees form at shear boundaries—usually between the avalanche deposit and valley sides or at the margins of unconfined avalanches (Siebert et al. 1995;

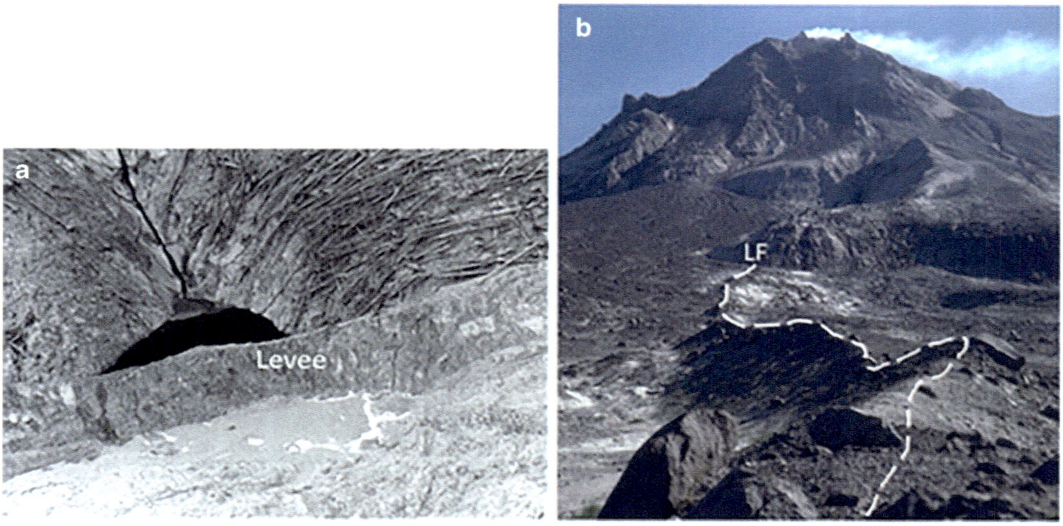

Fig. 8 Volcanic debris avalanche (VDA) deposit lateral levees. **a** Valley-margin lateral levee of the 1980 Mount St. Helens (Washington, USA) avalanche, where it blocks and dams a small tributary stream. The volume of the lake is about 50 million cubic meters. Logs on the ground were felled by the lateral blast of the eruption and are 30–50 m long. Photo modified from Glicken (1996). **b** Lateral levee of unconfined 1883 Burr Point VDA deposit at Augustine Volcano (Alaska) (Siebert et al. 1989, 1995). Levee (dashed line tracing crest) is about 2 km from base of lava flow (LF) at center of photo to near pre-1883 shoreline; in 1988 had a maximum height of ca. 25 m. Avalanche deposits to left of levee are overlain by post-avalanche block-and-ash-flow deposits from 1883 and later eruptions. Smithsonian photo by L. Siebert

Waitt and Begét 2009). Occasionally, they form within avalanche deposits (Francis et al. 1985; Glicken 1996). Their height can mark the fully dilated thickness of part of the VDA during flow or that of valley margin deposits pushed aside by the moving avalanche (Shea and van Wyk de Vries 2008). In unconfined areas they are generally higher than the main body of the deposit once movement stops. Levees commonly occur as single, narrow, linear ridges, but they can also form as wide linear zones of irregular positive topography. Levees of VDAs in confined valleys are typically steeper-sided and taller than levees at the margins of unconfined avalanche deposits. At their position along the margin of a deposit, levees are typically the part of the deposit that dams tributary valleys (Fig. 8). Due to their relatively low compaction and often narrow cross-sectional profiles, levee dams are especially susceptible to rapid failure if a lake forms in a tributary valley behind the levee. Large lakes formed by VDA debris blockages, such as at Bandai volcano (Japan) (Sekiya and Kikuchi 1889), Mount St. Helens (Scott 1988b), and the Colima volcanic complex (Capra and Macías 2002; Cortés et al. 2010) have failed and triggered massive lake-breakout floods and lahars composed of remobilized VDA deposits.

A variety of ridges, faults, and patterns of lineation can develop on the interior surfaces of VDA deposits (for example, Fig. 2). Irregular transverse compressional ridges develop on some VDA deposits, giving the deposit surface a wrinkled appearance, particularly near flow margins. Longitudinal flow-parallel ridges and longer and thinner flow bands can also develop and be prominent in some debris avalanches (Fig. 9), such as at Llullaillaco volcano (Chile/Argentina) (Francis and Wells 1988; Richards and Villeneuve 2001), Tutupaca volcano (Peru) (Valderrama et al. 2018), and the 1964 VDA deposit of Shiveluch volcano (Russia) (Belousov et al. 1999). Flow-parallel longitudinal ridges appear to be formed by a variety of processes, as inferred from field observations on deposits and from analogue modeling (Inokuchi 1985;

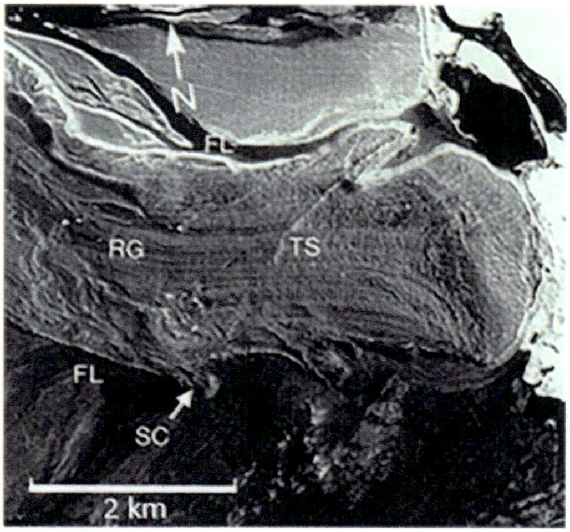

Fig. 9 Satellite image of Llullaillaco (Chile/Argentina) volcanic debris avalanche (VDA) deposit with prominent flow-parallel ridges. FL: Flow levees; RG: Ridges and grooves. SC marks a scarp that underlies the VDA deposit in direction of white arrow but does not interrupt VDA flow below location TS. Image modified from Francis and Wells (1988)

Pouliquen et al. 1997; Shea and van Wyk de Vries 2008; Dufresne and Davies 2009; Johnson et al. 2012; Valderrama et al. 2016, 2018). An unusual chevron or "herring-bone" pattern of short linear v-shaped ridges (v apexes pointing uphill) has also been observed on the surface of the Lastarria VDA deposit (Chile), which was inferred to reflect by high-velocity emplacement (Naranjo and Francis 1987).

Owing to their weight and momentum, large debris avalanches tend to plow into unconsolidated deposits, pushing them ahead of the avalanche front like a bulldozer and producing thrust faults and soft-sediment deformation in terrain adjacent to the deposit. Distal hummocks and ridges sometimes consist primarily or entirely of substrate materials that have been entrained (Belousov et al. 1999; Clavero et al. 2004; Paguican et al. 2012, 2021). Ballistic impacts from large gravel-size clasts tossed hundreds of meters beyond the steep distal fronts of some avalanche deposits imply that deceleration of debris avalanches during emplacement can be very abrupt (Plafker and Eriksen 1978; Siebe et al. 1992).

2.2 Lithofacies

Lithofacies characterization (see Sect. 2.2 in Chap. "Deposit Identification and Interpretation") of VDA deposits incorporates features of landscape-scale deposit architecture and morphology (discussed above), as well as outcrop-scale and particle-scale features that are discussed in more detail in Sects. 3 and 4. VDA deposit facies are based on the relative proportions of debris-avalanche blocks, solid gravel-size clasts, and matrix material. *Block facies* and *mixed facies* are the main lithofacies types of VDA deposits (Figs. 10, and 11), although the relative proportion of their components and their sedimentary textures vary among proximal, medial, distal, and marginal zones as block sizes and degrees of mixing differ with distance from source (Glicken 1991; van Wyk de Vries and Delcamp 2015). Contacts between facies are typically gradational, although contacts between juxtaposed debris-avalanche blocks and between debris-avalanche blocks and mixed material can be sharp.

The *block facies* (Fig. 10) is dominated by debris-avalanche blocks that are packed tightly together; individual blocks may or may not be deformed by internal shear or bulk consolidation. Where avalanches can spread and rapidly thin during emplacement, the block facies is generally concentrated along the thick central axis of the deposit and closest to source. Debris-avalanche blocks tend to be smaller at the base and lateral margins of deposits than those in the upper parts of deposits along central axes of flow (Takarada et al. 1999). Marginal zones are also progressively enriched in mixed material (Palmer et al. 1991). In thick valley-confined deposits, the block facies (still dominantly proximal) typically fills the valley from side to side (Palmer et al. 1991; Glicken 1996). Matrix and gravel-size clasts in the block facies are found both within fractured, faulted, and partially disrupted debris-avalanche blocks (intrablock mixed material) and between different debris-avalanche blocks (interblock mixed material). Bernard et al. (2021) distinguished between block-facies material derived from the edifice and that incorporated from the substrate.

Fig. 10 Block facies of volcanic debris avalanche (VDA) deposits in outcrop. **a** Tightly packed fragments of a moderately fractured and slightly deformed clast (about 1-m diameter) in block facies of VDA deposit on west flank of Cotopaxi volcano (Ecuador) (Vezzoli et al. 2017). Rock hammer at bottom left shows scale. **b** Two large light-colored debris-avalanche blocks (large fragments of an old dacite lava dome that was part of the pre-failure edifice) in the 1980 avalanche deposit from Mount St. Helens (Washington. USA). The block on the right had been injected by two thin black basalt dikes that are highly irregular in shape. People at bottom center show scale. USGS photos by T. Pierson

The *mixed facies* in VDA deposits (Fig. 11), originally termed "matrix facies" (Crandell et al. 1984; Crandell 1989; Glicken 1986, 1996), is composed of mixed material (gravel-size hard clasts plus matrix material) that typically contains some debris-avalanche blocks totally suspended in mixed material. The mixed facies is more homogeneous in color and texture than the block facies, and its constituent debris-avalanche blocks and gravel-size clasts tend to be smaller, more widely spaced, and slightly less angular than in the block facies. These characteristics become more pronounced with increasing distance from source, and the ratio of mixed facies to block facies also increases with distance from source (Ui et al. 1986b; Palmer et al. 1991; Siebert 1996; Takarada et al. 1999). Deformation within debris-avalanche blocks (faulting, folding, shearing) also increases with distance from source (Ui 1983; Ui et al. 1986b).

Some VDA deposit facies classifications and terminology include additional facies types (e.g., Glicken 1991; Palmer et al. 1991; Ui et al. 2000; van Wyk de Vries and Davies 2015; Bernard et al. 2021), as summarized by Dufresne et al. (2021b). Some authors distinguish an additional intermediate facies that contains fewer and generally smaller blocks than in the block facies. Van Wyk de Vries and Delcamp (2015) and van Wyk de Vries and Davies (2015) distinguished between monomictic matrix (produced by milling of debris-avalanche blocks of one rock type) and polymictic matrix, and they also recognized a basal facies, where the lowest meter or so of the deposit is characterized by finer-grained and abraded material with sheared lenses of incorporated substrate. Concentrated lahar deposits on or adjacent to VDA deposits have also been identified as a separate VDA lithofacies (Palmer et al. 1991).

Large debris-avalanche blocks undergo relatively little shear deformation during transport, despite their pervasive fracturing. Thus, original volcanic structures and stratigraphy, such as lava flows, dikes, and tephra beds remain more or less intact and recognizable (Fig. 12). Hydrothermal alteration commonly imparts intense color to originally black or gray edifice rocks, and minimally deformed debris-avalanche blocks in avalanche deposits will preserve such color. Alteration hues commonly include white, yellow, orange, red, and purple, but unaltered lava flows and pyroclastic deposits may also provide black,

Fig. 11 Mixed facies of volcanic debris avalanche (VDA) deposits in outcrop. **a** Mixed facies (mixed material) of the 7000 yr BP volcanic debris avalanche (VDA) deposit on the southwest flank of the Colima volcanic complex (Mexico), exposed in the canyon wall near El Remate (Cortés et al. 2019). **b** Highly fractured dark debris-avalanche block suspended in light-gray mixed material in the mixed facies of the 1980 Mount St. Helens VDA deposit (Washington, USA). **c** Light-colored mixed material surrounding and injected into a darker debris-avalanche block in the mixed facies of the Acajutla VDA deposit, Santa Ana volcano (El Salvador) (Siebert et al. 2004). Rock hammer (circled) for scale. USGS photo **a** by T. Pierson; Smithsonian photos **b, c** by L. Siebert

gray, white, or oxidized pink or red tones. Colors can be uniform or mottled; boundaries between colors may be diffuse but they are commonly very sharp. They can have chromas (color-saturation values) as high as '6' on the Geological Society of America Rock Color Chart (Rock Color Chart Committee 1991) and '8' on the Munsell Soil Color Charts (1975).

2.3 Erosion and Effects on Vegetation

Debris avalanches are highly erosive and catastrophically destructive to all vegetation in their flow paths, including large-diameter trees (Glicken 1996; Takarada et al. 1999; Arce et al. 2008; Cortés et al. 2010). Entire forests can be completely stripped from parts of hillsides, and valley floors can be buried under hundreds of meters of debris. Trees are uprooted and broken into smaller, highly abraded logs and splintered wood fragments. Lateral flow boundaries show abrupt transitions from undisturbed vegetation to barren razed ground (Fig. 13a). While some of this wood debris is scattered throughout the avalanche deposit, much of it can be concentrated at the flow front or in jumbled hummocks of woody debris (Glicken 1986, 1996) (Fig. 13b). In addition, trees along the margins of debris avalanches may be severely abraded and damaged by flying rocks (Crandell and Fahnestock 1965).

Fig. 12 Stratigraphy from source volcanic edifices preserved in large debris-avalanche blocks and exposed by mass wasting and fluvial erosion. **a** Original pre-collapse stratigraphy of bedded lapilli-and-ash layers showing brittle deformation (faulting), Acajutla volcanic debris avalanche (VDA) deposit from Santa Ana volcano (El Salvador) (Siebert et al. 2004). Scale bar is 2 m long. **b** Debris-avalanche blocks in 1980 VDA deposit at Mount St. Helens (Washington, USA). Blocks are pervasively fractured (hundreds of fractures per meter). Cliff faces are about 12–20 m high. **c** Roadcut exposure of Escuintla VDA deposit from Fuego-Acatenango complex (Guatemala) (Vallance et al. 1995). Outcrop shows extensive normal faulting and shear, boudinage of smaller dark-colored lava flow segments (upper center) and both brittle and ductile (top of outcrop) deformation. Dashed lines show selected normal faults that offset thick lava flow segments (LF); person (circled) for scale. **d** Hummock margin dominated by extensional normal faulting in VDA deposit 21 km from Mount Shasta volcano (California, USA) (Crandell 1989). Outcrop is about 10–15 m tall. Smithsonian photos **a, c, d** by L Siebert; USGS photo **b** by T. Pierson

Fig. 13 Volcanic debris avalanche (VDA) effects on forest vegetation. **a** Valley side (on right) and toe of ridge (on left) stripped of forest vegetation by the 1984 VDA about 7 km downstream from source at Ontake volcano (Japan). VDA volume estimated to be 32–36 × 10^6 m^3 (Endo et al. 1989). This photo (September 8, 1985) shows structures being built on valley floor following the VDA as part of a sediment-retention dam. Avalanche deposit on valley floor here has been buried by post-avalanche sediments. USGS photo by T. Pierson. **b** Jumbled piles of woody debris at distal end of 1980 VDA deposit at Mount St. Helens (Washington, USA). Note people at lower left for scale. Smithsonian photo by L. Siebert

VDAs can carve striations, grooves, and flutes into bedrock substrates, which can closely resemble glacial abrasion marks (Fig. 14), and debris avalanches can erode deeply on surfaces underlain by unconsolidated deposits. Large pieces of valley floor bedrock and sediment can be eroded and entrained in debris avalanches, as well (Takarada et al. 1999; Cortés et al. 2010). However, in other cases, VDAs have flowed over soft surfaces with little erosion. Note that there are differences between VDA-striated bedrock and glacially striated bedrock.

Debris avalanches move at velocities much faster than glaciers but only very briefly. The underlying rock can be gouged but not polished. Pronounced smoothing and polishing of

Fig. 14 Parallel striations carved into bedrock along failure plane of 1980 debris avalanche on a previously unglaciated location on the north flank of Mount St. Helens (Washington, USA). Such striations resemble those eroded into bedrock by glaciers, but the eroded surface is rough. Ice axe for scale. USGS photo by T. Pierson

substrate rock is the hallmark of the prolonged grinding action of glaciers, not VDAs.

3 Outcrop-Scale Characteristics of Deposits

Debris-avalanche deposits are characteristically massive, ungraded, and extremely poorly sorted diamicts devoid of primary bedding features. At outcrop scale, mixed facies of VDA deposits commonly resemble deposits of other types of mass flows, such as lahars and pyroclastic density currents (PDCs) (particularly, block-and-ash flows). They can also resemble glacial tills. Deposits may contain exotic gravel-size clasts (such as well-rounded stream cobbles or boulders), particularly near the lower boundary of a deposit (Takarada et al. 1999). However, VDA deposits have outcrop-scale characteristics that distinguish them from other types of mass-flow and glacial deposits. Outcrop-scale characteristics diagnostic of VDA deposits are commonly not visible on surficial exposures, making cross-sectional views in riverbanks, coastal cliffs, road cuts, and quarries essential. Even these exposures may be partially obscured by slope wash or weathering, and scraping with shovels or other tools may be necessary to reveal features needed for deposit identification.

3.1 Stratification

Volcanic debris-avalanche deposits are inherently massive and lack primary bedding structures, yet bedding may be visible within debris-avalanche blocks (e.g., tephra layers, lava flows, and PDC or lahar deposits). Such strata are inherited from a failed edifice. If an outcrop is small and a debris-avalanche block is large, examination of multiple outcrops is necessary. However, shear structures that resemble bedding may be present in outcrops—subhorizontal shear zones and juxtaposition of different textures across shear planes. Such pseudo-bedding is most likely to be found near the base of the deposit and along its lateral margins (Glicken 1996, 1998), although multiple shear zones at the base of separate debris-avalanche blocks transported within the VDA may reflect complex relative horizontal and vertical motion during progressive fragmentation (Roverato et al. 2015).

3.2 Gravel-Size Clast Framework

Sediment transport by VDAs offers only limited opportunity for effective interaction between individual particles and a fluid phase, so any internal arrangements of coarse particles resembling clast- or matrix-supported internal structures in VDA deposits, particularly those in proximal or medial settings, are likely the result of preexisting sedimentary textures and structures inherited from pyroclastic deposits originating from the failed edifice and transported *en masse* by the avalanche. However, particle–fluid interactions occur in distal settings within the mixed facies, but we know of no characteristic internal structures in these deposits.

3.3 Fabric

Rock fragments in VDA deposits typically do not show preferred orientations. Transported logs, however, can have distinctly preferred orientations, with long axes aligning themselves with flow direction (Takarada et al. 1999). As with stratification, fabric may be inherited from intact pieces of a failed volcanic edifice or from sedimentary units entrained from the valley floor.

The spatial distribution of gravel-size clasts in outcrop faces is commonly clustered, which is a unique characteristic of VDA deposits. Clasts commonly are found in monomictic clusters (Fig. 15), due to limited shearing and partial disaggregation of debris-avalanche blocks that results in little mixing of clasts with in the surrounding mixed material (Glicken 1986; 1996). Coarsely fractured clasts can be sheared and stretched out to the point where the original gravel-size is no longer recognizable, but the resultant clasts still form a grouping or domain with individual clasts separated but

Fig. 15 Clustering of gravel-size clasts in outcrop. **a** Clast cluster in the 1984 Ontake (Japan) volcanic debris avalanche (VDA) deposit (Endo et al. 1989). The cluster formed by progressive disaggregation of a debris-avalanche block, prior to the deposit coming to rest. Rock hammer for scale. **b** Gravel-size clast cluster surrounded by brown matrix with few large clasts in Okinajima VDA deposit from Bandai volcano (Japan) (Nakamura 1978). People along road for scale. Smithsonian photos by L. Siebert

not dispersed. Gravel-size distributions typically become more random with distance from source, but gravel-size clast clusters may be present in distal areas of even very long-runout deposits (Siebert et al. 2004; Roverato et al. 2015).

VDA deposits have bulk densities that are distinctly less than their pre-failure source material. The 1980 Mount St. Helens deposit had an average bulk density of 1850 kg/m^3 (range 1440 to 2180) compared to the calculated mean bulk density of 2310 kg/m^3 for the rock in the volcano's flank before the avalanche occurred (Glicken 1986, 1996). This approximately 20-percent increase in bulk volume is caused by disaggregation during flow. Similarly, the 1984 Ontake avalanche deposit had bulk densities ranging from 1500 to 1700 kg/m^3, although the mean bulk density of the pre-failure material (which was known to have several porous pumice beds) was not known (Suwa et al. 1985).

3.4 Grading

Grading is seldom apparent in small outcrops of debris-avalanche deposits, but it can be found in the basal shear zone of VDA deposits (Takarada et al. 1999). On the scale of the entire deposit, normal grading of gravel-size clasts and inverse grading of woody debris is present in some thick vertical sections (Takarada et al. 1999), although inverse grading can occur in nonvolcanic and volcanic landslides and is attributed to stronger basal shearing and fragmentation or to clast segregation in a moving mass (e.g., Shea and van Wyk de Vries 2008; Procter et al. 2009; Paguican et al. 2021). Grading relationships can be complicated by overlapping lobes from retrogressive landslides.

3.5 Fluid-Escape Structures and Deformation

Fluid-escape structures have not been reported in VDA deposits. However, localized dewatering of the 1980 Mount St. Helens debris avalanche did occur beginning about 5 h after deposition, resulting in the expulsion of liquefied slurry that coalesced to become a large lahar (Janda et al. 1981; Fairchild 1985, 1987). It is possible that deep dissection of this avalanche deposit may someday reveal fluid-escape structures.

Fig. 16 Irregular yellowish-brown clastic dike intruding debris-avalanche deposit from Raung volcano (Indonesia) (Siebert 2002). Orientation of the dike is irregular to near horizontal above this point. Scale bar is 2-m long. Smithsonian photo by L. Siebert

Clastic dikes form by injection of mixed material into fractures in debris-avalanche blocks due to internal pressure from rapid loading. The dikes are typically centimeters to a meter or more in width, with lengths of meters to tens of meters. Orientations of clastic dikes range from vertical to highly irregular, due to block geometries and deformation patterns during or after emplacement (Fig. 16). The homogeneous color of injected mixed material contrasts with the more variable colors of the debris-avalanche blocks they intrude. Clastic dikes are commonly intruded along fault zones within large debris-avalanche blocks (van Wyk de Vries and Davies 2015). Clastic dikes are rare in VDF deposits but may be present in VHF deposits. Dikes of mafic lava may also be preserved in debris-avalanche blocks (Fig. 10).

3.6 Shear Deformation

Brittle deformation structures in debris-avalanche blocks can include block elongation, conjugate reverse faults, and a variety of minor faults (e.g., Takarada et al. 1999; Roberti et al. 2017; Makris et al. 2023b). Fault offsets range from millimeters to many meters, observable at debris-avalanche block margins or in bedding contacts within the blocks. Observed faults may represent primary faults within transported segments of the pre-failure edifice or faults produced during transport or from post-avalanche settling. Faults during transport usually affect only the debris-avalanche blocks, although they may present but not as apparent in the surrounding mixed material; faults that penetrate both blocks and mixed material either reflect bulk brittle behavior during movement (more sliding than flow) or post-depositional consolidation. Such faults can be used to differentiate deposits of debris avalanches from those of lahars; syn-depositional faulting does not occur in lahar deposits—the material is too fluid or plastic. Displacement may occur along discrete fault planes in coherent materials found within broader shear zones in looser, more brecciated materials (Shea and van Wyk de Vries 2008). Shearing of extensively and finely fractured debris-avalanche blocks commonly results in plastic deformation that can range from slight bending or elongation to extreme elongation and contortion (Fig. 17).

Intense shear at the lower boundary layer of a moving avalanche (within about 5 m of lower contact) can result in extreme "smearing out" of debris-avalanche blocks to form a type of crude layering (Siebert et al. 1996). Basal contacts between VDA deposits and underlying substrate are typically smooth, even if the substrate shows evidence of shear extending below the contact and substrate material is entrained by the avalanche (Fig. 18) (Belousov et al. 1999; Clavero et al. 2002; Dufresne et al. 2010; Keigler et al. 2011). Bernard and van Wyk de Vries (2017) and Perinotto et al. (2015) noted increased elongation (or flattening) of basal VDA deposit particles with their long axes aligning parallel to the direction of maximum shear strain. Schneider and Fisher (1998) used anisotropy of magnetic susceptibility (AMS) measurements at Monts Dore in France to evaluate basal shear processes. Shape Preferred Orientation (SPO) analysis of deposits emplaced with basal shear at Peruvian volcanoes (Bernard and van Wyk de Vries 2017; Bernard et al. 2019) has shown that gravel-size clast distributions correlate with increased fracturing, with cataclasis data informing syn-emplacement processes including flow transformation implications.

3.7 Material Consistency

Outcrop-scale consistency of VDA deposits is a function of the part of the deposit being sampled, and it can vary greatly between disaggregated debris-avalanche blocks and homogenized mixed material. This wide variation alone can be useful in distinguishing VDA deposits from lahar deposits, PDC deposits, or tills, which tend to have more uniform consistency.

Rock material within debris-avalanche blocks that have been fractured (but not crushed) is typically loose to friable under moist field conditions, because not enough fines have been created during fracturing to help hold the debris-avalanche block together. When the rock material has been intensely crushed (sub-millimeter fracture spacing), fines are present to provide cohesion, and consistency may be firm to very firm when moist and slightly plastic to plastic when wet. Intact parts of debris-avalanche blocks may remain as fully lithified rock. Overall, debris-avalanche blocks may range from soft to well-indurated, depending on the degree of fragmentation and crushing.

Mixed material in debris avalanches typically contains a fine-grained clastic matrix that may have been subjected to overburden pressures exerted by deposits hundreds of meters thick. Thus, a wide variation in consistency may be

Fig. 17 Deformed volcanic debris avalanche (VDA) blocks. **a** Normal fault (dashed line) with ~ 3-m offset within Acajutla VDA deposit, Santa Ana volcano (El Salvador) (Siebert et al. 2004). **b** Highly deformed debris-avalanche blocks composed of highly fractured andesite (darker areas) within lighter-colored mixed material of the 18,500 yr BP Nevado de Colima (Mexico) VDA deposit (Cortés et al. 2019). **c** Shear zones extending diagonally across photo create schlieren of gray and oxidized red layers within Riobamba VDA deposit from Chimborazo volcano (Bernard et al. 2008). Smithsonian photos **a, c** by L. Siebert; USGS photo **b** by T. Pierson

Fig. 18 Basal contacts of volcanic debris avalanche (VDA) deposits. **a** Flat-lying basal contact of a 28–30 km^3 VDA deposit above yellow–brown paleosol on south flank of Popocatépetl volcano (Mexico) about 15 km from present summit (Siebe et al. 2017). Smithsonian photo by L. Siebert. **b** Top 10–15 cm of paleosol (in lower part of photo) at same location as in **a**, which contains shear zones parallel to contact with the base of the avalanche deposit (slightly overhanging surface above upper end of rock hammer handle). USGS photo by T. Pierson

encountered, depending on abundance of fines, their grainsize and mineralogy, and position within the deposits. In addition, varying degrees of localized post-depositional cementation can result in differences in hardness. Case hardening where induration of exposed surfaces contrasts with more friable deposit interiors is common in river-bank, coastal, or other exposures. Generally, mixed material within fresh debris-avalanche deposits is very firm to extremely firm (when moist) and hard to very hard (when dry), whether rich or poor in clay. Yet when clay-poor material is dug or scraped with a tool, it tends to break off as individual grains. Material relatively rich in hydrothermal alteration minerals breaks off in chunks like clay-rich lahar deposits.

3.8 Evidence of Heat

Most volcanic debris-avalanche deposits are emplaced at relatively low temperatures, even when accompanied by magmatic eruptions (Siebert 2002). However, if an avalanche deposit contains quenched fragments of an intruding cryptodome, as was the case in the 1980 Mount St. Helens VDA, gravel-size clasts from the rapidly cooled cryptodome are likely to exhibit prismatic jointing. In the Ferrar Large Igneous Province (Antarctica), (Reubi et al. 2005) noted basaltic bodies with chilled rims that were hot and plastic during transport in deposits of an avalanche inferred to have been triggered by emplacement of sub-surface intrusions. If entrained hot fragments are large enough, it might be possible for gas-escape pipes or highly localized charring of woody debris to develop. Some deposits contain pyroclasts, fragments of lava flows, or charred wood from PDC deposits of the pre-collapse edifice, which could retain evidence of high temperature from earlier eruptions (Siebert 2002). Clement et al. (1993) found paleomagnetic evidence in lithic gravel-size clasts for variable primary temperatures up to 350–580 °C in medial parts of a VDA deposit from the Colima volcanic complex.

Some deposits of both nonvolcanic and volcanic landslides show evidence of frictional heating, even partial melting of rock along slide planes at temperatures > 1500 °C, forming glass layers (hyalomylonites) that some authors distinguish from pseudotachylites formed during seismic or impact events (Erismann 1979; Weidinger et al. 2014). Glass layers termed pseudotachylites or frictionites have been reported at the base of VDA deposits from Pichu-Pichu volcano (Peru) (Legros et al. 2000; Bernard et al. 2019; Hughes et al. 2020), in the French Massif Central (Bernard and van Wyk de Vries 2017), and on a larger scale within the megaslides of the Markagunt Plateau in Utah (Hacker et al. 2014; Biek et al. 2019).

4 Particle-Scale Characteristics of Deposits

4.1 Particle Composition

VDA deposits originating from composite volcanoes create mainly polymictic sediment mixtures that are derived from a variety of rock types and degrees of hydrothermal alteration from source edifices. However, small outcrop faces may display strikingly monomictic mixtures if exposure is limited to a single debris-avalanche block of uniform rock type. Hydrothermally altered rock in gravel-size clasts and debris-avalanche blocks is commonly juxtaposed with unaltered volcanic clasts, and debris-avalanche deposits are well known for containing debris-avalanche blocks exhibiting sharp contacts between units of strongly contrasting colors and textures.

Debris-avalanche deposits may contain exotic gravel-size clasts of various types. Accidental clasts picked up from the bed tend to be concentrated near the base of the deposit (Takarada et al. 1999); this is also true for some PDC and lahar deposits (Walker 1971; Pierson et al. 1990). Mixed material also commonly contains splintered fragments of wood that are often concentrated toward the front and lateral margins of the deposit (Glicken 1986, 1996). Such wood fragments generally are not carbonized. However, the presence of exotic clasts and woody debris are not unique to debris-avalanche deposits.

4.2 Particle Roundness, Shape, Fracturing, and Surface Texture

The shape of gravel-size clasts is highly irregular and depends on how the failed edifice breaks up and on the degree of shear deformation experienced within the avalanche. Some shattered

Fig. 19 Contrasting lithologies (different colors) within a small debris-avalanche block in the Osceola lahar deposit, Mount Rainier (Washington, USA), show high degrees of plastic deformation (Vallance and Scott 1997). This debris-avalanche block was rafted far downstream after the initial avalanche transformed to a lahar. Pen at top center for scale. USGS photo by K. Scott

debris-avalanche blocks subjected to high shearing forces can be contorted and irregular in shape (Fig. 19). Features of both brittle and ductile deformation are present, sometimes in the same outcrop, the latter perhaps reflecting localized areas of higher water content (Delcamp et al. 2016). The shapes of hard gravel-size clasts depend on fracture orientations within source debris-avalanche blocks and most clasts are approximately equant in shape unless dominantly parallel fractures produce more tabular clasts.

Gravel-size clasts within highly fractured debris-avalanche blocks tend to be very angular because little differential movement occurred. Clasts with this degree of angularity are seldom found in other mass-flow or flood deposits. Boulder-size clasts within mixed material where shearing and significant differential movement has occurred are generally angular but may show some rounding, though typically not as much rounding as is seen in boulders in VDF deposits. In some cases, subrounded to rounded gravel-size clasts have been reported in VDA deposits (Palmer et al. 1991), and such rounded clasts may be found adjacent to highly angular clasts (Fig. 20).

Temporal and spatial factors influence particle rounding and degree of fracturing in VDA deposits (Fig. 21). Typical fracture densities range from 1 to 70 per linear meter of outcrop exposure at Mount Shasta (California, USA) and Taranaki volcano (New Zealand) (Ui and Glicken 1986; Ui et al. 1986b). Some debris-avalanche blocks or parts of blocks, however, are found to be pervasively shattered rock, crushed to a friable granular state with fracture densities estimated at more than 1000 per linear meter (Komorowski et al. 1991). Analysis of fractal dimensions and the circularity of jigsaw-cracked gravel-size clasts at Piton de Neiges volcano (Reunion Island) by (Perinotto et al. 2015) demonstrated that these parameters did not vary significantly with distance from the source, implying that the jigsaw fractures (see discussion below) were created close to their depositional site. Alternatively, they found that isolated gravel-size clasts showed increased roundness with travel distance due to abrasion by surrounding mixed material. This abrasion also increased the proportion of matrix material.

Most micro- and macro-scale fracturing of debris-avalanche blocks in VDA deposits has been attributed to proximal compression-rarefaction waves early in the collapse process (Glicken 1986; Komorowski et al. 1991) and to rapid unloading of lithostatic pressure at the source by sudden removal of overburden (Belousov et al. 1999). Micro-texture studies (Caballero and Capra 2011; Roverato et al. 2015; Procter et al. 2021) have further demonstrated that high-pressure particle contacts cause clast comminution both during failure and during subsequent VDA transport (Dufresne et al. 2021b).

Fig. 20 Angular to subrounded gravel-size clasts in the 1640 volcanic debris avalanche (VDA) deposit from Komagatake volcano (Japan) (Yoshimoto and Ui 1998). VDA deposit is overlain by pumice-fall deposit from the 1640 eruption. Scale bar is 2-m high. Smithsonian photo by L. Siebert

Pervasive, fine-scale (mm to cm) fracturing of hard rock clasts with little to no shear displacement is commonly found at a variety of scales within VDA deposits (Ui 1983; Komorowski et al. 1991; Takarada et al. 1999; Brideau and Procter 2015) (Fig. 22). These fractures are termed *jigsaw cracks or jigsaw fractures* from the resemblance of the interlocking fragments to puzzle pieces. Jigsaw-fractured clasts, when surrounded by unfractured matrix material, are unique to VDA deposits. The "puzzle pieces" may (1) be in contact with each other, (2) have narrow gaps between them (which accentuate the jigsaw-puzzle appearance), or (3) have crushed intrablock or intraclast matrix material injected into the cracks. Where fracturing is extreme, a clast may at first glance appear not to be fractured at all, but when examined closely it reveals tightly spaced (<1 mm) fractures that can cause it to disaggregate into angular, sand-size or finer fragments when it is removed from the outcrop. Jigsaw cracks may be of variable density within the same fractured clast, and gaps in jigsaw fractures can also increase with distance from source (Ui 1983; Ui et al. 1986b). Note that jigsaw-fit textures also occur in a more localized context in hyaloclastites and peperites generated primarily by in situ non-explosive quench fragmentation at the margins of subaqueous lava flows, domes or sills (e.g., Scutter et al. 1998; Schmidt and Schmincke 2002; Skilling et al. 2002).

Jigsaw fracturing and localized crushing of particles is apparent to microscopic scale. After the pioneering scanning electron microscopy of Komorowski et al. (1991), additional work at a number of other volcanoes has expanded understanding of microscopic textures and their interpretive value, namely at Shiveluch volcano (Belousov et al. 1999), Nevado del Huila volcano, Colombia (Pulgarín et al. 2004), Piton de Neiges volcano (Perinotto et al. 2015); Nevado de Toluca and Colima volcanic complex, Mexico (Caballero and Capra 2011; Roverato and Capra 2013), Taranaki volcano (Roverato et al. 2015; Procter et al. 2021), Monts Dore (Bernard and van Wyk de Vries 2017), and Pichu-Pichu volcano (Bernard et al. 2019; Hughes et al. 2020). Images of particles

Fig. 21 Types, degrees, and scales of fracturing in gravel-size clasts. **a** Relatively coarse fracture pattern in unaltered dacite clast in the 1792 Mayu-yama (Japan) volcanic debris avalanche (VDA) deposit (Unzen volcanic complex) (Ohta 1969). Left scale bar shows 10 cm. **b** Coarse jigsaw fracturing of an undeformed clast surrounded by unfractured mixed material in the Barilles VDA deposit at Barú volcano (Panama) (Herrick et al. 2013). Machete (~1-m long) at left for scale. **c** Moderate fracturing in a clast composed of porphyritic dacite in the 1792 Mayu-yama VDA deposit. **d** Two highly crushed debris-avalanche blocks distinguished by pinkish color of oxidized (top) and gray color of unoxidized material (bottom) juxtaposed along uneven contact (dashed line) in the 1792 Mayu-yama VDA deposit. USGS photos **a**, **c** by T. Pierson, **b** by J. Herrick, used with permission, **d** by L. Siebert, Smithsonian

0.5–4.0 mm reveal: (1) microcracks with offsets, separation, jigsaw patterns and broken crystals, just as observed in gravel-size VDA clasts; (2) hackly surface textures on protruding parts of grains that reflect crushing during repeated high-pressure collisional contacts between sand-sized grains during flow; (3) percussion marks; (4) linear raised lips extending across clasts; (5) staircase geometry; and (6) parallel ridges and grooves, the latter two features being more prominent in basal parts of deposits (Caballero and Capra 2011) (Fig. 23). Virtually all matrix-size grains examined exhibit some combination of these features.

4.3 Particle-Size Distribution and Sedimentary Texture

Particle-size (grain-size) distributions in VDA deposits display considerable variation and can differ from those of other deposits of volcanic particles due to the singular presence of large domains of both block-facies and mixed-facies material in variable proportions. Analysis of particle size in VDA deposits involves determination of the size distribution of only the hard, coherent particles within the mixed material, not fractured pieces of debris-avalanche blocks.

Fig. 22 Jigsaw fractures in gravel-size clasts in volcanic debris avalanche (VDA) deposits. **a** Fractures in boulder-size clast, most of which are not showing separation, in VDA deposit at Cofre de Perote volcano (Mexico) (Carrasco-Núñez et al. 2010). Rock hammer for scale. Smithsonian photo by L. Siebert. **b** Fractures showing separation between pieces of a disaggregated gray gravel-size clast suspended in brown matrix material in the 1980 VDA deposit at Mount St. Helens (Washington, USA). Fingers touching outcrop at lower right for scale. USGS photo by T. Pierson

VDA mixed material samples are typically very poorly to extremely poorly sorted using the criteria of (Folk 1974), and size distributions range from unimodal to polymodal. The 1980 Mount St. Helens VDA deposit was the first to be studied in detail (Voight et al. 1981, 1983; Glicken 1986, 1996; Fisher et al. 1987; Komorowski et al. 1991). Data across facies show, on average, 45% gravel (range 22.0 to 92.3%), 45% sand (range 5.6 to 65.4%), and 10% silt and clay (range 1.1 to 21.2%). Reported Inman sorting coefficients (Inman 1952) of samples from the deposit range from 2.2 to 13.0φ (Voight et al. 1983; Glicken 1986, 1996). Average values appear to be dependent on analytical technique (Glicken 1986, 1996), and range from about 3.5 to 7.1φ for samples containing clasts up to 10 cm. The largest solid clast found in the Mount St. Helens VDA deposit has an intermediate diameter of 8 m (actual size $15 \times 8 \times 5.2$ m) and the finest material in the matrix is clay size (finer than 0.004 mm). Most sample size distributions (block and mixed facies) are bimodal, with a coarse peak maximum at −5φ or −4φ (32 or 16 mm) and a fine peak maximum at 0φ or 1φ (1.0 or 0.5 mm). The fine peak likely results from an abundance of feldspar, amphibole, and pyroxene crystals released from source rocks by crushing during avalanche movement (Glicken 1986, 1996), and the coarse peak likely reflects the average fracture or joint spacing in the source rock. Unimodal size distributions are both coarse- and fine-tailed. Cumulative probability and Rosin-law plots of sample data show that the size distributions differ from both the ideal normal and Rosin-law distributions, although they are closer to a normal distribution

Fig. 23 Scanning electron microscope (SEM) images of microtextures on grains in volcanic debris avalanche (VDA) deposits. **a** Fracture generation with separation of cracks and small-scale displacements. **b** Jigsaw fractures in part of a lithic grain. **c** Shattered plagioclase crystal (pl) at upper right, adjacent to glass at left along basal shear zone. **d** Hackly texture on projecting parts of glass-covered pyroxene crystal from highly irregular breakage by repeated high-pressure grain contacts. **e** Pair of lip structures wider and deeper to top of image. **f** Grooves and scratches of variable size and depth. **g** Percussion marks (P) from grain-to-grain impact. **h** Truncated crystal with staircase geometry. Modified image sources: **a** Colima volcanic complex (Mexico) (Roverato and Capra 2013), **c** Pichu-Pichu volcano (Peru) (Hughes et al. 2020), **d** Mount St. Helens (Washington, USA) 1980 VDA deposit (Komorowski et al. 1991), **e** Nevado de Toluca volcano (Mexico) (Caballero and Capra 2011), Images **b, f, g, h** from Taranaki volcano (New Zealand) (Roverato et al. 2015). Image collection modified from Siebert and Reid (2023) ▶

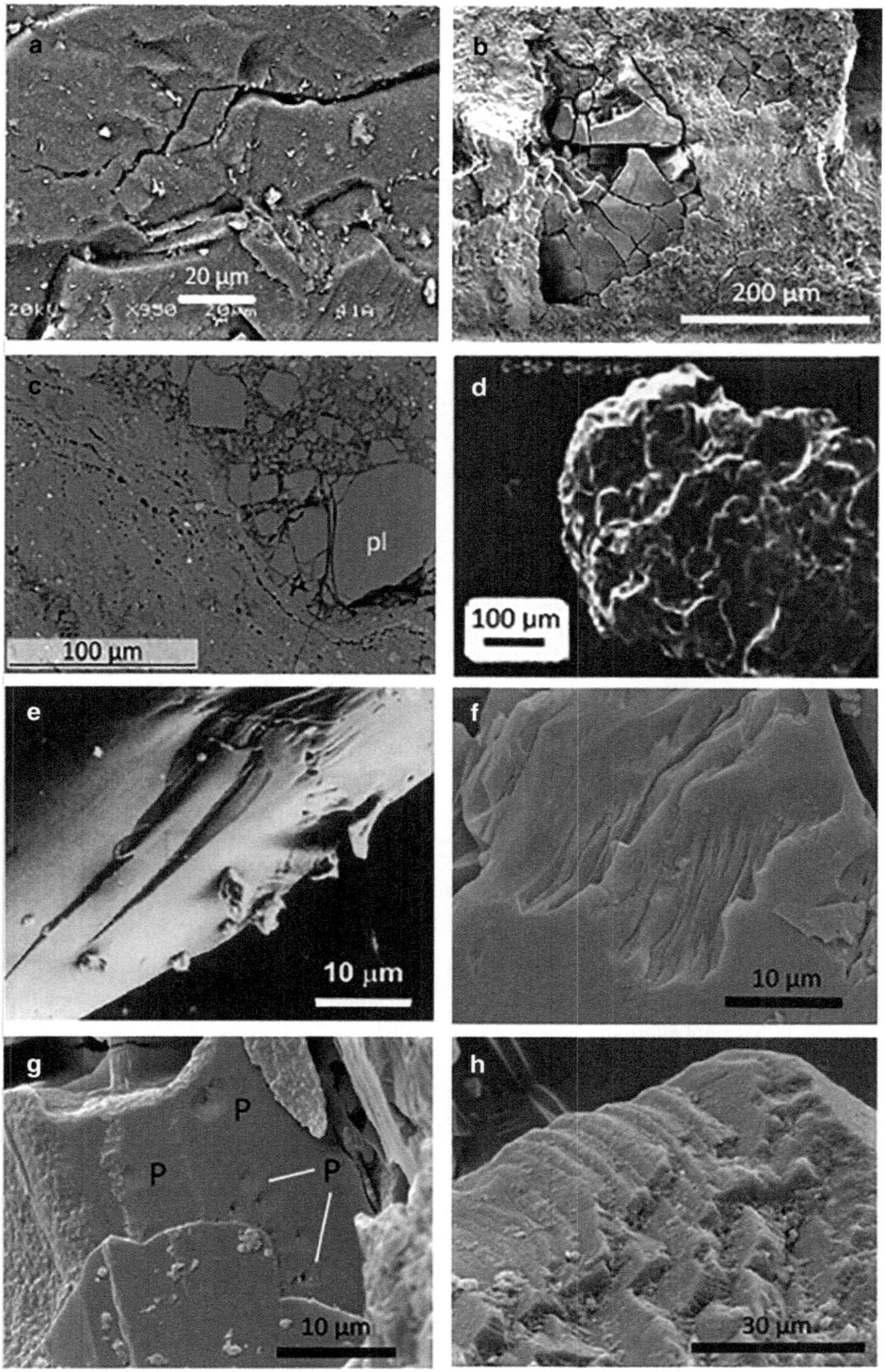
a
20 µm
b
200 µm
c
pl
100 µm
d
100 µm
e
10 µm
f
10 µm
g
P
P
P
10 µm
h
30 µm

(Glicken 1986; 1996). Rosin-law distributions are typically obtained for artificially crushed material and might be expected in such a large landslide deposit.

VDA deposits reflect increased mixing and homogenization with distance traveled (Glicken 1986, 1996; Ui 1986). Most fractured rock particles were either inherited from an already fractured edifice (Hausback and Swanson 1990; Glicken 1998) or produced in the earliest stages of slope failure and initial mass acceleration (Glicken 1986, 1996; Komorowski et al. 1991), but some comminution of gravel-size clasts and production of fines occurs during transport due to high-pressure particle contacts (Perinotto et al. 2015; Dufresne et al. 2021b). At the same time, some depletion of fines may occur in flowing avalanches as the result of elutriation. A thin layer of elutriated sediment was deposited by an air blast from 1963 VDA from Little Tahoma Peak at Mount Rainier (Washington, USA) (Crandell and Fahnestock 1965). However, the proportion of matrix material generally increases with increasing distance traveled by a VDA. Matrix material proportion increased to 50% at Taranaki volcano, with distal deposits also showing a decrease in mean grain size (Roverato et al. 2015). Dufresne et al. (2021b) further noted that fines often increase overall with distance but with variations due to lithology differences. Degree of particle breakage and sorting also varies with degree of valley confinement (Takarada et al. 1999; Paguican et al. 2021) and was noted on a smaller scale between flow-parallel ridge crests and their margins on a VDA deposit incorporating fragments of a collapsed hot lava dome (Valderrama et al. 2016).

Following the pioneering VDA particle-size distribution studies of Glicken (1986, 1996) at Mount St. Helens, subsequent particle-size distribution studies, primarily at collapses from other stratovolcanoes, were summarized by Dufresne et al. (2021b). Full-deposit size and sorting data from VDA deposits overlap considerably with size data from deposits of other volcanic particle–fluid flows, particularly lahars and pyroclastic density currents. VDA deposits are often as coarse or coarser than the pyroclastic-flow field of (Walker 1971) and more poorly sorted than Walker's pyroclastic-fall field (Siebert et al. 1995, 2004; Dufresne et al. 2021b). Size and sorting data collected after (Walker 1971) show considerable overlap between pyroclastic-flow and pyroclastic-fall fields, and pyroclastic-surge data overlap broadly with both fields. Fisher and Schmincke (1984) noted that these size and sorting diagrams should be used with caution. Statistical parameters of particle-size distribution, including mean particle size, sorting, kurtosis, and skewness, can be determined for VDA deposits, but varying analytical and sampling techniques complicate direct data comparison, and particle-size-distribution data alone may not be indicative of the process that emplaced a deposit. Deposit morphology and outcrop- and microscopic-scale textures are essential additional tools in VDA deposit identification.

4.4 Vesicles and Intergranular Voids

Voids commonly are found in the gaps of open fractures within gravel-size clasts. Cylindrical voids can also be found where wood fragments have rotted away. Vesicles and other primary voids, however, are extremely uncommon within mixed material in VDA deposits, although vesicles have been reported within localized and marginal zones of saturated sediment (Siebert 1984; Bernard et al. 2009) or where full transformation to saturated lahar-facies material has occurred (see Chap. "Lahar Deposits—Debris-Flow Type").

5 Key Characteristics of Volcanic Debris Avalanche (VDA) Deposits

VDA deposits can resemble other types of volcaniclastic sediments, particularly some VDF and volcano-glacial deposits. A summary table of all relevant characteristics of VDA deposits, together with an identification key, is found in Chap. "Deposit Identification: Key and Summary Tables", and a gallery of outcrop photographs is in Chap. "Annotated Outcrop Photographs: Volcanic Wet-Flow Deposits and Other Unconsolidated Volcanic Deposits". Some of the most useful identification criteria are listed below and are described in detail in the sections above. However, any one characteristic can be found in at least one other type of deposit. Therefore, no single deposit characteristic is entirely sufficient for emplacement process identification. Field observations and physical characteristics of sediment must be evaluated together for an accurate assessment of a deposit's origin.

The characteristics listed below and in Chap. "Deposit Identification: Key and Summary Tables" are based on evaluation of fresh deposits. Fresh deposits and freshly impacted flow paths provide the best opportunities for identifying emplacement process, prior to the complicating effects of weathering, mineral alteration, chemical staining, and cementation of the sediment particles themselves; erosion of landforms; and decay of affected vegetation.

- **Landscape-scale criteria**
  (1) Surface topography commonly contains irregularly shaped and distributed hills (hummocks) that range from about a meter to more than a kilometer across.
  (2) Closed depressions between hummocks commonly host small ponds or lakes.
  (3) Broad flat-surfaced areas with few if any hummocks may be found in medial to distal parts of deposits.
  (4) Unusually thick (commonly > 100 m) and unstratified deposits are common in proximal to medial parts of deposits.
  (5) Large debris-avalanche blocks are common in VDA deposits; they are composed of highly fractured, minimally deformed, unconsolidated to semi-consolidated segments of the failed source volcanic edifice and commonly retain their original primary stratification.
  (6) Deposits are typically bounded by steep, sharply defined, lateral and frontal margins (where not buried or where deposit did not transform to a lahar).
  (7) Longitudinal or transverse ridges (or both) occur on some deposit surfaces; flow levees along deposit margins are common.

- **Outcrop-scale criteria**
  (1) Abrupt color changes between deposit components are common and diagnostic of VDA deposits; adjacent fragments of unaltered and hydrothermally altered volcanic rock in debris-avalanche blocks have sharply contrasting intense colors (black, white, tan, red, orange, yellow, purple) and are commonly juxtaposed. These contrasting colors (common in block facies) stand out against the neutral brown and gray tones of the mixed facies.
  (2) Jigsaw fracturing of particles ranging from sand size to boulder size is common and diagnostic of VDA deposits.
  (3) Very angular gravel-size clasts are commonly clustered (i.e., not randomly or uniformly distributed spatially).
  (4) Plastic deformation of debris-avalanche blocks suspended in matrix material is common and includes elongation, folding, smearing, and contorting.
  (5) Inverse or normal grading and well-developed fabrics are largely absent.
  (6) Mixed-facies deposits are massive (unstratified), although original stratification of deposits from the pre-failure edifice may be preserved in some debris-avalanche blocks; concentrated shear within deposits can produce subhorizontal pseudo-bedding.
  (7) Clastic dikes may be injected into debris-avalanche blocks during VDA failure and transport; they are preserved due to minimal deformation of the blocks. Clastic dikes commonly have irregular margins and variable orientations. Fluid-escape structures are rare or absent.

(8) Mafic lava dikes may be preserved in debris-avalanche blocks containing coherent pieces of the edifice.
(9) Homogeneous, unstratified mixed material (gravel-size clasts and matrix material) is extremely poorly sorted and becomes more abundant (in proportion to debris-avalanche blocks with distance from source.

- **Particle-scale criteria**
  (1) Sediment in VDA mixed facies deposits is polymictic; fragmental material within single-lithology debris-avalanche blocks is monomictic.
  (2) Gravel-size clasts in VDA deposits are largely angular to very angular in shape, particularly in proximal parts of deposits; partly rounded clasts, if present, are found in medial to distal parts of deposits in mixed material.
  (3) Gravel-size clasts may be extensively fractured yet still relatively coherent.
  (4) Bubble vesicles are absent, except in deposits of lahars formed from transformation of VDAs.

References

Andrade SD, van Wyk de Vries B (2010) Structural analysis of the early stages of catastrophic stratovolcano flank-collapse using analogue models. Bull Volcanol 72:771–789. https://doi.org/10.1007/s00445-010-0363-x

Arce JL, Macías R, García Palomo A, Capra L, Macías JL, Layer P, Rueda H (2008) Late Pleistocene flank collapse of Zempoala volcano (Central Mexico) and the role of fault reactivation. J Volcanol Geotherm Res 177(4):944–958. https://doi.org/10.1016/j.jvolgeores.2008.07.015

Belousov A, Belousova M, Voight B (1999) Multiple edifice failures, debris avalanches and associated eruptions in the Holocene history of Shiveluch volcano, Kamchatka, Russia. Bull Volcanol 61:324–342. https://doi.org/10.1007/s004450050300

Belousov A, Belousova M (1996) Large scale landslides on active volcanoes in the 20th century—examples from the Kurile-Kamchatka region (Russia). In: Senneset (ed) Landslides. Balkema, Rotterdam

Bernard K, van Wyk de Vries B (2017) Volcanic avalanche fault zone with pseudotachylite and gouge in French Massif Central. J Volcanol Geotherm Res 347:112–135

Bernard B, van Wyk de Vries B, Leyrit H (2009) Distinguishing volcanic debris avalanche deposits from their reworked products: the Perrier sequence (French Massif Central). Bull Volcanol 71:1041–1056. https://doi.org/10.1007/s00445-009-0285-7

Bernard B, Takarada S, Andrade SD, Dufresne A (2021) Terminology and strategy to describe large volcanic landslides and debris avalanches. In: Roverato M, Dufresne A, Procter J (eds) Volcanic debris avalanches, vol 6. Advances in Volcanology. Springer, Cham, pp 51–73. https://doi.org/10.1007/978-3-030-57411-6_3

Bernard, B, van Wyk de Vries B, Barba D, Leyrit H, Robin C, Alcaraz S, Samaniego P (2008) The Chimborazo sector collapse and debris avalanche: deposit characteristics as evidence of emplacement mechanisms. J Volcanol Geotherm Res 176(1):36–43. https://doi.org/10.1016/j.jvolgeores.2008.03.012

Bernard K, van Wyk de Vries B, Thouret J-C (2019) Fault textures in volcanic debris-avalanche deposits and transformations into lahars: the Pichu Pichu thrust lobes in south Peru compared to worldwide avalanche deposits. J Volcanol Geotherm Res 371:116–136

Biek RF, Rowley PD, Hacker DB (2019) The gigantic Markagunt and Sevier gravity slides resulting from mid-Cenozoic catastrophic mega-scale failure of the Marysvale volcanic field, Utah, USA. Geol Soc Am Field Guide 56:1–121

Branney MJ, Gilbert JS (1995) Ice-melt collapse pits and associated features in the 1991 lahar deposits of Volcán Hudson, Chile: criteria to distinguish eruption-induced glacier melt. Bull Volcanol 57:293–302

Brideau MA, Procter JN (2015) Discontinuity orientation in jigsaw clasts from volcanic debris avalanche deposits and implications for emplacement mechanism. GeoQuébec 2015:20–23

Bronto S (1989) Volcanic geology of Galunggung, West Java, Indonesia. Dissertation, University of Canterbury, 490 pp

Caballero L, Capra L (2011) Textural analysis of particles from El Zaguán debris avalanche deposit, Nevado de Toluca volcano, Mexico: evidence of flow behavior during emplacement. J Volcanol Geotherm Res 200:75–82

Capra L, Macías JL (2002) The cohesive Naranjo debris-flow deposit (10 km^3): a dam breakout flow derived from the Pleistocene debris-avalanche deposit of Nevado de Colima Volcano (México). J Volcanol Geotherm Res 117(1–2):213–235. https://doi.org/10.1016/S0377-0273(02)00245-7

Carrasco-Núñez G, Siebert L, Díaz-Castellón R, Vázquez-Selem L, Capra L (2010) Evolution and hazards of a long-quiescent compound shield-like volcano: Cofre de Perote, Eastern Trans-Mexican Volcanic belt. J Volcanol Geotherm Res 197(1–4):209–224. https://doi.org/10.1016/j.jvolgeores.2009.08.010

Clavero JE, Sparks RSJ, Huppert HE, Dade WB (2002) Geological constraints on the emplacement mechanism of the Parinacota debris avalanche, northern Chile. Bull Volcanol 64:40–54

Clavero J, Polanco E, Godoy E, Matthews S, Sparks RS (2004) Substrata influence in the transport and emplacement mechanism of the Ollagüe debris avalanche (northern Chile). Acta Vulkanol 16(1/2):1–18

Clement BM, Connor CB, Draper G (1993) Paleomagnetic estimate of the emplacement temperature of the long-runout Nevado de Colima volcanic debris avalanche deposit, Mexico. Earth Planet Sci Lett 120:499–510

Cortés A, Macías JL, Capra L, Garduño-Monroy VH (2010) Sector collapse of the SW flank of Volcán de Colima, México—the 3600 yr BP La Lumbre-Los Ganchos debris avalanche and associated debris flows. J Volcanol Geotherm Res 197:52–66

Cortés A, Komorowski J-C, Macías J, Capra L, Layer P (2019) Late Pleistocene-Holocene debris avalanche deposits from Volcán de Colima, Mexico: portrait of a persistently hazardous volcano. https://doi.org/10.1007/978-3-642-25911-1_4

Crandell DR, Miller CD, Glicken HX, Christiansen RL, Newhall CG (1984) Catastrophic debris avalanche from ancestral Mount Shasta volcano, California. Geology 12:143–146

Crandell DR (1989) Gigantic debris avalanche of Pleistocene age from ancestral Mount Shasta volcano, California, and debris-avalanche hazard zonation. US Geol Surv Bull 1861:32

Crandell DR, Fahnestock RK (1965) Rockfalls and avalanches from Little Tahoma Peak on Mount Rainier Washington. US Geol Surv Bull 1221-A:30

Delcamp A, Roberti G, van Wyk de Vries B (2016) Water in volcanoes: evolution, storage and rapid release during landslides. Bull Volcanol 78:87. https://doi.org/10.1007/s00445-016-1082-8

Dufresne A, Davies TR (2009) Longitudinal ridges in mass movement deposits. Geomorphology 105:171–181

Dufresne A, Geertsema M (2020) Rockslide–debris avalanches: flow transformation and hummock formation, examples from British Columbia. Landslides 17(1):15–32

Dufresne A, Salinas S, Siebe C (2010) Substrate deformation associated with the Jocotitlán edifice collapse and debris avalanche deposit. Central México. J Volcanol Geotherm Res 197:133–148. https://doi.org/10.1016/j.jvolgeores.2010.02.019

Dufresne A, Siebert L, Bernard B (2021a) Distribution and geometric parameters of volcanic debris avalanche deposits. In: Roverato M, Dufresne A, Procter J (eds) Volcanic debris avalanches, vol 6. Advances in Volcanology. Springer, Cham, pp 75–90. https://doi.org/10.1007/978-3-030-57411-6_4

Dufresne A, Zernack A, Bernard K, Thouret JC, Roverato M (2021b) Sedimentology of volcanic debris avalanche deposits. In: Roverato M, Dufresne A, Procter J (eds) Volcanic debris avalanches. Advances in Volcanology. Springer, Cham 6:175–210 https://doi.org/10.1007/978-3-030-57411-6_8

Dykes AP, Bromhead EN (2018) The Vaiont landslide: re-assessment of the evidence leads to rejection of the consensus. Landslides 15(9):1815–1832. https://doi.org/10.1007/s10346-018-0996-y

Endo K, Sumida M, Machida M, Furuichi M (1989) The 1984 collapse and debris avalanche deposits of Ontake Volcano, central Japan. In: Latter JH (ed) Volcanic hazards. Springer, Berlin, pp 210–229. https://doi.org/10.1007/978-3-642-73759-6_14

Erismann TH (1979) Mechanisms of large landslides. Rock Mech 12:15–46

Fairchild LH (1985) Lahars at Mount St. Helens, Washington. Dissertation, University of Washington, 374 pp

Fairchild LH (1987) The importance of lahar initiation process. In: Costa JE, Wieczorek GF (eds) Debris flows/avalanches: process, recognition, and mitigation. Geol Soc Am Rev Eng Geol 7:51–62

Fisher RV, Glicken HX, Hoblitt RP (1987) May 18, 1980, Mount St. Helens deposits in South Coldwater Creek, Washington. J Geophys Res 92:10267–10283

Fisher RV, Schmincke H-U (1984) Pyroclastic rocks. Springer, Berlin, p 472

Folk RL (1974) Petrology of sedimentary rocks. Hemphill, Austin, 184 pp

Francis PW, Wells GL (1988) Landsat thematic mapper observation of debris avalanche deposits in the Central Andes. Bull Volcanol 50:258–278. https://doi.org/10.1007/BF01047488

Francis PW, Gardeweg M, Ramirez CF, Rothery DA (1985) Catastrophic debris avalanche deposit of Socompa volcano, northern Chile. Geology 13:600–603. https://doi.org/10.1130/0091-7613(1985)13<600:CDADOS>2.0.CO;2

Glicken H (1986) Rockslide-debris avalanche of May 18, 1980, Mount St. Helens Volcano, Washington. Dissertation, University of California, Santa Barbara, 303 pp

Glicken H (1991) Sedimentary architecture of large volcanic debris avalanches. In: Fisher RV, Smith GA (eds) Sedimentation in volcanic settings, vol 45. Society Sediment Geology Special Publication, pp 99–106

Glicken H (1996) Rockslide debris avalanche of May 18, 1980, Mount St. Helens Volcano, Washington. U S Geol Surv Open-File Rep 96–677, 90 pp and 5 plates.

Glicken H (1998) Rockslide-debris avalanche of May 18, 1980, Mount St. Helens volcano, Washington. Bull Geol Surv Jpn 49:55–106

Hacker DB, Biek RF, Rowley PD (2014) Catastrophic emplacement of the gigantic Markagunt gravity slide, southwest Utah—implications for hazards associated with sector collapse of volcanic fields. Geology 42(11):943–946. https://doi.org/10.1130/G35896.1

Hausback BP, Swanson DA (1990) Record of prehistoric debris avalanches on the north flank of Mount St. Helens volcano, Washington. Geosci Can 17(3):142–145

Herrick JA, Siebert L, Rose WI (2013) Large-volume Barriles and Caisán debris avalanche deposits from Volcán Barú, Panama. Geol Soc Am Spec Pap 498:1–22. https://doi.org/10.1130/2013.2498(09)

Hughes A, Kendrick JE, Salas G, Wallace PA, Legros F, Di Toro D, Lavallée Y (2020) Shear localisation, strain partitioning and frictional melting in a debris avalanche generated by volcanic flank collapse. J Struct Geol 140:104132. https://doi.org/10.1016/j.jsg.2020.104132

Inman DL (1952) Measures of describing the size distribution of sediments. J Sediment Petrol 22:125–145

Inokuchi T (1985) The Ontake rockslide and debris avalanche caused by the Naganoken-Seibu earthquake, 1984. In: Proceedings of the IVth international conference and field workshop on landslides, Tokyo, pp 329–338

Janda RJ, Scott KM, Nolan KM, Martinson HA (1981) Lahar movement, effects, and deposits. In: Lipman PW, Mullineaux DR (eds) The 1980 eruptions of Mount St. Helens, Washington, vol 1250. US Geology Survey Professional Paper, pp 461–478

Johnson CG, Kokelaar BP, Iverson RM, Logan M, LaHusen RG, Gray JMNT (2012) Grain-size segregation and levee formation in geophysical mass flows. J Geophys Res Earth Surf 117:F01032. https://doi.org/10.1029/2011JF002185

Keigler R, Thouret J-C, Hodgson KA, Neal VE, Lecointre JA, Procter JN, Cronin SJ (2011) The Whangaehu formation: debris-avalanche and lahar deposits from ancestral Ruapehu volcano, New Zealand. Geomorphology 133:57–79. https://doi.org/10.1016/j.geomorph.2011.06.019

Komorowski J-C, Glicken HX, Sheridan MF (1991) Secondary electron imagery of microcracks and hackly fracture surfaces in sand-size clasts from the 1980 Mount St. Helens debris—avalanche deposit Implications Particle-Particle Interactions. Geology 19:261–264. https://doi.org/10.1130/0091-7613(1991)019<0261:SEIOMA>2.3.CO;2

Legros F, Cantagrel J-M, Devouard B (2000) Pseudotachylyte (frictionite) at the base of the Arequipa volcanic landslide deposit (Peru): implications for emplacement mechanisms. J Geol 108:601–611. https://doi.org/10.1086/314421

Lipman PW, Mullineaux DR (eds) (1981) The 1980 eruptions of Mount St. Helens, Washington. US Geol Surv Prof Pap 1250:844. https://doi.org/10.3133/pp1250

Makris S, Roverato M, Lomoschitz A, Cole P, Manzella I (2023b) The propagation and emplacement mechanisms of the Tenteniguada volcanic debris avalanche (Gran Canaria): field evidence for brittle fault-accommodated spreading. J Volcanol Geotherm Res 435:107773. https://doi.org/10.1016/j.jvolgeores.2023.107773

Moore JG, Clague DA, Holcomb RT, Lipman PW, Normark WR, Torresan ME (1989) Prodigious submarine landslides on the Hawaiian Ridge. J Geophys Res 94:17465–17484

Munsell Soil Color Charts (1975) Munsell color (Kollmorgen Corp), Baltimore

Nakamura Y (1978) Geology and petrology of Bandai and Nekoma volcanoes. Tohoku Univ Sci Rep Ser 3(14):67–119

Naranjo JA, Francis P (1987) High velocity debris avalanche at Lastarria volcano in the north Chilean Andes. Bull Volcanol 49:509–514

Ohta K (1969) Study on the collapse of Mayu-Yama: on the mechanism for collapse. Sci Shimabara Inst Volcanol Balneol, Fac Sci Kyushu Univ (in Japan with English Abstract)

Paguican EMR, van Wyk de Vries B, Lagmay, A (2014) Hummocks: how they form and how they evolve in rockslide-debris avalanches. Landslides 11:67–80. https://doi.org/10.1007/s10346-012-0368-y

Paguican EMR, Roverato M, Yoshida H (2021) Volcanic debris avalanche transport and emplacement mechanisms. In: Roverato M, Dufresne A, Procter J (eds) Volcanic debris avalanches, vol 6. Advances in Volcanology. Springer, Cham, pp 143–173. https://doi.org/10.1007/978-3-030-57411-6_7

Paguican EMR, van Wyk de Vries B, Lagmay A (2012) Volcano-tectonic controls and emplacement kinematics of the Iriga debris avalanches (Philippines). Bull Volcanol 74:2067–2081. https://doi.org/10.1007/s00445-012-0652-7

Palmer BA, Alloway BV, Neall VE (1991) Volcanic debris-avalanche deposits in New Zealand—lithofacies organization in unconfined, wet-avalanche flows. In: Fisher RV, Smith GA (eds) Sedimentation in volcanic settings, vol 45. SEPM Special Publication, pp 89–98

Perinotto H, Schneider J-L, Bachèlery P, Le Bourdonnec FX, Famin V, Michon L (2015) The extreme mobility of debris avalanches: a new model of transport mechanism. J Geophys Res Solid Earth 120:8110–8119. https://doi.org/10.1002/2015JB011994

Pierson TC, Janda RJ, Thouret JC, Borrero CA (1990) Perturbation and melting of snow and ice by the 13 November 1985 eruption of Nevado del Ruiz, Colombia, and consequent mobilization, flow, and deposition of lahars. J Volcanol Geotherm Res 41:17–66

Pierson TC, Siebert L, Harpel CJ, Scott KM (2018) Geologic field-trip guide of volcaniclastic sediments from snow- and ice-capped volcanoes—Mount St. Helens, Washington, and Mount Hood, Oregon. US

Geol Surv Sci Invest Rep 2017–5022-F:1–97. https://doi.org/10.3133/sir20175022F

Plafker G, Ericksen GE (1978) Nevados Huascarán avalanches, Peru. In: Voight B (ed) Rockslides and avalanches, 1st edn. Elsevier, Amsterdam, pp 277–314

Pouliquen O, Delour J, Savage SB (1997) Fingering in granular flows. Nature 386:816–817. https://doi.org/10.1038/386816a0

Prager C, Zangerl C, Nagler T (2009) Geological controls on slope deformations in the Köfels rockslide area (Tyrol, Austria). Austrian J Earth Sci 102(2):4–19

Procter JN, Cronin SJ, Zernack AV (2009) Landscape and sedimentary response to catastrophic debris avalanches, western Taranaki, New Zealand. Sediment Geol 220(3–4):271–287. https://doi.org/10.1016/j.sedgeo.2009.04.027

Procter JN, Zernack AV, Cronin SJ (2021) Computer simulation of a volcanic debris avalanche from Mt. Taranaki, New Zealand. In: Roverato M, Dufresne A, Procter J (eds) Volcanic debris avalanches, vol 6. Advances in Volcanology. Springer, Cham, pp 281–310. https://doi.org/10.1007/978-3-030-57411-6_11

Pulgarín B, Capra L, Cepeda H, Macías JL (2004) Late Pleistocene deposits associated with a southern flank collapse of the Nevado del Huila volcanic complex (Colombia). Acta Vulcanol 16(1–2):37–58

Qi S, Xu Q, Zhang B, Zhou Y, Lan H, Li L (2011) Source characteristics of long runout rock avalanches triggered by the 2008 Wenchuan earthquake. China. J Asian Earth Sci 40(4):986–906. https://doi.org/10.1016/j.jseaes.2010.05.010

Reiche P (1937) The toreva-block—a distinctive landslide type. J Geol 45:538–548

Reubi O, Ross P-S, White JDL (2005) Debris avalanche deposits associated with large igneous province volcanism: an example from the Mawson formation, Central Allan Hills, Antarctica. Geol Soc Am Bull 117:1615–1628

Richards JP, Villeneuve M (2001) The Llullaillaco volcano, northwest Argentina: construction by Pleistocene volcanism and destruction by sector collapse. J Volcanol Geotherm Res 105:77–105. https://doi.org/10.1016/S0377-0273(00)00245-6

Roberti G, Friele P, van Wyk de Vries B, Ward B, Clague JJ, Perotti L, Giardino M (2017) Rheological evolution of the Mount Meager 2010 debris avalanche, southwestern British Columbia. Geosphere 13(2):369–90. https://doi.org/10.1130/GES01389.1

Rock Color Chart Committee (1991) Rock color chart. Geol Soc Am, Boulder, CO

Rodríguez I, Páez J, Maxmillian S, van Wyk de Vries B, Godoy B (2020) Dynamics and physical parameters of the Lastarria debris avalanche, Central Andes. J Volcanol Geotherm Res 402:106990. https://doi.org/10.1016/j.jvolgeores.2020.106990

Roverato M, Capra L (2013) Características microtexturales como indicadores del transporte y emplazamiento de dos depósitos de avalancha de escombros del volcán de Colima. Rev Mex Cien Geol 30:512–525

Roverato M, Cronin S, Procter J, Capra L (2015) Textural features as indicators of debris avalanche transport and emplacement, Taranaki volcano. Geol Soc Am Bull 127:3–18. https://doi.org/10.1130/B30946.1

Roverato M, Dufresne A, Procter J (eds) (2021) Volcanic debris avalanches. Advances in Volcanology. Springer, Cham, 358 pp. https://doi.org/10.1007/978-3-030-57411-6

Salinas S, López-Blanco J (2010) Geomorphic assessment of the debris avalanche deposit from the Jocotitlán volcano, central Mexico. Geomorphology 123:142–153

Schmidt R, Schmincke H-U (2002) From seamount to oceanic island, Porto Santo, central East-Atlantic. Int J Earth Sci 91(4):594–614. https://doi.org/10.1007/s00531-001-0243-x

Schneider JL, Fisher RV (1998) Transport and emplacement mechanisms of large volcanic debris avalanches: evidence from the northwest sector of Cantal Volcano France. J Volcanol Geotherm 83:141–165

Scott KM (1988b) Origin, behavior, and sedimentology of prehistoric catastrophic lahars at Mount St. Helens, Washington. In: Clifton HE (ed) Sedimentologic consequences of convulsive geologic events, vol 229. Geology Society America Special Paper, pp 23–36. https://doi.org/10.1130/SPE229-p23

Scott KM, Vallance JW, Pringle PP (1995) Sedimentology, behavior, and hazards of debris flows at Mount Rainier, Washington, vol 1547. US Geology Survey Professional Paper, 56 pp

Scott KM, Macías JL, Naranjo JA, Rodríguez S, McGeehin JP (2001) Catastrophic debris flows transformed from landslides in volcanic terrains: mobility, hazard assessment, and mitigation strategies. US Geology Survey Professional Papep 1630, 59 pp

Scutter CR, Cas RA, Moore CL, De Rita D (1998) Facies architecture and origin of a submarine rhyolitic lava flow-dome complex, Ponza, Italy. J Geophys Res Solid Earth 10.103(B11):27551–66. https://doi.org/10.1029/98JB01121

Sekiya S, Kikuchi Y (1889) The eruption of Bandai-san. Tokyo Imp Univ Coll Sci J 3(2):91–172

Shea T, van Wyk de Vries B (2008) Structural analysis and analogue modeling of the kinematics and dynamics of rockslide avalanches. Geosphere 4(4):657–686

Siebe C, Komorowski J-C, Sheridan MF (1992) Morphology and emplacement of an unusual debris-avalanche deposit at Jocotitlán volcano, central Mexico. Bull Volcanol 54:573–589. https://doi.org/10.1007/BF00569941

Siebe C, Salinas S, Arana-Salinas L, Macías JL, Gardner J, Bonasia R (2017) The ~23,500 y 14C BP White Pumice Plinian eruption and associated debris avalanche and Tochimilco lava flow of Popocatépetl volcano, México. J Volcanol Geotherm

Res 333–334:66–95. https://doi.org/10.1016/j.jvolgeores.2017.01.011

Siebert L (1984) Large volcanic debris avalanches: characteristics of source areas, deposits, and associated eruptions. J Volcanol Geotherm Res 22:163–197. https://doi.org/10.1016/0377-0273(84)90002-7

Siebert L (1996) Hazards of large volcanic debris avalanches and associated eruptive phenomena. In: Scarpa R, Tilling RI (eds) Monitoring and mitigation of volcano hazards. Springer, Berlin, pp 541–572

Siebert L, Glicken H, Ui T (1987) Volcanic hazards from Bezymianny- and Bandai-type eruptions. Bull Volcanol 49:435–459. https://doi.org/10.1007/BF01046635

Siebert L, Glicken H, Kienle J (1989) Debris avalanches and lateral blasts at Mount St Augustine volcano, Alaska. Nat Geogr Res 5(2):232–249

Siebert L, Begét JE, Glicken H (1995) The 1883 and late-prehistoric eruptions of Augustine volcano, Alaska. J Volcanol Geotherm Res 66:367–395. https://doi.org/10.1016/0377-0273(94)00069-S

Siebert L (2002) Landslides resulting from structural failure of volcanoes. In: Evans SG, DeGraff JV (eds) Catastrophic landslides: effects, occurrence, and mechanisms, vol 15. Geology Society America Review Engineering Geology, pp 209–235. https://doi.org/10.1130/REG15-p209

Siebert L, Reid ME (2023) Lateral edifice collapse and volcanic debris avalanches: a post-1980 Mount St. Helens perspective. Bull Volcanol 85(61). https://doi.org/10.1007/s00445-023-01662-z

Siebert L, Roverato M (2021) A historical perspective on lateral collapse and volcanic debris avalanches. In: Roverato M, Dufresne A, Procter J (eds) Volcanic debris avalanches, vol 6. Advances in Volcanology. Springer, Cham, pp 11–50. https://doi.org/10.1007/978-3-030-57411-6_2

Siebert L, Kimberly P, Pullinger CR (2004) The voluminous Acajutla debris avalanche from Santa Ana volcano, western El Salvador, and comparison with other Central American edifice-failure events. In: Rose WI, Bommer JJ, López DL, Carr MJ, Major JJ (eds) Natural hazards in El Salvador, vol 375. Geology Society America Special Paper, pp 5–23. https://doi.org/10.1130/0-8137-2375-2.5

Siebert L, Alvarado GE, Vallance JW, van Wyk de Vries B (2006) Large-volume volcanic edifice failures in Central America and associated hazards. In: Rose WI, Bluth GJS, Carr MJ, Ewert J, Patino LC, Vallance J (eds) Volcanic hazards in Central America, vol 412. Geology Society America Special Paper, pp 1–26. https://doi.org/10.1130/2006.2412(01)

Skilling IP, White JD, McPhie J (2002) Peperite: a review of magma–sediment mingling. J Volcanol Geotherm Res 114(1–2):1–17. https://doi.org/10.1016/S0377-0273(01)00278-5

Suwa H, Okunishi K, Okuda S, Takahashi H, Hasegawa H, Takada M, Takaya S (1985) Depositional characteristics of the debris avalanche at Mt. Ontake in 1984. Annu Disaster Prev Res Inst, Kyoto Univ 28(B-1):505–518 (in Japan)

Takarada S, Ui T, Yamamoto Y (1999) Depositional features and transportation mechanism of valley-filling Iwasegawa and Kaida debris avalanches, Japan. Bull Volcanol 60:508–522. https://doi.org/10.1007/s004450050248

Ui T (1983) Volcanic dry avalanche deposits-identification and comparison with nonvolcanic debris streams deposits. J Volcanol Geotherm Res 18:135–150. https://doi.org/10.1016/0377-0273(83)90006-9

Ui T, Glicken H (1986) Internal structural variations in a debris-avalanche deposit from ancestral Mount Shasta, California, USA. Bull Volcanol 48:189–194. https://doi.org/10.1007/BF01087673

Ui T, Kawachi S, Neall VE (1986b) Fragmentation of debris avalanche material during flowage—evidence from the Pungarehu Formation, Mount Egmont, New Zealand. J Volcanol Geotherm Res 27:255–264

Ui T, Takarada S, Yoshimoto M (2000) Debris avalanches. In: Sigurdsson H, Houghton B, McNutt SR, Rymer H, Stix J (eds) Encyclopedia of volcanoes. Academic Press, San Diego, pp 617–626

Ui T (1985) Debris avalanche deposits associated with volcanic activity. Proceedings IVth international conference and field workshop on landslides, 23–31 Aug, 1985, Tokyo, Japan Landslide Society, pp 405–410

Ui T (1989) Discrimination between debris avalanche and other volcaniclastic deposits. In: Latter JH (ed) Volcanic hazards, vol 1. International Association Volcanology Chemical Earth's Interaction Proceeding Volcanology, pp 201–209

Ui T, Yamamoto H, Suzuki-Kamata K (1986a) Characterization of debris avalanche deposits in Japan. In: Kushiro I (ed) M Sakuyama and H Fukuyama memorial volume, vol 29, pp 231–243. https://doi.org/10.1016/0377-0273(86)90046-6. J Volcanol Geotherm Res

Valderrama P, Roche O, Samaniego P, van Wyk de Vries B, Bernard K, Marino J (2016) Dynamic implications of ridges on a debris avalanche deposit at Tutupaca volcano (Southern Peru). Bull Volcanol 78:14. https://doi.org/10.1007/s00445-016-1011-x

Valderrama P, Roche O, Samaniego P, van Wyk de Vries B, Araujo G (2018) Granular fingering as a mechanism for ridge formation in debris avalanche deposits: laboratory experiments and implications for Tutupaca volcano, Peru. J Volcanol Geotherm Res 349:409–418. https://doi.org/10.1016/j.jvolgeores.2017.12.004

Vallance JW, Scott KM, Pringle PT (1995) Sedimentology, behavior, and hazards of debris flows at Mount Rainier, Washington, vol 1547. US Geology Survey Professional Paper, 56 pp

Vallance JW, Scott KM (1997) The Osceola Mudflow from Mount Rainier: sedimentology and hazard implications of a huge clay-rich debris flow. Geol Soc Am Bull 109:143–163. https://doi.org/10.1130/0016-7606(1997)109<0143:TOMFMR>2.3.CO;2

van Wyk de Vries B, Delcamp A (2015) Volcanic debris avalanches. In: Shroder JF, Davies T (eds) Landslide hazards, risks and disasters. Academic Press, San Diego, pp 131–157. https://doi.org/10.1016/B978-0-12-396452-6.00005-7

Vezzoli L, Apuani T, Corazzato C, Uttini A (2017) Geological and geotechnical characterization of the debris avalanche and pyroclastic deposits of Cotopaxi Volcano (Ecuador). A contribute to instability-related hazard studies. J Volcanol Geotherm Res 332:51–70

Voight B, Sousa J (1994) Lessons from Ontake-san: a comparative analysis of debris avalanche dynamics. Eng Geol 38:261–297. https://doi.org/10.1016/0013-7952(94)90042-6

Voight B, Janda RJ, Glicken H, Douglass PM (1983) Nature and mechanics of the Mount St. Helens rockslide-avalanche of 18 May 1980. Geotechnique 33:243–273. https://doi.org/10.1680/geot.1983.33.3.243

Voight B, Glicken H, Janda RJ, Douglass PM (1981) Catastrophic rockslide avalanche of May 18. In: Lipman PW, Mullineaux DR (eds) The 1980 eruptions of Mount St. Helens, Washington, vol 1250. US Geology Survey Professional Paper, pp 347–377

Wadge G, Francis PW, Ramirez CF (1995) The Socompa collapse and avalanche event. J Volcanol Geotherm Res 66:309–336. https://doi.org/10.1016/0377-0273(94)00083-S

Waitt RB, Begét, JE (2009) Volcanic processes and geology of Augustine Volcano, Alaska. US Geol Surv Prof Pap 1762:1–78. 2 plates, scale 1:25,000

Walker GPL (1971) Grain-size characteristics of pyroclastic deposits. J Geol 79:696–714

Weidinger JT, Korup O, Munack H, Altenberger U, Dunning SA, Tippelt G, Lottermoser W (2014) Giant rockslides from the inside. Earth Planet Sci Lett 389:62–73. https://doi.org/10.1016/j.epsl.2013.12.017

van Wyk de Vries B, Davies TR (2015) Landslides, debris avalanches, and volcanic gravitational deformation. In: Sigurdsson H (ed) The encyclopedia of volcanoes, 2nd edn. Elsevier, Amsterdam, pp 665–685. https://doi.org/10.1016/B978-0-12-385938-9.00038-9

Yoshida H (2014) Hummock alignment in Japanese volcanic debris avalanches controlled by pre-avalanche slope of depositional area. Geomorphology 223:67–80. https://doi.org/10.1016/j.geomorph.2014.06.024

Yoshida H, Sugai T, Ohmori H (2012) Size-distance relationships for hummocks on volcanic rockslide-debris avalanche deposits in Japan. Geomorphology 136:76–87. https://doi.org/10.1016/j.geomorph.2011.04.044

Yoshimoto M, Ui T (1998) The 1640 sector collapse of Hokkaido Komagatake volcano, northern Japan. Bull Volcanol Soc Jpn 43:137–148 (in Japan with English abstract)

Volcanic Water Floods—Process Overview

Abstract

Floods of water originating from volcanically disturbed landscapes—volcanic water floods (VWFs)—differ from floods in most other types of drainage basins in the world in significant ways. They flow at higher discharges (flow rates), have higher flood peaks, and have shorter flow durations relative to watershed area than most other types of floods. They can have suspended-sediment concentrations that approach those of hyperconcentrated flow, and perhaps most importantly, they can transport such enormous amounts of sediment (suspended load and bed load) that rivers cannot transport all of it farther downstream. Resulting sediment storage in river valleys can result in channel aggregation and burial of valley floors by meters of sediment within hours or tens of meters within a few years. Small, short-duration VWFs in steep terrain often resemble dilute lahars—muddy floods moving as highly turbulent, very muddy water in multiple waves or surges down a channel. Larger, less concentrated VWFs behave more like muddy nonvolcanic floods but with greater-than-normal sediment transport rates. VWFs can be triggered by high-intensity or long-duration rainfall, rapid melting of snow and ice by volcanic eruptions, lake breakout floods, and glacial outburst floods. In highly disturbed watersheds, large-volume VWFs are likely to transform to lahars.

1 Definitions and Context

Volcanic water floods (VWFs) are water floods from volcanoes, supercharged with both suspended-load and bed-load sediment eroded from volcanically disturbed landscapes. They differ from normal water floods in undisturbed terrain by their capacity for higher discharges (flow rates) relative to the size of the contributing watershed, their typically shorter durations, and their high to extreme suspended-sediment concentrations[1] that can approach concentrations in hyperconcentrated flows. Nonetheless VWFs behave hydraulically as Newtonian fluids, like water floods everywhere. Small, short-duration VWFs in steep terrain often resemble dilute lahars—muddy floods moving as highly turbulent, very muddy water in multiple waves or

[1] Sizes of suspended normal density rock particles in VWFs include grains in the clay, silt, and fine sand-size fractions. Clay minerals may be present in streams draining hydrothermally altered volcanic edifices. Where pumice and scoria are present, gravel-size particles (lapilli) are commonly transported in suspension.

T. C. Pierson et al., *Deposits of Volcanic Wet Flows*, Advances in Volcanology,
https://doi.org/10.1007/978-3-031-66574-5_9

Fig. 1 Examples of high-concentration water floods caused by landscape disturbance. **a** Surge in a volcanic water flood (VWF) coming down Smith Creek (frontal bore flowing toward camera) in the blast zone during first major rain after the May 18, 1980, eruption of Mount St. Helens (Washington, USA). Peak sediment concentration of this flood, measured about 15 km farther downstream, was 178,000 mg/L (5.7 vol% solids) (Dinehart 1986). USGS photo by H. Martinson, Nov 6, 1980. **b** VWF in North Fork Toutle River in a wide braided-channel reach downstream of Mount St. Helens near the downstream end of the 1980 debris-avalanche deposit on November 7, 2006. Note active erosion of bank on left, which is about 1 m high. Flow was toward camera. USGS photo by J. Major. **c** Sediment-rich water flood on Hood River on afternoon of November 7, 2006, at Hood River Railroad bridge about 40 km downstream of the summit of Mount Hood (Oregon, USA). Extremely heavy rain from an atmospheric river storm triggered landslides at the head of this basin that triggered the high sediment transport rate. Flow depth about 3 m; sediment concentration here was not measured. Flow was left to right. Photo © by Darryl Lloyd, used with permission. **d** VWF on Toutle River about 73 km downstream of Mount St Helens at the Tower Road gaging station on Dec 9, 2015. Measured suspended sediment concentration 15 min before this photo was taken was 19,600 mg/L (about 0.8 vol%), with about 45% of sample being silt and clay. Flow is toward camera (viewed from a bridge) and at least several meters deep; flooded right riverbank is visible in upper left. USGS photo by T. Pierson

surges down a channel (Fig. 1). Larger, less concentrated VWFs behave more like nonvolcanic floods but with greater-than-normal sediment transport rates. VWFs typically transport significantly more sediment than floods of equivalent size in undisturbed nonvolcanic watersheds, commonly resulting in unusually high sediment yields for water floods.

Because of their high sediment loads, VWFs can have significant geomorphic impacts on river channels tens to hundreds of kilometers downstream of source volcanoes. Generally, they are not as destructive as lahars at equivalent discharges, but they pose significant hazards to valleys downstream of recently active volcanoes because of their tendency to promote higher than expected flood peaks and water volumes, lateral channel shifting (undercutting of banks and terraces) and large-scale channel aggradation. VWFs have many similarities with flash floods in arid nonvolcanic terrain (Lekach and Schick 1982; Reid and Frostick 1987; Laronne and Reid 1993; Reid et al. 1998; Cohen and Laronne 2005). They also have similarities with high-concentration floods in volcanic and nonvolcanic watersheds that have been recently disturbed—not by volcanic eruptive activity but by other natural and human-caused disturbances such as logging, wildfire, intensive agriculture, surface mining, and landslides (Schumm and Rea 1995; Meyer and Wells 1997; Miller and Benda 2000; Hoffman and Gabet 2007; Reneau et al. 2007; 2009).

VWFs in typical volcanic terrains transport up to 11–15 vol% suspended sediment while maintaining the properties of Newtonian fluids (Dinehart 1986; Pierson et al. 2005). The upper concentration limit for Newtonian fluid flow depends on the particle-size distributions and particle compositions found in the erodible sediment in a particular landscape. Sediment mixtures rich in clay minerals would have limits lower than 11 vol% (Hampton 1975), whereas those with atypically low fines contents would have limits higher than 15 vol% (Fei 1983; Wan and Wang 1994).

2 Initiation and Mobilization

VWFs may be generated by eruptive processes such as eruptions through crater lakes (Cronin et al. 1996; Kilgour et al. 2010) or rapid melting of snow and ice by hot volcaniclastic particles (Pierson et al. 1990; Dorava and Meyer 1994; Trabant et al. 1994; Pierson 1999), but most are generated by secondary processes following active eruption activity—heavy rainfall runoff, lake outbreaks, and glacier outbursts (e.g., McGimsey et al. 1994; Waythomas et al. 1996; Guðmundsson et al. 1997; Dinehart 1998; Manville et al. 1999, 2005; Maria et al. 2000; O'Connor et al. 2001; Hodgson and Nairn 2005; Major and Mark 2006; Kataoka et al. 2008; Kataoka 2011; Pierson et al. 2013). Although sudden releases of large water volumes in volcanic terrain commonly transform from VWFs to lahars before traveling more than a few kilometers (e.g., Walder and Driedger 1994b; Pierson 1999; Procter et al. 2010), lahars typically transform back to VWFs farther downstream (Pierson and Scott 1985; Pierson 1999; O'Connor et al. 2001).

2.1 Rainfall-Runoff Floods

The magnitude, timing, and sediment load of rainfall-generated floods are greatly influenced by the type, degree, extent, and recency of volcanic disturbance to a watershed (Pierson and Major 2014). Large explosive eruptions producing widespread and fine-grained tephra fall (TF) and pyroclastic density current (PDC) deposits have huge hydrologic impacts on watersheds. During and shortly following such eruptions, heavy rainfall runoff mainly produces lahars, but VWFs typically become the dominant sediment-transporting process and downstream hazard once much or most of the readily available and highly erodible volcaniclastic source sediments have been removed and watersheds start to return to a state of geomorphic stability (e.g., Major et al. 1996; Pierson et al. 1996; Rodolfo et al. 1996; Hayes et al. 2002; Gran et al. 2011). After some eruptions, however, rainfall runoff may produce few if any lahars, and subsequent hydrologic responses are limited to VWFs, whether occurring soon after disturbance or delayed for years, decades, or possibly centuries (e.g., White et al. 1997; Pierson et al. 2011, 2013). VWFs (rather than lahars) are commonly the main result of posteruptive rainfall runoff where (a) low-energy explosive eruptions produce TF deposits that are coarse or have limited areal distributions, (b) rainfall is of low intensity or short duration, (c) runoff magnitude is large relative to the volume of sediment available for erosion, or (d) nonexplosive eruptions produce lava flows or domes but little tephra.

The combined disturbance effects of vegetation loss and decrease in soil infiltration capacity

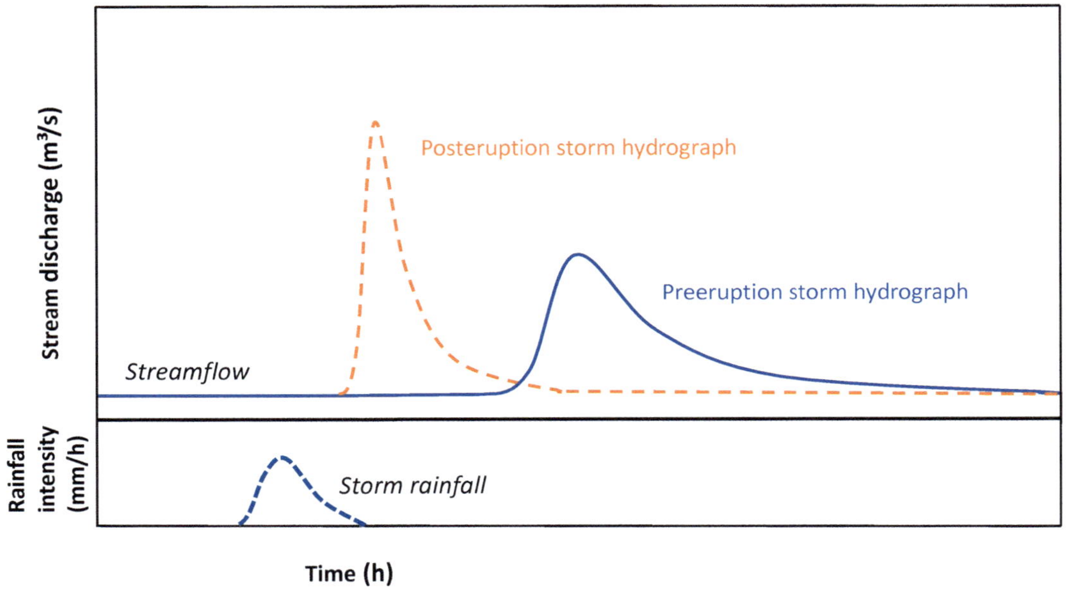

Fig. 2 Schematic representation of the typical effects of volcanic disturbance (fine ash deposition and vegetation removal) in a highly disturbed watershed on a hypothetical runoff-generated flood hydrograph. Time scale and vertical axis units are arbitrary. Runoff occurs sooner and peak discharges are higher when fine-grained tephra blankets impede infiltration of rainfall and when lack of vegetation allows all the rain to reach the ground instead of some proportion of it being intercepted by tree and understory leaves and stems

due to deposition of fine-grained tephra blankets have a major impact on watershed hydrology and flood generation (Pierson and Major 2014) (see Sect. 3 in Chap. "Scaling and Timing of Volcanic Wet Flows in Response to Eruptions"). Vegetation loss allows more rain to reach the ground surface due to rain no longer being intercepted by plant leaves and stems, and the additional rainfall reaching the ground runs off faster due to decreased soil infiltration capacity and decreased surface roughness (Pierson and Major 2014). Both effects can persist for decades—the interval that largely defines the posteruptive period (Suwa and Yamakoshi 1999; Major et al. 2000, Major and Yamakoshi 2005; Gran et al. 2011) (see Chap. "Scaling and Timing of Volcanic Wet Flows in Response to Eruptions"). These changed watershed conditions following volcanic disturbance allow posteruptive rainfall runoff to produce VWFs that have greater water volumes, higher peak discharges, steeper hydrograph rises, and shorter lags between rainfall peaks and flood peaks in comparison to pre-disturbance floods in the same watersheds (Fig. 2) (Smith 1991a, b; Major and Mark 2006; Pierson and Major 2014).

After a volcanic disturbance to a landscape, drainage basins typically return, over time, to a relatively stable geomorphic equilibrium condition that can be defined by some combination of flood magnitude and frequency; stability of channel dimensions, patterns, and profile elevations; vegetation patterns; and sediment loads (Pierson and Major 2014; Major et al. 2021) (see Chap. "Scaling and Timing of Volcanic Wet Flows in Response to Eruptions"). This transition is termed the *geomorphic recovery* of a drainage basin, and it typically follows a rate law in the form of a negative exponential function (Graf 1977) (Fig. 6) in Chap. "Scaling and Timing of Volcanic Wet Flows in Response to Eruptions", although there may not be a single path or single outcome that defines recovery (Phillips 2009)—a new equilibrium condition may be like or quite unlike the preeruption condition. The time required for the transition back to some normative state of geomorphic stability can vary significantly depending on the type and degree of disturbance (Major 2004; Manville et al. 2009b; Pierson and Major 2014).

Degree of volcanic disturbance can vary among different drainage basins affected by the same eruption. Watersheds with minor to moderate levels of disturbance may return to relatively stable conditions after only a few years (Major et al. 2000; Pierson et al. 2013; Iroumé et al. 2020). In contrast, prolonged recoveries can persist for decades, centuries, or even millennia, due either to continued reworking of long-lingering reservoirs of eruption-generated sediment (Kuenzi et al. 1979; Major et al. 2000; Hayes et al. 2002; Manville 2002; Major 2004; Manville and Wilson 2004; Manville et al. 2005, 2009b; Pierson et al. 2011; Gran et al. 2011) or to the inherent instability of many volcanic edifices and the excessively high rates of sediment production there by normal geomorphic and glacial processes (Czuba et al. 2010, 2012). In cases where eruptive activity is frequent, however, drainage basins may not recover or reach stable conditions for extended periods due frequent or ongoing eruptive disturbance or edifice instability that continually interrupts system recovery (Major et al. 2021).

2.2 Meltwater Floods

Hot PDCs and ballistic ejecta can scour, displace, entrain, and turbulently mix with snow and ice in river valleys, on volcano flank snowpacks, on alpine glacier surfaces, in craters, and on summit ice caps to produce large volumes of meltwater within minutes (Waitt et al. 1983; Pierson and Scott 1985; Pierson et al. 1990; Thouret 1990, Dorava and Meyer 1994; Pierson 1999; Walder 2000a, b; Waythomas et al. 2013). Meltwater from such sources is commonly already mixed with volcaniclastic particles (and commonly also with ice fragments) as it is produced. Further vigorous entrainment of loose sediment during descent down steep, debris-covered slopes nearly always results in transformation of meltwater floods to lahars. VWFs may not transform, however, if meltwater releases are relatively small or if a succession of meltwater-producing events has already removed some of the sediment available for erosion (e.g., Dorava and Meyer 1994).

2.3 Lake Breakout Floods

Lakes are commonly impounded on or near volcanic edifices—either within volcanic craters and calderas or behind debris blockages, lava flows, or glaciers acting as natural dams. Debris blockages can be formed by rapidly emplaced deposits of PDCs, TFs, lahars, and volcanic debris avalanches (VDAs), as well as by slowly aggrading alluvial fans and channel fills and by glacial moraines (Costa 1988; Costa and Schuster 1988; Manville et al. 1999, 2007; O'Connor et al. 2013). Sudden release of water impounded by natural dams in volcanic terrain can produce enormous (up to 20 km^3) water floods that are among the most far-reaching volcanic wet-flow hazards (McGimsey et al. 1994; Waythomas et al. 1996; Manville et al. 1999; 2005; Pierson 1999; Maria et al. 2000; Hodgson and Nairn 2005; Manville and Cronin 2007; Manville et al. 2007; Kataoka et al. 2008; Kataoka 2011), particularly when the ratio of sediment volume available for erosion to water volume is low. If available sediment volumes (or ice-particle volumes) are large relative to flood water volume, transformation to lahars is common (e.g., Pierson and Scott 1985; Scott 1988b, 1989; Cronin et al. 1996; Manville and Cronin 2007; Waythomas et al. 2013).

Most natural dams impounding lakes at volcanoes and elsewhere ultimately fail without engineering intervention (Schuster 2000), and the majority fail within a year of formation (Costa and Schuster 1988). This instability results from the uncompacted nature of the material forming debris blockages and from uncontrolled rises in lake levels. Lake levels typically rise steadily because inflow of water from precipitation and snowmelt usually exceeds losses from seepage and evaporation. Failure of natural dams occurs when lake levels rise to

Fig. 3 Breaching in 1981 of small lake on the Mount St. Helens (Washington, USA) 1980 volcanic debris avalanche (VDA) deposit due to overtopping (observed) of the debris blockage impounding the water. Breach widened and deepened noticeably in only minutes as outflow eroded the unconsolidated VDA deposit. Estimated width of breach in photo is 10–15 m. USGS photo taken from helicopter by T. Pierson

critical levels, and dam failure is typically very rapid (Costa 1988; Costa and Schuster 1988; Schuster 2000) (Fig. 3). Flood discharges resulting from lake breakouts are a function of release volume and breach formation rate (Walder and O'Connor 1997).

In contrast to the common occurrence of unstable natural dams that readily fail after impounding lakes, a smaller percentage of lake-forming debris blockages remain stable for many years, if not indefinitely. The size, shape, permeability, degree of consolidation, and coarse gravel content) of such blockages can affect resistance to failure by overtopping or by sapping erosion. Low dams tend to have low-gradient overtopping channels and impound shallow lakes that can only produce relatively low hydrostatic seepage pressures. Dams with broad crests provide long seepage pathways and greater blockage mass. Clay-rich dams have low permeability to slow seepage. Bouldery, clay-rich, and well-consolidated blockage materials resist erosion when overtopping occurs and form relatively stable drainage channels. Good examples of lake-impounding debris blockages with long-term stability include Silver Lake, downstream of Mount St. Helens (Washington, USA), and Lake Kapowsin, downstream of Mount Rainier (Washington, USA). Both lakes are dammed by broad blockages of clay-rich lahar deposits (Scott 1988b, 1989; Scott et al. 1995).

3 Glacier-Outburst Floods

Glacier-outburst floods (also known by the Icelandic term *jökulhlaups*) are sudden releases of large volumes of liquid water accumulated in reservoirs beneath or within glaciers (Richardson 1968; Björnsson 1975; Guðmundsson et al. 2015). Water for these floods is produced through melting of ice and snow by subglacial volcanic eruptions, geothermal heating around fumaroles, or melting at the glacier surface (ablation) during warm or rainy weather. If an eruption is not involved, accumulations of water within or beneath glaciers may occur slowly and can go undetected prior to reservoir release.

Relatively small outburst floods from temperate glaciers on volcano flanks commonly

Fig. 4 Eruption-triggered glacier outburst flood (*jökulhlaup*) in progress from the Vatnajökull ice cap in a very wide braided channel in SE Iceland on November 5, 1996. View is looking upstream toward the ice cap. Note surviving span of highway bridge in center of photo. Discharge at time photo was taken was about 15,000 m^3/s. Photo © by Magnús T. Guðmundsson, used with permission

occur during hot weather in late summer or early autumn when glacier meltwater production is especially high and when early autumn rain can augment summer melting (Richardson 1968; Driedger and Fountain 1989; Walder and Driedger 1994a, b, 1995). Glaciers on Mount Rainier, a volcano with frequent glacier outburst floods in recent decades, release up to 10^6 m^3 of water per flood (Richardson 1968). Most outburst floods at Mount Rainier transform to small lahars (volcanic debris flows), owing to flow routes through steep, narrow canyons with near-vertical walls of unconsolidated debris. Estimated peak discharges for a few of these flows have ranged from 10^2 to10^3 m^3/s (Richardson 1968).

Eruptive activity beneath glaciers or ice caps can produce much larger volumes of meltwater than is possible by rapid surface ablation. For example, the 1996 Gjálp volcano eruption in Iceland melted and accumulated a huge volume of water beneath the Vatnajökull ice cap (Guðmundsson et al. 1997; Russell and Knudsen 1999; Maria et al. 2000). The basaltic andesite fissure eruption occurred beneath 500–750 m of glacier ice over a period of 13 days. For 5 weeks, 15–20 °C meltwater flowed 5–10 km through a channel at the glacier bed into a subglacial lake in the caldera of the Gjálp vent's primary volcano, Grimsvötn, raising the lake level at a rate of 15–20 m/day. Eventually the roof of ice over the growing lake collapsed in several places, indicating the location of the reservoir. Slightly more than a month after the eruption began, the deepening lake produced hydrostatic pressures high enough to lift the ice dam from a gap in the eastern wall of the caldera, bursting out as a catastrophic jökulhlaup (Fig. 4). This flood reached a peak discharge of 52,000 m^3/s approximately 12 h after onset, and more than 3 km^3 of water were released over two days, flowing mainly as a VWF. Ice blocks the size of multi-story buildings and ~ 1.8×10^8 m^3 of sediment were transported by the flood, resulting in an average of 4 m of net aggradation (minimum 9 m near the ice margin) on the glacial outwash plain downslope of the glacier, and a net addition of a 7 km^2 delta to the coastline.

Not all subglacial eruptions produce glacial outburst floods, however. The slow 2004–2008 dome-building eruption of Mount St. Helens occurred beneath about 120 m of crater-filling

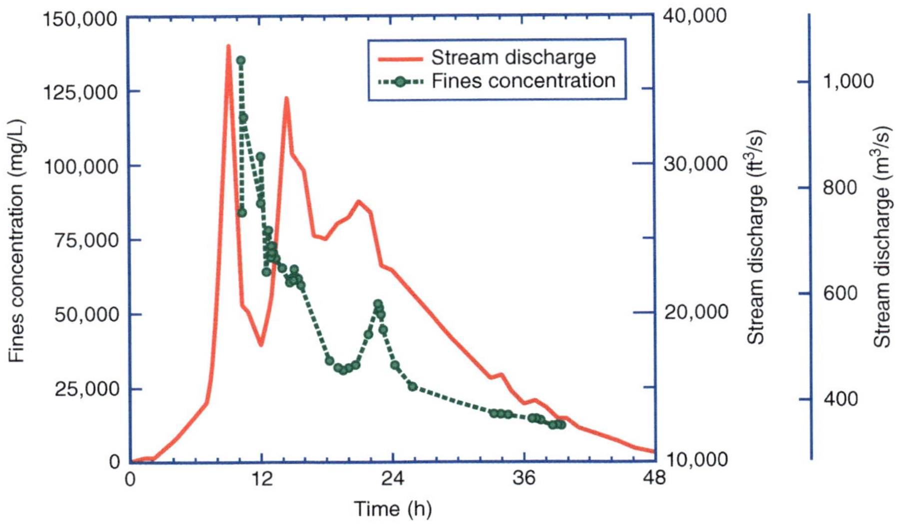

Fig. 5 Hydrograph of rainfall-runoff volcanic water flood (VWF), recorded on the lower Toutle River, 82 km downstream of Mount St. Helens (Washington, USA) on February 20–21, 1982. Drainage above this gaging station is 1326 km^2 in area. Roughly ¾ of the watershed area above this station was disturbed by the 1980 eruption. Peak suspended-sediment concentration at this gaging station was about 5 vol% (modified from Dinehart 1998)

glacier ice, and roughly 10^7 m^3 of ice was melted during the 3.3-year eruption (Walder et al. 2007, 2008; Dzurisin et al. 2015). Yet little or no meltwater accumulated beneath or within the glacier and no outburst floods occurred. Instead, most of the meltwater percolated downward into pervious rock debris on the crater floor, recharging the groundwater/hydrothermal system, while the remainder flowed out of the crater via two surface streams at steady low discharges.

4 Flow Behavior

VWFs from recently disturbed watersheds typically move as highly concentrated, high discharge, turbulent, unsteady water flows with high rates of sediment transport; event durations are relatively short, ranging from hours to days, depending on rainfall duration and basin size (Dinehart 1998; Hayes et al. 2002; Alexander et al. 2010). Such floods characteristically have sharply peaked hydrographs and overall high sediment-transport rates (both suspended and bed load). Hydrograph rising limbs can be very steep; falling limbs are somewhat less steep (Fig. 5). Flood flow fronts may be marked by a relatively gradual transition from low-discharge clear-water flow to high-discharge muddy streamflow, by concentrations of floating logs at the flow front, or by discrete frontal bores in small basins (Fig. 1a).

Mean flow velocities of VWFs generally range from 2 to 5 m/s in steep but relatively shallow channels on and immediately downstream of volcano flanks. These velocities are lower than those of lahars in the same channels at equivalent depth and slope (Pierson and Scott 1985; Lavigne and Suwa 2004). When unconfined by canyons, VWFs create and flow within shallow braided channels, a channel configuration that tends to keep velocities relatively low, even for large floods. For example, the largest recorded water flood in Iceland had flow depths of only a few meters and a peak average flow velocity to 2–3 m/s for a peak discharge of about 52,000 m^3/s (Maria et al. 2000) (Fig. 4).

VWFs commonly entrain sediment and transform to lahars in recently disturbed drainage basins, thereby increasing in volume, peak discharge, flow velocity, erosional capacity, and hazard to life and property. Floods triggered by high-intensity rainfall on fresh pyroclastic deposits, lake breakouts through breached natural dams, and glacier outburst floods are all capable of rapidly entraining sufficient sediment for that transformation to occur (Richardson 1968; O'Connor et al. 2001; Walder and Driedger 1994b, 1995; Pierson 1999; Marren et al. 2009; Procter et al. 2010). Channels particularly susceptible to rapid sediment entrainment are steeper than about 8°, constricted enough for flow to impinge on one or both banks, and incised deeply into unconsolidated material, with banks at angle of repose or steeper (Pierson 1999; O'Connor et al. 2001).

4.1 Material Properties and Sediment Transport

Sediment is transported in all water floods by a combination of *wash load* (sediment particles continually in suspension above the bed regardless of flow conditions), *suspended load* (sediment particles kept in suspension whenever forces applied by the flow are sufficient to keep them in suspension), and *bed load* (sediment particles that move mainly along the bed by saltation, rolling, or sliding). In VWFs, however, the wash load is typically made up of greater than normal amounts of silt- and clay-size particles (commonly abundant in disturbed volcanic terrain), making the flood water quite muddy. In addition, larger low-density pumice particles are commonly included in the wash load, if available, because they can float. Normal-density sand-size and fine gravel-size particles move either as suspended or bed load, depending on flood hydraulics and the concentration of fines in the wash load. Coarser dense gravel particles generally move only as bed load, and bed-load transport by VWFs may include formation of basal traction carpets (Todd 1989; Sohn 1997) and incipient basal granular flows (Manville and White 2003).

All water floods transport sediment, but VWFs typically transport more sediment and do so more effectively than water floods from undisturbed watersheds in older volcanic terrain or in nonvolcanic areas (Major et al. 2000, 2021, 2004; Hayes et al. 2002; Major and Mark 2006; Gran et al. 2011; Pierson et al. 2013; Major et al. 2021). The key reason for this is that volcanically disturbed ground surfaces directly expose huge volumes of loose and highly erodible sediment to falling rain (see Sect. 2.1). In addition, VWFs transport sediment more efficiently due to reconfigured channels that are commonly straighter (and therefore steeper) and hydraulically smoother than before disturbance. High rates of sediment input to rivers following eruptions typically result in relatively smooth riverbeds composed mainly of medium to coarse sand (Kuenzi et al. 1979; Dinehart 1998; Montgomery et al. 1999; Hayes et al. 2002; Gran et al. 2006; Pierson et al. 2011). Such smooth, sandy channel beds are highly mobile, and rates of bed-load sediment transport are high, because the effective boundary shear stress on a smooth bed is high, while the critical shear stress (the shear stress required to get sand-size bed material moving) is low (Montgomery et al. 1999). As sand inputs to rivers decline over time, riverbeds become progressively armored with gravel and more stable (Gran and Montgomery 2005; Gran et al. 2006).

Type and severity of disturbance in a particular drainage basin affects peak sediment concentration in VWFs following a volcanic eruption. This was clearly the case at Mount St. Helens in 1980 (Table 1). Peak sediment concentrations at eight gaging stations in three different drainage basins at Mount St. Helens ranged from 0.9 to 11.3 vol% in the first year following disturbance (not considering the high value in the table caused by riverbank modification). For comparison, 28 large floods from undisturbed volcanic terrain in the western Cascade Range of Oregon and Washington in 1964–1965 (same climatic and ecological region as Mount St. Helens and in drainage basins up to several hundred square kilometers in area) never carried

Table 1 Annual maximum suspended-sediment concentrations in volcanic water floods (VWFs) at U.S

River	Gaging station (station no.)	Contributing drainage area (km^2)	Types of disturbance	Peak sediment concentration (vol%)		
				WY 1981	WY 1982	WY 1983
North Fork Toutle	At Kid Valley (14,241,100)	736	VDA, PDC, lahar, TF	11.3	7.2	6.2
Toutle	At Tower Road (14,242,580)	1285	VDA, PDC, lahar, TF	No data	3.4	4.2
Toutle	At Highway 99 (14,242,690)	1326	VDA, PDC, lahar, TF	14.6†	9.5†	1.3
Cowlitz	At Castle Rock (14,243,000)	5796	VDA, PDC, lahar, TF	1.7	1.6	1.2
South Fork Toutle	At Camp 12 (14,241,490)	303	Dilute PDC, lahar	1.3	1.0	1.1
Muddy	Near Cougar (14,216,500)	218	Dilute PDC, lahar	5.7	3.1	1.8
Pine Creek	Near Cougar (14,216,900)	67	Lahar	4.0	2.6	0.9
Green	Above Beaver Creek (14,240,800)	334	Dilute PDC, tephra fall	0.9	0.2	0.1

† Higher than expected values, probably due to erosion of dredge spoils placed along bank of this river reach several kilometers upstream of gaging station in 1980

Geological Survey gaging stations on rivers draining Mount St. Helens (Washington, USA) in the three years following the May 18, 1980, eruption (Dinehart 1986). The Cowlitz River station is just a few kilometers downstream of the Toutle River—Cowlitz River confluence, and its values show the effect of dilution of Toutle River flood water with Cowlitz River water. Disturbance types include volcanic debris avalanche (VDA), pyroclastic density current (PDC), lahar, and tephra fall (TF) (Major et al. 2000). Dates preceded by WY indicate that these are "water years," which extend from October 1 of the previous year to September 30 of that year. Additional gaging-station data can be found at https://waterdata.usgs.gov/nwis/sw?

suspended-sediment concentrations exceeding 0.1 vol% solids (Waananen et al. 1971). A concentration of 11 vol% is close to the transition to hyperconcentrated flow for sediment mixtures with a small amount of clay, as is the case at Mount St. Helens (Pierson 2005b). Where eruptive products are finer grained and rainfall tends to be more intense, such as at Mount Pinatubo (Philippines) following its large explosive eruption in 1991, numerous rainfall-triggered floods transformed to lahars during the early posteruptive period (Pierson et al. 1996; Rodolfo et al. 1996; Umbal and Rodolfo 1996; Hayes et al. 2002).

Distance from source also affects sediment load. Rivers draining disturbed watersheds on volcanoes generally entrain much of their sediment from the abundant and readily erodible sediment sources in their steeper headwaters near the volcano, but erosion and deposition can occur irregularly along any channel, depending on localized conditions. Decreased sediment entrainment with distance from source plus increased dilution from inflowing tributary streams typically result in decreasing suspended-sediment concentration with distance from source (Table 2).

Annual sediment yields from basins severely disturbed by volcanic processes can range from 10^3 to 10^7 Mg/km^2, and yields are highest in the first year or two following volcanic disturbance (Major et al. 2000; Gran et al. 2011; Pierson et al. 2013) (Fig. 12 in Chap. "Scaling and Timing of Volcanic Wet Flows in Response to Eruptions"). This contrasts with yields of about 10^2 Mg/km^2 for undisturbed forested basins in the same volcanic terrain. Posteruption sediment

Table 2 Changing peak suspended-sediment concentration with distance from main sediment source for one volcanic water flood (VWF)—the February 19–20, 1982, rainfall-runoff flood on the Toutle River, downstream of Mount St. Helens (Washington, USA) (Dinehart 1986)

Gaging station	Distance from Toe of VDA deposit (sediment source) (km)	Peak sediment concentration (vol%)	Peak discharge (m^3/s)
North Fork Toutle River at Kid Valley (no. 14241100)	24	7.2	408
Toutle River at Tower Road (no. 14242580)	48	3.4	804
Toutle River at Highway 99 (no. 14242690)	56	9.5†	680
Cowlitz River at Castle Rock (no. 14243000)	62	1.6	1552

† Higher than expected value, probably due to erosion of dredge spoils placed along bank of this river reach several kilometers upstream of gaging station in 1980
The primary sediment source, the 1980 volcanic debris avalanche (VDA) deposit, fills the uppermost 25 km of the river valley. Additional gaging-station data can be found at https://waterdata.usgs.gov/nwis/sw?

yields vary with degree and nature of the volcanic disturbance to the basins (Lehre et al. 1983; Leavesley et al. 1983; Collins and Dunne 1986; Montgomery et al. 1999; Major et al. 2000; Hayes et al. 2002; Gran and Montgomery 2005; Major and Yamakoshi 2005; Swanson and Major 2005; Gran et al. 2006; Major and Mark 2006; Pierson et al. 2011). Basins experiencing numerous posteruption lahars in addition to VWFs (typical in the tropics where high-intensity rainfall is common) are likely to have extreme sediment yields.

Despite such high sediment loads and rates of sediment transport documented in rivers draining volcanically disturbed drainage basins, sediment inputs can be so great to rivers draining such disturbed landscapes that the rivers simply cannot transport all of it—there is insufficient sediment transport capacity. The sediment that cannot be transported and flushed out of the basins goes into storage within basins on channel beds, floodplains, and valley floors, resulting in *channel aggradation*—the net accumulation of sediment that causes the elevation of riverbeds to be raised (Pierson and Major 2014) (see Sect. 6).

5 Magnitude and Frequency

VWFs occur on a variety of scales, from discharges barely high enough to overtop the banks of a stream channel to catastrophic deluges, particularly glacier outburst floods triggered by subglacial eruptions and large lake-breakout floods (Björnsson 1975; Donnelly-Nolan and Nolan 1986; Waythomas et al. 1996; Guðmundsson et al. 1997; Manville et al. 1999, 2007; Macías et al. 2004; Hodgson and Nairn 2005; Kataoka et al. 2008; Kataoka 2011). Flood magnitude and frequency depend on the type of initiation mechanism, climatic factors, and the type, severity, and recency of volcanic disturbance. In the first few decades following the May 18, 1980, eruption at Mount St. Helens, rare rainfall-triggered 100-year floods transported up to 50% of the posteruptive annual sediment load in a single day, yet the greatest amount of suspended sediment transported from the disturbed areas around the volcano (60–95% of annual load) was mainly mobilized by moderate-sized river discharges—mean annual floods and 2-year floods (Major 2004). In the tropics, large sediment-transporting flood flows can occur on a nearly daily basis during the monsoon season (Martinez et al. 1996; Pierson et al. 1996; Rodolfo et al. 1996; Tuñgol and Regalado 1996; Lavigne and Suwa 2004). Small glacier outburst floods at Mount Rainier can occur at an average rate of several per decade (Richardson 1968). Large, eruption-triggered glacier-outburst floods in Iceland (*jökulhlaups*) occur at the rate of one to several per century (Björnsson 1975).

6 Sediment Deposition

In-channel deposition by water floods largely involves accretion of bed-load sediment on channel beds; suspended-load deposition occurs more in areas of current stagnation—on floodplains or in backwater areas. Low-concentration streamflow in channels with high rates of sediment input commonly transports and deposits bed load in mobile sheet-like accumulations of sand and gravel only a few particles thick—*bed-load sheets*, which are dragged along by shear stress imposed by the turbulent overpassing streamflow (Whiting et al. 1988; Gao 2008; Venditti et al. 2008). High-concentration VWFs with higher bed shear stress are thought to transport sand and gravel bed load in similar but thicker and more continuous mobile *traction carpets* of highly concentrated bed-load particles, depositing particles at the interface between the traction carpet and the channel bed (Lowe 1982; Hanes and Bowen 1985; Sohn 1997). Rapid suspension fallout of suspended particles from the turbulent, muddy flood water adds particles to the traction carpet from above, whereas accretion of particles to the bed removes particles from the traction carpet from below. Movement of particles within traction carpets involves progressively more dense packing of particles closer to the bed, with a downward transition from collisional to frictional particle–particle interactions. Particle motion mainly involves scattered saltating, colliding particles not in continuous contact with each other at the top of the traction carpet; sliding of thin sheets of contiguous particles in the middle; and dense particle–fluid flow at the base of the carpet, possibly inertially driven and decoupled from the overriding flood flow (Hanes and Bowen 1985; Todd 1989; Sohn 1997; Druitt 1998; Manville and White 2003).

In highly disturbed basins, VWF deposition occurs on broad braid plains often extending from valley side to valley side (Montgomery et al. 1999; Hayes et al. 2002; Gran and Montgomery 2005; Gran et al. 2006). During intereruptive periods, fluvial systems partly recovered from volcanic disturbance exhibit in-channel bed-load deposition combined with overbank suspended-load deposition on low terraces adjacent to single-thread or anabranching channels (e.g., Pierson et al. 2011). Despite sometimes copious amounts of silt and clay in the suspended and wash load, the turbulence that is characteristic of VWFs in moderate to steep channels ensures that the silt + clay fraction stays largely in suspension. Thus, in-channel VWF deposits are typically composed of sand and gravel with almost no fines (silt or clay) mixed in.

The dominant geomorphic response to rapid and excessive sedimentation by VWFs (or by a combination of VWFs and lahars) is channel aggradation (meters to tens of meters), bank erosion and channel widening (hundreds of meters), and channel avulsion—direct consequences of volcanically impacted river basins having more sediment input than output (Kuenzi et al. 1979; Vessell and Davies 1981; Meyer et al. 1986; Smith 1987a; Newhall and Punongbayan 1996; Hayes et al. 2002; Segschneider et al. 2002; Manville et al. 2005; Pierson et al. 2011) (Fig. 6). Maximum aggradation levels can be reached within a day or two, within in a few years, or possibly after a century or more (Kuenzi et al. 1979; Vessell and Davies 1981; Czuba et al. 2010, 2012; Pierson et al. 2011, 2013; Pierson and Major 2014) (Fig. 6). Furthermore, aggradation in main valleys can (1) dam tributary valleys, leading to back-flooding and the potential for later lake-breakout floods (e.g., (Kuenzi et al. 1979; Umbal and Rodolfo 1996) (Fig. 7), and (2) drastically reduce flood conveyance capacity in river channels. Thus, structures built on floodplains

Fig. 6 Hazardous impacts of channel aggradation caused in part by volcanic water floods (VWFs) in drainage basins affected by recent volcanic disturbance. **a** Bluff of sandy volcano-fluvial deposits about 20 m high resulting mainly from channel aggradation from multiple posteruptive VWFs that occurred over twelve years, about 70 km downstream of Mount Hood (Oregon, USA). A forested low terrace was adjacent to the preeruption river channel and then buried (paleoterrace surface at the level of the man's waist). One lahar deposit and multiple VWF sand deposits buried the forest between 1781 and 1793, which coincided with active growth and repeated collapse of a lava dome near the volcano summit (Pierson et al. 2011). USGS photo by T. Pierson. **b** Town of Chaitén (Chile) that was inundated by ~ 1 m of VWF overbank deposits in 2008 following 7 m of channel aggradation caused first by a lahar and then by a VWF over a period of 48 h (Pierson et al. 2013). Town was also bisected by a new channel of the Chaitén (Blanco) River that avulsed from the sediment-filled abandoned channel visible in the upper right to a rapidly incised and widened new channel that initially followed a street. Town is on a low alluvial fan about 10 km downstream of Chaitén volcano. Open-source photo (IMG 086, June 3, 2008) by SERNAGEOMIN (Chilean volcano-monitoring agency). **c** Depiction of rapid posteruptive river channel aggradation by lahars and VWFs on Sacobia River, about 30 km downstream of vent at Mount Pinatubo (Philippines) and about 10 km downstream of sediment source (thick valley-filling pyroclastic density–current deposits) (Scott 1996a, 1996b; Pierson et al. 1996)

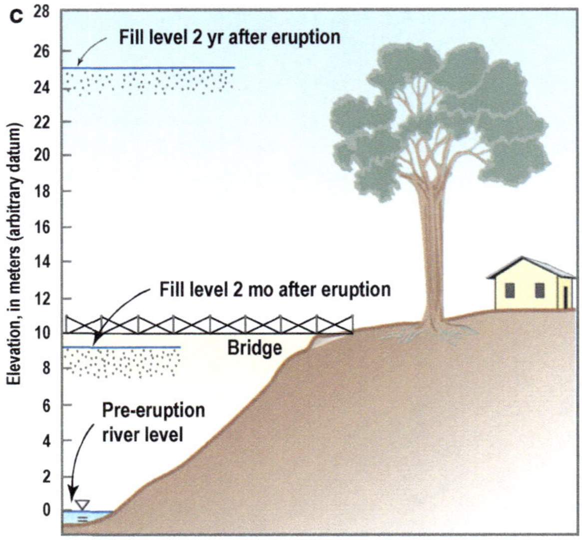

Fig. 6 (continued)

Fig. 7 Lake formed by backwater flooding of tributary to the Mirimla River on the northeast side of Mount Pinatubo (Philippines). Lahar and volcanic water flood (VWF) deposits aggraded the Mirimla channel (visible in distance), blocking the mouth of this tributary stream. USGS photo by T. Pierson

and river terraces in communities downstream of active volcanoes can be exposed to significant VWF hazards for many years following eruptions. Lingering excess sedimentation hazards in river valleys draining active volcanoes are not always recognized in volcano hazard assessments.

References

Alexander J, Barclay J, Susnik J, Loughlin SC, Herd RA, Darnell A, Crosweller S (2010) Sediment-charged flash floods on Montserrat: the influence of synchronous tephra fall and varying extent of vegetation damage. J Volcanol Geotherm Res 194:127–138

Björnsson H (1975) Subglacial water reservoirs, jökulhlaups, and volcanic eruptions. Jökull 25:1–14

Cohen H, Laronne JB (2005) High rates of sediment transport by flashfloods in the Southern Judean Desert, Israel. Hydrol Process 19:1687–1702. https://doi.org/10.1002/hyp.5630

Collins BD, Dunne T (1986) Erosion of tephra from the 1980 eruption of Mount St. Helens. Geol Soc Am Bull 97:896–905

Costa JE (1988) Floods from dam failures. In: Baker VR, Kochel RC, Patton PC (eds) Flood geomorphology. Wiley, New York, pp 439–463

Costa JE, Schuster RL (1988) The formation and failure of natural dams. Geol Soc Am Bull 100:1054–1068

Cronin SJ, Neall VE, Lecointre JA, Palmer AS (1996) Unusual snow slurry lahars from Ruapehu volcano, New Zealand, September 1995. Geology 24:1107–1110

Czuba JA, Czuba CR, Magirl CS, Voss FD (2010) Channel-conveyance capacity, channel change, and sediment transport in the lower Puyallup, white, and carbon rivers, western Washington, vol 2010-5240. US Geology Survey Science Investigation Report, 104 pp. http://pubs.er.usgs.gov/publication/sir20105240

Czuba JA, Magirl CS, Czuba CR, Curran CA, Johnson KH, Olsen TD, Kimball HK, Gish CC (2012) Geomorphic analysis of the river response to sedimentation downstream of Mount Rainier, Washington, vol 2012–1242. US Geology Survey Open-File Report, 134 pp

Dinehart RL (1986) Sediment data for streams near Mount St. Helens, Washington—volume 2. Water years 1981–83, vol 85–632. US Geology Survey Open-File Report, 438 pp

Dinehart RL (1998) Sediment transport at gaging stations near Mount St. Helens, Washington, 1980–1990: data collection and analysis, vol 1573. US Geology Survey Professional Paper, 105 pp

Donnelly-Nolan JM, Nolan KM (1986) Catastrophic flooding and eruption of ash-flow tuff at Medicine Lake volcano, California. Geology 14:875–878

Dorava JM, Meyer DF (1994) Hydrologic hazards in the lower Drift River basin associated with the 1989–1990 eruption of Redoubt Volcano, Alaska. In: Miller TP, Chouet BA (eds) The 1989–1990 eruptions of Redoubt Volcano, Alaska; J Volcanol Geotherm Res 62:387–407

Driedger CL, Fountain AG (1989) Glacier outburst floods at Mount Rainier, Washington State, U.S.A. Annu Glaciol 13:51–55

Druitt TH (1998) Pyroclastic density currents. Geol Soc Lond Spec Publ 145:145–182. https://doi.org/10.1144/GSL.SP.1996.145.01.08

Dzurisin D, Moran SC, Lisowski M, Schilling SP, Anderson KR, Werner C (2015) The 2004–2008 dome-building eruption at Mount St. Helens, Washington: epilogue. Bull Volcanol 77:89. https://doi.org/10.1007/s00445-015-0973-4

Fei X (1983) Grain composition and flow properties of heavily concentrated suspensions, in Proceedings of the second international symposium on river sedimentation, english summary. Water Resources and Electrical Power Press, Nanjing, China, pp 307–308

Gao P (2008) Transition between two bed-load transport regimes: saltation and sheet flow. J Hydraulic Eng 134(3). https://doi.org/10.1061/(ASCE)0733-9429(2008)134:3(340)

Graf WL (1977) The rate law in fluvial geomorphology. Am J Sci 277:178–191

Gran KB, Montgomery DR (2005) Spatial and temporal patterns in fluvial recovery following volcanic eruptions: channel response to basin-wide sediment loading at Mount Pinatubo, Philippines. Geol Soc Am Bull 117:195–211

Gran KB, Montgomery DR, Sutherland DG (2006) Channel bed evolution and sediment transport under declining sand inputs. Water Resour Res 42:W10407

Gran KB, Montgomery DR, Halbur JC (2011) Long-term elevated post-eruption sedimentation at Mount Pinatubo, Philippines. Geology 39(4):367–370. https://doi.org/10.1130/G31682.1

Guðmundsson MT (2015) Hazards from lahars and jökulhlaups. In: Sigurdsson H, Houghton B, McNutt SR, Rymer H, Stix J (eds) The encyclopedia of volcanoes, 2nd edn. Elsevier, Amsterdam, pp 971–984

Guðmundsson MT, Sigmundsson F, Björnsson H (1997) Ice-volcano interaction of the 1996 Gjálp subglacial eruption, Vatnajökull, Iceland. Nature 389:954–957

Hampton MA (1975) Competence of fine-grained debris flows. J Sediment Petrol 45:834–844

Hanes DM, Bowen AJ (1985) A granular-fluid model for steady intense bed-load transport. J Geophys Res 90:9149–9158

Hayes SK, Montgomery DR, Newhall CG (2002) Fluvial sediment transport and deposition following the 1991 eruption of Mount Pinatubo. Geomorphology 45:211–224

Hodgson KA, Nairn IA (2005) The ca. AD 1315 syn-eruption and AD 1904 post-eruption breakout floods from Lake Tarawera, Haroharo caldera, North Island, New Zealand. New Zeal J Geol Geophys 48:491–506
Hoffman DF, Gabet J (2007) Effects of sediment pulses on channel morphology in a gravel-bed river. Geol Soc Am Bull 119(1–2):116–125. https://doi.org/10.1130/B25982.1
Iroumé A, Paredes A, Garbarino M, Morresi Ramon D, Batalla J (2020) Post-eruption morphological evolution and vegetation dynamics of the Blanco River, southern Chile. J S Am Earth Sci 104:102809. https://doi.org/10.1016/j.jsames.2020.102809
Kataoka KS (2011) Geomorphic and sedimentary evidence of a gigantic outburst flood from Towada caldera after the 15 ka Towada-Hachinohe ignimbrite eruption, northeast Japan. Geomorphology 125:11–26
Kataoka KS, Urabe A, Manville V, Kajiyama A (2008) Breakout flood from an ignimbrite-dammed valley after the 5 ka Numazawako eruption, northeast Japan. Geol Soc Am Bull 120:1233–1247
Kilgour G, Manville V, Della Pasqua F, Graettinger A, Hodgson KA, Jolly GE (2010) The 25 September 2007 eruption of Mount Ruapehu, New Zealand: directed ballistics, surtseyan jets, and ice-slurry lahars. J Volcanol Geotherm Res 191:1–14. https://doi.org/10.1016/j.jvolgeores.2009.10.015
Kuenzi WD, Horst OH, McGehee RV (1979) Effect of volcanic activity on fluvial-deltaic sedimentation in a modern arc-trench gap, southwestern Guatemala. Geol Soc Am Bull 90:827–838
Laronne JB, Reid I (1993) Very high rates of bedload sediment transport by ephemeral desert rivers. Nature 366:148–150
Lavigne F, Suwa H (2004) Contrasts between debris flows, hyperconcentrated flows and stream flows at a channel of Mount Semeru, East Java, Indonesia. Geomorphology 61:41–58
Leavesley GH, Lusby GC, Lichty RW (1983) Infiltration and erosion characteristics of selected tephra deposits from the 1980 eruption of Mount St. Helens, Washington, USA. Hydrol Sci J 34:339–353
Lehre AK, Collins BD, Dunne T (1983) Post-eruption sediment budget for the North Fork Toutle River drainage, June 1980-June 1981. In: Okuda S, Netto A, Slaymaker O (eds) Extreme land forming events, vol 46. Zeitschrift Geomorphologie, Supplement Bd, pp 143–163
Lekach J, Schick AP (1982) Suspended sediment in desert floods in small catchments. Israel J Earth Sci 31:144–156
Lowe DR (1982) Sediment gravity flows; II, Depositional models with special reference to the deposits of high-density turbidity currents. J Sediment Res 52(1):279–297
Macías JL, Capra L, Scott KM, Espíndola JM, García-Palomo A, Costa JE (2004) The 26 May 1982 breakout flows derived from failure of a volcanic dam at El Chichón, Chiapas, Mexico. Geol Soc Am Bull 116:233–246. https://doi.org/10.1130/B25318.1
Madej MA, Ozaki V (2009) Persistence of effects of high sediment loading in a salmon-bearing river, northern California. In: James LA, Rathburn SL, Whittecar GR (eds) Management and restoration of fluvial systems with broad historical changes and human impacts, vol 451. Geology Society America Special Paper, pp 43–55
Major JJ, Mark LE (2006) Peak flow responses to landscape disturbances caused by the cataclysmic 1980 eruption of Mount St. Helens, Washington. Geol Soc Am Bull 118:938–958. https://doi.org/10.1130/B25914.1
Major JJ, Pierson TC, Dinehart RL, Costa JE (2000) Sediment yield following severe volcanic disturbance—a two-decade perspective from Mount St. Helens. Geology 28:819–822
Major JJ (2004) Posteruption suspended sediment transport at Mount St. Helens: decadal-scale relationships with landscape adjustments and river discharges. J Geophys Res Earth Surf 109:F01002. https://doi.org/10.1029/2002JF000010
Major JJ, Yamakoshi T (2005) Decadal-scale change of infiltration characteristics of a tephra-mantled hillslope at Mount St. Helens, Washington. Hydrol Process 19:3621–3630. https://doi.org/10.1002/hyp.5863
Major JJ, Janda RJ, Daag AS (1996) Watershed disturbance and lahars on the east side of Mount Pinatubo during the mid-June 1991 eruptions. In: Newhall CG, Punongbayan RS (eds) Fire and mud: eruptions and lahars of Mount Pinatubo, Philippines. Philippines Institute Volcanology Seismological, Quezon City, and University of Washington Press, Seattle, pp 895–919
Major JJ, Spicer KR, Mosbrucker AR (2021) Effective hydrological events in an evolving mid-latitude mountain river system following cataclysmic disturbance—a saga of multiple influences. Water Resour Res 57(2):e2019WR026851. https://doi.org/10.1029/2019WR026851
Manville VR (2002) Sedimentary and geomorphic responses to ignimbrite emplacement: readjustment of the Waikato River after the A.D. 181 Taupo eruption. New Zealand. J Geol 110:519–541
Manville V, Cronin SJ (2007) Breakout lahar from New Zealand's Crater Lake. Eos, Trans Am Geophys Union 88(43):441–442
Manville VR, White JDL (2003) Granular mass flows at the base of sediment-laden floods, and the roles of flow competence and flow capacity in the deposition of stratified boulder sands. Sediment Geol 155:157–173
Manville V, Wilson CJN (2004) The 26.5 ka Oruanui eruption, New Zealand: a review of the roles of volcanism and climate in the post-eruptive sedimentary response. New Zeal J Geol Geophys 47:525–547

Manville V, White JDL, Houghton BF, Wilson CJN (1999) Paleohydrology and sedimentology of a post-1.8 ka breakout flood from intracaldera Lake Taupo, North Island, New Zealand. Geol Soc Am Bull 111:1435–1447

Manville V, Newton EH, White JDL (2005) Fluvial responses to volcanism: resedimentation of the 1800a Taupo ignimbrite in the Rangitaiki River catchment, North Island, New Zealand. Geomorphology 65:49–70

Manville V, Hodgson KA, Nairn IA (2007) A review of break-out floods from volcanogenic lakes in New Zealand. New Zeal J Geol Geophys 50:131–150

Manville V, Segschneider B, Newton E, White JDL, Houghton BF, Wilson CJN (2009b) Environmental impact of the 1.8 ka Taupo eruption, New Zealand: landscape responses to a large-scale explosive rhyolite eruption. Sediment Geol 220:318–336

Maria A, Carey S, Sigurdsson H, Kincaid C, Helgadóttir G (2000) Source and dispersal of jökulhlaup sediments discharged to the sea following the 1996 Vatnajökull eruption. Geol Soc Am Bull 112:1507–1521

Marren PM, Russsel AJ, Rushmer EL (2009) Sedimentology of a sandur formed by multiple jökulhlaups, Kverkfjöll, Iceland. Sediment Geol 213:77–88

Martinez MML, Arboleda RA, Delos Reyes PJ, Gabinete E, Dolan MT (1996) Observations of 1992 lahars along the Sacobia-Bamban River system. In: Newhall CG, Punongbayan RS (eds) Fire and mud: eruptions and lahars of Mount Pinatubo, Philippines. Philippines Institute Volcanology Seismology, Quezon City, and University of Washington Press, Seattle, pp 1033–1044

McGimsey RG, Waythomas CF, Neal CA (1994) High stand and catastrophic draining of intracaldera Surprise Lake, Aniakchak volcano, Alaska. In: Geologic studies in Alaska by the U.S. Geological survey, 1993, vol 2017. US Geology Survey Bulletion, pp 59–71. https://doi.org/10.3133/70180217

Meyer DF, Janda RJ, Keller S (1986) Sedimentation downstream from the 18 May 1980 North Fork Toutle River debris avalanche deposit, Mount St. Helens, Washington. Eastern Washington University Press, Cheney, pp 68–86

Meyer GA, Wells SG (1997) Fire-related sedimentation events on alluvial fans, Yellowstone National Park, U.S.A. J Sediment Res 67:776–791

Miller DJ, Benda LE (2000) Effects of punctuated sediment supply on valley-floor landforms and sediment transport. Geol Soc Am Bull 112:1814–1824

Montgomery DR, Panfil MS, Hayes SK (1999) Channel-bed mobility response to extreme sediment loading at Mount Pinatubo. Geology 27:271–274

Newhall CG, Punongbayan RS (1996) Fire and mud: eruptions of Mount Pinatubo, Philippines. Philipp Inst Volcanol Seismol, Quezon City, and University of Washington Press, Seattle, 1126 pp

O'Connor JE, Hardison JH III, Costa JE (2001) Debris flows from failures of Neoglacial-age moraine dams in the three sisters and Mt. Jefferson Wilderness Areas, Oregon. US Geol Surv Prof Pap 1606:1–93

O'Connor JE, Clague JJ, Walder JS, Manville V, Beebee RA (2013) Outburst floods. Treatise Geomorphol 9:475–510

Phillips JD (2009) Changes, perturbations, and responses in geomorphic systems. Prog Physical Geogr 33:17–30

Pierson TC (2005) Hyperconcentrated flows—transitional process between water flow and debris flow (Chapter 8). In: Jakob M, Hungr O (eds) Debris-flow hazards and related phenomena. Praxis-Springer, Berlin, pp 159–202

Pierson TC, Major JJ (2014) Hydrogeomorphic effects of explosive volcanic eruptions on drainage basins. Annu Rev Earth Planet Sci 42:469–507. https://doi.org/10.1146/annurev-earth-060313-054913

Pierson TC, Scott KM (1985) Downstream dilution of a lahar: transition from debris flow to hyperconcentrated streamflow. Water Resour Res 21:1511–1524

Pierson TC, Janda RJ, Thouret JC, Borrero CA (1990) Perturbation and melting of snow and ice by the 13 November 1985 eruption of Nevado del Ruiz, Colombia, and consequent mobilization, flow, and deposition of lahars. J Volcanol Geotherm Res 41:17–66

Pierson TC, Pringle PT, Cameron KC (2011) Magnitude and timing of downstream channel aggradation in response to a dome-building eruption at Mount Hood, Oregon. Geol Soc Am Bull 123:3–20

Pierson TC, Major JJ, Amigo Á, Moreno H (2013) Acute sedimentation response to rainfall following the explosive phase of the 2008–2009 eruption of Chaitén volcano, Chile. Bull Volcanol 75:723. https://doi.org/10.1007/s00445-013-0723-4

Pierson TC (1999) Transformation of water flood to debris flow following the eruption-triggered transient-lake breakout from the crater on March 19, 1982. In: Pierson TC (ed) Hydrologic consequences of hot-rock/snowpack interactions at Mount St. Helens volcano, Washington 1982–84, vol 1586. US Geology Survey Professional Paper, pp 19–36

Pierson TC, Daag AS, Delos Reyes PJ, Regalado MTM, Solidum RU, Tubianosa BS (1996) Flow and deposition of posteruption hot lahars on the east side of Mount Pinatubo, July–October 1991. In: Newhall CG, Punongbayan RS (eds), Fire and mud: eruptions and lahars of Mount Pinatubo, Philippines: Philippines Institute Volcanology Seismological, Quezon City, and University of Washington Press, Seattle, pp 921–950

Procter J, Cronin SJ, Fuller IC, Lube G, Manville V (2010) Quantifying the geomorphic impacts of a lake-breakout lahar, Mount Ruapehu, New Zealand. Geology 38:67–70

Reid I, Frostick LE (1987) Flow dynamics and suspended sediment properties in arid zone flash floods. Hydrol Process 1:239–253

Reid I, Laronne JB, Powell DM (1998) Flash-flood and bedload dynamics of desert gravel-bed streams. Hydrol Process 12:543–557

Reneau SL, Katzman D, Kuyumjian GA, Lavine A, Malmon DV (2007) Sediment delivery after a wildfire. Geology 35:151–154

Richardson D (1968) Glacier outburst floods in the Pacific Northwest, vol 600-D. US Geology Survey Professional Paper, pp 79–86

Rodolfo KS, Umbal JV, Alonso RA, Remotigue CT, Paladio-Melosantos ML, Salvador JHG, Evangelista D, Miller Y (1996) Two years of lahars on the western flank of Mount Pinatubo: initiation, flow processes, deposits, and attendant geomorphic and hydraulic changes. In: Newhall CG, Punongbayan RS (eds), Fire and mud: eruptions and lahars of Mount Pinatubo, Philippines. Philippines Institute Volcanology Seismological, Quezon City, and University of Washington Press, Seattle, pp 989–1013

Russell AJ, Knudsen Ó (1999) Controls on the sedimentology of the November 1996 jökulhlaup deposits, Skeiðarársandur, Iceland. In: Smith ND, Rogers J (eds) Fluvial sedimentology VI. International association sedimentology, pp 315–329

Schumm SA, Rea DK (1995) Sediment yield from disturbed earth systems. Geology 23:391–394

Schuster RL (2000) Outburst debris-flows from failure of natural dams. In: Wieczorek GF, Naeser ND (eds) Debris-flow hazards mitigation: mechanics, prediction, and assessment. Balkema, Rotterdam, pp 29–42

Scott KM (1988b) Origin, behavior, and sedimentology of prehistoric catastrophic lahars at Mount St. Helens, Washington. In: Clifton HE (ed) Sedimentologic consequences of convulsive geologic events, vol 229. Geology Society America Special Paper, pp 23–36. https://doi.org/10.1130/SPE229-p23

Scott KM (1989) Magnitude and frequency of lahars and lahar-runout flows in the Toutle-Cowlitz River system, vol 1447-B. US Geology Survey Professional Paper, 33 pp

Scott KM, Vallance JW, Pringle PP (1995) Sedimentology, behavior, and hazards of debris flows at Mount Rainier, Washington, vol 1547. US Geology Survey Professional Paper, 56 pp

Scott WE, Hoblitt RP, Torres RC, Self S, Martinez MML, Nillos T (1996a) Pyroclastic flows of the June 15, 1991, climactic eruption of Mount Pinatubo. In: Newhall CG, Punongbayan RS (eds) Fire and mud: eruptions and lahars of Mount Pinatubo, Philippines. Philippines Institute Volcanology Seismology, Quezon City, and University of Washington Press, Seattle, pp 545–57

Scott KM, Janda RJ, de la Cruz EG, Gabinete E, Eto I, Isada M, Sexon M, Hadley KC (1996b) Channel and sedimentation responses to large volumes of 1991 volcanic deposits on the east flank of Mount Pinatubo. In: Newhall CG, Punongbayan RS (eds) Fire and mud: eruptions and lahars of Mount Pinatubo, Philippines. Philippines Institute Volcanology Seismology, Quezon City, and University of Washington Press, Seattle, pp 971–988

Segschneider B, Landis CA, Manville V, White JDL, Wilson CJN (2002) Environmental response to a large, explosive rhyolite eruption: sedimentology of post-1.8 ka pumice-rich Taupo volcaniclastics in the Hawke's Bay region, New Zealand. Sediment Geol 150:275–299

Smith GA (1987a) The influence of explosive volcanism on fluvial sedimentation: the Deschutes formation (Neogene) in central Oregon. J Sediment Petrol 57:613–629

Smith RCM (1991b) Post-eruption sedimentation on the margin of a caldera lake, Taupo Volcanic Centre, New Zealand. Sediment Geol 74:89–138

Smith GA (1991a) Facies sequences and geometries in continental volcaniclastic sediments. In: Fisher RV, Smith GA (eds) Sedimentation in volcanic settings. SEPM (Soc Sediment Geol) Spec Publ 45:109–121

Sohn YK (1997) On traction-carpet sedimentation. J Sediment Res 67:502–509

Suwa H, Yamakoshi T (1999) Sediment discharge by storm runoff at volcanic torrents affected by eruption. Zeitschrift Geomorphologie, Suppl Bd 114:63–88

Swanson FJ, Major JJ (2005) Physical events, environments, and geological—ecological interactions at Mount St. Helens: March 1980–2004. In: Dale VH, Swanson FJ, Crisafulli CM (eds) Ecological responses to the 1980 eruption of Mount St. Helens. Springer, New York, NY. https://doi.org/10.1007/0-387-28150-9_3

Thouret J-C (1990) Effects of the November 1985 eruption on the snowpack and ice cap of Nevado del Ruiz volcano, Colombia. J Volcanol Geotherm Res 41:177–201

Todd SP (1989) Stream-driven, high-density gravelly traction carpets: possible deposits in the Trabeg Conglomerate formation, SW Ireland and some theoretical considerations of their origin. Sedimentology 36(4):513–530. https://doi.org/10.1111/j.1365-3091.1989.tb02083.x

Trabant DC, Waitt RB, Major JJ (1994) Disruption of Drift glacier and origin of floods during the 1989–1990 eruptions of Redoubt Volcano, Alaska. J Volcanol Geotherm Res 62:369–385

Tuñgol NM, Regalado MTM (1996) Rainfall, acoustic flow monitor records, and observed lahars of the Sacobia River in 1992. In: Newhall CG, Punongbayan RS (eds) Fire and mud: eruptions and lahars of Mount Pinatubo, Philippines. Philippines Institute Volcanology Seismology, Quezon City, and University of Washington Press, Seattle, pp 1023–1032

Umbal JV, Rodolfo KS (1996) The 1991 lahars of southwestern Mount Pinatubo and evolution of the lahar-dammed Mapanuepe Lake. In: Newhall CG,

Punongbayan RS (eds) Fire and mud: eruptions and lahars of Mount Pinatubo, Philippines. Philippines Institute Volcanology Seismology, Quezon City, and University of Washington Press, Seattle, pp 951–970

Venditti JG, Nelson PA, Dietrich WE (2008) The domain of bedload sheets. Marine and river dune dynamics, Leeds, United Kingdom

Vessell RD, Davies KK (1981) Nonmarine sedimentation in an active fore arc basin. In: Ethridge FG, Flores RM (eds) Nonmarine depositional environments: models for exploration, vol 31. Society economic paleontologie mineral special publication, pp 34–45

Waananen AO, Harris DD, Williams RC (1971) Floods of December 1964 and January 1965 in the far western states, Part 1. Description, vol 1866-A. US Geology Survey Water-Supply Paper, 265 pp

Waitt RB, Pierson TC, MacLeod NS, Janda RJ, Voight B, Holcomb RT (1983) Eruption-triggered avalanche, flood, and lahar at Mount St. Helens—effects of winter snowpack. Science 221:1394–1397

Walder JS (2000a) Pyroclast/snow interactions and thermally driven slurry formation. Part 1: theory for monodisperse grain beds. Bull Volcanol 62:105–118

Walder JS (2000b) Pyroclast/snow interactions and thermally driven slurry formation. Part 2: experiments and theoretical extension to polydisperse tephra. Bull Volcanol 62:119–129

Walder JS, Driedger CL (1994a) Geomorphic change caused by outburst floods and debris flows at Mount Rainier, Washington, with emphasis on Tahoma Creek valley. US Geol Surv Water-Resour Invest Rep 93–4093:1–93

Walder JS, Driedger CL (1994b) Rapid geomorphic change caused by glacial outburst floods and debris flows along Tahoma Creek, Mount Rainier, Washington, USA. Arctic Alpine Res 26:319–327

Walder JS, O'Connor JE (1997) Methods for predicting peak discharge of floods caused by failure of natural and constructed earthen dams. Water Resour Res 33:2337–2348

Walder JS, LaHusen RG, Vallance JW, Schilling SP (2007) Emplacement of a silicic lava dome through a crater glacier: Mount St. Helens, 2004–06. Annu Glaciol 45:14–20

Walder JS, Driedger CL (1995) Frequent outburst floods from South Tahoma Glacier, Mount Rainier, U.S.A.; relation to debris flows, meteorological origin and implications for subglacial hydrology. J Glaciol 41(137):1–9. https://doi.org/10.3189/S0022143000017718

Walder JS, Schilling SP, Vallance JW, LaHusen RG (2008) Effects of lava-dome growth on the Crater Glacier of Mount St. Helens, Washington. In: Sherrod DR, Scott WE, Staufer PH (eds) A volcano rekindled: the renewed eruption of Mount St. Helens, vol 1750. US Geology Survey Professional Paper, pp 257–276

Wan Z, Wang Z (1994) Hyperconcentrated flow. Balkema, Rotterdam, p 290

Waythomas CF, Walder JS, McGimsey RG, Neal CA (1996) A catastrophic flood caused by drainage of a caldera lake at Aniakchak Volcano, Alaska, and implications for volcanic hazards assessment. Geol Soc Am Bull 108(7):861–871

Waythomas CF, Pierson TC, Major JJ, Scott WE (2013) Voluminous ice-rich lahars generated during the 2009 eruption of Redoubt Volcano, Alaska. J Volcanol Geotherm Res 259:389–413

White JDL, Houghton BF, Hodgson KA, Wilson CJN (1997) Delayed sedimentary response to the A.D. 1886 eruption of Tarawera, New Zealand. Geology 25(5):459–452

Whiting PJ, Dietrich WE, Leopold LB, Drake TG, Shreve RL (1988) Bedload sheets in heterogeneous sediment. Geology 16(2):105–108. https://doi.org/10.1130/0091-7613(1988)016<0105:BSIHS>2.3.CO;2

Volcanic Water-Flood Deposits

Abstract

Deposits of volcanic water floods (VWFs) commonly comprise a significant proportion of the volcanic deposits on the lower edifice flanks and in downstream river valleys at many volcanoes. Following explosive eruptions, the availability of abundant fresh and loose volcaniclastic sediment typically leads to extreme sediment loading in rivers and thick sequences of deposits emplaced by multiple high-concentration VWFs in river valleys downstream of source volcanoes. VWF deposits are commonly interbedded with deposits of lahars, volcanic debris avalanches (VDAs), pyroclastic density currents (PDCs), and tephra fall (TF) on volcano flanks and debris aprons. The proportion of VWF deposits (relative to deposits of other types of flows) increases with distance away from the source volcano. Deposits of individual VWFs are commonly 1–5 m thick, whereas sequences dominated by VWF deposits from multiple floods can be up to hundreds of meters thick. Key features of VWF deposits include distinct plane-parallel bedding (commonly with other preserved bed forms), mainly sandy beds with gravel lenses, loose consistencies (friable/soft sand), minimal interstitial fines preserved in sandy units, mixing of juvenile particles with remobilized older bed material, and moderate sorting.

1 Overview

Deposits of volcanic water floods (VWFs) are a category of fluvial deposits—*volcano-fluvial deposits*—and they comprise a significant proportion of the secondary volcaniclastic deposits at many volcanoes. Following explosive eruptions, the availability of abundant volcaniclastic material in humid climatic environments typically leads to extreme sediment loading in rivers and thick sequences of deposits emplaced by multiple high-concentration water floods in river valleys downstream of source volcanoes. Such excessive amounts of sediment rapidly added to river channels produce a cascade of geomorphic responses—rapid downstream channel aggradation (i.e., infilling with sediment and rise of channel-bed elevation) and consequent changes in river channel pattern, form, and position within valleys (Smith 1991a; Montgomery et al. 1999; Hayes et al. 2002; Gran and Montgomery 2005; Manville et al. 2005; Pierson et al. 2011; see Chap. "Volcanic Water Floods—Process Overview").

Rivers overwhelmed with sediment following volcanic disturbance typically develop braided channel forms, which tend to be particularly well developed in gravel-bedded channels (Kuenzi et al. 1979; Germanoski and Schumm 1993; Collinson 1996; Segschneider et al. 2002; Gran and Montgomery 2005; Pierson and Major 2014). Where actively aggrading rivers

T. C. Pierson et al., *Deposits of Volcanic Wet Flows*, Advances in Volcanology,
https://doi.org/10.1007/978-3-031-66574-5_10

are confined by valley sides, entire valley floors may be submerged during floods, and such rivers do not have floodplains in the classic sense. Instead, they have braid plains covered by active and abandoned channels and bars. Total aggradation in confined channels during a single eruptive cycle can be at least 20 to 30 m (e.g., Pierson et al. 2011), with short-term aggradation rates as high as 2–5 m/day (Scott et al. 1996; Pierson et al. 2013) and up to about 20 m/year (Rodolfo et al. 1996). Over time, multiple cycles of aggradation and degradation result in the formation of alluvial fill terraces at various elevations above the active channels (Germanoski and Schumm 1993; Pierson et al. 2011) (Fig. 1). Rivers unconfined by valley walls (on ring plains and alluvial fans/plains) typically display less total aggradation and more lateral shifting of channels with sediment deposition over broad areas (e.g., Kuenzi et al. 1979; Smith 1987a, b; Pierson et al. 1996; Donoghue and Neall 2001; Zernack et al. 2009, 2011). Resulting deposit geometries, lithofacies, and areal extents are of types and scales seldom seen in non-volcanic flood deposits (Smith 1987a, 1991a, b).

2 Landscape-Scale Characteristics of Deposits

VWF deposits are commonly interbedded with deposits of lahars (volcanic debris flows [VDFs] and volcanic hyperconcentrated flows [VHFs]), volcanic debris avalanches (VDAs), pyroclastic density currents (PDCs), and tephra fall (TF) in proximal and medial zones. The proportion of VWF deposits (relative to PDC and lahar deposits) increases with distance away from the source volcano (see Sect. 2.1). VDF deposits seldom extend as far from source as VWF

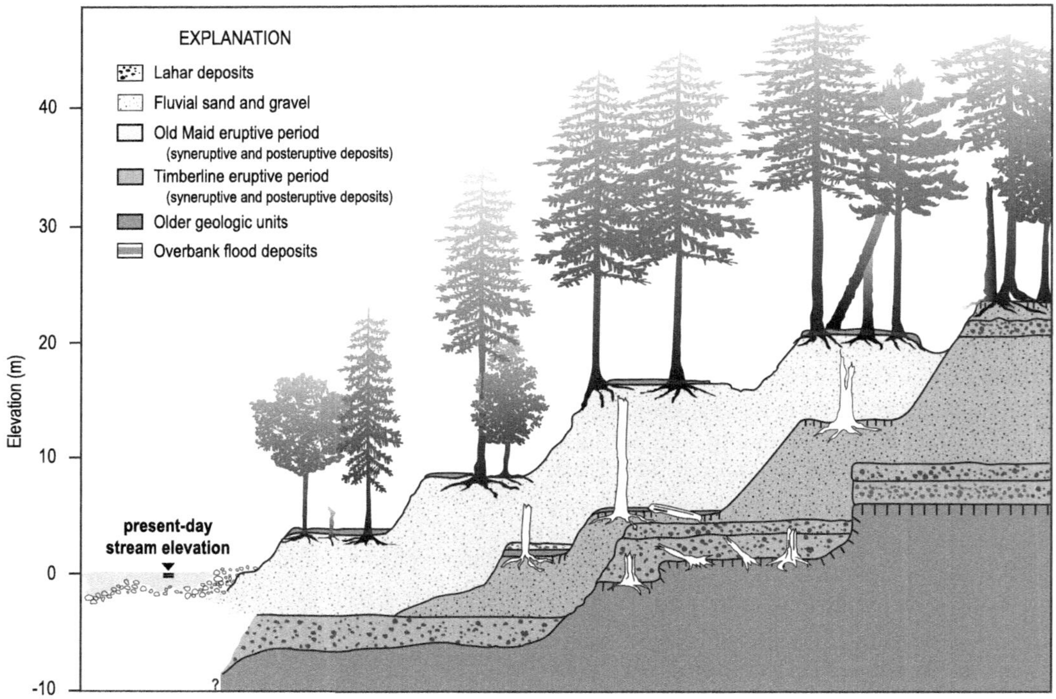

Fig. 1 Remnants of eruption-induced aggradational fill deposits, dominantly volcanic water flood (VWF) deposits, preserved as exposed and buried alluvial terraces on lower Sandy River, about 70 km downstream of Mount Hood (Oregon, USA) (Pierson et al. 2009). At least two posteruptive aggradation–degradation cycles are preserved in this sequence: those of the Old Maid eruptive period (1781–1801?) and the Timberline eruptive period (~300–600). White tree trunks shown are buried in overlying deposits and locally exposed by bank erosion

deposits, due to the potential for clay-poor flow mixtures to be diluted at higher-order stream confluences and transform to VWFs. Clay-rich VDFs, on the other hand, do not readily mix with stream water and may travel long distances (see Sect. 4.2 in Chap. "Lahars—Process Overview").

Deposits of individual VWFs are commonly 1–5 m thick (Nakayama and Yoshikawa 1997; Kataoka and Nakajo 2002; Manville et al. 2005, 2009; Umazano et al. 2008; Pierson et al. 2013). Sequences dominated by VWF deposits from multiple floods, sometimes from multiple eruption cycles, can be up to hundreds of meters thick (Smith 1987a, b), although it can be difficult to discern which VWF deposits were emplaced by individual floods. This is because composition of source material sediment and sedimentary textures can be similar from one flood to the next within a single eruption cycle, and evidence of non-deposition (e.g., buried soils or layers of organic debris) may be absent due to insufficient time for soil formation, bioturbation, or erosion on floodplains (Smith 1991b; Pierson et al. 2011).

2.1 Deposit Architecture and Morphology

2.1.1 Sediment-Body Shape

VWF sediment bodies in actively aggrading systems form downstream-tapering wedges of clastic sediment. These can be confined within incised river valleys, forming sediment fills from one valley side to the other (Fig. 2a), or sediment can spread broadly and unconfined across alluvial fans or plains (Fig. 2b). The downstream extent of VWF sediment bodies can exceed 100 km from the source volcano (Kuenzi et al. 1979; Vessell and Davies 1981; Smith 1987a, b, 1991a; b; Newhall and Punongbayan 1996). Thick sequences of aggraded sediment resulting from excessive sediment input to rivers (e.g., Fig. 1) are commonly preserved as alluvial terraces following channel incision of valley sediment fills. Such sediment fills represent in-channel storage of excess sediment that is emplaced over time in response to drainage network geometry and limits to sediment-transport capacity (Miller and Benda 2000; Lisle 2008; Pierson et al. 2011; Pierson and Major 2014; Gran and Czuba 2017).

VWFs flowing from highly confined point sources, such as the exit point from a narrow canyon, produce alluvial fan deposits with common characteristics: (1) deposit thickness and particle-size decay approximately exponentially in the direction of flow; (2) minimal deposition occurs at the fan apex where flow velocities are high; and (3) coarsest bed-load deposition occurs closest to the apex and suspended-load deposition occurs more distally (van Wagoner et al. 2003).

2.1.2 Large-Scale Sedimentary Structures

Bars are positive-relief, three-dimensional accumulations of gravel and sand that develop by vertical accretion within a river channel or by lateral accretion along the margin of a channel in flows transporting significant volumes of bed load (Miall 1996). Bars formed by VWFs are most commonly longitudinal *midchannel bars*, which are a prominent feature in braided rivers carrying exceptional volumes of coarse bed load during syneruptive and posteruptive periods (Fig. 3a–c). Dimensions are highly variable, but large lake-outbreak floods can build longitudinal gravel bars up to 10 m high, 0.9 km wide, and 2.5 km long (Hodgson and Nairn 2005). Midchannel bars tend to migrate during high flows, gradually moving downstream due to erosion of coarse sand and gravel (generally up to cobble size) from their upstream ends and redeposition of that material at their downstream ends. Cross-bedding that reflects this mode of bed-form migration is commonly preserved in bar sediments. Coarser gravel clasts tend to be concentrated on bar surfaces, but in many cases, these coarse-grained surfaces are only one or two particles thick over finer and more poorly sorted bed material. A characteristic of fluvial braided channels is the juxtaposition of coarse

a

b

◀ **Fig. 2** Sedimentary bodies produced by volcanic water flood (VWF) sedimentation in response to volcanic disturbance: **a** Valley-confined VWF deposition in the aggrading Rayas River channel (January 2010 photo) about 25 km downstream of Chaitén volcano (Chile), which is visible in upper right of photo. It erupted explosively in May 2008. **b** Aerial view of unconfined VWF deposition overflowing from the sediment-filled Bamban River channel (long arrow marking original channel position) onto a broad alluvial braid plain about 40 km downstream of Mount Pinatubo (Philippines) in August 1991. The volcano erupted explosively in June 1991, and deposit thickness on this alluvial plain was 1–2 m at this time. Largest island of dark vegetation in middle distance (short arrow) shelters a still-occupied small village temporarily protected from inundation by sandbags. USGS photos by T. Pierson

Fig. 3 Midchannel longitudinal bars. **a** Sandy longitudinal midchannel bar in the Chaitén (Blanco) River downstream of Chaitén volcano (Chile). View is looking upstream. **b** Gravelly longitudinal midchannel bars in a narrow aggrading reach of the Rayas River 5 km downstream of Chaitén volcano. View is looking downstream. **c** Large and small midchannel longitudinal bars in channel of the Rayas River, 19 km downstream of the volcano. View is downstream toward river mouth. USGS photos by T. Pierson, 2010

gravelly longitudinal bar surfaces with shallow channels—braid channels, wash-over channels, and chute cutoffs—bedded with deposits that are typically well-sorted medium to coarse sand. Such abrupt changes in particle size and sorting, visible in outcrops as scour-fill lenses of sand adjacent to truncated lenses of coarse gravel, are characteristic of VWF deposits in volcanic settings. They are uncommon in dilute lahar deposits and nonexistent in concentrated lahar and VDA deposits.

Berms, flat-topped accumulations of typically sandy sediment attached to the lateral margins of a channel in or near alcoves and reentrants along the bank, are also common sedimentary structures formed by VWFs (Fig. 4). They develop primarily along channel margins or in backwater eddies from fallout of suspended sediment, where velocity and turbulence are diminished. Tops of berms are commonly close to the level of maximum flood stage (the flood's *high-water mark*).

Fig. 4 Flood berms formed by volcanic water floods with high suspended-sediment loads. **a** Flood berm composed of cross-bedded medium sand, deposited in November 2006 in a back-water eddy along the upper Sandy River, about 15 km downstream of the summit of Mount Hood (Oregon, USA). Weakly rooted trees were tilted upstream by the reverse flow in the eddy. Main flow direction in channel was left to right. Deposit is about 1.5 m thick. **b** Flood berm of laminated medium to fine sand in a shallow alcove along bank of lower Sandy River, about 80 km downstream of Mount Hood, deposited by the same flood. Flow direction was toward camera, and the boundary between sand and green grass in the upper left marks the approximate level of maximum flood stage. Deposit thickness varied from centimeters to about one meter. USGS photos by T. Pierson

Outsized clasts—clasts significantly larger than median bed material—occur on and in VWF deposits emplaced by high-discharge flows. They occur either singly or as small groupings or clusters of such clasts. Outsized clasts are most common in extreme-discharge VWF deposits such as lake-breakout floods, which are capable of moving boulders meters in diameter (Waythomas et al. 1996; Manville et al. 1999; Hodgson and Nairn 2005).

2.1.3 Deposit Margins

VWFs do not form flow levees, as lahars commonly do. Instead, lateral margins of flood deposits are typically marked by strand lines of flotsam (logs, twigs, leaves, pumice clasts etc.), features that are typically not preserved in older deposits because they are readily remobilized by later floods. Deposits emplaced near lateral margins and on overtopped terraces are typically fine grained (very fine sand and silt) and thinly laminated.

2.1.4 Bounding Surfaces

Braided channel systems develop rapidly in actively aggrading river channels that are overwhelmed with excessive volumes of bedload sediment (Smith and Smith 1984; Schumm 1985), and the upper surfaces of VWF deposits typically retain the morphologic characteristics of braided channels (e.g., Fig. 2) (Kuenzi et al. 1979, Vessell and Davies 1981; Smith 1987a, b; Montgomery et al. 1999; Hayes et al. 2002; Gran and Montgomery 2005). Braided-channel surface features (Fig. 5a and c), including (1) anastomosing networks of numerous, shallow (0.2–1 m), low-sinuosity channels, (2) scour holes in finer sediment around outsized clasts on deposit surfaces, (3) dune bed forms in areas with sandy facies, and (4) abrupt juxtapositions of sediments of different particle size, are diagnostic of VWF deposits and not found on sediment body surfaces emplaced by lahars. Braided channel systems commonly extend over broad areas (braid plains) when not confined by valley walls, and braided channels downstream of volcanoes typically continue from proximal to distal depositional environments, where braiding may persist for decades following eruptive activity due to sediment reworking from farther upstream (Umazano et al. 2008). Because VWFs more commonly deposit sediment than erode it,

Fig. 5 Surface features of volcanic water flood (VWF) deposits: **a** Surface of the North Fork Toutle River floodplain about 6 km downstream of the toe of the 1980 volcanic debris-avalanche deposit (primary sediment source) and about 33 km downstream of Mount St. Helens (Washington, USA) immediately after a VWF in 1999. Flow direction was toward camera. Coin for scale on sand surface is 23 mm in diameter. **b** Abrupt change in mean particle size on a bar surface upstream of a logjam in the Sandy River about 70 km downstream of Mount Hood (Oregon, USA). Volcanic disturbance to this watershed occurred in 1781–1801(?). View is across the channel; flow was right to left. **c** Surface of VWF deposits in the active floodplain of the North Fork Toutle River in 2007. Terrace on far side of channel is also composed mainly of VWF deposits. Entire channel is incised into the 1980 Mount St. Helens volcanic debris-avalanche deposit (its hummocks stand behind the trees on the terrace). Note the two people next to the channel for scale. USGS photos by T. Pierson

the lower surfaces of sediment bodies usually reflect pre-flood topography.

Armoring of bar surfaces—the development of a surface layer of gravel (Fig. 5b on right side of photo) occurs on VWF channel bed surfaces, as it does on channel beds in non-volcanic terrain, in response to the waning sediment supply. Degree of armoring in channels recovering from volcanic disturbance reflects time since initial disturbance, and it is one measure of geomorphic recovery of a drainage basin (see Sect. 2.2 in Chap. "Scaling and Timing of Volcanic Wet Flows in Response to Eruptions"). As sediment loading of rivers wanes over time following cessation of volcanic activity, channels re-incise their aggradational sediment deposits and re-establish features characteristic of undisturbed fluvial systems: cobbly to bouldery armored beds and more simplified channel patterns (single-thread and anabranching channels) (Gran and Montgomery 2005; Gran et al. 2006).

VWF deposits emplaced by glacial outburst floods may, in some cases, contain large blocks of ice. When such blocks melt, shallow depressions may form on the surface of the deposit where an ice block used to be or above where it used to be if the block had been buried. Such depressions can resemble glacial kettles or secondary explosion pits.

2.2 Lithofacies

VWF deposits exhibit a large variety of lithofacies (see Sect. 2.2 in Chap. "Deposit Identification and Interpretation" for definitions) that vary over short distances both horizontally and vertically—an important characteristic of these deposits. Key aspects of VWF lithofacies include: (1) distinct stratification; (2) a variety of bed forms—plane-parallel bedding, low-angle cross-bedding, and high-angle cross-bedding; and (3) moderate to good sorting (Davies et al. 1978; Mathisen and Vondra 1983; Smith 1987a). Sediment particle sizes and textures can vary tremendously; these are dependent on localized flow energy at the time of deposition. Three generalized depositional facies (lithofacies associations) have been defined for VWF deposits: (1) *sheetflood facies*, (2) *channel facies*, and (3) *overbank facies* (Smith 1987a) (Fig. 6).

Sheetflood-facies deposits in volcanic settings are composed mainly of plane-parallel-bedded strata of poorly sorted, medium- to coarse-grained sand or gravelly sand, commonly with beds, lenses, or stringers of sandy pebble gravel (Fig. 6a). They occur both in broad shallow channels and as spillover onto unchannelized surfaces and are commonly found in sequences 10 m or more thick (Smith 1987a, 1991a; b; Manville et al. 2009; Pierson et al. 2011). Individual beds are commonly 5–25 cm in thickness, typically separated by low-angle erosion surfaces showing < 10 cm relief (Smith 1987a). Low-angle (antidune) cross-bedding and scour-fill cross-bedding may alternate with planar strata, but planar-tabular (dune form) cross-bedding is rare (Smith 1987a). Although this facies association may extend over 100 m laterally and longitudinally, individual beds rarely extend more than 1–2 m laterally. In non-volcanic terrain, recognition of sheetflood facies occurrence has been largely restricted to steep (2–25°) debris/alluvial fans at the mouths of canyons, deposited by shallow, unconfined, high discharge water flows (flash floods) that spread out as they flow downslope (e.g., Blair and MacPherson 1994).

The sheetflood facies association is characteristic of extreme rates of sediment transport (bed load and suspended load) by high-energy, shallow, upper-regime water flow in hydraulically smooth sand-bedded channels during syneruptive and posteruptive periods (Smith 1987a, 1991a, b; Gran and Montgomery 2005; Manville et al. 2009). Rapidly shifting, shallow, braided channel complexes are the depositional setting (Smith 1987a, 1991a, b; Gran and Montgomery 2005). We observed the requisite conditions for sheetflood facies deposition in a braided channel as far as 35 km downstream of Mount St. Helens following the 1980 eruption. Fluvial deposits are only a minor component of volcaniclastic sequences on debris aprons and ring plains in proximal volcanic settings, but where present they are typically sheetflood facies deposits.

Fig. 6 Examples of three major volcanic water flood (VWF) lithofacies: **a** Sheetflood facies—plane-parallel-bedded coarse sand with gravel lenses in broad channel about 15 km downstream of Michinmahuida volcano (Chile). Pattern reflects rapidly shifting shallow channels. Pencil is 15 cm long. **b** Proximal channel facies—plane-parallel-bedded fine to medium gravel with imbrication of gravel-size clasts—derived from locally reworked block-and-ash flow deposits in Chaitén (Blanco) River, 7 km downstream of Chaitén volcano (Chile); flow was from left to right. Trenching shovel is 58 cm long. **c** Distal channel facies—cross-bedded medium to coarse sand and fine gravel—in Sandy River about 70 km downstream of Mount Hood (Oregon, USA). Flow was from left to right. Pencil is about 15 cm long. **d** Overbank facies deposit (same location as photo **c** but lower in stratigraphic section); it is composed of laminated fine sand and silt. Deposit overlies a gray (A horizon) and orange (B horizon) buried soil with sediment-filled macropores on surface of buried paleoterrace. Note buff-colored, slightly deformed, silty cap layers that are separating overbank deposits of individual floods. Flow was approximately left to right. Width of curved scrapper-handle shaft is 5 mm. USGS photos by T. Pierson

VWF channel facies in rivers draining stratovolcanoes with coarse and poorly sorted sediment sources vary in their characteristics depending on channel width and gradient, which are functions of distance downstream of the source volcano (Smith 1987a). In steep proximal channels deposits may be dominated by massive or crudely bedded, poorly sorted gravels up to boulder size, where imbrication of some clasts is usually evident and long axes of clasts may be transverse to flow direction, owing to tractional transport on the bed (Smith 1987a) (Fig. 6b). Farther downstream, channel-facies deposits become enriched in greater amounts of medium to coarse sand with progressively fewer gravel interbeds and lenses (Fig. 6c). Bedding structures include flat planar bedding (subhorizontal), high-angle planar-tabular cross-bedding (formed by dune migration), high-angle trough cross-bedding, and low-angle antidune and scour-fill cross-bedding. Gravel in layers and lenses is typically imbricated. Juxtaposed beds with contrasting sedimentary textures may result from sediment transport at different times and discharges (Smith 1987a). In contrast to the facies described above, deposits emplaced by large lake-breakout or glacier outburst floods can exhibit additional types and spatial distributions of channel facies (e.g., Maria et al. 2000; Manville et al. 1999, 2005; Kataoka et al. 2008). Exceptionally coarse channel facies emplaced by extremely large floods can include boulders many meters in diameter (Manville et al. 1999; Hodgson and Nairn 2005). Bedding characteristics indicate that channel-facies deposits are emplaced by shallow to moderate-depth floods that are flowing mostly in the lower flow regime, even while still transporting high sediment loads (Manville et al. 2009). This facies is emplaced by high energy flows in aggrading channels with anabranching or single-thread channel forms, as opposed to braided channels (Smith 1991a; b; Pierson et al. 2011).

The VWF overbank facies association (also called floodplain facies) is represented by massive to finely laminated silts and silty fine sands, typically only centimeters thick for individual flow units, that have been draped on raised depositional surfaces by slowly moving spillover flow from adjacent channels (Smith 1987a; Pierson et al. 2011) (Fig. 6d). Laminations are commonly < 1 mm thick and difficult to see unless the outcrop face has been roughened by wind erosion. Bedding structures may include ripple-drift cross-bedding, but planar bedding is the dominant bedding type, which may be wavy due to having been draped over uneven microtopography on the depositional surface. Massive bedding in fine sand units is also common in overbank facies deposits where suspension fallout deposition produces good sorting without abrupt changes in grain size. Fine overbank deposits commonly contain impressions of leaves, stems, and seeds, as well as molds of small logs and woody debris, and if older than a year or so they may exhibit traces of burrows and roots. Fine-grained tephra-fall deposits can sometimes conformably mantle overbank facies surfaces and appear to be part of the fluvial sequence. Thicker accumulations of silt and silty fine sand (up to several meters) may represent deposition in ponds formed on the irregular surfaces of floodplains or terraces (Smith 1987a).

VWF overbank deposits reflect the shallow flow depths and slow velocities at the lateral margins of VWFs, where only the fine sediment being transported in suspension at high flood stages is carried out onto floodplains and low terraces (Tunbridge 1981; Smith 1987a; Pierson et al. 2011). In gradually aggrading posteruptive channels, successions of seasonal floods may leave a sequence of stacked thin overbank facies deposits that are eventually overlain by channel facies deposits, reflecting the point at which the aggrading bed reaches the level of the floodplain or terrace (Pierson et al. 2011).

2.3 Erosion and Effects on Vegetation

VWFs transporting high sediment loads tend to aggrade their beds and erode their banks (i.e., channels widen) when flow is not laterally constrained within incised valleys, although this response may be nonlinear and include multiple pulses of sediment (e.g., Hayes et al. 2002; Gran and Montgomery 2005; Manville et al. 2005; Pierson et al. 2011). Channel incision normally returns once sediment inputs decline below critical levels. However, vertical downcutting may occur when sediment loads are still high if channel width is constrained and channel gradient is steep (e.g., Meyer and Martinson 1989). Evidence of high-discharge VWFs escaping established channels may include (1) eroded valley-side soils, older deposits, or talus aprons; (2) abandoned channels; (3) streamlined longitudinal ridges; and (4) erosion surfaces with boulder lags (Waythomas et al. 1996; Manville et al. 1999; Hodgson and Nairn 2005).

VWFs have less impact on vegetation than lahars. The upstream sides of tree trunks can be abraded or even locally debarked, but the damage is usually confined to (1) the lower 50 cm or so of trunks—the zone of abrasion by saltating bed load, and (2) a zone about 1 m above and below the level of peak stage, where trees may be struck by floating debris (Pierson 2005). Branches of trees and bushes are commonly bent but not broken and act as "trash racks" for light floating debris. Tree stems can become wrapped by thin, flexible debris (branches, long grass, rope, etc.). Grass on bar surfaces and banks is often flattened but not eroded. Crenulations in tree bark may collect sand particles and smooth stems submerged by VWFs may display very thin silty coatings, but mud coatings thicker that a fraction of a millimeter typically do not adhere to tree trunks during VWFs (Pierson 2005).

3 Outcrop-Scale Characteristics of Deposits

3.1 Stratification and Bed Forms

Stratification in VWF deposits can be massive or faint if deposits are coarse grained and were emplaced by sustained high-energy flows (Fig. 7). More generally, however, bedding in VWF deposits is distinct (see Chap. "Deposit Identification and Interpretation"). Flat, subhorizontal bedding parallel to the channel bed, *plane-parallel bedding*, is common in VWF deposits (Fig. 8), but two general types of cross-bedding are produced by different bed forms: *Low-angle cross-bedding* deviates $<15°$ from the bed plane and dips both upstream and downstream; *high-angle cross-bedding* dips only downstream and is steeper than 15° (Miall 1977, 1996; Smith 1986) (Fig. 9) (see

a **b**

Fig. 7 Massive to very crudely bedded coarse volcanic water flood (VWF) deposits that reflect transport by high-energy flows. **a** Massive, slightly imbricated, clast-supported bouldery gravel along the Klickitat River about 20 km downstream of Mount Adams (Washington, USA); flow was right to left. Note person at bottom center for scale. **b** Crudely plane-parallel-bedded fluvial gravel and coarse sand derived from an adjacent lithic pyroclastic density current (PDC) deposit along the Chaitén (Blanco) River, about 4 km downstream of Chaitén volcano (Chile), the PDC source. Left scale on card is in centimeters. USGS photos by T. Pierson

Fig. 8 Subhorizontal plane-parallel bedding in volcanic water flood (VWF) deposits: **a** Coarse sandy plane-parallel beds in a Sandy River terrace, about 70 km downstream of Mount Hood (Oregon, USA). All grains are normal density dacite or andesite and somewhat similar in size. Flow was obliquely right to left toward camera. **b** Plane-parallel-bedded coarse sand and fine normal-density gravel in Sandy River at same location as photo **a**. Flow was from right to left. Note low-angle truncation of sandy beds. The pencil is about 15 cm long. **c** Plane-parallel bedding that is poorly defined where segregated coarser pumice grains are absent, about 10 km downstream of Chaitén volcano (Chile) in Chaitén (Blanco) River. Flow was toward camera. Pencil is about 15 cm long. USGS photos by T. Pierson

Sect. 3.2 in Chap. "Deposit Identification and Interpretation"). Most plane-parallel bedding encountered in steep volcanic terrain and low-angle cross-bedding are characteristic of flows transitional to the upper flow regime—shallow, turbulent, rapid flow (Smith 1986, 1987a; Dinehart 1992; Manville et al. 2009). Low-angle cross-bedding records either the migration of antidunes or shallow scour and fill of the bed (Smith 1986). High-angle cross-bedding forms mainly in flow moving in the lower flow-regime—relatively slow, turbulent, deep flow;

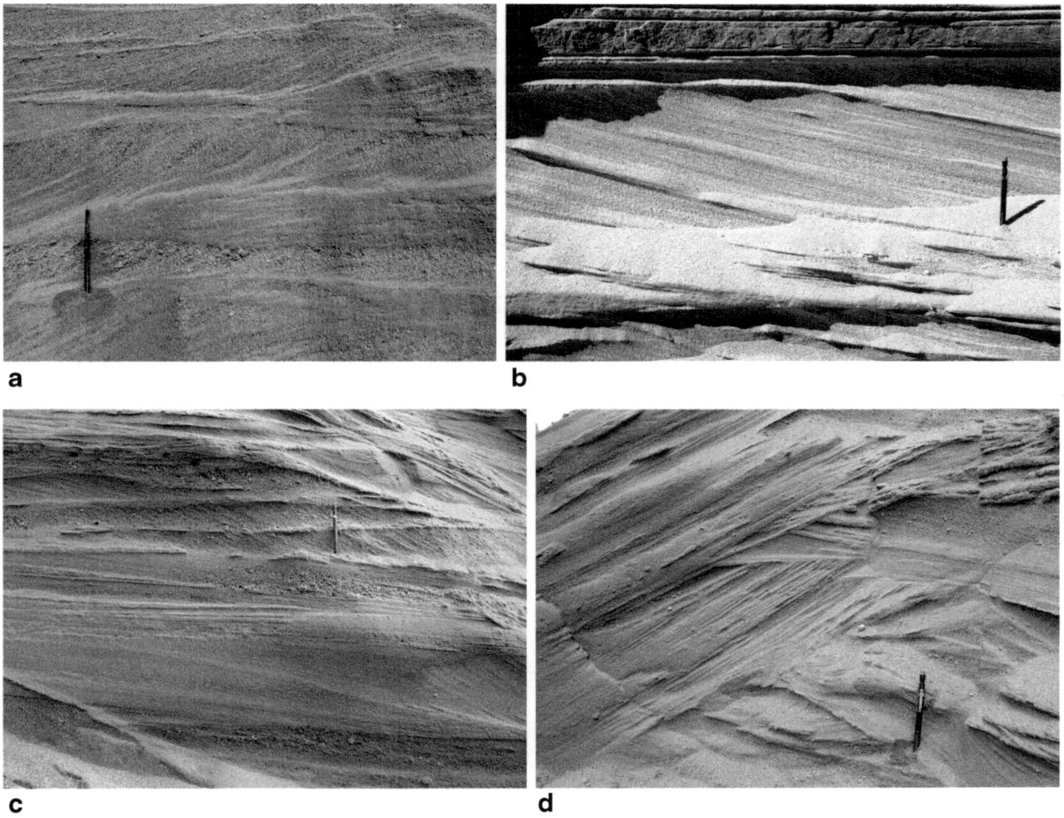

Fig. 9 Cross-bedding in volcanic water flood (VWF) deposits: **a** Wavy cross-beds in slightly gravelly coarse sand in a syneruptive fluvial sequence along the Sandy River downstream of Mount Hood (Oregon, USA), probably the result of migrating 3-dimensional dunes. Pencil is 15 cm long. **b** Planar cross-bedded medium to coarse sand formed by two-dimensional (2-D) dunes migrating downstream along the bank of the Negro River. Flow was left to right. Subhorizontal plane-parallel-bedded sand caps the cross-bedding. Abundant light-colored loose sand (in which the pencil is sticking) indicates how friable the sand is at this outcrop face. Pencil is 15 cm long. **c** Medium to coarse sand with different types of cross-bedding within an aggradational sequence in Sandy River, 70 km downstream of Mount Hood. Flow was from left to right. Pencil is 15 cm long. Planar cross-bedding from downstream migration of 2-D dunes visible in center of photo. **d** Trough cross-bedded medium to coarse sand in Sandy River formed by migration of 3-D dunes, 70 km downstream of Mount Hood. Flow was approximately toward the camera. Pencil is 15 cm long. USGS photos by T. Pierson

it records the downstream migration of ripples (if<5 cm in amplitude) or dunes (if>5 cm in amplitude) (Miall 1977, 1996; Smith 1986), but it also occurs as backset beds formed in upper-regime flow by chute-and-pool and cyclic-step bed forms and in scour-and-fill structures. Two-dimensional bed forms produce planar cross-bedding that is best seen in outcrops parallel to flow direction (i.e., flow direction of the formative flow and not necessarily the current flow direction). Three-dimensional dune forms and scours produce curved steep cross-bedding with various orientations and angles relative to formative flow direction.

Beds in posteruptive VWF deposits typically have only limited lateral extent longitudinally or transversely relative to the channel axis (Smith 1987a)—a key feature to help with identification. This is because sediment is deposited mainly by many small, shallow, and intersecting channels on braid plains, which are continuously shifting and truncating preexisting beds.

Truncated strata seldom exceed a meter or two in lateral extent. Abrupt changes in grain size are also common over vertical distances of centimeters to tens of centimeters in this depositional setting. Deeper flows in valley-confined single channels (after sedimentation rates have declined) can emplace deposits that are more laterally continuous.

3.2 Gravel-Size Clast Framework

Gravel is typically abundant in volcanic settings, and gravel clasts in beds emplaced by VWFs are almost always clast supported. Matrix-supported structures are rare, likely limited to cases where gravel clasts are rolled on a sand bed and enveloped by aggrading sand. Thick beds of coarse gravel virtually always have an openwork structure with voids only partially filled with finer sediment.

3.3 Fabric

Gravel clasts having platy or tabular shapes characteristically develop an imbricate fabric in response to bed shear stress imposed by the flow. Long *a*-axes lie transverse to the direction of flow, and larger flat surfaces formed by the *a*-*b* plane dip upstream. Imbricated fabrics can be strongly developed on channel-bed and bar surfaces (Fig. 10).

3.4 Grading

Grading appears to be uncommon within individual beds in most posteruptive VWF deposits (e.g., Figs. 7, 8 and 9). Where it appears to be present (e.g., in the lower gravelly bed in Fig. 8b), it can be difficult to assess whether it is occurring within a single depositional unit. Contacts between old and new beds can be diffuse or sharp, and sharp contacts are commonly produced by sub-horizontal erosional surfaces. Scour of the channel bed may produce bedding contacts at angles quite different from the plane of the channel bed (Visher 1972).

3.5 Fluid-Escape Structures and Deformation

VWF bed-load deposits tend not to trap fines together with sand during deposition or

Fig. 10 Imbrication of gravel clasts on bars, with clast a-b planes dipping upstream. **a** Broad bar surface in the Sandy River, about 70 km downstream of Mount Hood (Oregon, USA). Flow was from right to left. Pencil resting on the large cobble in foreground is 15 cm long and points downstream. **b** Gravel bar in a mountain streambed in the Scottish Highlands following a recent flood (note grass flattened on far bank by the flood). Flow was from left to right. Largest boulder on right is about 0.6 m wide. USGS photos by T. Pierson

consolidate significantly under loading, so pore water can escape quickly from sandy or gravelly volcano-fluvial deposits without excess pore pressures developing. Thus, fluid-escape structures tend not to form and are not normally reported in such deposits. However, fine-grained VWF sedimentary layers emplaced in backwater areas by suspension fallout may remain saturated for a time and may experience undrained loading and deform during rapid deposition or earthquakes.

3.6 Material Consistency

Moderately to well-sorted, coarse VWF sediments usually contain little silt or clay, unlike volcanic debris flow (VDF) and volcanic hyperconcentrated flow (VHF) deposits which typically contain noticeable amounts of fines. Therefore, consistency typically is loose to very friable when dry, and loose to soft when moist. There is virtually no coherence between grains. Sand and fine gravel particles ravel easily from dry outcrop faces when disturbed by a tool (or even touched). Sand that is relatively "clean" (free of fines) is a key characteristic of VWF deposits.

4 Particle-Scale Characteristics of Deposits

4.1 Particle Composition

Fluvial volcanic sediment typically includes some mixture of (1) primary volcaniclastic particles reworked from original deposits of unconsolidated pyroclastic particles, (2) secondary volcaniclastic particles remobilized from stream beds or volcano-fluvial deposits, and (3) epiclastic particles from eroded lithified pyroclastic units, lavas, or bedrock. Pulses of volcanic sediment are delivered to rivers following volcanic disturbance, and hence their deposits are typically extremely rich in fresh volcaniclastic particles (Smith 1991a; b). Such oligomictic or monomictic sediment mixtures generally include mostly recently erupted rock fragments in the gravel size fraction and whole crystals or crystal fragments, glass fragments (shards or fragments of microvesicular pumice), and minor lithic fragments in the sand fraction (McPhie et al. 1993). Deposits emplaced in intereruptive periods are more typically polymictic, although they can be oligomictic in lithologically uniform terrains, such as on basaltic islands like Hawaii. All glass fragments in deposits are prone to rapid chemical alteration and devitrification, and basaltic hyaloclastite shards are especially prone to alteration, even in young deposits, where sideromelane is commonly replaced by palagonite (McPhie et al. 1993). See Sect. 4.1 in Chap. "Deposit Identification and Interpretation" for definitions of degrees of particle mixing.

4.2 Particle Roundness, Shape, Fracturing, and Surface Texture

Volcanic terrains typically contribute texturally immature sediment to rivers, so grain shape and roundness of particles transported by VWFs are a function of sediment source type and location, particle hardness, degree of weathering, and transport history. Fresh dense-rock volcaniclasts in the pebble to cobble size range, being transported for the first time by geomorphic processes and for only a few kilometers from their source, typically remain very angular to angular in shape (Fig. 11). However, boulder-size clasts of hard lithologies transported in high-discharge water floods can become noticeably more rounded after only a few kilometers of transport (Scott and Gravlee 1968), so particle size (implying high particle-to-particle collision forces) is also an important factor in particle rounding. Soft pumice clasts, on the other hand, may or may not be rounded during water-flood transport (depending on degree of suspension), but they will be well rounded if entrained from PDC flow deposits. Older, more

Fig. 11 Mostly very angular to angular pebble- and cobble-size dense-rock bed-material particles of glassy rhyolite (viewed vertically from above), emplaced by volcanic water floods (VWF) on the surface of a longitudinal bar in the Chaitén (Blanco) River channel (Chile). Distance transported was 3–9 km from the probable source—two valley-filling pyroclastic density current (PDC) deposits (lithic block-and-ash flow deposits) that formed most of the vertical streambanks along the channel upstream of this site, along with reworked deposits of the same material. This site is 10 km downstream of Chaitén volcano, the source of the PDCs in 2008 and 2009 (Major et al. 2013). A few sub-rounded older clasts are also visible. Flow direction was from top to bottom of the photo. USGS photo in 2011 by T. Pierson

rounded, gravel-size clasts of dense-rock lithologies may be picked up from streambeds and streambanks, where they may have endured thousands of years of particle–bed collisions, intermittent reworking, and both chemical and physical weathering processes. These older and rounder clasts mix with fresh angular volcaniclastic particles during VWFs. Sand grains are different, however. Normal-density sand-size rock particles require prolonged transport distances (presumably thousands of kilometers) to acquire significant rounding (Twenhofel 1945; Frings 2008), due to mainly suspended-load transport during high-discharge VWFs that precludes strong particle–particle collisions. In any case, we have observed that fresh (first-cycle) sand-size particles of dense volcaniclastic rock types are always very angular in samples of VWF deposits, even in distal locations.

Volcaniclastic particles composed of dense-rock lithologies or crystals are, for the most part, roughly equant in shape, although fragments from highly fractured lava can be non-equant. Crystals are commonly rod- or blade-shaped, and fresh glass shards are highly angular and have blocky, wedge-like, or splintery shapes (McPhie et al. 1993). More mature volcano-fluvial sands do not contain glass shards, and lithic and crystal fragments tend to be more rounded, smooth, and with no characteristic fracture or etch patterns (Visher 1972).

Large gravel-size clasts transported by high-discharge water floods are seldom fractured (Scott and Gravlee 1968), probably because fracturing likely results in clasts breaking apart during transport. Transport of smaller clasts by lower-discharge floods probably results in collisional impacts that are less intense than during transport by VDFs and VDAs. Surface textures of gravel-size clasts are typically rough in proximal settings, in contrast to the smoother-surfaced clasts that are typical farther downstream in volcanic drainages or in non-volcanic streams.

4.3 Particle-Size Distribution and Sedimentary Texture

Sediment deposited by VWFs in proximal areas is dominated by coarse sand and gravel, often mixed in different proportions. Fines content is minimal but typically slightly higher than in fluvial sediments from nonvolcanic terrains. Sandy VWF deposits typically have unimodal size distributions and moderate sorting values in the range 0.5 to 0.9φ (Scott 1988a; Letsinger 1994). Sorting in gravelly deposits can have higher values, but VWF deposits are significantly better sorted than lahar deposits.

Particle-size distributions can have distinctly negative skewness and be mesokurtic to extremely leptokurtic (Auer et al. 2012). Very coarse water-flood deposits in steep channels—*debris-flood* deposits (Church and Jakob 2020)—are probably rare in most volcanic arc settings, because enough fines are usually present to cause water floods in steep channels to transform quickly to lahars. Where very coarse water-flood deposits have been studied in non-volcanic settings, sorting values are poor to very poor (~1.0 to 3.0φ) (Scott and Gravlee 1968; Glancy and Harmsen 1975).

Moderately sorted sandy VFW deposits may show relatively little variation in particle-size distribution in deposits emplaced by different floods. For example, of variations in particle-size distribution in distal VWF deposits at a study site 65 km downstream of Mount St. Helens (Washington) (Letsinger 1994) can be seen in Table 1. These posteruptive VWF deposits were emplaced in a braided river channel at the end of a period of rapid channel aggradation (10 m of aggradation between 1980 and 1983) in a river valley that was heavily impacted by the South Fork Toutle River lahar on May 18, 1980 (Janda et al. 1981; Scott 1988a). The three VWF deposits all had sorting values between 0.7 and 0.9φ. The flood deposits in some outcrops

Table 1 Size-distribution characteristics of three volcanic water flood (VWF) deposits on the South Fork Toutle River downstream of Mount St. Helens (Washington, USA), emplaced during a period of elevated posteruptive sedimentation (Letsinger 1994). Lithofacies codes are defined in Sect. 2.2 in Chap. "Deposit Identification and Interpretation"

Deposit	Mean thickness in centimeters (Range)	Mean grain size in phi units	Sorting	Comments
1	12 (4–15)	1.8φ (medium sand)	0.88φ (moderately sorted)	Deposited at head and tail of midchannel gravel bar; erosive basal contact; sediments generally fining upward; lithofacies *St*
2	12 (5–25)	2.6φ (fine sand)	0.74φ (moderately sorted)	Deposited on abandoned vegetated bar; non-erosive basal contact; generally coarsening upward; lithofacies *Sh, Sp*
3	13.5 (3–25)	1.7φ (medium sand)	0.89φ (moderately sorted)	Deposited in channel; non-erosive/erosive basal contacts; contains lenses of coarse sand and pebbly sand; lithofacies *Sh, Ss*

appeared to be massive when first cleaned off, but subtle stratification could be discerned by tracing ferromagnesian minerals and sand-sized pumice grains.

4.4 Vesicles and Intergranular Voids

Fluvial sediments do not normally contain vesicles. Their better sorting and near absence of fines allow any ingested air to escape rather than be trapped as bubbles when depositional units are emplaced. Coarse fluvial gravels, however, commonly contain irregular intergranular voids between gravel clasts, especially in thick gravel lenses. When such voids are abundant, deposits have distinct openwork structures.

5 Key Characteristics of Volcanic Water-Flood (VWF) Deposits

VWF deposits can resemble several other types of volcaniclastic sediments, particularly some VHF, PDC, and tephra-fall (TF) deposits. A summary table of relevant characteristics for VWF deposits is found in Chap. “Deposit Identification: Key and Summary Tables”, and a gallery of outcrop photographs is in Chap. “Annotated Outcrop Photographs”. Some of the most useful identification criteria are listed below and are described in detail in the sections above. However, any one characteristic can be found in at least one other type of deposit. Therefore, no single deposit characteristic is entirely sufficient for emplacement process identification. Field observations and physical characteristics of sediment must be evaluated together for an accurate assessment of a deposit's origin.

The characteristics listed below and in Chap. “Deposit Identification: Key and Summary Tables” are based on evaluation of fresh deposits. Fresh deposits and freshly impacted flow paths provide the best opportunities for identifying emplacement process prior to the complicating effects of weathering, mineral alteration, chemical staining, and cementation of the sediment particles themselves; erosion of landforms; and decay of affected vegetation.

- **Landscape-scale criteria**
  - (1) Deposit surfaces can be flat or terraced, commonly with bars and bed forms.
  - (2) Deposit surfaces exhibit patches of sediment with abrupt changes in average grain size (e.g., pebble gravel adjacent to sand).
  - (3) Little abrasion damage occurs to standing trees in flow paths, but if present, the damage occurs just above the bed (caused by bed-load transport abrasion) and at the level of peak stage (caused by abrasion from floating debris).
  - (4) No sediment coatings are found on standing tree trunks (though possibly some sand grains trapped in crenulations of tree bark).
  - (5) No lateral flow levees.
- **Outcrop-scale criteria**
  - (1) Sediment is mainly stratified, with sharp or diffuse contacts between beds; some gravel deposits can be massive.
  - (2) Erosional truncation of beds (from shifting channels) is common, and therefore there is limited lateral continuity of beds. Lenses of gravel are common.
  - (3) Abrupt changes in average grain size between beds are common.
  - (4) Bedding commonly exhibits bed forms. The presence of high-angle (>15°) planar cross-bedding (formed by ripples and two-dimensional dunes) is virtually always restricted to alluvial deposits, including VWF deposits. Steep curvilinear cross-beds are common in VWF deposits where paleo-flow directions were oblique to outcrop faces.

(5) Both plane-parallel bedding and cross-bedding are common, often alternating in vertical sequences.
(6) Gravel bars are common (appear as gravel lenses in vertical outcrop faces).
(7) Fabric—imbrication of elongate or flat gravel clasts is common.
(8) Loose packing—deposits are friable to very friable when dry; soft when moist. Sandy outcrop faces ravel easily when scraped.
(9) No dewatering structures are present.
(10) Voids between gravel-size clasts are common.

- **Particle-scale criteria**
  (1) Minimal fines are preserved in the sandy matrix of channel-facies deposits; some fines (or mainly fines) may be present in overbank-facies deposits.
  (2) Deposits emplaced in volcanically disturbed basins during or shortly after eruptions are likely to be monomictic to oligomictic.
  (3) Slight rounding of fresh, normal-density, pebble- and cobble-size clasts may occur; these clasts are more typically very angular to angular. Older volcaniclastic clasts may show some rounding (possibly subangular to well-rounded); these older clasts are commonly mixed with fresh angular clasts. Fresh boulder-size clasts may be partially rounded. Sand-size particles are very angular.
  (4) Sorting is moderate within individual beds—mainly in the range 0.5–0.9φ, with values commonly 0.7–0.9φ.
  (5) Bubble vesicles are absent.

References

Auer A, White JDL, Manville V (2012) Cryptic eruption of Mt. Ruapehu revealed by deposits of sediment-laden streamflow in a steep mountain valley: the 4 ka Kiwikiwi Formation, Whangaehu Valley, New Zealand. J Volcanol Geotherm Res 243–244:45–58. https://doi.org/10.1016/j.jvolgeores.2012.07.003

Blair TC, McPherson JG (1994) Alluvial fans and their natural distinction from rivers based on morphology, hydraulic processes, sedimentary processes, and facies assemblages. J Sediment Res 64(3):450–489

Church M, Jakob M (2020) What is a debris flood?. Water Resour Res 56(8):e2020WR027144. https://doi.org/10.1029/2020WR027144

Collinson JD (1996) Alluvial sediments. In: Reading HG (ed) Sedimentary environments: processes, facies and stratigraphy, 3rd edn. Blackwell Science, Oxford, pp 37–82

Davies DK, Vessell RK, Miles RC, Foley GG, Bonis SB (1978) Fluvial transport and sediment modifications in an active volcanic region. In: Miall AD (ed) Fluvial sedimentology, vol 5. Canadian Society Petroleum Geology Memory, pp 61–84

Dinehart RL (1992) Evolution of coarse gravel bed forms; field measurements at flood stage. Water Resour Res 28:2667–2689

Donoghue SL, Neall VE (2001) Late quaternary constructional history of the southeastern Ruapehu ring plain, New Zealand. New Zeal J Geol Geophys 44(3):439–466. https://doi.org/10.1080/00288306.2001.9514949

Frings RM (2008) Downstream fining in large sand-bed rivers. Earth-Sci Rev 87(1–2):39–60. https://doi.org/10.1016/j.earscirev.2007.10.001

Germanoski D, Schumm SA (1993) Changes in braided river morphology resulting from aggradation and degradation. J Geol 101:451–466

Glancy PA, Harmsen L (1975) A hydrologic assessment of the September 14, 1974, flood in Eldorado Canyon, Nevada. US Geol Surv Prof Pap 930:1–28

Gran KB, Czuba JA (2017) Sediment pulse evolution and the role of network structure. Geomorphology 277:17–30. https://doi.org/10.1016/j.geomorph.2015.12.015

Gran KB, Montgomery DR (2005) Spatial and temporal patterns in fluvial recovery following volcanic eruptions: channel response to basin-wide sediment loading at Mount Pinatubo, Philippines. Geol Soc Am Bull 117:195–211

Gran KB, Montgomery DR, Sutherland DG (2006) Channel bed evolution and sediment transport under declining sand inputs. Water Resour Res 42:W10407

Hayes SK, Montgomery DR, Newhall CG (2002) Fluvial sediment transport and deposition following the 1991 eruption of Mount Pinatubo. Geomorphology 45:211–224

Hodgson KA, Nairn IA (2005) The ca. AD 1315 syn-eruption and AD 1904 post-eruption breakout floods from Lake Tarawera, Haroharo caldera, North Island, New Zealand. New Zeal J Geol Geophys 48:491–506

Janda RJ, Scott KM, Nolan KM, Martinson HA (1981) Lahar movement, effects, and deposits. In: Lipman PW, Mullineaux DR (eds) The 1980 eruptions of Mount St. Helens, Washington, vol 1250. US Geology Survey Professional Paper, pp 461–478

Kataoka K, Nakajo T (2002) Volcaniclastic resedimentation in distal fluvial basins induced by large-volume explosive volcanism. The Ebisutoge—Fukuda tephra, Plio-Pleistocene boundary, central Japan. Sedimentology 49:319–334

Kataoka KS, Urabe A, Manville V, Kajiyama A (2008) Breakout flood from an ignimbrite-dammed valley after the 5 ka Numazawako eruption, northeast Japan. Geol Soc Am Bull 120:1233–1247

Kuenzi WD, Horst OH, McGehee RV (1979) Effect of volcanic activity on fluvial-deltaic sedimentation in a modern arc-trench gap, southwestern Guatemala. Geol Soc Am Bull 90:827–838

Letsinger S (1994) Late-stage syneruption sedimentation in the South Fork Toutle River, Toutle, Washington. Thesis, University of Idaho, 38 pp

Lisle TE (2008) The evolution of sediment waves influenced by varying transport capacity in heterogeneous rivers. In: Habersack H, Piégay H, Rinaldi M (eds) Gravel-bed rivers VI: from process understanding to river restoration. Elsevier, Amsterdam, pp 443–469

Major JJ, Pierson TC, Hoblitt RP, Moreno H (2013) Pyroclastic density currents associated with the 2008–2009 eruption of Chaitén volcano (Chile)—forest disturbances, deposits, and dynamics. Andean Geol 40(2):324–358

Manville V, White JDL, Houghton BF, Wilson CJN (1999) Paleohydrology and sedimentology of a post-1.8 ka breakout flood from intracaldera Lake Taupo, North Island, New Zealand. Geol Soc Am Bull 111:1435–1447

Manville V, Newton EH, White JDL (2005) Fluvial responses to volcanism: resedimentation of the 1800a Taupo ignimbrite in the Rangitaiki River catchment, North Island, New Zealand. Geomorphology 65:49–70

Manville V, Segschneider B, Newton E, White JDL, Houghton BF, Wilson CJN (2009) Environmental impact of the 1.8 ka Taupo eruption, New Zealand: landscape responses to a large-scale explosive rhyolite eruption. Sediment Geol 220:318–336

Maria A, Carey S, Sigurdsson H, Kincaid C, Helgadóttir G (2000) Source and dispersal of jökulhlaup sediments discharged to the sea following the 1996 Vatnajökull eruption. Geol Soc Am Bull 112:1507–1521

Mathisen ME, Vondra CF (1983) The fluvial and pyroclastic deposits of the Cagayan Basin, Northern Luzon, Philippines—an example of non-marine volcaniclastic sedimentation in an interarc basin. Sedimentology 30:369–392

McPhie J, Doyle M, Allen R (1993) Volcanic textures—a guide to the interpretation of textures in volcanic rocks. Centre for Ore Deposit and Exploration Studies, University of Tasmania, Hobart, p 196

Meyer DF, Martinson HA (1989) Rates and processes of channel development and recovery following the 1980 eruption of Mount St. Helens, Washington. Hydrol Sci J 34:15–127

Miall AD (1977) A review of the braided river depositional environment. Earth Sci Rev 13:1–62

Miall AD (1996) The geology of fluvial deposits. Springer, Berlin, p 582

Miller DJ, Benda LE (2000) Effects of punctuated sediment supply on valley-floor landforms and sediment transport. Geol Soc Am Bull 112:1814–1824

Montgomery DR, Panfil MS, Hayes SK (1999) Channel-bed mobility response to extreme sediment loading at Mount Pinatubo. Geology 27:271–274

Nakayama K, Yoshikawa S (1997) Depositional processes of primary to reworked volcaniclastics on an alluvial plain; an example from the Lower Pliocene Ohta tephra bed of the Tokai Group, central Japan. Sediment Geol 107:211–229. https://doi.org/10.1016/S0037-0738(96)00022-X

Newhall CG, Punongbayan RS (1996) Fire and mud: eruptions of Mount Pinatubo, Philippines. Philippines Institute Volcanology Seismology, Quezon City, and University of Washington Press, Seattle, 1126 pp

Pierson TC (2005) Distinguishing between debris flows and floods from field evidence in small watersheds. US Geol Surv Fact Sheet 2004–3142, 4 pp

Pierson TC, Major JJ (2014) Hydrogeomorphic effects of explosive volcanic eruptions on drainage basins. Annu Rev Earth Planet Sci 42:469–507. https://doi.org/10.1146/annurev-earth-060313-054913

Pierson TC, Pringle PT, Cameron KC (2011) Magnitude and timing of downstream channel aggradation in response to a dome-building eruption at Mount Hood, Oregon. Geol Soc Am Bull 123:3–20

Pierson TC, Major JJ, Amigo Á, Moreno H (2013) Acute sedimentation response to rainfall following the explosive phase of the 2008–2009 eruption of Chaitén volcano. Chile. Bull Volcanol 75:723. https://doi.org/10.1007/s00445-013-0723-4

Pierson TC, Daag AS, Delos Reyes PJ, Regalado MTM, Solidum RU, Tubianosa BS (1996) Flow and deposition of posteruption hot lahars on the east side of Mount Pinatubo, July–October 1991. In: Newhall CG, Punongbayan RS (eds), Fire and mud: eruptions and lahars of Mount Pinatubo, Philippines: Philippines Institute Volcanology Seismology, Quezon City, and University of Washington Press, Seattle, pp 921–950

Pierson TC, Scott WE, Vallance JW, Pringle PT (2009) Eruption-related lahars and sedimentation response downstream of Mount Hood: field guide to volcaniclastic deposits along the Sandy River, Oregon. In: O'Connor JE, Dorsey RJ, Madin IP (eds) Volcanoes to vineyards: geologic field trips through the dynamic landscape of the Pacific Northwest. Geol Soc Am Field Guide 15:229–244

Rodolfo KS, Umbal JV, Alonso RA, Remotigue CT, Paladio-Melosantos ML, Salvador JHG, Evangelista D, Miller Y (1996) Two years of lahars on the western flank of Mount Pinatubo: initiation, flow processes, deposits, and attendant geomorphic and hydraulic changes. In: Newhall CG, Punongbayan

RS (eds), Fire and mud: eruptions and lahars of Mount Pinatubo, Philippines. Philippines Institute Volcanology Seismology, Quezon City, and University of Washington Press, Seattle, pp 989–1013

Schumm SA (1985) Patterns of alluvial rivers. Annu Rev Earth Planet Sci 13:5–27

Scott KM (1988a) Origins, behavior, and sedimentology of lahars and lahar-runout flows in the Toutle-Cowlitz River system. US Geology Survey Professional Paper 1447-A, 74 pp

Scott KM, Gravlee GC Jr (1968) Flood surge on the Rubicon River, California—hydrology, hydraulics and boulder transport. US Geology Survey Professional Paper 422-M, 40 pp

Scott KM, Janda RJ, de la Cruz EG, Gabinete E, Eto I, Isada M, Sexon M, Hadley KC (1996) Channel and sedimentation responses to large volumes of 1991 volcanic deposits on the east flank of Mount Pinatubo. In: Newhall CG, Punongbayan RS (eds) Fire and mud: eruptions and lahars of Mount Pinatubo, Philippines. Philippines Institute Volcanology Seismology, Quezon City, and University of Washington Press, Seattle, pp 971–988

Segschneider B, Landis CA, Manville V, White JDL, Wilson CJN (2002) Environmental response to a large, explosive rhyolite eruption: sedimentology of post-1.8 ka pumice-rich Taupo volcaniclastics in the Hawke's Bay region, New Zealand. Sediment Geol 150:275–299

Smith GA (1986) Coarse-grained nonmarine volcaniclastic sediment: terminology and depositional process. Geol Soc Am Bull 97:1–10

Smith GA (1987a) The influence of explosive volcanism on fluvial sedimentation: the Deschutes formation (Neogene) in central Oregon. J Sediment Petrol 57:613–629

Smith RCM (1991b) Post-eruption sedimentation on the margin of a caldera lake, Taupo Volcanic Centre, New Zealand. Sediment Geol 74:89–138

Smith ND, Smith DG (1984) William River: an outstanding example of channel widening and braiding caused by bed-load addition. Geology 12:78–82

Smith GA (1987b) Sedimentology of volcanism-induced aggradation in fluvial basins: examples from the Pacific Northwest, U.S.A. In: Ethridge FG, Flores RM, Harvey MD (eds) Recent developments in fluvial sedimentology, vol 39. Society economic paleontology mineral special publication, pp 217–228

Smith GA (1991a) Facies sequences and geometries in continental volcaniclastic sediments. In: Fisher RV, Smith GA (eds) Sedimentation in volcanic settings, vol 45. SEPM (Society sedimentary geology) Special Publication, pp 109–121

Tunbridge IP (1981) Sandy high-energy flood sedimentation—some criteria for recognition, with an example from the Devonian of S.W. England. Sediment Geol 28(2):79–95

Twenhofel WH (1945) The rounding of sand grains. J Sediment Res 15(2):59–71. https://doi.org/10.1306/D4269224-2B26-11D7-8648000102C1865D

Umazano AM, Bellosi ES, Visconti G, Melchor RN (2008) Mechanisms of aggradation in fluvial systems influenced by explosive volcanism: an example from the Upper Cretaceous Bajo Barreal Formation, San Jorge Basin, Argentina. Sediment Geol 203:213–228

Vessell RD, Davies KK (1981) Nonmarine sedimentation in an active fore arc basin. In: Ethridge FG, Flores RM (eds) Nonmarine depositional environments: models for exploration, vol 31. Society Economic Paleontology Mineral Special Publication, pp 34–45

Visher GS (1972) Physical characteristics of fluvial deposits. In: Rigby JK, Hamblin WK (eds) Recognition of ancient sedimentary environments, vol 16. Society Economic Paleotol Mineral Special Publication, pp 84–97

Van Wagoner JC, Hoyal DCJD, Adair NL, Sun T, Beaubouef RT, Deffenbaugh M, Dunn PA, Huh C, Li D (2003) Energy dissipation and the fundamental shape of siliciclastic sedimentary bodies. Search and Discovery (AAPG Data Pages Inc) Article #40080, 9 pp

Waythomas CF, Walder JS, McGimsey RG, Neal CA (1996) A catastrophic flood caused by drainage of a caldera lake at Aniakchak Volcano, Alaska, and implications for volcanic hazards assessment. Geol Soc Am Bull 108(7):861–871

Zernack AV, Cronin SJ, Neall VE, Procter JN (2011) A medial to distal volcaniclastic record of an andesite stratovolcano: detailed stratigraphy of the ring-plain succession of south-west Taranaki, New Zealand. Int J Earth Sci (Geol Rundsch) 100:1937–1966. https://doi.org/10.1007/s00531-010-0610-6

Zernack AV, Procter JN, Cronin SJ (2009) Sedimentary signatures of cyclic growth and destruction of stratovolcanoes: a case study from Mt. Taranaki, New Zealand. Sed Geol 220(3–4):288–305. https://doi.org/10.1016/j.sedgeo.2009.04.024

Other Processes Depositing Volcanic Sediment

Abstract

Pyroclastic density currents (PDCs), tephra falls (TFs), and glaciers deposit volcanic sediment, including juvenile volcaniclastics, in the same depositional zones as volcanic wet flows—on the flanks of volcanic edifices and in downstream river valleys. These processes can emplace deposits that closely resemble those of volcanic wet flows, and these other types of deposits are commonly interbedded with wet-flow deposits. This chapter gives an overview of the emplacement processes and physical characteristics of deposits emplaced by PDCs, TFs, and glaciers. It does not attempt to provide comprehensive treatments of pyroclastic or glacial deposits. However, it does emphasize the characteristics of subaerial PDC, TF, and glacial deposits that differ from those of volcanic wet flows. Relevant key characteristics of PDC, TF, and glacial deposits (primarily tills) are given for comparison, and numerous evaluation criteria are presented that can be used to differentiate volcanic wet-flow deposits (those of lahars, volcanic debris avalanches, and volcanic water floods) from deposits of these other processes.

1 Introduction

Pyroclastic density currents (PDCs), tephra falls (TFs), and glaciers can deposit volcanic sediment, including juvenile volcaniclastics, in the same depositional zones as volcanic wet flows—on the flanks of volcanic edifices and in downstream river valleys—as well as outside these zones. Sediments deposited by pyroclastic flows and falls are commonly called *tuff*, *lapilli tuff*, or *tuff breccia*, and sediment emplaced directly by glaciers and unreworked by flowing water is termed *till*—the type of glacial deposit most encountered on volcanoes. Glacial deposits emplaced by flowing glacial meltwater are termed *glacial outwash* (Flint 1971; Ehlers 1996; Neuendorf et al. 2005); these deposits are mainly found in low-gradient valleys downstream of glaciated volcanoes. PDC, TF, and till or outwash deposits are commonly interbedded with volcanic wet-flow deposits and can closely resemble them. Thus, in field studies of unconsolidated deposits on volcanoes, it is critical to be able to distinguish between these other deposits and the deposits of volcanic debris flows (VDFs), volcanic hyperconcentrated flows (VHF), volcanic debris avalanches (VDAs), and volcanic water floods (VWFs).

T. C. Pierson et al., *Deposits of Volcanic Wet Flows*, Advances in Volcanology,
https://doi.org/10.1007/978-3-031-66574-5_11

This chapter gives an overview of the emplacement processes and physical characteristics of deposits emplaced by PDCs, TFs, and glaciers. It emphasizes how the characteristics of their subaerial deposits differ from those of volcanic wet flows. It does not attempt to provide comprehensive treatments of pyroclastic or glacial deposits. For this reason, discussion of these "other processes" is largely limited to the kinds of pyroclastic and glacial deposits most likely to be encountered on the flanks and downstream of stratovolcanoes and lava dome complexes—the principal types of volcanoes hosting the volcanic wet flows that are the focus of this book. This subset of pyroclastic and glacial deposits includes (1) mainly deposits of small-volume PDCs (as opposed to large ignimbrites); (2) proximal and relatively coarse TF deposits (as opposed to distal, thin, and very fine-grained tephra layers); and (3) tills emplaced by alpine glaciers on volcanoes (as opposed to continental ice sheets or deposits other than till). In volcanic settings, glacial outwash deposits are mainly emplaced by VWFs occurring during glacial periods, so we make no attempt to differentiate between glacial and nonglacial VWF deposits. These three subsets of "other deposits" from stratovolcanoes and dome complexes are particularly similar in both scale and sedimentology to volcanic wet-flow deposits. However, some deposits emplaced by eruptions at maars and tuff rings are included to show features found more broadly in volcanic settings.

Explosive eruptions create most of the volcanic rock particles transported and deposited by processes other than volcanic wet flows, and there are three basic types of explosive eruptions. *Magmatic eruptions* derive their explosive power from the rapid exsolution of dissolved gases in magma. *Phreatomagmatic eruptions* derive their power from a combination of magmatic gas exsolution and the rapid vaporization of pressurized groundwater during the interaction of ascending magma with shallow aquifers. *Phreatic eruptions* are driven solely by steam explosions caused by the heating of confined shallow groundwater by a magmatic heat source. Eruption type and magma volume, composition, temperature, and gas content are controlling factors in determining eruption magnitude, style, and degree of explosivity. They also determine the size and texture of pyroclastic particles and the volume, thickness, and area of PDC and TF deposits. Both quantitative criteria and qualitative descriptors are used to categorize explosive eruptions (Table 1).

There is a wide and comprehensive literature treating these "other deposits." For processes and deposits of both PDCs and TFs, a partial list of references includes: Fisher (1961, 1966b), Walker (1971), Wright et al. (1980, 1981), Fisher and Schmincke (1984), Rowley et al. (1985), Cas and Wright (1987), Carey (1991), Branney and Kokelaar (1992, 2002), Boudon et al. (1993), McPhie et al. (1993), Druitt (1998), Miyabuchi (1999), Freundt et al. (2000), Houghton et al. (2000a, b, 2015), Valentine and Fisher (2000), Sigurdsson et al. (2000, 2015), Druitt et al. (2002), Saucedo et al. (2004), Schmincke (2004), White and Houghton (2006), Lube et al. (2007, 2020), Hildreth and Fierstein (2012), Sulpizio et al. (2014), Brown and Andrews (2015), Dufek et al. (2015), Houghton and Carey (2015), Macorps et al. (2018), and Pollock et al. (2019). Comprehensive treatments of glacial processes and deposits may be found in, among others, Boulton and Deynoux (1981), Eyles et al. (1983), Benn (1995), Ehlers (1996), Zielinski and van Loon (1996), Benn and Evans (2010), and Iverson (2017).

2 Pyroclastic Density Current

2.1 Process Overview

Pyroclastic density currents (PDCs) are dry (mostly), expanded, generally ground-hugging and density-stratified flows of heterogeneous particle–fluid mixtures. They are typically extremely hot, driven mainly by gravity (in some cases, assisted by the propulsive forces of lateral explosions), and range across a solids-concentration continuum from highly

Table 1 Ways of rating and describing degree of explosivity and magnitude in volcanic eruptions (modified from Newhall and Self 1982; Siebert et al. 2010)

VEI	0	1	2	3	4	5	6	7	8
Eruption description	Gentle	Effusive	Explosive						
Magnitude of explosive component	Non-explosive	Small	Moderate	Moderate to large	Large	Very large	"Cataclysmic"		
Tephra volume ejected (m^3)	<10^4	10^4–10^6	10^6–10^7	10^7–10^8	10^8–10^9	10^9–10^{10}	10^{10}–10^{11}	10^{11}–10^{12}	>10^{12}
Eruption style	Hawaiian →	→				← Ultra-Plinian →	→	→	→
		Strombolian →	→		← Plinian →	→	→	→	
			← Vulcanian →	→	→				
Eruption column height (km)	<0.1 ac	0.1–1.0 ac	1–5 ac	3–15 asl	10–25 asl	>25 asl			

VEI is the Volcano Explosivity Index (Newhall and Self 1982). Eruption column heights are measured either "above crater" (ac) or "above sea level" (asl). Boiling-over-type eruptions can occur during different styles and magnitudes of eruptions. During boiling-over eruptions, a dense froth of solidifying pyroclasts pours over a crater rim or through a notch in a crater, creating a lithic PDC

concentrated to very dilute (Kieffer 1981a, b; Branney and Kokelaar 2002; Sulpizio et al. 2014; Cole et al. 2015; Dufek et al. 2015; Girolami et al. 2015); see Chap. "Deposits Emplaced by Volcanic Wet Flows" for more detail (Fig. 1). They move as highly mobile hot dispersions of rock fragments, mixed with a fluid phase consisting mainly of entrained air plus exsolving magmatic gases, and they generally leave behind landscapes scoured or buried, scorched, and devoid of life. Gravity segregation of particles during flow commonly leads to the development of stratified flow with a dense, highly concentrated, ground-hugging basal flow zone that grades upward into a dilute and highly turbulent upper flow zone. These two progressively evolving zones or parts of the original flow can eventually become decoupled and follow slightly different flow trajectories downslope (Fisher 1995). Initially, PDCs can be locally erosive, but as slope gradient decreases and flows spread out, PDCs progressively deposit solid particles as they flow, becoming more dilute with distance from source (Fisher 1979; Sheridan 1979; Fisher and Schmincke 1984; Cas and Wright 1987; Druitt 1998; Freundt et al. 2000; Valentine and Fisher 2000; Wilson and Houghton 2000; Branney and Kokelaar 2002; Sarocchi et al. 2011; Sulpizio et al. 2014; Dufek et al. 2015).

The dominant fluid phase in all PDCs is dusty gas, although in some volcanic systems cooler PDCs may have a fluid phase composed of a mixture of dusty gas and water droplets

Fig. 1 Small pyroclastic density current descending north flank of Mount St. Helens (Washington, USA), August 7, 1980. USGS archive photo

or films on grains (Fisher and Schmincke 1984; Wilson and Houghton 2000; Brown and Andrews 2015). This gaseous pore fluid, commonly with sustained high fluid pressures, contributes to the high mobility of dense PDCs (Wilson 1980, 1984; Carey 1991; Branney and Kokelaar 2002; Roche 2012). Only subaerial PDC deposits are considered here, although some PDCs can flow for kilometers on the surface of water bodies or as submarine flows and emplace subaqueous or submarine deposits (e.g., Sparks et al. 1980; Carey et al. 1996; Allen and Cas 2001; Trofimovs et al. 2006).

Flow velocities can range from meters to hundreds of meters per second (Moore and Sisson 1981; Tanguy 1994; Valentine and Fisher 2000; Komorowski et al. 2013; Lube et al. 2020), with high velocities and long runout distances presumed to be facilitated by fluidization through ingestion and heating of air (Wilson 1980, 1985; Wilson and Houghton 2000; Girolami et al. 2015) or by momentum initially imparted by explosions (Kieffer 1981a, b; Esposti Ongaro et al. 2012).

Emplacement temperatures commonly range from <100 to 550 °C but can reach 1000 °C for extremely large silicic PDCs (Hoblitt and Kellogg 1979; Voight and Davis 2000; Charbonnier and Gertisser 2008; Bryan et al. 2010). To retain liquid water, flows must be emplaced at temperatures <100 °C (Moore 1967; Fisher and Waters 1970; Lorenz 1974).

Flow runout distances can range from <1 km for small-volume, low-energy currents, to >200 km for flows from large caldera-forming eruptions (Wilson et al. 1995; Druitt 1998; Druitt et al. 2002; Bryan et al. 2010; Brown and Andrews 2015). At most stratovolcanoes and dome complexes, however, PDCs typically travel from 1 to 15 km from source (e.g., Saucedo et al. 2004; Komorowski et al. 2013; Macorps et al. 2018). Flow trajectories and travel distances can vary where PDCs become highly density stratified and high-concentration basal flows decouple from their dilute overriding currents, which is common in complex topography (Fisher and Schmincke 1984; Boudon et al. 1993; Yamamoto et al. 1993; Freundt et al. 2000; Charbonnier and Gertisser 2008; Sarocchi et al. 2011).

Nomenclature for PDCs has been evolving over the years (Chapin and Elston 1979; Fisher and Schmincke 1984), with the terms *pyroclastic flow* and *pyroclastic surge (ash-cloud surge)* being widely adopted in the mid-twentieth century to describe distinct types of flows of high and low particle concentrations, respectively. But further research has shown that dry flows of particle–gas dispersions occur on a continuum of behaviors that (1) grade from one to another, (2) vary with particle concentration, and (3) evolve with time and with position within a flow. Thus, the more inclusive terms *pyroclastic density current* (or *pyroclastic current*) has now been widely adopted to include the whole range of flow processes that occur along this continuum, recognizing that PDCs can evolve continuously during flow through the actions of particle erosion and entrainment, particle deposition and elutriation, entrainment and heating of atmospheric air, turbulent mixing, and gravity segregation (Branney and Kokelaar 2002; Sulpizio et al. 2014; Dufek et al. 2015; Lube et al. 2020). The bulk particle-concentration threshold separating concentrated PDCs from dilute PDCs appears to lie with the range of about 10–30 vol%, depending on defining criteria (Druitt 1998; Branney and Kokelaar 2002; Burgisser and Bergantz 2002; Bursik et al. 2005; Dufek et al. 2015; Lube et al. 2020). Fully turbulent upper parts of dilute currents likely contain < 1 vol% solids (Mele et al. 2015; Brosch and Lube 2020; Dellino et al. 2020; Lube et al. 2020). Terminology can be further subdivided by variations in particle componentry, as particle composition plays a role in flow mechanics. For example, flows composed mainly of pumice or scoria particles are termed *pumiceous or scoriaceous PDCs*. Those containing mainly fragments of previously formed rock (from collapsed lava domes or from some boiling-over eruptions) are termed *lithic PDCs* or *block-and-ash flows*.

Small- to moderate-volume PDCs are generated in four basic ways as listed below. Large to extremely large pumiceous PDCs are generated by another mechanism, *caldera collapse*—a process in which huge volumes of vesiculating magma can be ejected rapidly from multiple vents and fissures (e.g., Bacon 1983; Wilson 1985; Brown and Branney 2004; Hildreth and Fierstein 2012)

(1) **Fallback of ejected particles on the outer flanks of volcanoes during eruptions**. Fall-back scenarios range from gravitational collapse of high-energy explosive eruption columns to the pyroclastic fountaining and "boiling over" of coarse-particle mixtures onto outer edifice flanks during lower-energy eruptions (Carey and Bursik 2000, 2015; Cioni et al. 2000, 2015; Freundt et al. 2000; Branney and Kokelaar 2002; Dufek et al. 2015; Rader et al. 2015). PDCs formed by column collapse typically include mostly highly vesicular glassy pyroclasts; such flows are also termed *ash flows*, *pumice flows*, or *scoria flows*. However, PDCs formed by low-energy fall-back of explosively ejected dome rock, low-energy fountains or jets during Strombolian eruptions, or even lower-energy boiling-over eruptions are typically composed of low to moderately vesicular rock fragments and generate lithic PDCs (Freundt et al. 2000; Belousov et al. 2020).
(2) **Gravitational collapse of lava domes or lava flows** (Mellors et al. 1988; Sato et al. 1992; Miyabuchi 1999; Druitt et al. 2002; Sarocchi et al. 2011; Komorowski et al. 2013; Brown and Andrews 2015; Dufek et al. 2015). Collapse is sometimes accompanied by explosive fragmentation, and these lithic PDCs are commonly termed *block-and-ash flows*.
(3) **Laterally directed explosions ejecting fragmental material at high velocity across the terrain** (Gorshkov 1959; Gorshkov and Dubik 1970; Hoblitt et al. 1981; Kieffer 1981a, b; Moore and Sisson 1981; Hoblitt 2000; Belousov et al. 2007; Esposti Ongaro et al. 2012; Dufek et al. 2015; Houghton et al. 2015; Zimanowski

et al. 2015). Such PDCs may evolve from one or more explosions resulting from the abrupt depressurization of near-surface magma bodies or from magma interacting with groundwater or surface water. Where focused laterally in one direction by failure geometry, they are termed *directed blasts*. Explosions resulting from magma interacting with water produce cool, dilute, and usually wet and laterally directed PDCs that are traditionally termed *base surges* (e.g., Moore 1967; Waters and Fisher 1971; Valentine and Fisher 2000). In PDCs initiated by explosions, the initial explosion-driven momentum of the current wanes within tens of seconds and gravity flow takes over (Esposti Ongaro et al. 2012).

(4) **Gravitational failure of fresh PDC deposits**. Secondary generation of PDCs by gravitational failure and remobilization of freshly emplaced and still-hot and very loosely consolidated PDC deposits on steep terrain are termed *secondary PDCs* (Brantley and Waitt 1988; Scott et al. 1996; Torres et al. 1996; Branney and Kokelaar 2002; Risica et al. 2022). This is a less common mechanism than the others listed above.

The reigning conceptual model for emplacement of PDC deposits is *progressive aggradation* in which incremental accretion of particles takes place along the lower flow boundary as the flow travels away from source (Fisher 1966a; Branney and Kokelaar 1992, 2002; Legros and Martí 2001; Sulpizio et al. 2014; Girolami et al. 2015; Pollock et al. 2019; Lube et al. 2020). Aggradation may be continuous, discontinuous, or stepwise over the duration of the current. Gradients in mixture concentration, particle characteristics (size, shape, density), particle transport mode (suspension versus traction), and shear rate at the lower flow boundary largely determine how particles lose energy, cross the lower flow boundary of the current, and come to rest as part of the deposit. The mechanism of *edge-friction damming*, a minor depositional mechanism demonstrated in concentrated lahars (see Sect. 6 in Chap. "Lahars—Process Overview"), may also be operative locally within decelerating high-concentration PDCs at highly frictional coarse flow fronts and margins (e.g., Calder et al. 2000). Strictly considered, *en masse* deposition (simultaneous "freezing" of the entire vertical section) probably does not occur (Branney and Kokelaar 2002), but very rapid aggradation of rapidly decelerating high-concentration currents may, nonetheless, provide a snapshot of flow structure in the final stages of movement (Roche 2012).

Progressive aggradation implies that a depositional flow unit must be constructed over time, from the base upward, as the overriding current lays down particles sequentially as the PDC passes over—the same mechanism operative in lahar deposition. Therefore, a single vertical section in an outcrop does not record an instantaneous snapshot of the flow at the moment of deposition. Instead, it records the progressive accretion of particles at that place along the lower flow boundary. Thus, stratification and lithofacies variations within a PDC deposit record the non-uniformity and unsteady flow of a PDC in both time and space (Branney and Kokelaar 2002; Sulpizio et al. 2014).

Highly dilute PDCs have an additional depositional mechanism—gravity fall through air. This occurs when horizontal motion of a highly dilute PDC stalls and the current's density becomes equal to or less than the surrounding air, due to the continued entrainment and expansion of air. This causes the dilute PDC to cease forward motion, become buoyant, and loft upward to form a co-PDC plume (Sulpizio et al. 2014; Dufek et al. 2015). When this happens, deposition of fine particles remaining in suspension changes to gravity fallout of individual particles or ash pellets—now fall deposits instead of progressive-aggradation deposits (Brown et al. 2010), and rate of deposition is controlled entirely by the settling velocity of the particles or pellets.

2.2 Pyroclastic Density Current (PDC) Deposits

The wide variety of PDC deposits that may be encountered in the field reflects differences in magma chemistry, volatile content, temperature,

a b

Fig. 2 Pyroclastic density current (PDC) deposit geometries: **a** broad, relatively thin fan of PDC deposits (the Pumice Plain) from multiple pumiceous flow events from different eruptions in 1980 on the north flank of Mount St. Helens (Washington, USA). Thickness locally exceeds 40 m. USGS photo by T. Pierson in 2006. **b** Flat-topped, eroded remnants of a thick, narrow, valley-fill PDC deposit, emplaced in the upper White River valley of Mount Hood (Oregon, USA) by repeated dome-collapse PDCs during the last two eruptions of Mount Hood approximately 240 and 1500 years ago. USGS photo by T. Pierson

and crystallinity; eruption style; and vesicularity of the erupted lava fragments (Brown and Andrews 2015). These differences can manifest within eruptions, as well as between eruptions. Fresh subaerial PDC deposits can be welded (lithified) or nonwelded (unconsolidated). Welding is a post-depositional effect of retained heat and compaction, and some welded tuffs have properties not present in nonwelded deposits (see below). However, discussion in this section focuses mainly on nonwelded deposits. The term *ignimbrite* has been used by some authors to refer to the deposit of any PDC (Neuendorf et al. 2005), while others have restricted it (by definition or implication) to deposits of PDCs composed of vesiculated juvenile particles (mainly pumice) (e.g., Sparks et al. 1973; Cas and Wright 1987; Dade and Huppert 1996; Legros and Martí 2001; Branney and Kokelaar 2002).

2.2.1 Landscape-Scale Characteristics of Deposits

Deposit Architecture and Morphology

The geometry and features of PDC deposits are determined by bulk solids concentration of the particle–fluid mixture, degree of density segregation within a flow, PDC energetics, and the topography onto which deposits are emplaced. Scale and extent depend on the volume of erupted material.

Broad overlapping depositional fans are built by PDCs on undissected lower flanks of volcanoes where high-concentration currents repeatedly sweep down, spread out, and are emplaced. Where volcano flanks are dissected, deposits of concentrated flows accumulate as narrow, valley-confined depositional bodies (e.g., Rowley et al. 1985; Boudon et al. 1993; Miyabuchi 1999; Saucedo et al. 2004; Lube et al. 2007; Douillet et al. 2013; Major et al. 2013) (Fig. 2).

Deposits of small-volume high-concentration flows commonly display long and sinuous planforms (sometimes overlapping), even where unconfined by valley sides. On open slopes, flows are prevented from spreading out by lateral flow levees, and such deposits typically terminate in steep lobate flow fronts. Both the levees and flow fronts are coarse, highly frictional, and more fines-depleted than deposit interiors, commonly displaying openwork accumulations of gravel-size clasts (Sheridan 1979; Rowley et al. 1981; Fisher and Schmincke 1984; Cas and Wright 1987) (Fig. 3a, b). Deposits typically exhibit complexly digitate margins (Fig. 3c) where elongate lobes may overlap or crosscut

Fig. 3 Architectural features of deposits from small-volume, high-concentration pyroclastic density currents (PDCs). **a** Flow levee on lateral margin of pumiceous PDC on north flank of Láscar volcano (Chile), emplaced in 1993 on the surface of an older scoriaceous PDC deposit. Flow was toward camera. Note coarseness of clasts composing the levee and the shear zone separating the levee from the interior of the flow. Collapsed trekking pole for scale (arrow). USGS photo by T. Pierson. **b** Coarse openwork flow front of a lobe of the October 17, 1980, PDC on the north flank of Mount St. Helens (Washington, USA), taken one day after emplacement. USGS photo by T. Leighley. **c** Aerial view of highly digitate lobes of a PDC deposit emplaced during the June 12, 1980, eruption of Mount St. Helens. USGS photo by P. Rowley, June 13, 1980. **d** Multiple PDC flow lobes on north flank of Mount St. Helens, 1980. USGS photo (archival). **e** Toe of light-gray 1993 PDC debris fan, built on darker reddish-brown fan of scoria-flow deposits on the north flank of Láscar volcano. Note overlapping of individual flow deposits, digitate nature of flow fronts, and light-colored single-lane road in lower right for scale. USGS photo by T. Pierson. **f** Dissected pyroclastic debris fan (Zigzag Canyon) on southwest flank of Mount Hood (Oregon, USA), composed of dome-collapse PDC deposits and minor lahar deposits. Fan thickness above the lava flow near the bottom of the canyon (white arrow) is about 150 m, accumulated, mostly during one eruptive period about 1500 years ago. Note dark remnant of lava dome in upper right of photo (black arrow). USGS photo by T. Pierson

each other (Fig. 3d, e). Even though individual deposits are generally thin (< 1–10 m) and small in volume (< 0.2 km^3), large fans and thick valley fills can accumulate from repeated emplacement of hundreds to thousands of flows, particularly PDCs repeatedly emplacing deposits downslope of growing lava domes (Yamamoto et al. 1993; Miyabuchi 1999; Herd et al. 2005) (Fig. 3f). Relations between thickness and distance from source or thickness and slope angle can be complex (Boudon et al. 1993; Freundt et al. 2000; Saucedo et al. 2004). Unconfined larger-volume pumiceous flow units can produce broad sheet-like deposits with tapering margins (Cole et al. 2002).

PDC deposits from large eruptions (particularly caldera eruptions) commonly fill broad valleys and bury preexisting topography (Fig. 4); the largest may radiate in all directions from source vents and can cover areas of more than 10^5 km^2 (Sheridan 1979; Fisher and Schmincke 1984; Cas and Wright 1987; Freundt et al. 2000; Bryan et al. 2010; Hildreth and Fierstein 2012). When ponding in valleys or filling caldera depressions, deposit thicknesses from single eruptions can exceed 200 m (Scott et al. 1996; Bryan et al. 2010). Large pumiceous PDC deposits commonly decrease exponentially in thickness with distance from source (Freundt et al. 2000), and thickness in valleys is much thicker than on adjacent slopes and ridges (Brown and Andrews 2015).

Low-concentration PDCs, in comparison to valley-confined, high-concentration flows, tend to emplace terrain-mantling deposits a few centimeters to several meters thick, which in high-relief terrain are thicker and coarser-grained in valley bottoms and thinner and finer on hillslopes (e.g., Hoblitt et al. 1981; Bull and Cas 2000). In low-relief terrain, low-concentration currents form relatively thin sheets of typically uneven thickness (pinching and swelling) (Wilson and Houghton 2000). When associated with gravity-segregated basal underflows, deposits of low-concentration currents tend to form thin tongues of finer-grained material that extend beyond or on different trajectories from the associated basal-current deposits (Moore and Sisson 1981; Yamamoto et al. 1993; Miyabuchi 1999; Valentine and Fisher 2000) (Fig. 5). Dilute PDC deposits are commonly fringed by zones of singed vegetation, where tree leaves have been killed by heat but not burned (Winner and Casadevall 1981). Dilute PDCs generated by eruptions at stratovolcanoes may extend >25 km from source and cover hundreds of square kilometers in radiating fan-shaped patterns (Moore and Sisson 1981; Bursik et al. 1998; Valentine and Fisher 2000; Belousov et al. 2007). Dilute currents produced by caldera eruptions may emplace much more extensive deposits (Dade and Huppert 1996).

Upper surfaces of PDC deposits can be flat, slightly convex, or undulating, with patterns of lobes or flow-parallel furrows and ridges marking otherwise smooth cross-valley profiles (Charbonnier and Gertisser 2008). Pristine surfaces are commonly modified by fluvial channel incision to varying degrees within days to months of emplacement (Fig. 6). Some deposit surfaces may exhibit localized networks of polygons (tens of meters across) created by deeply penetrating, intersecting, decimeter-scale, orthogonal fractures that may be infilled with wind-blown ash (Whelley et al. 2012; Robinson et al. 2017) (Fig. 7). Smaller-scale polygonal fracturing (and formation of columnar jointing) has also been observed in thick hot PDC deposits (Wright et al. 2011). Formation of the large polygons is attributed to initial deaeration, subsequent post-depositional compaction, and thermal contraction, and they may focus fumarolic degassing of thick deposits (Whelley et al. 2012; Hildreth and Fierstein 2012). Large polygonal fracture patterns have not been reported in other types of volcaniclastic deposits, although they could feasibly form in polar environments (ice-wedge polygons) or evaporite basins (Kocurek and Hunter 1986; Wales et al. 2020). Surfaces of dilute PDC deposits composed of mainly ash-size particles may exhibit primary bed forms (dunes or antidunes) produced during emplacement with orientations related to flow direction (Hoblitt et al. 1981; Douillet et al. 2013), although similar eolian features having different orientations can be superimposed on such

Fig. 4 Aerial views of large valley-filling partially welded pumiceous pyroclastic density current (PDC) deposit—the "Valley of Ten Thousand Smokes" ignimbrite sheet, emplaced by the 1912 Novarupta-Katmai eruption in southwest Alaska (USA) (Hildreth and Fierstein 2012). **a** Middle part of the deposit, viewed from western margin; **b** western margin of deposit. Stream channel estimated to be on the order of 10 m wide. USGS photos by T. Pierson

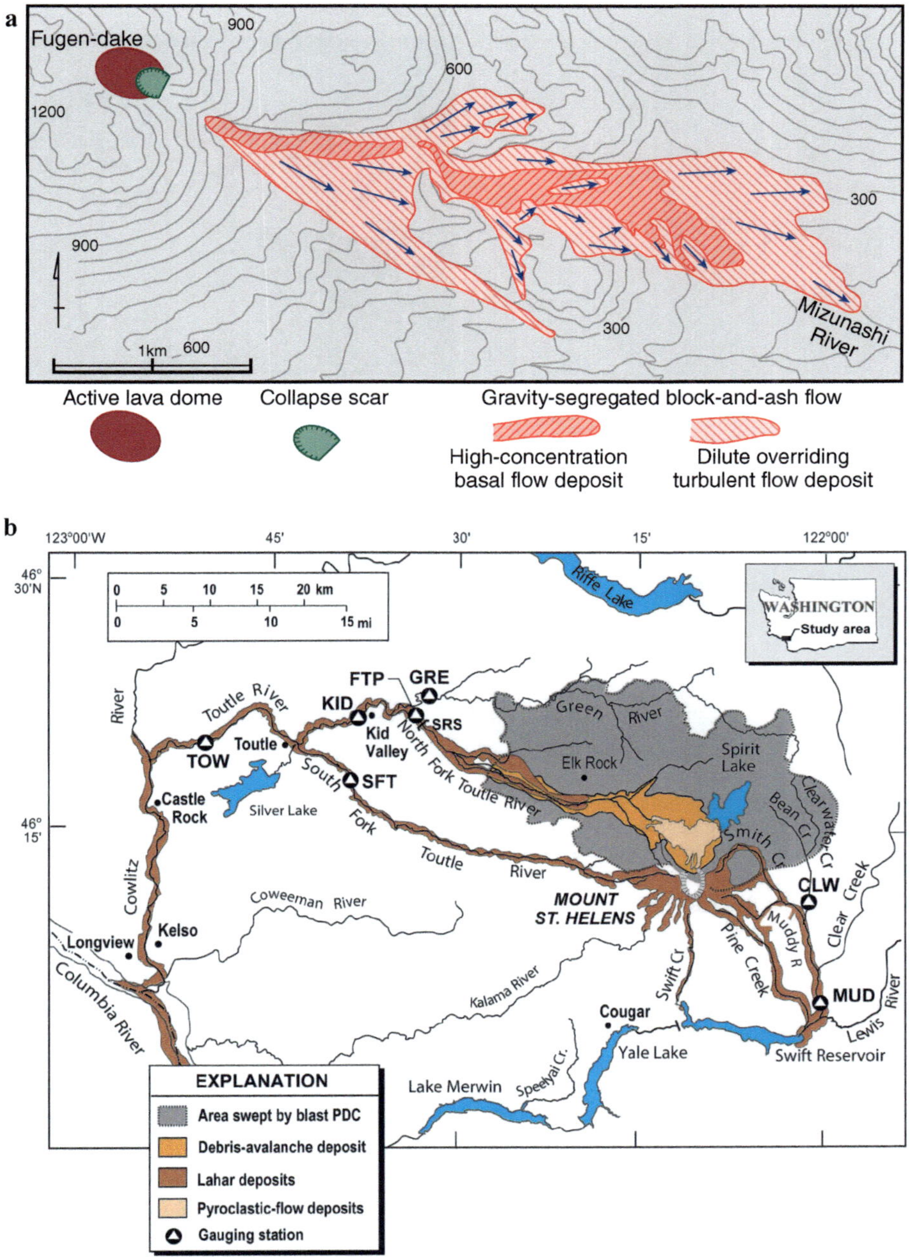

Fig. 5 Lateral extent of concentrated and dilute pyroclastic density current (PDC) deposits. **a** Deposits of dense basal currents and associated dilute currents from several lithic PDCs produced by collapses from the growing Unzen lava dome (Japan), 1991–1995 (modified from Yamamoto et al. 1993). **b** Deposit of the PDC associated with the directed blast during the May 18, 1980, eruption at Mount St. Helens (Washington, USA)—here labeled the "blast PDC," and the zone of accumulation of concentrated PDC deposits emplaced by column-collapse PDCs during the main Plinian phase of the eruption (here labeled "pyroclastic-flow deposits") (Lipman Mullineaux 1981). The "blast PDC" was deposited by progressively more dilute flow phases (Esposti Ongaro et al. 2012). These two depositional areas contrast with the areas affected by the passage and deposition of a large debris avalanche and several major lahars. Widths of lahar channels are exaggerated to make them visible at the scale of this map. Gauging stations shown were established to record volcanic water flood (VWF) discharge and sediment concentration following the 1980 eruption (Dinehart 1998)

Fig. 6 Upper surfaces of pyroclastic density current (PDC) deposits showing varying degrees of post-emplacement channel incision: **a** headward gully erosion on the minimally dissected surface of the partially welded "Valley of Ten Thousand Smokes" ignimbrite sheet from the 1912 Novarupta-Katmai eruption (Alaska, USA), more than 100 years after deposit emplacement. Eolian dunes on the deposit surface were noted in some locations. **b** Shallowly eroded surface of the nonwelded 2009 lithic PDC deposit in Chaitén (Blanco) River valley, about 6 km downstream from its lava dome source (steaming in background) at Chaitén volcano (Chile), about one year after emplacement. Dead trees in left foreground were engulfed in the deposit and killed, but not knocked down. **c** Rilled and gullied surface of the nonwelded 1991 pumiceous PDC deposit in Sacobia River valley about 2 months after emplacement, lower east flank of Mount Pinatubo (Philippines). Deposit filled the valley up to about 200 m deep in places (Scott et al. 1996); white bluff in foreground is where still-hot and dry deposit was being actively eroded by the river. **d** Deeply eroded 1991 pumiceous PDC deposit in upper Abacan River valley about 2 years after emplacement, east flank of Mount Pinatubo (Philippines). Stream channel in bottom of canyon is estimated to be 10–15 m wide. Photographed while standing on remnant of original fill surface. USGS photos by T. Pierson

surfaces (Smith and Katzman 1991). Large-scale flow-parallel furrows have also been noted on surfaces swept by high-velocity directed blasts (Moore 1967; Kieffer and Sturtevant 1988; Belousov et al. 2007).

Lithofacies

PDC deposits exhibit a wide range of lithofacies types and associations that typically vary both vertically and laterally within individual flow units (Branney and Kokelaar 2002). Vertical variation in a deposit reflects changes in processes and conditions over time at the lower flow boundary of the PDC—changes in the steadiness of current, magma composition, and eruption rate; horizontal variation reflects the nonuniformity of the current due to topography and evolution of the current with distance traveled (Branney and Kokelaar 2002). Lithofacies may grade into one another (e.g., Calder et al. 2000).

Lithofacies classifications for PDC deposits (e.g., Branney and Kokelaar 2002) tend to

a b

Fig. 7 Large-scale polygonal fracture patterns on pyroclastic density current (PDC) deposit surfaces: **a** surface of 1993 pumiceous PDC deposit on the north side of Láscar volcano (Chile), in the highly arid Altiplano region of northern Chile where little rain has fallen since emplacement. Person is standing at the intersection of two partially infilled fractures. **b** Aerial photo of fracture polygons on surface of large PDC valley fill (Valley of Ten Thousand Smokes), emplaced in 1912 by the Novarupta eruption. Sides of polygons are tens of meters long. USGS photos by T. Pierson

use pyroclast-specific terminology for particles and sedimentary textures, which are not always directly equivalent to the simplified sedimentological classification used in this book to facilitate comparison between wet and dry volcaniclastics (see Table 2 in Chap. "Deposit Identification and Interpretation"). Under the sedimentological classification, concentrated PDC deposits mainly have lithofacies categorized as massive and graded matrix-supported diamicts (Dmm, Dmg) and clast-supported diamicts (Dcm, Dcg), massive or stratified gravels or sandy gravels (Gm, Gh, Gw, Gs), and massive or stratified sands or gravelly sands (Sm, Sh. Sw, Ss). These lithofacies are also found in VDF and VHF deposits. In fact, lahar deposits can be composed entirely of PDC component materials if remobilized from a freshly emplaced PDC deposit (e.g., Pierson 1985; Pierson et al. 1996). Dilute PDC deposits tend to have sandy lithofacies, with the possible addition of massive or laminated sandy mud (Fm, Fl).

An early conceptualization of lithofacies associations for PDC deposits was genetic, assuming an idealized type-section that resulted from an assumed specific sequence of events or processes during an eruption. The interpretation of this idealized association included (a) an initial tephra-fall deposit at the base of the sequence; (b) a ground-surge deposit[1] (layer 1); (c) a thick pyroclastic-flow unit composed with a thin fine-grained basal layer (layer 2a) and a thicker, coarser, and massive layer; (d) an overlying ash-cloud deposit (layer 3); and in some cases (e) a capping fallout deposit and possibly a lava flow on top of the whole sequence (Sparks et al. 1973; Sheridan 1979; Wright et al. 1981; Wilson and Walker 1982; Fisher and Schmincke 1984). More recent work (Branney and Kokelaar 1992, 2002; Legros and Martí 2001; Sulpizio and Dellino 2008) has emphasized the variety in PDC deposit facies and the importance of nongenetic lithofacies descriptions for interpreting changing depositional conditions during the flow. Branney and Kokelaar (2002) especially focused on ways that different PDC lithofacies can form through progressive aggradation as flow conditions and properties change.

Erosion and Effects on Vegetation

PDCs of varying concentration can, in general, be highly erosive (Wilson and Walker 1981; Sparks et al. 1997; Grunewald et al.

[1]A ground-surge deposit (Layer-1 in the idealized sequence) is a fine-grained, variably thin, cross-bedded surge deposit formed by deposition at the dilute leading edge of the PDC.

2000; Brand et al. 2014) (Fig. 8). Erosion is most intense near vents, on steep slopes, and downslope of hydraulic drops. Scouring can occur on snow and ice surfaces, on substrates of unconsolidated deposits, and in the walls of bedrock canyons. On steep slopes (> 25°) underlain by volcaniclastic debris, PDCs can entrain enough non-juvenile particles to make up as much as 40–50% of flow material volume (Bernard et al. 2014). Non-juvenile clasts in deposits commonly reflect erosion of underlying fragmental deposits (Wilson and Walker 1981; Bryan et al. 1998a).

PDCs are especially erosive on snow and ice, as well as on soft rock and unconsolidated deposits (Sparks et al. 1997; Walder 2000a, b) (Fig. 8). In the May 18, 1980, eruption of Mount St. Helens (Washington, USA), lithic PDCs (boil-over type) flowing on the southeast flank of the volcano, followed by pumiceous column-collapse PDCs later in the day, incised channels up to 10 m in glacier ice (Brugman and Meier 1981). Later in the year, multiple pumiceous PDCs from post-May 18 eruptions from late May through October 1980 cut channels up to 40-m deep in unconsolidated deposits at the mouth of the breached crater (Rowley et al. 1981). At Redoubt Volcano (Alaska, USA), multiple concentrated PDCs during the December 14–15, 1989, eruption eroded ice canyons up to 30 m wide and 40 m deep (Trabant et al. 1994). Large lithic PDCs from the 1783 eruption of Asama volcano (Japan) eroded a channel up to 2 km wide and 40 m deep in older pyroclastic deposits on the lower flank of the volcano (Aramaki 1956). Concentrated PDCs during the November 13, 1985, eruption of Nevado del Ruiz eroded broad swaths across glacier surfaces within minutes to produce about 10^7 m^3 of meltwater (Pierson et al. 1990; Thouret 1990). The highly energetic directed blast PDC at Mount St. Helens carved U-shaped furrows parallel to flow direction that had depths up to several meters and lengths up to several hundred meters; they formed as far as 9 km from the vent (Kieffer and Sturtevant 1988; Kieffer et al. 2021). Dilute PDCs can also be erosive (Moore 1967; Fisher 1977; Fisher and Schmincke 1984; Wilson and Houghton 2000).

PDCs are capable of burning, breaking, flattening, toppling, or burying trees and other vegetation along their flow paths, commonly entraining charred woody debris (e.g., Hoblitt et al. 1981; Blong 1984; Miyabuchi 1999; Major et al. 2013) (Fig. 9). Wood debris entombed in hot deposits may be partially to completely charred, including large logs, depending on the emplacement temperature of the deposit, duration of exposure to heat, and vertical position in a deposit relative to cooling surfaces (Boudon et al. 1993; Voight and Davis 2000). In energetic dilute flows, the lateral force, suspended particles, and intense heat can scorch, severely abrade, knock down, break tops from, de-limb, or defoliate trees along their flow paths, with damage gradients decreasing with distance away from source (Blong 1984; Swanson et al. 2013). Zones of singed, standing dead trees also are common around the peripheries of dilute PDC flow paths (Kieffer 1981a, b; Winner and Casadevall 1981; Miyabuchi 1999). Trees broken-off or toppled by PDCs are aligned with localized flow direction (Kieffer 1981a, b; Swanson et al. 2013; Gardner et al. 2017, 2018) (Fig. 9a–c), although trees killed and desiccated by heat from partial burial may break off later with more random alignments (Major et al. 2013). Wood in some PDC deposits may not be charred, however, because low-concentration PDCs cool with distance traveled due to radiant heat loss and ingestion of air and other material along the flow path (Hoblitt et al. 1981; Moore and Sisson 1981; Voight and Davis 2000). Dilute, increasingly buoyant PDCs can lift off the ground when encountering topographic barriers or at the distal ends of their flow paths (Gardner et al. 2017, 2018). Where liftoff has raised a current above the ground surface, it can still break off the tops of trees yet leave the lower tree trunks and low vegetation minimally damaged (Kieffer 1981b; Moore and Sisson 1981; Pierson 1985).

Fig. 8 Erosion (scouring) of ice and rock by pyroclastic density currents (PDCs): **a** overlapping remnant PDC deposits containing ice clasts and refrozen meltwater on the summit ice cap of Nevado del Ruiz volcano (Colombia), three weeks after the Nov 13, 1985, eruption (Pierson et al. 1990; Thouret 1990). Most of the PDCs and meltwater transformed to lahars. Ice axe shaft is about 3.5 cm wide. USGS photo by R. Janda. **b** Aerial view of deep gouging and smoothing of glacier surface on the west flank of Nevado del Ruiz volcano (Colombia), caused 3 weeks earlier by PDCs erupted during the November 13, 1985, eruption. Meltwater from the massive scouring of ice and snow produced catastrophic lahars in multiple drainages. USGS photo by R. Janda. **c** Erosion (deep striations and plucking) of a canyon wall composed of welded tuffs on the north flank of Láscar volcano (Chile), caused by concentrated pumiceous PDCs channelize in a ravine during the 1993 eruption. This erosion is described in detail in Sparks et al. (1997). USGS photo by T. Pierson. **d** Nonvesicular rhyodacite boulders in a pyroclastic fan deposit that were planed off (30–50 cm) by abrasion from concentrated PDCs on Láscar volcano—same as those that formed striations in photo **c** but without distinct striations in the rock. Trekking poles is 1.3 m long. USGS photo by T. Pierson

◀ **Fig. 9** Effects of pyroclastic density currents (PDCs) on forest vegetation: **a** old-growth Douglas fir forest (1–3 m diameter trunks) at Fawn Lake, knocked down by a large, high-velocity, dilute PDC (lateral blast) during the May 18, 1980, eruption of Mount St. Helens (Washington, USA). For scale, two researchers are in a small boat in the middle of the lake. The top of the volcano can be seen on the left horizon about 15 km to the southeast. USGS photo by Lyn Topinka. **b** Total destruction area (tree removal zone) affected by a small low-temperature dilute PDC that spilled over the caldera rim (top of photo) during the explosive phase of the 2008 eruption of Chaitén volcano (Chile). Tree trunks and smaller stem and branches were not charred. Photo by F. Swanson, January 2011, used with permission. **c** Small-diameter trees abraded, broken, and toppled by a dilute low-temperature PDC (same flow as in **b**), 3–4 km downslope from rim. Burial of woody debris was <1 m. Even fine twigs and branches were not charred. Researchers can be seen (for scale) in the upper right on the skyline. USGS photo by T. Pierson. **d** Splintered trunks of trees sheared off by a concentrated PDC downstream of Chaitén volcano in the Chaitén River valley. USGS photo by T. Pierson. **e** Upright charred and partially charred tree and brush stems that had been buried up to several meters in a moderately hot concentrated PDC deposit downstream of Chaitén volcano in 2009. Stems were then sheared off by another flow at the top of the deposit that had buried them. Subsequent stream erosion exposed the truncated stems. A remnant of the original PDC deposit is visible in lower left foreground. This site was on the lateral margin of the PDC flow path; view is downstream. USGS photo by T. Pierson. **f** Tree kill along distal flow path of low-velocity, moderate-temperature 2009 PDC downstream of Chaitén volcano. PDC flowed among trees in the standing forest without charring wood. USGS photo by T. Pierson

2.2.2 Outcrop-Scale Characteristics of Deposits

Stratification and Bed Forms

Degree of stratification and presence of bed forms within individual deposits of PDCs is variable (Fisher and Schmincke 1984; Carey 1991; Branney and Kokelaar 2002) (Fig. 10). Deposits of PDCs can be massive, faintly/diffusely stratified, or well stratified. Bed surface roughness can play a role in development of strata (Brand et al. 2014). Local unconformities are common between flow units, and contacts between different units can be sharp or obscure (e.g. Sarocchi et al. 2011). Some contacts between flow units are planar and others may display undulations and flame structures formed by shear deformation (Pollock et al. 2019). Boundaries between flow units are usually defined by gradual to abrupt vertical changes in particle size or clast composition, by horizons free of gravel-size clasts (lapilli), or by layers or trains of lithic or juvenile clasts (e.g., Bryan et al. 1998a, b; Branney and Kokelaar 2002).

Stratification in pumiceous PDC deposits is highly variable, reflecting large differences in potential velocity, steadiness of flows, and solids concentration. These variations are reflected in deposits that can be massive with no obvious bed forms or well stratified with well-developed bed forms (Fisher and Schmincke 1984; Branney and Kokelaar 2002; Brown and Andrews 2015; Dellino et al. 2020). In deposits confined mainly to a volcano's flanks, longitudinal transitions occur from thick, coarse, massive deposits to thinner, finer-grained, bedded deposits over distances of hundreds to thousands of meters, which likely reflect decreases in flow concentration and average particle size due to selective deposition, continued fragmentation during flow, and a transition from concentrated-flow to dilute-flow depositional processes (Wright et al. 1981; Suzuki-Kamata et al. 2009). However, pumiceous PDCs also can deposit well-stratified deposits on steep proximal slopes that become more massive with distance from source and decreasing slope (Brand et al. 2016). Thus, pumiceous deposits may be massive, diffusely bedded, or distinctly bedded with sharp interbed contacts (Schmincke et al. 1973; Scott et al. 1996; Branney and Kokelaar 2002; Schmincke 2004).

Stratification in relatively small lithic PDC (block-and-ash-flow) deposits is generally limited (Moore and Melson 1969; Fisher and Schmincke 1984; Miyabuchi 1999; Saucedo et al. 2004; Charbonnier and Gertisser 2008; Sarocchi et al. 2011; Brown and Andrews 2015). Typically, there is a coarse, massive, gravel-rich basal depositional unit that was

Fig. 10 Varying degrees and types of stratification in pyroclastic density current (PDC) deposits: **a** massive concentrated PDC deposit (dome-collapse lithic PDC from the 1781 to 1793 Old Maid eruption) in White River valley, Mount Hood (Oregon, USA) (Scott and Gardner 2017). It contains variably vesiculated dacite fragments and charcoal. Small piece of carbonized wood to left of knife case, which is about 7 cm long. **b** Faint diffuse plane-parallel bedding within a matrix-supported pumiceous PDC deposit with faint cross-bedding (at top of photo) in the Abrigo ignimbrite, Tenerife (Canary Islands). Stratification is more easily apparent due to the presence of dark accidental lithic particles. **c** Stack of relatively thin massive density-graded PDC flow deposits in a layered sequence on Tenerife. The thin stratified fine-grained unit immediately below the pencil is interpreted as a water-reworked interbed. **d** Low-angle cross-bedding in PDC (base surge) deposit in a tuff ring sequence on Halla volcano (South Korea). Large blue marks on scale are 10 cm apart. USGS photos by T. Pierson

emplaced at low to moderate velocities by a highly concentrated phase of the current. This gravelly lithofacies may or may not grade upward to a finer-grained lithofacies (massive or bedded) that was emplaced by a more dilute, turbulent phase at the tail of the PDC. Contacts between the coarse and fine lithofacies emplaced by the concentrated and dilute flow phases can be gradational or sharp. However, in relatively distal depositional zones these flow phases can decouple to form two separate currents with different trajectories, leaving deposits in different (but often overlapping) areas (Fig. 5a). Deposit stratigraphy can be further complicated where vertical fallout of ash from co-PDC ash clouds caps the flow deposits.

Lithic PDC basal deposits can closely resemble VDF deposits at first glance, but intergranular voids are common in the PDC units and absent in VDF units. Basal zones depleted of gravel-size clasts are commonly observed in such lithic PDC deposits (and also in some pumiceous flow deposits), usually with gradational contacts into the overlying parts of the deposits (Fisher and Schmincke 1984; Branney and Kokelaar 2002). Such zones resemble sandy sole layers found in VDF deposits.

Lateral blasts are a type of highly erosive, high-velocity, and strongly density-stratified lithic PDC that emplace relative thin deposits over broad areas (Gorshkov 1959; Gorshkov and Dubik 1970; Hoblitt et al. 1981; Waitt 1981; Siebert et al. 1989; Bursik et al. 1998; Valentine and Fisher 2000; Belousov et al. 2007, 2020; Esposti Ongaro et al. 2012; Brown and Andrews 2015). Their deposits commonly contain several distinct strata with a combined overall normal grading that reflects progressive aggradation by currents that become more dilute with time and with distance traveled. Directed blast deposits typically have at least three layers: (1) a massive fines-depleted gravelly sand unit that commonly includes an unsorted basal layer of gravel and smashed woody vegetation (sometimes identified as a separate unit), (2) an intermediate poorly sorted, massive, and fines-rich sandy layer that may grade upward to small-scale bedding and cross-bedding, and (3) an upper, relatively fine-grained, normally graded (massive) layer of ash and accretionary lapilli emplaced by co-PDC tephra fall (Hoblitt et al. 1981; Waitt 1981; Belousov et al. 2007).

PDC bed forms are observed on deposit surfaces and interpreted from cross-bedding in vertical exposures. They are common, particularly in fine-grained deposits emplaced by low-concentration PDCs and reflect flow conditions in PDCs at the lower flow boundary during deposition. Wavy (lensoidal), complex, and low-angle (< 15°) cross-bedding observed in deposits of dilute PDCs—bedding types linked to antidunes, chute-and-pool structures, and cyclic steps in fluvial deposits—are suggestive of deposition by PDCs during critical to supercritical flow (e.g., Moore 1967; Waters and Fisher 1971; Schmincke et al. 1973; Fisher and Schmincke 1984; Brand et al. 2016; Douillet 2021). However, particle transport by dilute PDCs differs in various ways from that in sediment-laden VWFs, and the genetic implications of bed forms observed in PDC deposits may be different from those observed in fluvial deposits (Douillet et al. 2013; Douillet 2021). Four types of dune-like bed forms have been identified in planform in PDC deposits: (1) elongate (longitudinal), (2) transverse, (3) lunate, and (4) two-dimensional (crudely transverse and organized in trains) (Douillet et al. 2013). A distinctive aspect of PDCs generated by phreatomagmatic eruptions is that particles can be wet and cohesive, which can affect bed-form construction and can cause cohesive fine-grained deposits to be plastered onto nearly vertical surfaces (Fisher and Schmincke 1984; Houghton et al. 2000b) (Fig. 11).

PDC dune forms differ from those in fluvial deposits in several aspects (Valentine and Fisher 2000): (1) PDC dune forms are larger than could be formed by water flows of reasonable size, with wave lengths up to 20 m and heights up to 2 m; (2) backset angles of PDC dune forms are steeper than in fluvial dunes or antidunes; (3) PDC dune forms typically preserve both stoss-side and lee-side layering, whereas fluvial dune forms usually preserve only the lee-side strata; and (4) individual strata within PDC dune forms are relatively massive and poorly sorted compared to strata in fluvial dunes.

PDC dune forms are also distinctive when compared to eolian dunes sourced from reworked PDC or airfall tephra deposits. Dune forms in PDC deposits (1) display climbing cross-beds (eolian dunes do not), (2) preserve both lee- and stoss-side strata (eolian cross-beds are preserved only on the lee side), and (3) typically have low-angle foreset cross-beds (eolian foresets are typically high-angle) (Smith and Katzman 1991). Primary PDC deposits are also more poorly sorted than highly fines-depleted, wind-reworked PDC sediments.

Outsized clasts may occur in otherwise fine-grained PDC deposits (e.g., Bryan et al. 1998a, b), having been emplaced by flows or as ballistic bombs. If emplaced as part of the flow, they typically cause minimal disruption to the bedding below and around the clasts (Fisher

Fig. 11 Near vertically dipping, initially wet phreatomagmatic pyroclastic density current (PDC) deposit, plastered on upslope side of basalt boulder on flank of a tuff cone, Halla volcano (South Korea). Note pen for scale. USGS photo by T. Pierson

and Schmincke 1984) (Fig. 12a). Large clasts emplaced as ballistic blocks or bombs create small asymmetrical impact craters (bomb sags) in underlying deposits (Fig. 12b).

Gravel-Size Clast Framework

Large pumiceous PDC deposits typically have distinctively matrix-supported internal structures (Freundt et al. 2000; Branney and Kokelaar 2002) (Fig. 13a). Smaller-volume pumiceous PDC deposits, while being largely matrix-supported in the body of the deposit, commonly have upper surfaces, lateral margins, and flow fronts that are clast-supported with an openwork structure (Wilson and Head 1981). Zones within large pumiceous PDCs also can be clast supported (Fig. 13b). Lithic PDC deposits can be clast-supported or matrix-supported.

Fabric

Pumiceous and lithic PDC deposits exhibit strong particle fabrics of primary and secondary origin if non-equant clasts are present. Primary fabrics typically manifest as (1) bed-parallel alignment of clast long axes (a-axes) with flow (shear) direction for gravel-size clasts that are elongate (rod-shaped), (2) upslope-dipping imbrication of flat platy or blade-shaped gravel clasts on the aggrading bed surface, and (3) likely a combination of both for clasts with intermediate shapes (Schmincke et al. 1973; Kamata and Mimura 1983; Mimura 1984; Boudon et al. 1993; Bryan et al. 1998b; Hughes and Druitt 1998; Branney and Kokelaar 2002; Brown and Andrews 2015). However, localized ground interactions and thalweg directional variations, clast size and shape variations, and PDC dynamics can affect primary fabric development (Branney and Kokelaar 2002). Secondary fabrics can be formed shortly after deposition (and primary fabrics can be disrupted) by post-depositional deflation, degassing, and compaction of deposits, and by load flattening of hot pumice/scoria lapilli (fiamme) (Fisher and Schmincke 1984; Branney and Kokelaar 2002).

Grading and Density Segregation

Grading patterns within PDC deposits can be extremely varied and are influenced not only by flow dynamics and the sequence of deposition by progressive aggradation, but also by particle densities and gravity segregation during flow (Branney and Kokelaar 2002). In general, deposits show grading mainly with respect to the coarsest clasts present (i.e., coarse-tail grading); size distributions below a threshold

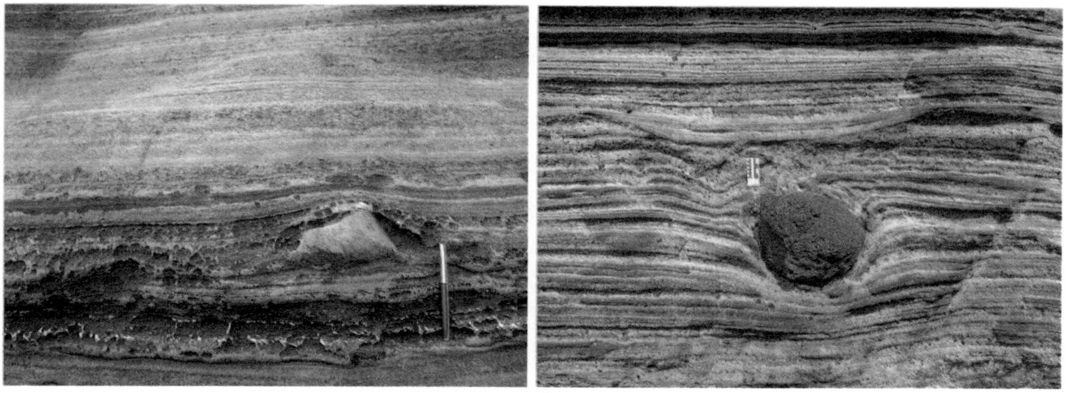

Fig. 12 Outsized clasts disrupting bedding in sand-size volcaniclastic deposits on outer flanks of a tuff ring, Halla volcano (South Korea): **a** pyroclastic density current (PDC)-transported clast resting on undeformed beds and buried by subsequent over-draping tephra fall (TF) and PDC deposits. Note pen for scale. **b** Ballistic clast resting at bottom of a bomb sag in bedded PDC and TF deposits, emplaced by ballistic impact. White scale card is about 14 cm long. USGS photos by T. Pierson

particle diameter tend to be identical throughout the deposit (Sparks 1976; Fisher and Schmincke 1984; Cas and Wright 1987). Pumiceous PDC deposits can be normally graded, ungraded, or inversely graded overall (Fisher and Schmincke 1984; Branney and Kokelaar 2002). Those containing both pumice clasts and denser lithic clasts may exhibit density-segregation grading with larger-than-average pumice particles concentrated near the top of the deposit and larger than average lithic clasts concentrated near the base (Figs. 10c and 14). Color contrasts between light-colored pumice and dark lithic fragments help make density-segregation grading obvious in outcrop. Lithic PDC deposits are usually either inversely graded or ungraded (Fig. 10a), occasionally showing normal grading (Moore and Melson 1969; Boudon et al. 1993; Francis 1993; McPhie et al. 1993; Miyabuchi 1999).

Gas-Escape Structures and Deformation

Gas-escape structures occur in many PDC deposits, primarily in those that are nonwelded (Walker 1971; Fisher and Schmincke 1984; Cas and Wright 1987; Freundt et al. 2000; Branney and Kokelaar 2002; Brown and Andrews 2015; Pacheco-Hoyos et al. 2020). They are particularly common in large, thick, pumiceous PDC deposits with prolonged durations of degassing (Fisher and Schmincke 1984; Cas and Wright 1987). These features have two basic components, which are not always differentiated: (1) the main gas-escape pathways—subvertical to vertical, pipe- or sheet-shaped, high-porosity elutriated conduits, through which gas escapes rapidly (Fig. 15a); and (2) cylindrical aureoles of mineralogically altered and cemented particles in the PDC deposit surrounding the conduits (Fig. 15b). Gas-escape structures in PDC deposits (conduits $\pm$ aureoles) are variously termed *gas-escape pipes, elutriation pipes, degassing pipes, gas segregation pipes, fossil fumarole pipes*, and *lapilli pipes*; aureoles are referred to as *alteration halos*, *fossil fumaroles*, *fumarolic mounds/pinnacles/ridges* (Sheridan 1970; Fisher and Schmincke 1984; Holt and Taylor 1998; Freundt et al. 2000). Soft-sediment deformation related to the fluidized condition of freshly emplaced PDC deposits has also been reported (Douillet et al. 2015). Syn-eruptive, soft-sediment deformation of deposits from dilute PDCs was documented at several volcanoes, with deformation interpreted to have been triggered by granular shear, dynamic pore pressure, ballistic impacts, and shock waves.

Fig. 13 Examples of matrix-supported and clast-supported internal structures within pumiceous pyroclastic density current (PDC) deposits. **a** Ungraded matrix-supported structure in ignimbrite, Tenerife (Canary Islands). Coin shown for scale is 22 mm in diameter. **b** Ungraded clast-supported zone within a thick PDF deposit that graded upward to a fine-grained facies, Tenerife. Pen is 14 cm long. **c** Variably graded matrix-supported pumiceous PDC deposits (person at base of outcrop for scale) on the lower north flank of Mount St. Helens (Washington, USA). Photos by B. Brand, used with permission

Gas-escape structures are most common in deposits of concentrated pumiceous PDCs and form in two ways. They can develop during and immediately after the incremental accretion of individual deposits when retained gases within the inflated deposit—both hot ingested air and exsolving magmatic gas escaping from cooling vesicular clasts—stream upward along preferred pathways and winnow out fines from within the deposit as it consolidates. They can also develop within a sequence of hot deposits that effectively cool, consolidate, and compact together (as a *cooling unit*) after emplacement (Druitt 1995; Branney and Kokelaar 2002; Brown and Andrews 2015; Pacheco-Hoyos et al. 2020). Preservation of gas-escape structures requires that there is little or no shearing of the aggrading deposit while gas is escaping. Although gas-escape pipes may be present in some lithic PDC deposits, the presence of these structures is generally attributed to PDC emplacement on wet ground or on buried vegetation, where escape of steam or heat-generated volatiles through the deposit produces the structures (Brown and Andrews 2015). Small pipes have been recorded in dilute PDC deposits, but rather than originating within those deposits, they are thought, instead, to reflect escape of gas from larger underlying PDC deposits (Fisher 1979). Degree of fines elutriation in degassing structures

Fig. 14 Well defined density segregation of light-colored pumice and denser, dark-colored lithic clasts in the middle pyroclastic density current (PDC) flow deposit, Tenerife (Canary Islands) (see also Fig. 10c). Where pumice clasts and accidental lithic particles are larger than the average particle size, density segregation leads to density-segregation grading in deposits. Upper unit here shows accumulation of dense clasts at its base (top of deposit not visible); lower unit shows no segregation of pumice clasts at its top. Pen is 14 cm long. USGS photo by T. Pierson

depends on the flux and velocity of escaping gas; minimal or low-velocity gas escape may only remove fine vitric ash, whereas high flux gas escape can remove almost all ash particles, leaving pipes defined by openwork lapilli. Gas flux in some pipes can form accretionary lapilli (Van Eaton and Wilson 2013).

Gas-escape pipes are best developed in the upper parts of deposits or cooling units (Fisher and Schmincke 1984). Pipes can be up to 10 m long and up to 30 cm in diameter, with subcircular to irregular cross-sections (Fisher and Schmincke 1984), but they are commonly only 1–2 m long and occur at different levels within a deposit, commonly pinching out before reaching the deposit or cooling-unit surface. Tubular pipes may be single, branching, or anastomosing in form, and they are generally sinuous, and initially nearly vertical in orientation. Late-stage shear or burial compaction may deform pipes (Branney and Kokelaar 2002).

The presence of pipe structures alone is not conclusive evidence of a PDC origin, because similarly shaped water-escape pipes can form in lahar deposits. However, intense oxidation of mineral grains within conduits, strong cementation of particles around pipes, and the presence of chemically altered aureoles are not observed in lahar water-escape pipes.

Material Consistency

Elutriation of fine ash during flow of dry pumiceous PDCs generally causes an enrichment of crystals and lithic particles and a depletion of fine-grained vitric particles, which can lead to an overall depletion of fines in some pumiceous deposits (Fisher and Schmincke 1984). However, fines depletion generally does not occur in wet deposits resulting from explosive magma-water interactions or in large ignimbrites (Fisher and Schmincke 1984). Overall fines depletion is common in concentrated

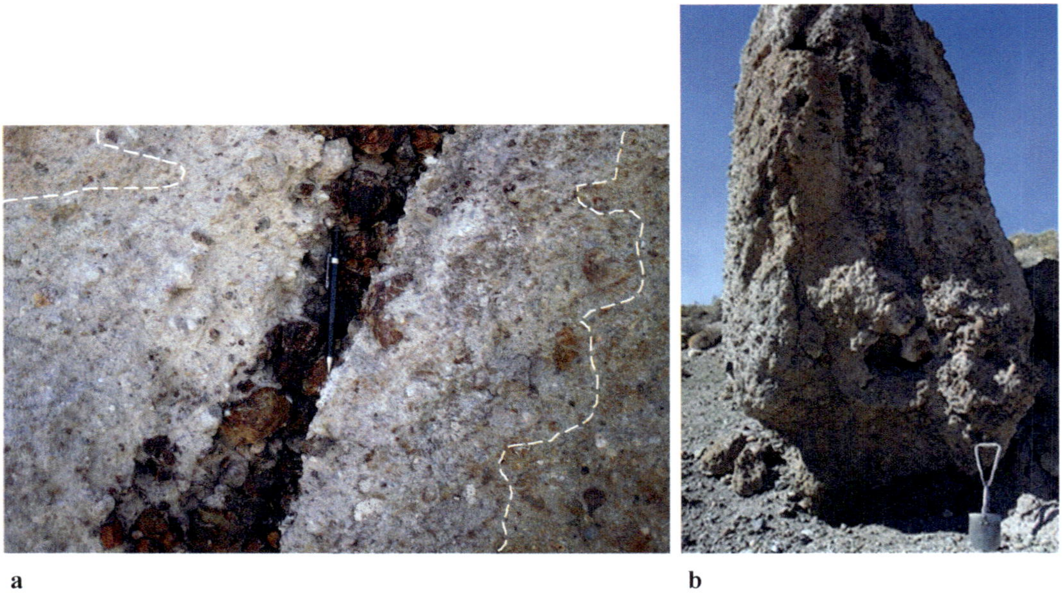

Fig. 15 Gas-escape pipes in a pyroclastic density current (PDC) deposit—the 27,000-year-old Soncor ignimbrite, west of Láscar volcano (Chile): **a** close-up view of a subvertical gas-escape pipe showing residual lapilli (matrix material removed by gas elutriation) that were cemented and stained in situ by vapor-phase mineral crystallization during gas escape. The outlined whitish zone around the pipe is an altered zone within the aureole. Black pencil in pipe is 14 cm long. **b** A hardened, erosionally resistant and chemically altered aureole (now exposed by erosion as a column) that formed around another gas-escape structure in the same deposit. USGS photos by T. Pierson

lithic PDC deposits (e.g., Boudon et al. 1993; Freundt et al. 2000). Our experience has been that matrix material in many small-volume PDC deposits (pumiceous and lithic types) is particularly depleted of extremely fine ash (silt- and clay-size particles), and this affects physical consistency of PDC deposits in comparison with wet-flow deposits.

Fines-depleted deposits from concentrated PDCs have a consistency that is typically loose to friable when dry, soft to slightly hard when moist, and nonsticky and nonplastic when wet (Table 4 in Chap. "Deposit Identification and Interpretation"). In general, fresh unconsolidated deposits of concentrated PDCs commonly look at first glance like concentrated VDF deposits, but they are less coherent, more prone to raveling, and easier to dig. Deposits of dilute PDCs tend to be firmer/harder than those of concentrated currents, and they can be similar in consistency to those of dilute VHFs. Both dilute PDC and VHF deposits are only partly fines-depleted, so other criteria are needed for differentiation.

Welding is the hardening (to near lithification) of very hot and thick PDC deposits that results from plastic deformation and sintering together of very hot glass fragments in the particle mixture (Fisher and Schmincke 1984; McPhie et al. 1993). The initially unconsolidated particles are formed into a solid mass by pressure and heat without melting of the glass to the point of being liquified. Welding can also occur in thick, hot, proximal, tephra-fall deposits. It is categorically diagnostic of emplacement by primary pyroclastic processes.

Welding in thick deposits commonly results in the formation of *fiamme*—pumice or scoria fragments compressed into dense, flattened, lenticular glassy clasts (usually black) that provide a bedding-parallel fabric or foliation to the unit (Fig. 16).

a b

Fig. 16 Compaction-deformed glassy clasts in thick, densely welded pumiceous pyroclastic density current (PDC) deposits exposed in the Las Cañadas caldera wall, Tenerife (Canary Islands): **a** lenticular fiamme. **b** Irregularly deformed glassy inclusion; origin is uncertain. USGS photos by T. Pierson

Zones of complete and partial welding are common in large-volume pumiceous deposits, particularly in the lower half of depositional units. However, hardening, densification, and increased cohesion of hot PDC deposits can also occur due to diffuse devitrification crystallization and vapor-phase crystallization of alteration minerals in the upper parts of thick PDC deposits without sintering (Wright et al. 2011). In addition, relatively rapid, low-temperature weathering reactions in humid tropical climates can also cause extensive induration of pumiceous PDC deposits. These reactions involve devitrification and hydration of glass particles and the subsequent cementation by crystallization of secondary minerals (Branney and Sparks 1990; Brown and Andrews 2015).

Evidence of Heat

Most PDC deposits show abundant evidence of very hot average emplacement temperatures—well above 100 °C, which is the upper limit for wet flows (Fisher and Schmincke 1984; Cas and Wright 1987; Freundt et al. 2000). Lithic PDCs commonly have emplacement temperatures of 400–600 °C (Brown and Andrews 2015), and welding requires sustained temperatures at or above 500–650 °C (Freundt et al. 2000). Banks and Hoblitt (1996) reported localized PDC-deposit temperatures up to 850 °C in a deposit emplaced during the May 18, 1980, eruption at Mount St. Helens. In general, PDC emplacement temperature depends on type of eruption (magmatic, phreatomagmatic, or phreatic); magma type and initial temperature; deposit thickness; degree of interaction with cooler groundwater, air, or ground surfaces during eruption and flow; PDC concentration; and distance traveled from source (McClelland and Druitt 1989).

Some PDCs have relatively cool emplacement temperatures and therefore may not produce strong outcrop-scale evidence of high heat. For example, a lithic PDC from Chaitén volcano (Chile) had an emplacement temperature <300 °C and charring of wood was limited to surficial scorching of only some wood surfaces—on logs buried in the deposits for longer periods than the uncharred wood (Major et al. 2013). As apparently was the case at Chaitén, initial temperature of a dome-collapse PDC may be relatively cool if a failed mass involves older cooler sectors of a lava dome, and all PDCs cool to some degree during flow by ingesting ambient-temperature air and entraining cool particles. PDCs from phreatomagmatic eruptions also cool from interaction with ground and surface water. Depending on initial flow temperature and on the amount of mixing with water and air, "cool" PDCs can have emplacement temperatures from <100° to about 350 °C (Moore 1967; Zanella et al. 2007; Porreca et al. 2008; Major et al. 2013). Furthermore, large roughness elements on ground surfaces in depositional zones, such as closely spaces buildings, can induce additional flow turbulence, air ingestion, and cooling (Zanella et al. 2007).

Very hot PDCs typically produce widely distributed carbonized wood (partially or completely charred logs or other wood fragments), abundant prismatically jointed and breadcrust clasts (particularly in lithic PDC deposits); pink coloration of deposit tops, thermoremanent magnetic orientation of clasts, phreatic explosion pit craters on deposit surfaces, and welding of particles in particularly hot and thick deposits. These are all reliable outcrop-scale indicators of high heat and a PDC origin, especially when multiple indicators are present in an outcrop (see Sect. 3.9 in Chap. "Deposit Identification and Interpretation"). However, the occurrence of only charcoal fragments, breadcrust clasts, or prismatically jointed blocks in a deposit does not necessarily preclude transport by a cold flow; they can be entrained from reworked pyroclastic deposits and redeposited by cold lahars or VWFs. Nevertheless, such heat indicators are much more abundant in PDC deposits than in lahar deposits (Smith and Lowe 1991).

2.2.3 Particle-Scale Characteristics of Deposits

Particle Composition

Most PDC deposits are composed of monomictic or oligomictic mixtures of rock particles. See Sect. 4.1 in Chap. "Deposit Identification and Interpretation" for definitions of these terms. Some PDC particle mixtures can be polymictic, such as PDCs initiated by (a) phreatomagmatic stratocone eruptions consisting largely of wall-rock lithic fragments (Vespermann and Schmincke 2000) and (b) phreatomagmatic maar explosions containing significant volumes of country-rock (wall rock or bedrock) fragments along with juvenile material (Cas and Wright 1987). Furthermore, some PDC deposits may be polymictic at their bases due to entrainment of various rock fragments by the flow heads but monomictic in the upper parts of their deposits, which were emplaced by the tails of the currents (e.g., Moore and Sisson 1981).

The minor amounts of lithic particles found in most PDC deposits are mainly stripped from conduit walls by explosive eruptions (cognate and accessory lithic particles). These include both older lava rock and bedrock fragments. Non-juvenile clasts up to 10 m in diameter have been reported in some PDC deposits from caldera-forming eruptions (Shawe and Snyder 1988). Accidental lithic fragments are scavenged from the overridden ground surface (Sparks 1976). Overall lithic contents can range from zero to more than 50 vol% (Sparks 1976; Bryan et al. 1998a), but in pumiceous PDC deposits they rarely exceed 5 vol% (Fisher and Schmincke 1984). Basal layers of large PDC deposits can be rich in woody debris as well as rock fragments (e.g., Waitt 1981).

Monomictic PDC deposits typically have matrix material and clasts that are the same color (e.g., Fig. 10b), which contrasts with typically multicolor polymictic deposits of most lahars, VDAs, and VWFs. Note that many outcrop faces in the field may appear to be uniformly colored due to a thin wash of matrix fines coating clasts; a few clasts may need to be broken open to ensure uniform composition. Oligomictic lithic PDC deposits with several different colors of clasts, on the other hand, can closely resemble VDF deposits (Fig. 17).

The componentry of pumiceous PDC deposits includes gravel-size clasts (particles >2 mm) composed almost entirely pumice fragments (together with minor lithic fragments) that typically have vesicularities over 50 vol% (Polacci 2005). The ash-size (sand-size) components of pumiceous PDC deposits include crystals, glass shards, pumice or scoria fragments, and dense-rock fragments (juvenile or lithic) in varying proportions, depending on rock type and eruption mechanism (Fisher and Schmincke 1984; Cas and Wright 1987). Proportions of these grain types in the matrix fraction can be helpful in reflecting flow type or origin (Yamamoto et al. 1993; Freundt et al. 2000; Carn et al.

Fig. 17 A dome-collapse lithic pyroclastic density current (PDC) deposit at Chaitén volcano (Chile) that is oligomictic and composed mainly of several different types, vesicularities, and colors of glassy rhyolite particles, as well as a small percentage of country-rock lithic particles. Camera lens cover is 5.8 cm in diameter. USGS photo by T. Pierson

2004; Herd et al. 2005; Major et al. 2013). Ash-size fractions of proximal pumiceous PDC deposits are typically rich in vitric shards due to abrasion of larger pumice clasts during flow. These small shards of broken volcanic glass appear as small platy or cuspate fragments that were broken from the walls of larger bubble vesicles; they are also termed bubble-wall fragments (Fisher 1963). The largest ash-size particles may include pumice fragments with intact vesicles. Deposits in medial to distal locations, where PDCs had prolonged runouts, typically contain an ash-size fraction enriched in crystals at the expense of vitric shards. This is due to preferential loss of lower density shards during elutriation (Walker 1972; Brown and Andrews 2015).

Lithic PDC deposits mainly contain dense to only moderately vesicular clasts, but clast populations can include some pumice or scoria (Charbonnier and Gertisser 2008). Dense fragments of dome rock commonly have a high vitric component, but fine particles lack the cuspate fragments of broken bubble walls originating from disintegrated pumice. Lithic PDC deposits may contain some cognate and accidental lithic fragments from their source vents, but they also entrain abundant exotic clasts along their flow paths (e.g., Major et al. 2013).

Ash pellets and *accretionary lapilli* are, respectively, unstructured and concentrically structured aggregates of ash particles that form in large, turbulent, density-stratified PDCs and are sometimes preserved in PDC deposits. Ash particles, turbulently suspended and frequently colliding, clump together due to electrostatic charges and surface tension of condensing water (Houghton et al. 2000b; Ross and White 2005; Brown et al. 2010; Van Eaton and Wilson 2013,

2015). They are especially common in eruptions involving magma-water interactions.

Structureless ash pellets form in lofted dilute co-PDC ash plumes and are usually deposited with co-PDC fall deposits beyond the margins of PDC deposits themselves (Brown et al. 2010). However, if ash pellets drop back into the denser and hotter parts of moving PDCs, they can be preserved in the upper parts of PDC deposits (Ross and White 2005). Accretionary lapilli, with multiple concentric layers of fine particles (Fig. 18), are thought to develop mainly in the highly turbulent, upper regions of PDCs where they can repeatedly recirculate, accrete additional layers, and be hardened by heat (Brown et al. 2010). If present, accretionary lapilli are generally preserved in the upper parts of PDC deposits (Brown et al. 2010). The key factor is prolonged suspension and recirculation, which is required for the accretion of concentric layers on accretionary lapilli.

Particle Roundness, Shape, Fracturing, and Surface Texture

Low-density pumice or scoria clasts in PDC deposits (including those emplaced by dilute flows) are commonly subrounded to well-rounded due to interparticle abrasion during flow, although clasts may break during transport, producing some angular clasts in the mix (Rowley et al. 1981; Fisher and Schmincke 1984; Schmincke 2004; Manga et al. 2011) (Fig. 19). Dense clasts in dome-collapse lithic PDC deposits tend to be angular, although

Fig. 18 Partially scraped vertical section of matrix-supported accretionary lapilli in a phreatomagmatic ignimbrite from the 25.4 ka Oruanui eruption of Taupo volcano (New Zealand). Left side of yellow tape (not scraped) shows weathered surface of the outcrop, with delicate lapilli structures preserved. Right side shows the freshly scraped surface, which reveals faint cross-sectional outlines of the centimeter-sized accretionary lapilli with concentric internal structures, floating in an extremely fine-grained ash matrix. Outcrop site is near the caldera margin. USGS photo by A. Van Eaton

subangular to subrounded boulders have been observed in these deposits only a few kilometers to tens of kilometers from source (Boudon et al. 1993). Some lithic PDC deposit outcrops show vertically upward improvement in smoothing and rounding of clasts, with breakage of large clasts commonly occurring at the base of the flow (Sarocchi et al. 2011). Glassy juvenile particles in hydrovolcanic deposits (dominantly ash and lapilli) have angular, fracture-controlled surfaces and blocky-equant shapes (Vespermann and Schmincke 2000). Abrasion between clasts in concentrated PDCs produces abundant ash, with much of it elutriated from basal flows to produce overriding dilute PDCs or lofted co-PDC plumes.

Slowly cooled juvenile clasts in PDC deposits (mainly lithic PDC deposits) commonly exhibit *breadcrust* fracturing and prismatic jointing (Fisher and Schmincke 1984). Prismatically jointed blocks form from cooling of already solid blocks and may be quite dense, whereas breadcrust surface textures form due to rapid cooling of solid outer rinds while cores are still vesiculating and expanding while ductile (Benage et al. 2014). Such fracturing is rarely observed in clasts within pumiceous or scoriaceous deposits, although mafic spatter clasts may show breadcrust surfaces. Breadcrust and prismatically jointed clasts may be eroded from lithic PDC deposits and transported by lahars, but the relative abundance of these types of clasts is likely lower in lahar deposits.

Particle-Size Distribution and Sedimentary Texture

Particle sizes over the full spectrum of PDC deposits range from micron-scale extremely fine ash to large blocks up to 10 m in diameter; size distributions are typically bimodal to polymodal (Walker 1971; Sparks 1976; Boudon et al. 1993; Freundt et al. 2000; Grunewald et al. 2000;

Fig. 19 Rounded to subrounded pumice and scoria clasts in a steep frontal lobe of the 1993 PDC deposit on the north side of Láscar volcano (Chile). Note the openwork structure (numerous intergranular voids) of this clast-supported deposit margin. Lens cap is 5.8 cm in diameter. USGS photo by T. Pierson

Sarocchi et al. 2011), although in some cases distributions can be unimodal (Hughes and Druitt 1998). Large volume pumiceous flow deposits can be rich in very fine and extremely fine ash (> 50 wt%) (e.g., Heming and Carmichael 1973; Druitt and Bacon 1986; Fierstein and Hildreth 1992), but the proportion of gravel-size clasts (those easily visible in outcrop) to matrix material varies widely. Size distributions are therefore highly variable—mixtures can be larger-clast-enriched, larger-clast-depleted, fines-enriched, and fines-depleted lithofacies. Despite these seeming differences in size distribution, most PDC particle-size distribution modes are typically in the range of medium to coarse ash (Valentine and Fisher 2000).

Average size of particles in pumiceous PDC deposits generally decreases with distance from vent. Maximum size of pumice clasts also varies with distance from vent but shows both decreases and increases, whereas maximum size of vent-derived lithic particles decreases exponentially with distance (Freundt and Schmincke 1992; Freundt et al. 2000). In deposits of highly dilute pumiceous currents, average particle sizes are finer than in concentrated PDCs. Median diameters range from -3.0 to $+4.0\varphi$ (8 mm–63 μm), and size distributions lack the pronounced fine and coarse tails of size distributions in deposits emplaced by concentrated basal flows (Valentine and Fisher 2000). Sorting values are commonly 1.5–4.5φ, and deposits of phreatomagmatic eruptions are particularly rich in fine ash and ash aggregates (ash pellets) (Brown and Andrews 2015).

Lithic PDC deposits are typically more strongly fines-depleted than most pumiceous deposits. This is probably because moving pumiceous flows are readily producing new fines through clast abrasion, even as other fines are lost through elutriation. Commonly, very fine and extremely fine ash contents in lithic PDC deposits (< 0.125 mm) are limited to 3–7 wt% (Kuntz et al. 1981; Boudon et al. 1993; Freundt et al. 2000; Sarocchi et al. 2011). In addition, lithic PDC deposits typically contain more gravel-size (non-matrix) clasts (i.e., clasts are more durable) than pumiceous flow deposits, and total gravel-size clast content may significantly exceed total ash content (Sarocchi et al. 2011). However, size distributions in some lithic PDC deposits from flows with relatively long runout distances appear to be a function of travel distance and reflect some disaggregation and diminution of the initially abundant large clasts. In these cases, field data show: (1) the proportion of fines increases with distance from source despite fines elutriation; (2) maximum clast size on deposit surfaces decreases with distance from source; (3) deposits show overall progressive fining with distance from source; and (4) ash content varies vertically within deposits, with fine ash commonly even more depleted near the upper surface of the deposit (Saucedo et al. 2004; Sarocchi et al. 2011). Where noted, fines depletion in unstratified deposits is an important clue in distinguishing PDC deposits from VDF deposits.

Published sorting coefficients (Inman criteria) in pumiceous PDC deposits range from about 1.5 to 5.0φ—well sorted to very poorly sorted (see Table 6 in Chap. "Deposit Identification and Interpretation") (Walker 1971; Kuntz et al. 1981; Valentine and Fisher 2000)—although some stratified lithofacies have slightly better sorting (Rowley et al. 1985; Wilson 1985; Scott et al. 1996). As a rule, increasing distance from the vent results in improved sorting (Freundt et al. 2000). Limited published data suggest that lithic PDC deposits have sorting coefficients ranging from about 2.0 to 4.0φ (Miyabuchi 1999; Freundt et al. 2000; Saucedo et al. 2004). Sorting coefficients in highly dilute flow deposits are intermediate between values for coarser flow deposits and tephra-fall deposits, ranging from about 1.0 to 4.0φ and averaging about 1.5φ (Walker 1971; Sparks 1976; Valentine and Fisher 2000). Cross-bedded base-surge deposits appear to be more poorly sorted than cross-bedded VWF deposits, which can otherwise look similar (Bull and Cas 2000). However, methodological differences in determining sorting coefficients make comparisons of coefficients between published PDC deposit data and published wet-flow deposit data difficult (see Sect. 4.3; Table 6 in Chap. "Deposit Identification and Interpretation").

Sedimentary textures of PDC deposits can closely resemble those of both lahar deposits and some water-reworked volcaniclastic deposits and volcanic flood deposits. Particle-size distribution alone is unlikely to provide an adequate basis for differentiation; other criteria are needed.

Vesicles and Intergranular Voids

Spherical gas-bubble vesicles from flow-entrained air or outgassed volatiles have not, to our knowledge, been found in subaerial settings in the matrix of fresh, dry, nonwelded, unconsolidated PDC deposits (either pumiceous or lithic). This lack of vesicles owes largely to the fact that gas can easily escape from these porous deposits during consolidation. However, primary vesicles can form (1) in PDC deposits emplaced by wet, fine-grained PDCs from hydrovolcanic eruptions, particularly those at tuff rings within a few kilometers from source (Fisher and Schmincke 1984; Vespermann and Schmincke 2000; Wilson 2001), (2) in welded rheomorphic ignimbrites (Andrews and Branney 2011), and (3) post-depositionally in deposits where infilling of pore spaces by eolian fines and infiltration of rainfall can trap air bubbles (Stieglitz and Inden 1969; Valentine and Harrington 2006).

Vesicles may be rare, but small irregularly shaped intergranular voids between coarse ash and fine lapilli particles are common in small-volume deposit interiors due to loose packing and fines depletion. Larger irregular voids are typically present between gravel-size clasts along deposit margins, at lobate flow fronts, on deposit surfaces, in lee-side pumice lenses, and within gas-escape pipes (Hoblitt et al. 1981; Rowley et al. 1981; Wilson 1985) (Fig. 19).

2.2.4 Key Characteristics of Pyroclastic Density Current (PDC) Deposits

Nonwelded PDC deposits, particularly those from small-volume currents, can closely resemble some volcanic wet-flow deposits, morphologically and texturally. The following features of PDC deposits are particularly useful in distinguishing them from volcanic debris flow (VDF) deposits, volcanic hyperconcentrated flow (VHF) deposits, volcanic debris avalanche (VDA) deposits, and volcanic water flood (VWF) deposits. A more complete list of characteristic properties is found in Chap. "Deposit Identification: Key and Summary Tables". Statements in italicized font draw comparisons with the different volcanic wet-flow deposits.

- **Landscape-Scale Criteria**
  - Relatively smooth or undulating upper surfaces may exhibit dunes (primary or secondary).
    Not reported on wet-flow deposit surfaces.
  - Thicknesses of dilute PDC deposits are commonly uneven.
    Not reported for VHF deposits.
  - Pronounced lobate deposit margins are common in deposits of relatively small high-concentration flows.
    Such lobes are also common in small-volume VDF deposits, but PDC lobes are commonly more numerous, more pronounced, and more digitated.
  - Flow levees and deposit margins with openwork structure are common in small-volume deposits.
    Levees are also found in small-volume VDF deposits but not in other wet-flow deposits.
  - Polygonal fracture networks (large-scale and small-scale types) may form locally on deposit surfaces (mainly from thermal contraction during cooling).
    Not present in wet-flow deposits.
- **Outcrop-Scale Criteria**
  - Deposits are commonly massive and graded; upper flow-regime bed forms are common in stratified dilute PDC deposits, but lower-flow regime forms are also reported.
    Scale and characteristics of bedding and bed forms may differ from those in VWF deposits or eolian deposits (PDC deposits reworked by wind).

- Coarse-tail grading is commonly pronounced, often combined with density-segregation grading of gravel-size clasts.
  Obvious density segregation of gravel-size clasts is absent in VDAs, rare in VDFs, and common in stratified VHFs and VWFs.
- Homogeneous color (clasts and matrix) and sedimentary texture are common in pumiceous and some lithic PDC deposits. Polymictic deposits can have multiple colors
  True color homogeneity is somewhat rare in volcanic wet-flow deposits once the surface wash of matrix fines is scraped away; exceptions are when primary volcaniclastic deposits are remobilized over short distances.
- Strong primary flow-parallel fabrics are commonplace in pumiceous and lithic PDC deposits; strong secondary subhorizontal layering of flattened glassy clasts is found in deposits that were originally very hot and thick.
  Similar flow-parallel fabrics are found in VDF and VHF deposits, but subhorizontal fabrics formed by load flattening only affect vegetation debris (not pumice clasts) in wet-flow deposits and tills.
- Gas-escape pipes are common in relatively thick deposits.
  Vapor-phase oxidation, cementation, and possible presence of accretionary lapilli differentiate PDC gas-escape conduits from water-escape structures in lahar deposits.
- Consistency is mainly loose/friable (soft) in fresh small-volume deposits; large, thick flow units emplaced hot are likely to be firmer/harder.
  Localized primary welding may occur in large, thick PDC deposits. *Welding does not occur in wet-flow deposits.*
  Secondary cementation from weathering reactions of glass components (devitrification) is common. *Secondary cementation also occurs in glass-poor wet-flow deposits due to cementation by iron oxide or amorphous silica precipitation from groundwater. Some hot lahar deposits can also be well cemented (cementing agents uncertain). VDA and VDF deposits are typically firmer/harder than PDC deposits.*
- Evidence of high heat is commonly abundant but not required.
  Carbonized wood (*found in smaller amounts in some lahar deposits also*).
  Prismatically jointed gravel-size clasts (*found in smaller amounts in some lahar deposits also*).
  Breadcrust-texture clast exteriors (*found in smaller amounts in some lahar deposits also*).
  Consistent thermoremanent magnetic orientation of clasts is common in eruption-generated PDCs but not common in flows derived from collapses of already erupted pyroclastic material (*sometimes present in hot lahar deposits, but only with largest clasts*).
  Pink coloration of deposit tops (*not in wet-flow deposits*).
  Phreatic explosion pit craters (*not in wet-flow deposits unless emplaced over hot PDC deposits; ice meltout pits can occur in some VDA, VDF, and glacial deposits—these can resemble explosion pit craters*).
  Welding (*not in wet-flow deposits*).

- **Particle-Scale Criteria**
  - Dominantly oligomictic or monomictic mixtures—composed mainly of the same juvenile pyroclastic or lava fragments but also may contain small fractions of nonjuvenile or exotic lithic fragments.
    Wet-flow deposits are mainly polymictic.
    Lithic PDC deposits sourced from different parts of lava dome complexes may be polymictic.
  - Sub-rounded to well-rounded clasts are common in pumiceous flow deposits; clasts are dominantly angular in short-runout lithic PDC deposits.
    In lahar deposits, particles finer than fine gravel tend to be angular unless they are entrained stream gravel particles; all particles in VDA deposits are angular.

- Matrix of small-volume deposits commonly depleted in fine to extremely fine ash (particularly in lithic PDC deposits). *VDF and VHF deposits commonly appear in outcrop to have similar sorting values as PDC deposits, but the fines depletion that is common in PDC deposits gives them a looser material consistency than is found in most lahar deposits.*
- Matrix vesicles absent in dry, unwelded deposits, but may be present in deposits emplaced by wet PDCs.

3 Tephra Fall

3.1 Process Overview

Subaerial tephra fall (TF) deposits (also called pyroclastic-fall deposits) are emplaced by the lofting of particles of fragmented volcanic rock, crystals, and glass (collectively called *tephra*) into the air during eruptions and their subsequent gravity-driven fallout through air onto land. This chapter focuses only on subaerial deposits because tephra deposits that accumulate through water in lacustrine or marine settings involve different depositional mechanisms and can have different deposit properties (e.g., White and Riggs 2001). Excellent summaries of subaerial tephra-fall processes and deposits are found in Fisher and Schmincke (1984), Cas and Wright (1987), Heiken and Wohletz (1991), Sparks et al. (1997), Houghton et al. (2000a), Cashman and Sparks (2013), Bonadonna et al. (2015), Brown and Andrews (2015), and Houghton and Carey (2015).

Mechanisms for lofting tephra particles in air by eruptions that are more or less explosive include some combination of (1) momentum-driven ejection of tephra and gas from a vent as jets or discrete explosions and (2) particle suspensions in hot eruption columns (or in ash clouds above hot PDCs) generated by turbulent particle entrainment and buoyancy in heated air (Fisher and Schmincke 1984; Cas and Wright 1987; Wilson and Houghton 2000; Houghton et al. 2000b; Carey and Bursik 2000, 2015). Atmospheric conditions, as well as magnitude and style of eruption, play an important role in the size and texture of tephra particles and in the thickness and areal extent of TF deposits (Watt et al. 2015; Van Eaton et al. 2015). TF deposits vary widely in grain size, from extremely fine ash microns in diameter to ballistic blocks meters in diameter.

Strong eruption plumes can carry ash several tens of kilometers above a volcano to or above the level of neutral buoyancy in the atmosphere and then spread radially outward to form an umbrella region, whereas weak plumes (relative to horizontal wind velocities) bend over in the downwind direction (Carey and Bursik 2015; Bonadonna et al. 2015). Suspended particles in strong plumes can drift with the wind for hundreds of kilometers, but wind direction can vary with altitude, sometimes leading to complex fallout patterns over extremely broad areas. Particle fallout occurs when updraft velocities can no longer keep individual particles (or aggregates of particles) in suspension, which typically results in finer-grained, better sorted deposits at increasing distances from the vent. The bulk of the ash from most Plinian eruptions (see Table 1 for definitions of eruption types) is emplaced at near-ambient temperatures due to cooling in the plume and during fallout from it. Close to the vent, large volumes of material from a collapsing Plinian eruption column may produce gravity-driven PDCs on the flanks of a volcano.

During lower-energy Hawaiian, Strombolian, and Vulcanian magmatic eruptions, as well as during some phreatomagmatic and phreatic eruptions, coarse-grained ballistic projectiles comprise an important subpopulation of tephra fallout particles. Coarse-grained ballistics (coarse lapilli, blocks, and bombs) may be solid or partially to wholly molten, depending on initial temperature and size-dependent cooler rate, and they are typically emplaced close to source—mainly within a few kilometers. Their trajectories, typically parabolic, are dictated mainly by ejection velocity and angle; gas jet velocity from the vent; mid-air particle collisions and fragmentation; and particle size, density, shape, rotation, and spin (Mastin 2001; Taddeucci et al. 2017). Flight paths of coarse-grained ballistic particles are minimally affected by thermal convection or wind. Accumulations of molten spatter around vents can form welded agglutinate carapaces with steep slopes that may become gravitationally unstable (Sumner 1998; Capaccioni and Cuccoli 2005). However, Vulcanian gas-driven explosions from growing lava domes or plugged volcanic vents eject ballistic projectiles more widely to distances up to 10 km (Houghton and Carey 2015; Taddeucci et al. 2017). Terminal fall velocities in air for particles >20 cm in diameter exceed 100 m/s (Fisher and Schmincke 1984). At these distances, unanticipated explosions and launching of blocks and bombs presents a serious hazard to nearby communities (Taddeucci et al. 2017) (Fig. 20).

3.2 Tephra Fall (TF) Deposits

TF deposits can be broadly distributed around and downwind of source volcanoes and can contain a variety of juvenile and wall-rock lithic particles (Fisher and Schmincke 1984; Heiken and Wohletz 1991; Houghton et al. 2000a; Houghton and Carey 2015). Transported particles are diverse, including:

- Meter-scale blocks and bombs of dense to moderately vesicular volcanic rock.
- Millimeter to centimeter-scale lapilli composed of pumice, scoria, dense rock or glass fragments, and pellets of aggregated ash.
- Millimeter to sub-millimeter-scale ash particles composed of dense rock; pumice or scoria; crystals and crystal fragments; aggregates of finer ash; and glass shards, droplets, and strands.

Particles with angular shapes (bounded by breakage surfaces) reflect their origin, ejection, and transport as brittle solids. Those with fluidal or plastically deformed shapes reflect their ejection while partially to fully molten. Cryptotephra layers—thin, extremely fine-grained (particles generally <125 μm), and far-traveled TF deposits that are invisible to the naked eye—are usually diluted by other sediment grains and have low concentrations of glass shards (Gehrels et al. 2008; Lane et al. 2014). They are detected by their magnetic susceptibility and chemical signatures in sediment profiles and are useful time-stratigraphic markers. However, they are not considered further here.

TF deposit thicknesses from single eruptions vary with eruption volume, eruptive energy, and distance from source. Close to the vent, accumulations of ballistic particles from low-energy eruptions and tephra from high-energy eruptions can exceed 10 m in thickness, whereas large-volume Plinian and ultra-Plinian eruptions can emplace >1 m of ash up to 100 km from source and measurable tephra thicknesses >1000 km downwind from source (Fisher and Schmincke 1984; Houghton et al. 2000a; Mastin et al. 2014). Measurable ash thicknesses from a single eruption can cover more than 10^6 km^2, and tephra layers from single eruptions become, on average, thinner and finer with distance away from source (Sulpizio 2005; Mastin et al. 2014; Bonadonna et al. 2015; Houghton et al. 2000a; Houghton and Carey 2015).

3.2.1 Landscape-Scale Characteristics of Deposits

Deposit Architecture and Morphology

The large-scale features of a TF layer are largely controlled by the style, energy, and volume of the eruption, as well as the atmospheric structure at the time of emplacement (Fisher and Schmincke 1984; Houghton et al. 2000a). Low-energy explosive basaltic eruptions with low magma-discharge rates and characterized by lava fountaining (Hawaiian- and Strombolian-type eruptions) tend to form relatively thick, potentially unstable, annular rings or aprons of lava spatter and bombs around a vent. These vent-localized deposits have the potential to form welded deposits of lava spatter (*agglutinates*). A major component of these deposits is heavy ballistic projectiles that are only slightly affected by wind during emplacement.

Fig. 20 Damage from ballistic projectiles to a school building (later abandoned) several kilometers from Usu volcano (Japan): **a** roof damage from ballistics ejected by an eruption in 2000. **b** Damage to siding of the same building during the same eruption. Photos by L. Siebert

Therefore, the areal distribution of these deposits is spatially limited and roughly symmetrical about the vent. Higher-energy explosive eruptions, usually of intermediate to silicic composition and having higher magma discharge rates (i.e., Plinian- and ultra-Plinian-type eruptions), produce mostly lapilli- and ash-size tephra particles that fall in widespread lobe-shaped or fan-shaped deposits, commonly with asymmetric and complex geometries that can be strongly affected by wind directions and aggregation of ash particles at different atmospheric levels (Carey and Sigurdsson 1982; Houghton et al. 2000a; Bonadonna et al. 2015; Major 2023).

A key landscape-scale feature of TF deposits is that they mantle topography on slopes <25–30° with deposits having relatively uniform thicknesses over limited areas, although on the scale of the whole depositional area, thickness decreases away from the vent and can be locally variable (Carey and Sigurdsson 1982; Fisher and Schmincke 1984). Valleys, hillslopes, and ridges within such restricted areas are draped with tephra beds of similar character and thickness, although slopewash over time can decrease bed thicknesses on hillslopes and increase it in valleys or depressions. This characteristic contrasts with PDC deposits that tend to be thinner and finer grained on ridge crests and hillslopes and thicker and coarser grained on valley floors (Cas and Wright 1987; Wilson and Houghton 2000). Landscape mantling TF deposits contrast with wet-flow deposits that are confined to valley floors, except where flow momentum may locally drive runup onto higher topography.

Over broad areas, deposit thickness generally decreases exponentially with distance from the vent (Fierstein and Nathenson 1992; Houghton et al. 2000a), although other models to describe thinning versus distance (e.g., power-law or Weibull functions) may provide better fits to decreasing thickness data (Houghton and Carey 2015). Secondary thickness maxima may occur far downwind of vents, typically attributed to fallout of ash aggregates that grew in the rising plume or downwind cloud (Carey and Sigurdsson 1982). Near-vent tephra thickness from a single Plinian eruption can exceed 100 m (Houghton et al. 2000a). Variation in wind direction and speed with altitude, as well as changes in eruption dynamics can result in (1) deposits having thickness maxima skewed to one side of the dispersal axis, (2) changes in azimuth of the dispersal axis, (3) displacement of maximum thickness away from the vent, or (4) multiple thickness maxima (Fisher and Schmincke 1984; Smith and Houghton 1995; Bryan et al. 2000).

Ash particles in eruption clouds tend to aggregate in the presence of liquid water originating from condensed atmospheric moisture, magmatic volatiles, or external water from aquifer, glacier, ocean, or lake sources (Brown et al. 2012; Van Eaton and Wilson 2013; Bonadonna et al. 2015; Houghton et al. 2015). Growth of secondary minerals like gypsum or halite may help cement these aggregates so that they are more likely to be preserved in the geologic record. In contrast, electrostatic attractions and mechanical interlocking tend to produce more loosely bound fluffy aggregates that readily collapse after deposition. Depending on their characteristics, aggregates may be referred to as accretionary lapilli, cored lapilli, ash clusters, or ash pellets. Their primary importance is that aggregates fall out sooner than their individual constituent particles otherwise would (Carey and Sigurdsson 1982; Fisher and Schmincke 1984; Brown et al. 2010, 2012; Van Eaton et al. 2015; Gudnason et al. 2018).

Lithofacies

Lithofacies in TF deposits can be highly variable in vertical sections on scales of millimeters to meters, depending on the sizes and sequences of eruptive pulses and distance from vents (Fisher and Schmincke 1984; Cas and Wright 1987; Houghton et al. 2000a), but vertical sequences can be laterally continuous for hundreds of meters, with virtually no change in bedding, componentry, or grain size. Note that this does not apply to ultra-proximal zones, where the transition from coarse ballistics to eruption-column fallout can be more abrupt. This lateral continuity of lithofacies is a hallmark of many TF deposits—a characteristic not shared by volcanic wet-flow or PDC deposits. However, common lithofacies in TF deposits may closely resemble those of VHF and VWF deposits, and thus a landscape-scale perspective is important. Commonly encountered lithofacies include massive or bedded lapilli units with variable proportions of ash-size material (Dmm, Dcm, Dms, Gmo, Gm, Gw, Gh) and ash-dominated units, with or without admixed lapilli (Sm, Sh, Sw, Sp, Ss, Sl, Fm, Fl)—see Table 2 in Chap. "Deposit Identification and Interpretation" for sedimentary lithofacies codes. Over large areas, downwind and lateral variations in lithofacies within discrete deposits are gradual and a function of atmospheric transport conditions and distance from the vent or the dispersal axis (Carey and Sigurdsson 1982; Bonadonna et al. 2015; Watt et al. 2015). These variations include decrease or increase in thickness, decrease in median and maximum particle size (total or among different grain-type populations), improvement in sorting, changes in degree of stratification, and changes in componentry.

Erosion and Effects on Vegetation

Physical erosion of the land surface directly by falling tephra is insignificant, being limited locally to impacts by ballistic blocks and by shallow sliding or sloughing of unstable tephra layers being emplaced on steep slopes. Physical effects of falling tephra on vegetation, however, can be substantial. Tephra fall can damage or kill vegetation (including trees) over broad areas by the action of (1) abrasion, perforation, or stripping of leaves (coarse particles falling near the vent); (2) overloading and breaking of branches through accumulation of fine tephra; (3) deposition of acidic aerosols with fine tephra on leaf surfaces; (4) interference with leaf respiration, stomatal heat regulation, and other

Table 2 Reported effects of tephra fall on vegetation (Seymour et al. 1983; Blong 1984; Swanson et al. 2013, 2016)

Thickness of tephra-fall deposit (mm)	Effects on vegetation
>2000	All vegetation killed
1000–1500	Most vegetation killed; some recovery
10–150	Severe damage to forest trees but tree mortality is highly variable and depends on species, physical damage during tephra fall, tree size/age, fine-ash coatings on foliage, and tephra thickness
<10–20	Threshold to severe damage to herbaceous crop plants

leaf functions by fine ash coatings on leaves; and (5) total or partial burial of plants (Blong 1984; Segura et al. 1994, 1995; Antos and Zobel 2005; Swanson et al. 2013) (Table 2). Vegetation responses to these effects extend over time and vary among plant species. In addition, stripped leaves and twigs, mosses, and epiphytes from trees can become incorporated into tephra-fall deposits (Swanson et al. 2013). In proximal zones during mafic eruptions, heat from accumulating layers of spatter and bombs causes vegetation damage and may start fires (Fig. 21).

3.2.2 Outcrop-Scale Characteristics of Deposits

Stratification and Bed Forms

Bedding in TF deposits is variable and principally a function of the steadiness of volumetric output of tephra from the vent (sustained versus unsustained discharge), eruptive energy, and erupting magma composition—variables that affect changes in grain size, color, texture, and sharpness of bedding contacts in the accumulating layers (Fisher and Schmincke 1984; Cas and Wright 1987; Houghton et al. 2000a; Houghton and Carey 2015) (Fig. 22). Deposition by single eruptions may result in (1) massive beds without internal stratification when eruptions are sustained at a relatively constant intensity; (2) alternating coarser and finer intervals (with only diffuse bedding contacts) when eruption intensity varies or pulsates over time or when partial column collapses episodically occur; and (3) sharply delineated beds when eruption bursts are spasmodic with intervening quiet intervals (Smith 1987; Houghton et al. 2000a; Houghton and Carey 2015). Sharp internal bedding contacts also may form when sudden changes occur in wind direction or speed, or when magma-water interaction abruptly changes eruption style or intensity (Fisher and Schmincke 1984). Partings of very fine to extremely fine ash imply breaks within spasmodic eruptions or the end of an eruptive episode (Houghton and Carey 2015). This is because it takes days to weeks for disaggregated very fine ash particles (< 30 μm diameter) to settle out of suspension (Rose and Durant 2011). Note that complete preservation of a TF depositional sequence of ash beds in areas receiving heavy rainfall cannot be assumed (Fig. 22e).

Bed forms are normally absent within or between tephra layers (Houghton et al. 2000a). However, the occurrence of strong winds during deposition can generate hybrid fall deposits with variable development of cross-bedding, with such deposits resembling low-concentration PDC (surge) deposits (Wilson and Hildreth 1998). Postdepositional reworking of fall-deposit surfaces by wind or water can also lead to the formation of ripples and dunes, which could be buried by subsequent tephra falls (e.g., Segerstrom 1950). Eolian reworking of tephra units can be discriminated using (1) distance from probable vents (relative to likely reach of low-concentration PDCs); (2) paleoflow direction (interpreted from bed-form orientations); (3) particle size (low lapilli content in eolian units); (4) particle sorting (better sorting and more symmetrical skewness with less fine ash in eolian units); (5) absence of very angular grains, especially pumice, in eolian units; and (6) dominantly high-angle cross-bedding (ripples and dunes, commonly erosionally truncated) in eolian units (Smith and Katzman 1991). Bed

Fig. 21 Examples of forest vegetation destruction by tephra fall (TF) at Chaitén volcano, southern Chile, following the explosive phase of the 2008–2009 eruption on May 1–8, 2008 (see Swanson et al. 2013). **a** Area of severe physical damage to temperate rainforest by near-vent lapilli TF. Photo by J. Jones, Jan 2010, used with permission. **b** Aerial view of moderate damage to trees (leaf kill in most trees) from a fall of about 50 cm of ash a few kilometers from the vent. USGS photo by T. Pierson, Feb 2010

forms resulting from postdepositional transport by flowing water are likely be restricted to channels eroded into tephra deposits (commonly broad and shallow on flatter gradients), although patterns of sheetwash-induced, intersecting, miniature alluvial fans ("parallelograms") that resemble ripples can form on flat surfaces (Segerstrom 1950). Any true fluvial bed forms would have orientations controlled by local topography and channel planform.

Tephra deposits typically form laterally extensive, plane-parallel laminae or beds that drape underlying topography in the fallout zone (*mantle bedding*); beds show little change in thickness over hundreds to perhaps thousands of meters. In steep terrain, tephra beds may terminate against steep slopes and may thicken at bases of slopes due to tephra sloughing off slopes farther uphill, usually from slopes steeper than 25–30° (Houghton et al. 2000a). Many coarse tephra beds on steep, unstable cinder cones similarly reflect tephra-bed instability. Secondary avalanching of tephra on slopes can result in thin mass-movement deposits that are laterally discontinuous, exhibit inverse grading, and interfinger with primary bedding (Cas and Wright 1987; Bryan et al. 2000). In humid regions tephra deposits can be rapidly reworked by rainfall runoff.

Gravel-Size Clast Framework

Gravel-size tephra particles (lapilli and blocks/bombs) occur in layers or in massive accumulations that are commonly clast supported with varying amounts of sand-size and finer material (ash) between the coarse clasts (Fig. 22a, d).

Fig. 22 Stratification in tephra fall (TF) deposits: **a** ash layers from multiple distinct eruptive pulses deposited several kilometers from the vent by the May 1–8, 2008, explosive eruption of Chaitén volcano (Chile). Dark organic soil visible at base of section is the preeruption soil surface. Some proportion of this tephra section was likely eroded by heavy rain. **b** Dominantly lapilli- and coarse ash-size scoriaceous tephra on north flank of Villarrica volcano (Chile); contacts between beds vary in degree of sharpness. **c** A phonolite TF deposit (white unit) is sandwiched between two dark basaltic scoria TF deposits, Tenerife (Canary Islands). **d** Close-up of massive phonolite pumice in **c**, with dark gray coin for scale (22 mm diameter) at center right of photo. **e** Remnant (~ 15 cm thick) of TF depositional unit (medium to coarse ash) deposited in May 2008 and photographed in Jan 2010. It was protected from raindrop impacts and rain-splash erosion (and preserved) beneath a shrub with dense foliage, about 10 km from Chaitén volcano. Trenching shovel rests on a bed of water-reworked tephra. USGS photos: **a** by R. Hoblitt (Jan 26, 2010), **b**, **c**, **d**, **e** by T. Pierson ▶

a
b
c
d
e

Openwork structures are common, particularly in coarse, near-vent accumulations of solidified pumice and scoria, such as on cinder cones.

Fabric

Fabric within clast-supported tephra beds is commonly isotropic because particles land in random orientations from airborne fallout, and generally have equant shapes (e.g., Fig. 22d). Bedding-parallel fabric may develop in deposits that contain elongate or platy crystals or shards or in near-vent welded fall deposits (Fisher and Schmincke 1984). Compaction of near-molten clasts in near-vent agglutinates and in thick deposits of hot airfall tephra also produces a bedding-parallel fabric (Fig. 23).

Grading

Tephra-fall deposits can have complicated grading patterns, showing both normal and inverse grading in various combinations, which result from changes in eruption rate and style, particle vesicularity, or wind speed and direction as the tephra is being deposited (Fisher and Schmincke 1984; Houghton et al. 2000a; Houghton and Carey 2015). Upward fining is common in TF deposits as eruptive energy wanes. Inverse grading at the base of a slope can also result from downslope raveling of coarser particles, common at the base of cinder cones (Fisher and Schmincke 1984).

Fluid-Escape Structures

Such structures generally are not present in tephra-fall deposits, but soft sediment deformation related to water escape during compaction has been observed in fine-grained phreatomagmatic TF deposits (Bryan et al. 1998b).

Material Consistency

Fresh unwelded tephra-fall deposits composed of medium ash (or coarser particles) are typically loose to soft when dry and loose to very friable when moist (see Sect. 3.8 in Chap. "Deposit Identification and Interpretation" for definitions). Fresh, dry, coarse ash/lapilli tephra beds are commonly the loosest, raveling easily from disturbed dry outcrop faces (deposit surfaces break up easily into individual particles when force or friction is applied). Ash particles may be tightly bound together if the tephra is moist (effect of water surface tension), if it contains more than ~ 10 vol% extremely fine ash (< 63 μm), or if it has undergone compaction or weathering reactions. Beds rich in very fine or extremely finer ash are firm (moist) or hard (dry), especially if the ash bed was deposited wet or after raindrop impacts or percolating surface water further packed the smallest particles into pore spaces between larger grains. These more consolidated beds are easy to dig with a tool but break off in chunks rather than ravel when disturbed. Hard and coherent tephra beds can be produced close to vents by welding (sintering of glass particles) or by agglutination (fusing of glass in partially molten bombs).

Evidence of Heat

Ash-size medial to distal tephra particles are typically cooled in the air during transport and thus are emplaced at ambient temperatures. In contrast, volcanic bombs and spatter falling close to source vents may retain sufficient heat for ignition of local fires, production of charcoal, or even for welding and rheomorphic flow in thick proximal deposits during Hawaiian and Strombolian eruptions (Taddeucci et al. 2015) (see Table 1 for descriptions of eruption type). High emplacement temperatures may be deduced from welding or fusing of pyroclasts, their coloration, surface textures, or clast morphology (see Sect. 3.9 in Chap. "Deposit Identification and Interpretation"). Dark-colored mafic clasts transported and emplaced while hot, including scoria, commonly have an intense brick-red color due to oxidation of abundant iron-bearing minerals on the surfaces of clasts exposed to air and water (Houghton and Carey 2015). Such oxidation is seen throughout the deposit and inside individual porous scoria clasts, but it appears only as a thin oxidized rind on dense pyroclasts. Light-colored silicic pyroclasts typically exhibit more subtle colors—orange, pink, red, purple, or even brown or black (Houghton and Carey 2015). These colors

Fig. 23 Bedding-parallel fabric in welded base of thick silicic tephra fall (TF) deposit with obsidian only at base, near Puerto de la Cruz, Tenerife (Canary Islands). Note that this welded tephra is nearly identical to welded and compaction-deformed glassy particles in a pumiceous pyroclastic density current (PDC) deposit in Fig. 16a. It is unclear whether this type of welding in thick TF deposits differs from similar welding in thick PDC deposits. Coin (22 mm diameter) for scale. USGS photo by T. Pierson

also tend to be pervasive throughout the deposit and be more intense in deposit interiors.

Rapid cooling in flight of partially molten ballistic projectiles commonly produces prismatic (radial) jointing and breadcrust surface textures with expanded surface fractures (Fig. 14a, b in Chap. "Deposit Identification and Interpretation") or cauliflower surface textures where surfaces are rapidly chilled by contact with water (Fig. 15d in Chap. "Deposit Identification and Interpretation"). Volcanic bombs that remain molten during at least part of their trajectories develop fluidal shapes (e.g., fusiform bombs, ribbon bombs, spherical bombs) and may deform plastically upon impact (see Sect. 3.2.3). Whatever the composition, larger and denser pyroclasts retain more heat during flight than do smaller and more vesicular clasts. The high temperature of ballistic blocks and bombs is commonly shown as incandescence in videos and photographs taken during Hawaiian, Strombolian, and Vulcanian eruptions at night.

3.2.3 Particle-Scale Characteristics of Deposits

Particle Composition

Tephra-fall deposits span the full compositional spectrum of magmas, although intermediate to silicic eruptions tend to produce greater TF volumes in proportion to total erupted volume than do mafic eruptions. In most cases, juvenile particles larger than ash size (> 2 mm) show some degree of vesicularity (Fisher and Schmincke 1984; Houghton et al. 2000a) (see Table 5 in Chap. "Deposit Identification and Interpretation" for definitions of tephra particle sizes). Ash-size juvenile components typically include some combination of individual crystals, dense vitric shards, pumice or scoria fragments, and ash aggregates. Ash aggregates can form rapidly in eruption plumes and scavenge much of the fine ash ejected in an eruption (Van Eaton et al. 2015). In addition, dense wall-rock lithic fragments can be found in many tephra-fall deposits.

Particle Roundness, Shape, Fracturing, and Surface Texture

Ejection of variably hot (and variably solid) pyroclastic particles into the air produces a variety of particle angularities, shapes, fracture patterns, and surface textures. These characteristics can be helpful in identifying tephra-fall deposits. Roundness and shape of particles in the coarse ash to medium-size lapilli size range in TF deposits provide especially helpful clues for deposit origin (Fisher and Schmincke 1984; Heiken and Wohletz 1991; McPhie et al. 1993; Liu et al. 2015).

Tephra particles that were solid when emplaced are typically very angular, although minor rounding of initially angular pumice particles can occur during fall and emplacement, principally through abrasion by particle collisions during fallback into vents prior to re-ejection (Houghton and Smith 1993). High-energy silicic eruptions produce abundant ash-size glass shards that are exceedingly angular, having sharp, acicular corners, curviplanar surfaces, and cuspate edges (Fig. 24a). These features are inherited from broken bubble walls from disintegrated pumice or scoria clasts. Thus, pumice beds containing mainly angular pumice lapilli and coarse ash (e.g., Fig. 22d) are likely to have a tephra-fall origin, whereas lapilli-size pumice particles in PDC and lahar deposits are quickly rounded by abrasion during flow (Bryan et al. 2000).

TF particles can also be very well rounded in certain situations. Round, mm- to cm-size ash and lapilli particles also form through aggregation of small ash grains during airborne transport, forming ash pellets and accretionary lapilli (Van Eaton and Wilson 2013; Houghton et al. 2015) (Fig. 24b). *Ash pellets*—internally structureless aggregations of ash particles—are, in TF deposits, generally confined to fall deposits from buoyant eruption plumes if enough water is present. They also occur in PDC deposits. In airborne tephra, ash pellets form from ash in dry eruption columns by electrostatic attraction or by acquisition of atmospheric moisture (condensation or ice-particle melting) as suspended particles rise and drift downwind. *Accretionary lapilli* form when small ash particles or pellets recirculate vertically in eruption columns and grow by repeated accretion of thin layers of fine ash particles. Neither ash pellets nor accretionary lapilli are likely to be preserved in volcanic wet-flow deposits, even if they were to drop onto those flows, because of their fragility.

Smooth glassy tephra particles, another type of well-rounded tephra, form by surface tension acting on small fragments of fully molten basaltic spatter during flight. When solidified, their shapes range from teardrop-shapes and nearly spherical glass beads (*Pele's tears*) to very elongate, cross-sectionally rounded, glass strands (*Pele's hair*) (Fig. 24c, d) (Heiken and Wohletz 1991).

Solid particles having dominantly equant shapes are produced by breakage of isotropic rock, either by fragmentation in the vent or by breakage upon impact from fall (Fisher and Schmincke 1984). Highly vesicular ash and lapilli particles tend to have irregular breakage surfaces but equant shapes when vesicles are undeformed (Fig. 25a). Fracture orientations are commonly random. Non-equant (elongate) shapes can be produced in ash- and lapilli-size particles when the relative size, symmetry, and abundance of vesicles imbue the rock with sufficient anisotropy (Fig. 25b) (Liu et al. 2015).

Coarser ballistic projectiles with partially molten interiors but solid exterior rinds chilled during flight commonly develop prismatic (radial) fracture patterns and breadcrust surface textures that can form non-equant clasts when

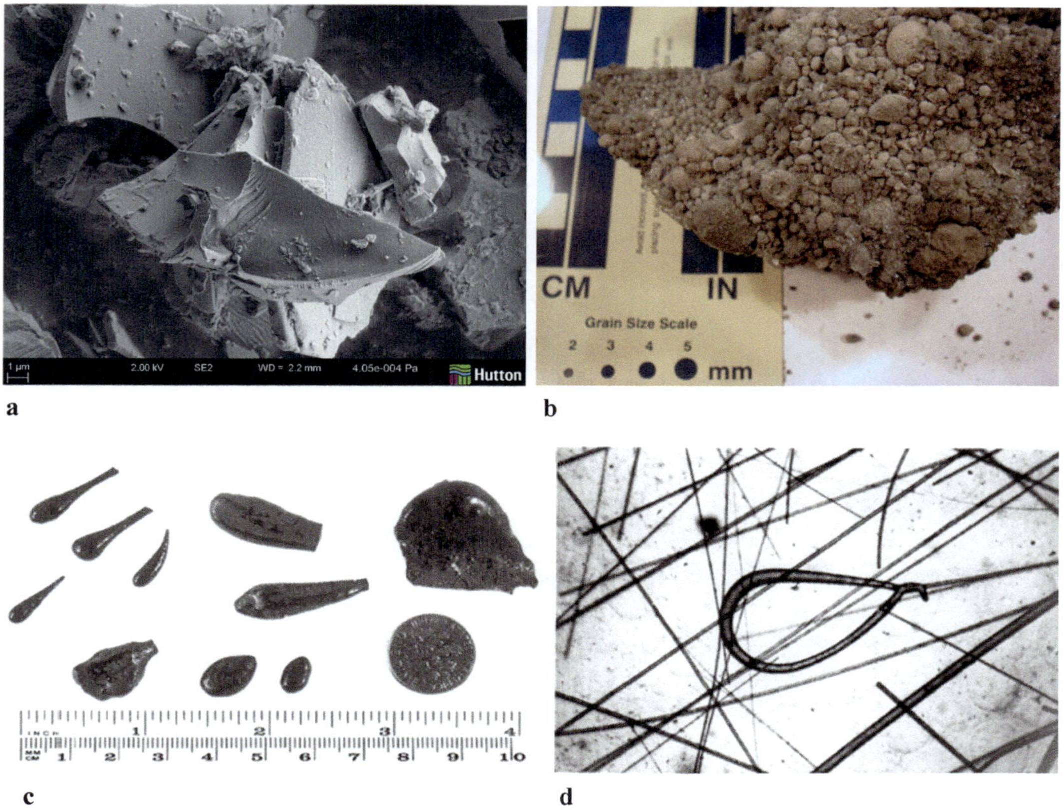

Fig. 24 Variably shaped and rounded particles of volcanic ash (see also Fig. 25a, b): **a** glass-shard ash particles (formed from broken bubble vesicle walls) in ash from Grímsvötn volcano (Iceland) that fell in Aberdeen, Scotland in May 2011 (scanning electron micrograph by Evelyne Delbos, The James Hutton Institute, used with permission). **b** Accretionary lapilli produced during the 2009 eruption of Redoubt Volcano (Alaska, USA), deposited in a water-rich fall environment about 12 km from source. USGS photo by K. Wallace. **c** Pele's tears—rounded basaltic glass spatter droplets from Mauna Ulu vent, Kilauea (Hawaii, USA) (USGS photo by J. D. Griggs). **d** Photomicrograph of fine smooth rods of volcanic glass ("Pele's hair") ejected from vent at Kilauea volcano, with diameter of coarsest strands about 0.1 mm. Photo from USGS photo archives

the ballistics break upon impact to form prism-shaped clasts (Fig. 25c). These fragments commonly have curviplanar and smooth breakage surfaces. Molten ballistic projectiles, however, tend to develop non-equant fluidal shapes during flight (Fig. 25d). Surface textures can be rough or relatively smooth. Fluidal volcanic bombs are named for their shapes—fusiform bombs (ellipsoidal, almond-shaped, or teardrop shaped) and ribbon bombs (ribbon or rod shaped), which may twist or bend during flight (Fig. 25e). Bombs still partially molten plastically deform (flatten) rather than break apart upon impact (Fig. 25f).

Particle-Size Distribution and Sedimentary Texture

Particle-size distributions of tephra-fall deposits emplaced by convective plume fallout from explosive magmatic eruptions are highly variable and depend on the type and magnitude of eruptions, on wind patterns and velocities, and on distance from source (Walker 1971; Fisher and Schmincke 1984; Cas and Wright 1987; Houghton et al. 2000a; Houghton and Carey 2015). Average particle size decreases with distance from the vent. Tephra from phreatic and phreatomagmatic eruptions is generally finer grained than tephra from magmatic eruptions,

Fig. 25 Shapes and degree of rounding and deformation of tephra fall (TF) particles transported either as fragmented solid particles or as partly or fully molten ballistics: **a** equant pumice particle with mainly spherical vesicles (USGS scanning electron micrograph). **b** Elongate pumice particles containing tubular vesicles (USGS scanning electron micrograph). **c** Prism-shaped fragments formed by breakage upon impact of the brittle glassy rind of a prismatically jointed volcanic bomb, Overo maar near Laguna Lejia, south of Láscar volcano (Chile). Exposed part of pencil is 7 cm long. USGS photo by T. Pierson. **d** Small, rounded, fluidal lapilli and volcanic bombs of basaltic composition. Coin is 19 mm in diameter (USGS photo archives). **e** Small basaltic ribbon bomb from southern Iceland (spatter from nearby vent). Pencil is 15 cm long. USGS photo by T. Pierson. **f** Flattened basalt "cowpie" bomb that was ejected as molten spatter in 1969 on Kilauea's upper East Rift Zone and which deformed upon impact. USGS photo by J. Lockwood

even in proximal locations (Houghton et al. 2000a; Morrissey et al. 2000).

Particle-size distributions of ballistic projectiles can differ significantly from those of fall from the convective eruption plumes, and the ballistic subpopulation of particles tends to show rapid thinning and fining away from the source vent (e.g., Houghton et al. 2017). Ballistic particles, generally much larger than particles falling out of convective plumes, can be numerous and large, exceeding a meter in diameter, reaching several kilometers from vents, and posing an important hazard for monitoring equipment, volcano tourists, and nearby communities (Breard et al. 2014; Fitzgerald et al. 2017). Distance traveled by ballistic projectiles generally increases with increasing explosivity (Fitzgerald et al. 2017).

Size distributions of tephra-fall deposits are usually unimodal or bimodal, with bimodality commonly better developed with distance from source (Walker 1971; Carey and Sigurdsson 1982; Fisher and Schmincke 1984; Gudnason et al. 2017, 2018), but transport mode (convective eruption plumes versus ballistic projectiles) greatly affects average particle size and sorting values of the subpopulations (Houghton et al. 2017). Addition of ash from simultaneous co-PDC plume fallout can further complicate size distributions (Eychenne et al. 2012).

TF deposits are commonly well sorted (1.0–2.0φ) when using Inman sorting criteria (σ_φ), as defined by Cas and Wright (see Table 6 in Chap. "Deposit Identification and Interpretation"). A broader range of sorting values (0.2–3.0φ) is encountered when considering the full range of eruption styles and magma/water ratios producing the tephra (Walker 1971; Houghton et al. 2000a). Density sorting commonly results from different fall velocities when ejecta contain both pumice and lithics (Fisher and Schmincke 1984). Sorting coefficients typically decline geometrically (degree of sorting improves) with distance from the vent (Fisher and Schmincke 1984; Houghton et al. 2000a). Tephra-fall deposits also tend to be strongly leptokurtic, with kurtosis values > 1.5 (Alloway et al. 2005).

Matrix Vesicles and Intergranular Voids

Trapped gas or air can escape easily from relatively well-sorted TF deposits of dry magmatic eruptions, which tend to be clast supported with little finer-grained intergranular matrix (Wilson and Houghton 2000). Under certain conditions, however, matrix vesicles can form: (1) when wind-transported fine ash is blown onto deposit surfaces and washed in by rainfall or (2) when falling fine-grained ash is wetted by condensed moisture in the air, by simultaneously falling rain, or by water erupted with the ash in hydrovolcanic eruptions. In these cases, air bubbles can be trapped within and between falling wet ash pellets ("mud rain") or between ash layers as wet ash accumulates on the ground (Fisher and Schmincke 1984). Secondary vesicles also may form in fine tephra after deposition due to entrapment of air bubbles by infiltrating rain. In addition, welded airfall tuffs may trap bubbles from exsolving gas (Fisher and Schmincke 1984).

Irregular void spaces between coarse-ash and lapilli particles, in contrast to vesicles, are common in uncompacted and relatively well-sorted tephra-fall deposits, which gives them a prominent openwork structure (e.g., Fig. 22d). This aids in distinguishing them from other types of volcaniclastic deposits.

3.2.4 Key Characteristics of Tephra Fall (TF) Deposits

Subaerial tephra fall (TF) deposits composed mainly of ash-size particles can mimic some volcanic wet-flow deposits, particularly volcanic hyperconcentrated flow (VHF) deposits and volcanic water flood (VWF) deposits; in some cases they can resemble thin, fine-grained volcanic debris flow (VDF) deposits. Properties that are particularly useful in distinguishing TF deposits from those volcanic wet-flow deposits are listed below, with contrasting characteristics of the wet-flow deposits italicized. A more complete list of characteristic TF deposit properties of all other deposit types is found in Chap. "Deposit Identification: Key and Summary Tables".

- **Landscape-Scale Criteria**
  - Topography-mantling beds exhibit uniform thickness on ridges, hillsides, and valley floors. Plane-parallel bedding conforms to underlying topography.
    Wet-flow deposits fill in topographic lows, and deposits are confined to valley floors except where they run up against topographic obstacles.
  - Lithofacies and bedding are laterally continuous for long distances (> 100 m).
    Not the case with VHF and VWF deposits; VDF deposits can be laterally continuous.
  - Forest vegetation may suffer widespread damage—physical and chemical effects.
    Damage from wet flows is highly localized and adjacent to channels or flow paths.
- **Outcrop-Scale Criteria**
  - Beds initially have uniform thickness and are laterally continuous in outcrop exposures (no thinning, scouring, or crosscutting unless reworked).
    Beds in VWF deposits and some VHF deposits are truncated by scouring.
  - Leaves and fine twigs can be vertically dispersed throughout the lower parts of lapilli tephra sections in forested terrain (vegetal debris progressively stripped from trees as tephra accumulates).
    Organic debris, if present, tends to be concentrated along contacts between different depositional units emplaced by separate flows or distinct surges within flows in VDF, VHF, and VWF deposits.
  - Bed forms are absent unless deposition is accompanied by strong wind or the beds are subsequently reworked by wind or surface water.
    Bed forms are common in VWF and some VHF deposits.
  - Grading can be exhibited over short vertical distances (centimeters) in vertical sections. Commonly beds are normally graded overall (finest ash takes longer to fall out), but complex grading patterns occur when eruptions are pulsatory.
    Grading is uncommon in VWF beds.
    Grading in VDF and VHF deposits is typically developed over larger vertical distances (tens of centimeters or meters).
  - Fresh deposits are relatively loose (soft) and easy to dig or scrape.
    Older deposits with a high glass content are highly susceptible to post-depositional cementation.
    Thick vent-proximal coarse deposits may be welded or fused (mainly from near-vent basaltic spatter and bombs).
    VDF and VHF deposits are typically firmer/harder than TF deposits, but VWF deposits have similar material consistencies.
  - Fluid-escape structures are absent.
    Fluid-escape structures are common in VHF deposits.
- **Particle-Scale Criteria**
  - Tephra from magmatic eruptions is oligomictic or monomictic—composed of a dominant juvenile rock type, commonly with small fractions of wall-rock lithic fragments.
    Phreatomagmatic TF deposits have high proportions of wall-rock lithics.
    TF deposits from phreatic eruptions, by definition, consist entirely of vent-clogging or wall-rock lithic particles.
    VDF, VWF, and VHF deposits are typically strongly polymictic.
  - Color of beds in a vertical sequence may change due to shifts in magma composition or vesicularity.
    Colors of VHF and VWF beds in a vertical sequence seldom show color changes, except subtle shading variation due to changes in particle size.
  - Particles of virtually all sizes are angular (subangular to very angular); angularity is more pronounced as grain size decreases.
    Angular pumice/scoria is not found in lahar or VWF deposits.
  - Tephra in individual tephra beds is commonly well sorted, as defined by Cas and Wright (1987) using Inman (1952) criteria.

VHF and VWF beds typically have poor to moderate sorting, as defined by Folk (1974).

- Openwork structure (voids between clasts) is common in lapilli (gravel)-size material.
  Equivalent openwork structure is absent in lahar deposits (except at deposit margins) and uncommon in VWF deposits (except in very coarse deposits).
- Matrix vesicles are rare (caused by secondary processes) in tephra from magmatic eruptions when tephra is coarser than about 0.1 mm (fine ash or coarser) and emplaced dry.
  Vesicles may be present in fine to extremely fine tephra when emplaced during rainfall, or when it falls out from phreatomagmatic or phreatic eruption plumes.
  Vesicles are common in VDF deposits but not in VHF or VWF deposits.
- Accretionary lapilli or ash pellets are common in tephra emplaced by phreatomagmatic eruptions.
  Ash aggregates are not present (not preserved) in volcanic wet-flow deposits.

4 Glacial Erosion and Deposition on Volcanoes

4.1 Process Overview

Glaciers are masses of ice, originally accumulated as snow in high-altitude or high-latitude settings, that flow downslope in response to gravity (Ehlers 1996). *Alpine glaciers* form in mountainous terrain, involve relatively small volumes of ice confined to individual valleys on mountainsides, and are currently active on the flanks of numerous volcanic edifices around the world. *Ice sheets* are larger ice masses that can cover entire mountain ranges and the intervening lowland terrain; large ice sheets on land are termed *continental glaciers*. Alpine glaciers on volcanoes (Fig. 26) erode, entrain, transport, and emplace volcanic rock debris in a variety of types of *volcano-glacial deposits*. When eruptive activity impacts glacier ice on volcanoes, hazardous volcanic wet flows can result (Björnsson 1975; Pierson et al. 1990; Guðmundsson et al. 1997; Huggel et al. 2007; Thouret et al. 2007; Waythomas et al. 2013). Alpine glaciers confined to the flanks of volcanoes (in contrast to larger ice sheets that flow across nonvolcanic terrain) leave deposits composed exclusively of volcanic rock particles, and alpine glacial deposits can be similar in both scale and sedimentology to volcanic wet-flow deposits. Thus, they are the focus of this section. Large ice sheets that were more extensive during past glacial periods have also emplaced deposits in valleys at the bases of volcanoes in many parts of the world. However, sediment deposits emplaced by these glaciers are likely to contain a high proportion of nonvolcanic clast lithologies.

Volcanoes in most parts of the world today host alpine glaciers, including those at tropical latitudes where summits exceed about 4500–5000 m above sea level (Benn et al. 2005). Glaciers are common on mid-latitude cones and ubiquitous on high-latitude volcanoes. Yet during the Pleistocene glacial periods, the average ice-mass altitudes and equilibrium-line altitudes (ELAs) of all glaciers were lower and ice masses were larger than they are today. Thus, Pleistocene glaciers deposited sediment higher on valley sides and farther down valleys than modern glaciers have. If glaciers are small or absent on a volcano now, it does not necessarily mean that outcrops of volcano-glacial deposits will not be found there.

Alpine glaciers may be found on all flanks of a volcano, although they form preferentially on sides with minimal solar insolation—north- and east-facing flanks in the Northern Hemisphere, and on south- and east-facing flanks south of the Equator. Small volcano glaciers start in depressions on edifice flanks that are then enlarged and deepened by glacier erosion—glacial *cirques*. Larger ice masses flow down from summit ice caps or out of volcanic craters. Alpine glacier ice movement (some combination of ice flow and basal sliding) occurs at surface velocities ranging from tens to a few hundreds of meters per year to over 1 m/day where basal sliding accounts for a large proportion of

Fig. 26 Alpine glaciers on stratovolcanoes in the Cascade Range (Washington, USA): **a** the Adams Glacier flowing down the north flank from the summit ice cap of Mount Adams on August 10, 2020. A lateral moraine is visible in front of the ice in the lower part of the photo. There is minimal rock debris visible on the glacier surface. USGS photo by T. Pierson. **b** Heavily debris-mantled Crater Glacier at Mount St. Helens on July 30, 2015. Ice flows out from the back of the crater, around both sides of the 1980–1986 and 2004–2008 lava domes, then northward into the crater breach. Terminus is descending Loowit Creek valley; ice thickness at terminus varies between 18 and 40 m. The thick layer of volcanic rock debris on glacier surface, has been emplaced by rockfalls from crater walls. The debris mantle helps insulate the ice from solar radiation and has allowed the glacier to advance over the last several decades when most glaciers in the region have been retreating. Glacier snout is eroding older volcaniclastic sediments. Photo by F. Barsotti, used with permission

movement (Ehlers 1996; Cuffey and Paterson 2010; Allstadt et al. 2015). Glacier surging can increase ice velocity by a factor of 100 but requires a mechanism for a buildup of high subglacial water pressure (Kamb et al. 1985).

Glaciers on volcanoes in mid or low latitudes are classified as temperate glaciers, which means that (1) liquid water is present on the glacier bed, and ice there remains near its pressure melting point year-round, and (2) some glacier motion may be accommodated by sliding on the bed (where ice is not frozen to the bed). The implications of these conditions are that temperate glaciers are powerfully erosive due to the erosive grinding action of sliding rock-studded ice and that subglacial meltwater streams transport abundant sediment (Ehlers 1996).

Glaciers erode and incorporate rock debris from their beds in two basic ways—plucking (quarrying) and abrasion (Iverson 1990; Ehlers 1996; Benn and Evans 2010). Plucking occurs where ice on the glacier bed, moving across an irregular bed surface, locally melts where it encounters a protruding knob of bedrock (pressure melting). This localized meltwater can then refreeze on the lee side of the knob where pressure is lower. If the rock on the lee side of the knob is jointed and the moving ice has refrozen to the bed, rock fragments can be plucked from the bedrock surface (Cuffey and Paterson 2010). Abrasion occurs where rock fragments entrained in the moving ice scrape against bedrock on the glacier bed, causing scoring and polishing of the bedrock surface. Plucking produces and incorporates relatively large rock fragments into the glacier ice, while abrasion produces abundant silt-size particles (glacial flour) that are the characteristic suspended sediment in glacial streams.

A third major way of adding rock debris to glaciers is by rockfall and rockslides from valley sides and cirque walls onto the glacier surface (*supraglacial rock debris*). Debris falling onto a glacier may later be buried by snow in the glacier's accumulation zone and become incorporated into the glacier ice (*englacial rock debris*). Downward melting of surface ice in the glacier's lower-altitude ablation zone further concentrates englacial rock fragments at the surface. Rockfall and melt-concentrated debris commonly produce a thick surface layer of rock fragments that insulates underlying ice from solar radiation and facilitates advancement of the glacier snout farther downslope than would otherwise occur (Fig. 26b).

4.2 Volcano-Glacial Deposits and Features

Glaciers deposit sediment in several ways to produce deposits with different characteristics (Flint 1971; Ehlers 1996; Benn and Evans 2010; Menzies et al. 2018). On volcanoes, deposition of volcano-glacial sediment is mainly accomplished (directly and indirectly) by relatively small alpine glaciers in steep terrain. Depositional processes can be grouped in two basic categories: (a) final transport and emplacement by flowing or stagnant ice through lateral shoving, mass movement of rock particles off glacier surfaces, or rock particles melting out of stagnant ice, and (b) final transport and emplacement by flowing water—glacial meltwater streams flowing off a glacier surface, flowing along the lateral margin of a glacier, or flowing out from meltwater streams within or beneath a glacier. Deposits emplaced by the first category of ice-dominant processes are broadly termed *till*. Deposits emplaced by the second category of water-dominant processes are broadly termed *glacial outwash* or *glaciofluvial deposits*. The interbedding of glacial deposits, particularly tills, with volcanic wet-flow and pyroclastic deposits on the flanks of volcanoes and in downstream valleys adds considerable complexity to the interpretation of the volcaniclastic stratigraphy (e.g., Scott et al. 2020). Although alpine glaciers are likely to be the main source of glacial deposits on volcanoes, arms of continental ice sheets may have flowed up valleys draining volcanoes in the past, also emplacing deposits, and further complicating deposit stratigraphy. Landforms and deposits emplaced by glaciers can closely resemble those volcanic debris avalanches (VDAs), volcanic debris flows (VDFs), volcanic water floods (VWFs), and some pyroclastic density currents (PDCs).

Till is the volcano-glacial deposit that most resembles volcanic wet-flow deposits due to its lack of stratification and poor sorting. Thus, till is the main focus of our discussion of volcano-glacial deposits. Tills can have different properties depending on their mode of formation. *Lodgment till* (also referred to as *basal till*) is glacial sediment transported or reworked by a glacier, deposited, and then overridden and highly compacted by the weight of the ice (Fig. 27a). Note that unconsolidated sediments beneath a lodgment till would also be compacted by the weight of overriding ice, but these deposits retain their original identity, as for example, a compacted fluvial outwash sand, and do not become part of the till. *Ablation till* is a broad category of sediment transported and deposited by a glacier but not overridden and therefore not compacted. Subtypes of ablation till include (a) rock debris within or on the ice that is conveyed to the lateral and terminal margins of the glacier by ice flowage or surface debris flowage (*flow tills*) and (b) rock debris on or within stagnant glacier ice that is lowered to the ground by ice melting to form a *melt-out till* deposit. Ablation-tills typically are deposited as *lateral moraines* and *terminal moraines* (Fig. 27b) or as loosely compacted, hummocky deposits on the former glacier bed—melt-out till. Melt-out till deposits (Fig. 27c) may inherit sedimentary textures and stratification from slumping, flowage, or fluvial reworking of sediment while still on the down-melting glacier surface or in ice-marginal stream channels (Fig. 27d).

The bulk of volcano-glacial outwash deposits are transported and emplaced by meltwater flows from glaciers downstream of glacier termini in broad, braided alluvial plains (*proglacial outwash*) (Fig. 28). Rate of sediment supply is typically high and deposit volumes are typically large, but glaciofluvial deposit facies differ very little from nonglacial braided fluvial systems (Miller 1996). Furthermore, glaciofluvial deposits in volcanic drainage systems are largely deposited relatively near

a b c d

Fig. 27 Types of till in the process of being emplaced. **a** Lodgment till actively forming/deforming beneath glacier ice at a lateral margin of Exit Glacier, Kenai Peninsula near Seward, Alaska. Access was in a gap between a lateral moraine and the glacier. Layer of lodgment till beneath ice is about 0.5 m thick (above the ice-marginal streambed), and glacier flow is from right to left. **b** Incipient small terminal moraine (ablation till) formed by a brief stillstand of the Exit Glacier terminus. Outflow from subglacial streams has cut through the moraine. The right lateral moraine (also ablation till) is visible beyond the ice. **c** Supraglacial debris on stagnant glacier ice in the process of forming melt-out till, just above toe of Nisqually Glacier, Mount Rainier (Washington, USA). Glacier here is about 170 m wide. **d** Right lateral margin of the Exit Glacier (looking upvalley) near the glacier terminus. A small ice-marginal stream is flowing between the glacier ice (on the right) and the ablation till of the right lateral moraine (on the left), providing some fluvial reworking of the coarse sediment. The loose boulders along the stream bed are up to about a meter in diameter. USGS photos by T. Pierson

advancing or retreating glacier termini, and such proximal outwash deposits are typically dominated by massive to crudely planar-bedded coarse gravels with some cross-bedding and sparse finer grained units (Miller 1996). VWF deposits, in contrast, tend to be dominated by sand and fine gravel (e.g., Pierson et al. 1996, 2013; see Chap. "Volcanic Water-Flood Deposits").

4.2.1 Landscape-Scale Characteristics of Volcano-Glacial Erosion and Deposition

Glaciers affect landscapes by a combination of large-scale erosion of terrain and deposition of sediment, and the erosional role in landscape modification is not always easy to separate from the depositional role. This section considers both in describing glacier impacts on landscapes.

Fig. 28 Glaciofluvial outwash plain deposit about 100 m downstream of the Exit Glacier terminus outside of Seward, Alaska. The largest clasts visible in foreground are about 40 cm in diameter. USGS photo by T. Pierson

Large-Scale Erosion and Deposit Architecture and Morphology

Glacially affected landscapes are mainly recognized by both their erosional and depositional landforms (Tables 3, 4; Fig. 29) (Ehlers 1996; Benn and Evans 2010; Menzies et al. 2018). Some are produced mainly by relatively large glaciers (Table 3), whereas others are formed mainly by relatively small glaciers—the size typically found on glaciated volcanoes (Table 4).

Ground moraine—formless accumulations of lodgment till, ablation till, or both—mantles the flow paths of past glaciers. In most volcanic terrain, these sediment sheets drape glaciated valley floors. Importantly, however, ground moraine can also be found on broad ridge crests if a volcano's ice cap during glacial periods was large enough to engulf that terrain. On valley floors ground moraine is relatively thin with respect to deposit area and somewhat uneven in thickness (Kjær et al. 2003). Till thickness commonly increases in the lee of bedrock knobs or small hills (Flint 1971; Boulton 1987), but generally lodgment till thickness beneath alpine glaciers ranges from centimeters to meters (Ehlers 1996; van der Meer et al. 1999; Kjær et al. 2003). Lodgment till sheets from continental glaciers can be hundreds of meters thick (Flint 1971). Furthermore, repeated advances of a glacier front can result in the emplacement of multiple, stratigraphically discrete lodgment tills (Fig. 29f). Ablation till thicknesses in ground moraines are highly variable and depend on the ease of accumulation of supraglacial debris on the glacier that emplaced it. Thus, thickness will be greater for narrow alpine glaciers confined by steep unstable valley sides than for larger, wider glaciers. Ablation till layers are generally thin but can range up to tens of

Table 3 Basic erosional and depositional landforms commonly produced by moderate-size to large glaciers (i.e., large alpine glaciers, montane ice caps, and lobes of continental glaciers) (adapted from Flint 1971; Ehlers 1996; Neuendorf et al. 2005; Benn and Evans 2010)

Landforms	Description
U-shaped valley	Valley in which glaciers have eroded bedrock and valley-fill deposits (and locally deposited sediment) to change the cross-valley profile from a V-shape (typical of valleys formed by stream incision) to a U-shape
Hanging valley	Tributary valley with valley floor altitude higher than the altitude of the main-valley floor, due to more intense erosion by the main-valley glacier
Drumlin	Elongate streamlined ridge composed primarily of till and aligned with direction of former ice flow, formed by molding of the underlying deposits. May be partly or wholly rock cored. Commonly hundreds of meters long, about half as wide, and tens of meters high. May be longitudinally symmetrical or have a steep stoss end and a gently sloping lee end. Often found in groupings (drumlin fields)
Esker	Elongate sinuous ridge composed of glaciofluvial sand and gravel, oriented mainly downvalley; they are formed by sediment deposition in subglacial stream-channel tunnels. Planforms are highly variable, ranging from single continuous ridges (uniform or variable height and width) to low ridges linking larger mounds (beaded eskers), to complex braided patterns
Fine-grained lake deposits in mouths of tributary valleys	If glacier ice melts out of a tributary valley before the ice in the main valley, an ice-dammed glacial lake forms where fine sediment is deposited. Following melting of the main-valley glacier, fine sediment typically forms a bench with steep bluffs in the mouth of the tributary valley
Flute (fluting)	Elongate streamlined ridge of sediment (till) aligned with direction of former ice flow, formed downslope of a boulder or bedrock knob by injection of wet sediment into the ice void created by the obstruction. Up to hundreds of meters long but only a few tens of centimeters to a few meters high and wide
Kame terrace	Narrow bench or ridge on a valley side (slope parallel to valley floor) composed of stratified (glaciofluvial) sand and gravel. Formed by deposition in the gap between melting ice and the adjacent valley wall or lateral moraine. Commonly pitted with kettles
Kettle (kettle hole)	Steep-sided, bowl-shaped depression formed by the melting out of ice blocks (detached stagnant ice) that had been buried in glacial outwash or ablation till. Typically up to tens of meters in diameter and meters deep. Commonly filled with water
Roche moutonnée (sheep-like rock)	Elongate streamlined knob of eroded bedrock, with long axis aligned with direction of former ice flow. Upglacier stoss end is smoothly rounded, striated, and gently inclined; downglacier lee end is steep, rough, and hackly, resulting from glacier plucking. Typically a few meters high and long

meters in thickness and can be distinctly hummocky in morphology (Flint 1971).

Rock glaciers (Fig. 30) are periglacial landforms defined as bodies of slowly moving or stagnant ice and rock debris that form either as well-insulated small glaciers heavily loaded with englacial and supraglacial rock debris or as bodies of ice-cemented rock talus that acquire enough intergranular ice to allow flowage. Large rock glaciers can resemble ground moraine, as well as small VDA or VDF deposits.

Lithofacies

Till is the type of volcano-glacial deposit most like volcano debris flow (VDF), volcanic debris avalanche (VDA), and lithic pyroclastic flow (PDC) deposits, and therefore till is the focus of discussion of volcano-glacial deposits for the remainder of this section. Glacial outwash

Table 4 Basic erosional and depositional landforms produced by glacial and periglacial processes that are commonly found on volcano edifices modified by small to moderate-size alpine glaciers

Landforms	Description
Lateral moraine (type of end moraine) (e.g., Fig. 29a)	Elongate ridge of rock debris that accumulates along the lateral margins of a glacier from rock debris that flows, falls, rolls, or slides off the glacier's surface. Commonly many tens of meters high, sharp crested (when fresh), and having angle-of-repose side slopes
Terminal moraine (type of end moraine) (e.g., Fig. 29b)	Irregular ridge of rock debris that accumulates at the downhill terminus of a glacier and is oriented mainly transverse to direction of glacier flow. It accumulates from rock debris that flows, falls, rolls, or slides off the glacier's terminus, marking the position of a prolonged still-stand of a glacier terminus. Typically occurs in a nested series of ridges marking the episodic retreat of a glacier terminus
Outwash plain (terraced valley fill if dissected by erosion) (e.g., Fig. 28)	Extensive sheets of coarse-grained glaciofluvial sediment (often preserved as terraces following stream incision) emplaced downstream of glacier termini. May have eskers and kettle holes. May have proglacial meltwater channels that are dry or host underfit streams (streams too small to have carved the channels)
Valley-side trim line (e.g., Fig. 29c)	Sharp line of demarcation, with undisturbed forest (or other vegetation) and soil above the line and bare scoured bedrock (with or without a plastering of till) below the line. Formed by the grinding of glacier ice against valley sides
Glacial cirque (e.g., Fig. 29d)	Erosional depression carved or enlarged by a small glacier on the flank of a mountain. If glacier is absent, depression may be occupied by a small lake dammed by the end moraine left by the glacier that formerly occupied the cirque

These features may also be found on terrain modified by large alpine glaciers, montane ice caps, and lobes of continental glaciers (adapted from Flint 1971; Ehlers 1996; Neuendorf et al. 2005; Benn and Evans 2010)

deposits are generally dominated by coarse gravel in proximal locations in contrast to the sand-dominated deposits of volcanic water floods (VWFs), but otherwise their lithofacies and deposit characteristics are the same as those for VWFs.

Lodgment and ablation tills are poorly sorted diamicts (Fig. 31) that are typically unstratified (massive) when deposited during a single glacier advance or retreat, but stratified deposits are found in some ice-marginal and melt-out tills. Facies descriptors and codes (Table 2 in Chap. "Deposit Identification and Interpretation") for commonly encountered tills (both lodgment and ablation types) include massive matrix-supported diamict (Dmm), massive clast-supported diamict (Dcm), stratified matrix-supported diamict (Dms), and stratified clast-supported diamict (Dcs). In general, tills exhibit lithofacies that are continuous laterally and vertically for tens to hundreds of meters, although vertical facies changes can be abrupt due to changes in particle size and the appearance or disappearance of structures (Benn and Evans 2010; Kjær et al. 2003).

Localized Erosion and Effects on Vegetation

Localized erosional effects of glacier flow can be readily seen on exposed bedrock surfaces. Rock clasts embedded in ice at the base of glaciers or in highly compressed layers of deforming subglacial sediment have points (asperities) that concentrate enormous stresses from clasts pressed onto small areas of bedrock at the glacier bed (Flint 1971; Iverson 1990; Clark 1991; Benn and Evans 2010). If the focused stress is sufficient to fracture the bedrock, the moving clasts can gouge linear grooves into the

a
b
Trimline
c
d
e
f

◀ **Fig. 29** Types of glacial landforms commonly found on or downstream of glaciated volcanoes. **a** Lateral moraines, viewed looking downvalley from Adams Glacier, on the north side of Mount Adams (Washington, USA); the person is standing on the crest of the right lateral moraine. The left lateral moraine is visible across the valley. USGS photo by T. Pierson. **b** Terminal moraine below toe of small glacier, Glacier National Park (Montana, USA). USGS photo archive. **c** Glacier trim line where glacier scoured soil and vegetation off the valley side during a previous period of glacial advance. USGS photo archive. **d** Glacial cirque occupied by the Carbon Glacier, Mount Rainier (Washington, USA). USGS photo by Austin Post, 1967. **e** Hummocky ground moraine (ablation till) in right-center part of photo, possibly ice-cored, and incipient small lateral moraines deposited in front of the retreating terminus of the Adams Glacier (seen in background) on north flank of Mount Adams. USGS photo by T. Pierson, 2021. **f** Layers of lodgment till of different ages on a ridge (ground moraine) at Bennett Pass on the south flank of Mount Hood (Oregon, USA) (Scott and Gardner 2017). USGS photo by T. Pierson

Fig. 30 Rock glacier flowing down from Sourdough Peak in Wrangell-St. Elias National Park (Alaska, USA). Due to their lobate form and lack of visible ice on their surfaces, rock glaciers can be confused with small VDA or VDF deposits. Photo by U.S. National Park Service, public domain

bedrock—*striations* (Fig. 32a). Note that VDAs, VDFs, and PDCs also can striate bedrock surfaces, but glacial striations tend to be longer and more continuous. Fine-grained rock particles in and beneath the glacier additionally smooth, sculpt, and polish bedrock to a greater degree than PDCs and volcanic wet flows can accomplish (Fig. 32b).

Plucking (quarrying) of bedrock by glaciers involves the fracture, detachment, and removal of large (> ~ 1 cm) rock fragments from the bedrock surface, which commonly occurs on the lee side of obstacles and at the tops of downward steps or depressions on the bed (Benn and Evans 2010). Quarrying results in small-scale chatter marks, crescentic fractures, conchoidal

a

b

◀**Fig. 31** Lodgment and ablation tills. **a** Highly compacted, unstratified lodgment till on the south flank of Mount Hood at Bennett Pass (Oregon, USA) (Scott and Gardner 2017). Tightly packed matrix does not separate easily from gravel-size clasts. Shovel shaft is 3.0 cm in diameter. **b** Unstratified ablation till (nonvolcanic origin) that closely resembles a volcanic debris flow (VDF) deposit. The outcrop is on the inside face of a terminal moraine at Wallowa Lake (Oregon, USA). This till contains rounded clasts of river gravel that were entrained by the glacier. USGS photo by T. Pierson

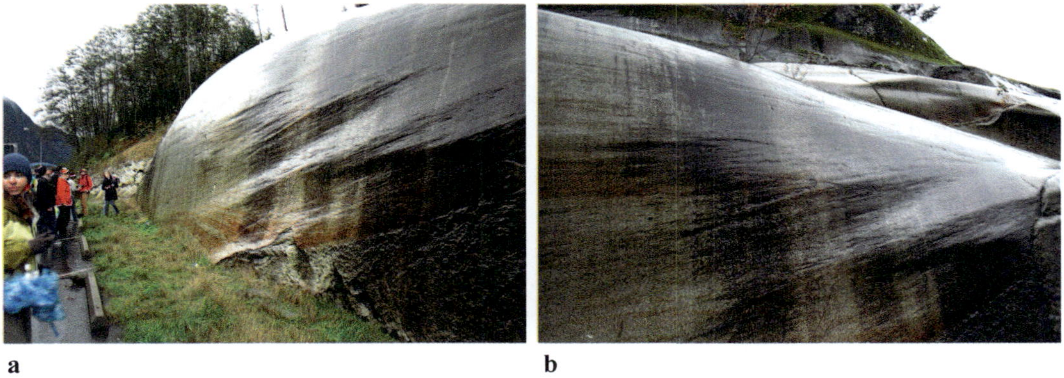

Fig. 32 Bedrock striated, sculpted, and polished by glacier flow. **a**, **b** Striations in bedrock (granodiorite), Coast Mountains (British Columbia, Canada). A smaller sculpted bedrock outcrop of this type is known as a *roche moutonnée* (sheep-like rock). The view frames in both photos are several meters wide. USGS photos by T. Pierson

fractures, and lunate fractures on otherwise smooth bedrock surfaces. Larger-scale erosion leads to the landforms listed in Table 3.

Advancing glaciers scour, crush, or bury any vegetation growing in their paths. Vegetation growing at the margins of a glacier is commonly buried by slumping supraglacial debris, snow avalanches, rockfall from valley sides, or ice-marginal fluvial sedimentation (Benn and Evans 2010). In addition, forest vegetation can become established on thick, supraglacial debris mantles on stagnant ice, parts of which could later be buried at random orientations in melt-out till (Porter and Carson 1971; Post and Streveler 1976). Glaciers also scour valley sides of all vegetation (and commonly all soil) to produce distinct trimlines (Flint 1971) (Fig. 29c). Glacier trimlines are typically higher above valley floors than trimlines produced by lahars or VDAs, and the trimlines remain consistently parallel to the valley floor, whereas lahar and VDA trimlines can vary in height above valley floors as they progress downvalley (e.g., Guthrie et al. 2012).

Glacial landforms and erosion on volcanoes can mimic some features produced by volcanic wet flows and PDCs. Distinguishing glacial landforms mainly depends on the scale of the features and on geometric details combined with sedimentary texture.

- Large VWFs can produce:
  - Broad alluvial plains of coarse fluvial sediment that resemble glacial outwash plains (e.g., Pierson et al. 1996). The main difference is the coarser average particle size in the outwash deposits that is due to higher average stream discharges during glacial periods.
  - Kettle-like meltout depressions in alluvial plains resulting from ice blocks transported by large glacier outburst floods (Fay 2002).
- Large VDFs and VDAs can produce:
  - Longitudinal bars that resemble drumlins but are typically about an order of magnitude smaller (Scott 1988).

- Lateral flow levees that resemble lateral moraines but are commonly smaller (Glicken 1996).
- Flutes on valley floors, bedrock striations, and valley-side trimlines (Pierson et al. 1990), but these glacial features are typically continuous over longer distances.
- Flank-collapse landslide scars at the sides of volcanoes that resemble glacial cirques, and some large failure scars may later be occupied by glaciers.
- Closed depressions on deposit surfaces (also from melting of entrained ice blocks) that resemble kettles (Glicken 1996).

• PDCs can produce kettle-like depressions that result from post-depositional phreatic explosions (Rowley et al. 1981; Hildreth and Fierstein 2012).

4.2.2 Outcrop-Scale Characteristics of Tills

Stratification

Single deposits of lodgment till are generally unstratified (Fig. 33a, b), but layering in lodgment till is sometimes observed. Stacking of multiple till sheets from different glacier advances or the overriding of stratified glaciofluvial deposits can produce layering of different till units with sharp contacts between them (Boulton and Deynoux 1981) (Figs. 29f and 33c). Furthermore, subhorizontal shear planes in lodgment till can resemble stratification. Ablation till may be massive or crudely stratified (Fig. 33d), although stratification results mainly from incremental deposition of numerous discrete mass flows produced by rock debris sliding off the glacier surface or from interbedding of mass-flow deposits with glaciofluvial deposits. Lack of stratification is also characteristic of VDF and VDA deposits.

Gravel-Size Clast Frameworks

Lodgment tills, especially those emplaced by large glaciers, typically have distinctively matrix-supported internal frameworks, commonly with gravel clasts widely spaced in highly compacted matrix material. Openwork frameworks in these deposits are highly unlikely. Ablation tills exhibit both clast-supported and matrix-supported frameworks. Openwork frameworks in these units are possible but rare.

Fabric

Lodgment tills that have been highly strained tend to have strong fabrics with long axes of gravel- and sand-size particles parallel to the direction of shear (more or less in the direction of glacier flow), commonly with a gentle fabric inclination (relative to macroscopic shear planes) that gives the deposit an upslope-dipping A-axis imbrication (Flint 1971; Mark 1974; Mills 1977a; Benn and Evans 2010; Kjær et al. 2003; Iverson 2017; Iverson et al. 2008). However, long axes can be oriented more transverse to flow direction in some cases, particularly where fabric inclination becomes steep and shear planes become tilted in response to glacier bed topography (Iverson 2017). Once developed, fabric does not weaken with continued strain (Iverson et al. 2008). Not all lodgment tills exhibit strong fabrics, however (Mark 1974; Mills 1977a; Benn and Evans 2010; Iverson et al. 2008). Rock-glacier and some PDC deposits tend to have strong flow-parallel fabrics, whereas ablation tills, melt-out tills, and some VDF deposits more typically have weak fabrics (Mills 1991).

Lodgment till subjected to extremely high overburden pressure and shearing may develop a type of subhorizontal foliation termed *fissility* (Flint 1971; Benn and Evans 2010). This texture occurs where fine-grained matrix particles become strongly oriented parallel to horizontal micro-shear planes. These small, discontinuous shear planes impart a lenticular, crinkly, or flaky appearance to till in a vertical outcrop face, and laminae typically bend around gravel clasts suspended in matrix material (Flint 1971) (Fig. 34). Fissility is not found in other types of volcaniclastic deposits.

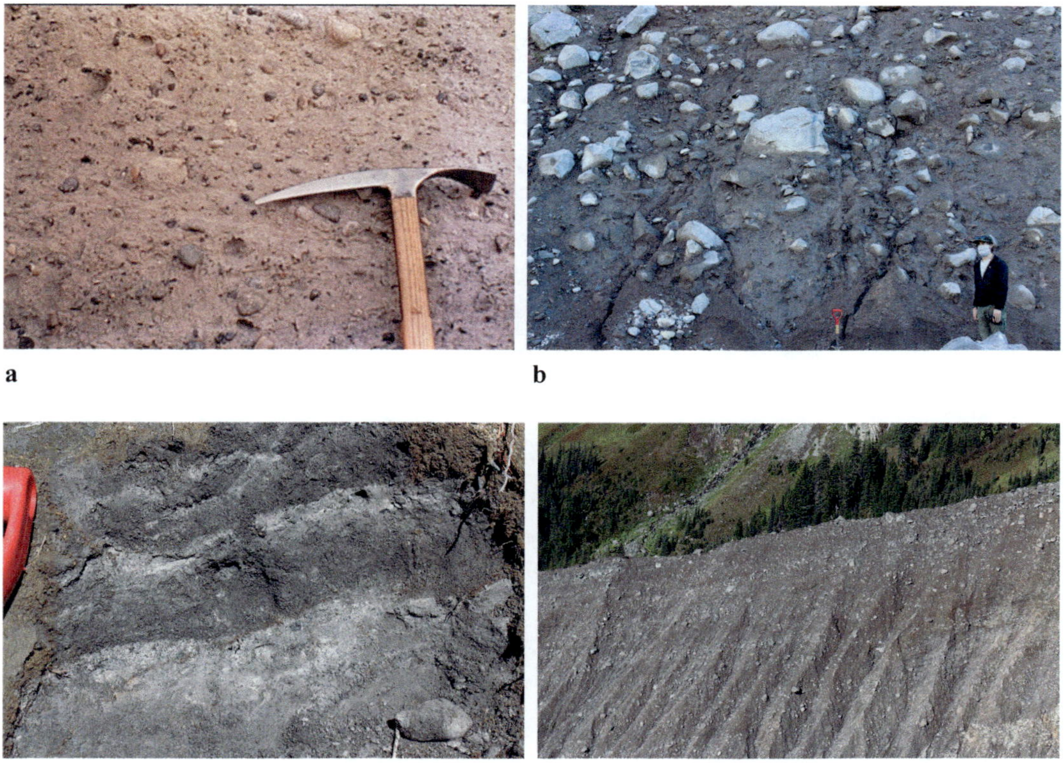

Fig. 33 Stratification in lodgment and ablation tills. **a** Unstratified and highly compacted massive lodgment till (nonvolcanic) emplaced by a Pleistocene continental ice sheet in the Puget Lowland near Port Townsend (Washington, USA). Ice-axe shaft is 35 mm wide. **b** Apparently unstratified and highly compacted massive lodgment till on a bench in Nisqually River valley at Mount Rainier (Washington, USA), emplaced by a large alpine glacier during Pleistocene glaciation. The outcrop surface is covered by a thin layer of rain-washed sand. **c** Layered, highly compacted, fine-grained lodgment till in Nisqually River valley at Mount Rainier. Shovel grip is 35 mm in diameter. **d** Crudely stratified ablation till on the inside face of the right lateral moraine, just downstream of the Nisqually Glacier terminus at Mount Rainier. Largest clasts are about 1 m in diameter. USGS photos by T. Pierson

Grading

We know of no reported instances of grading in glacial lodgment tills. Grading could be locally present in ablation tills in lateral and terminal moraines, in cases where some sorting and grading take place as debris slides or flows off glacier surfaces. Stagnant-ice melt-out tills are unlikely to be graded. Lack of grading is also typical of VDA deposits.

Fluid-Escape Structures

Tills normally do not contain fluid-escape structures. However, they have been reported resulting from unusual circumstances involving pressurized subglacial meltwater at Katla volcano (Iceland) (van der Meer et al. 1999; Kjær et al. 2003).

Shear Deformation

Lodgment till can be deformed by the huge normal stress of the overlying ice mass, combined with shear stress imposed by a glacier sliding over it. Degree of till deformation depends additionally on build-up of subglacial pore-water pressures, bed topography, granulometry, and strain history (Benn and Evans

Fig. 34 Fissile lodgment till. Example of a highly compacted, indurated, matrix-supported lodgment till with dark matrix of volcaniclastic sediment and weakly developed, discontinuous fissility (northern Patagonia, Chile). Fissle partings impart a subhorizontal "crinkly" appearance to outcrop face. Shovel blade is 16 cm wide. USGS photo by T. Pierson

2010; Kjær et al. 2003; Hart 2017). Some lodgment tills undergo brittle deformation and exhibit subhorizontal jointing with polished/striated joint surfaces, large-scale normal faulting, thrust faulting, subhorizontal shear planes, microshears, brecciation, boudins (lensoid structures), rotational structures, and crushed clasts and sand grains (Boulton and Deynoux 1981; Benn and Evans 2010; Hart 2017)—features also found in some large VDA avalanche blocks but not in lahar deposits. Type of shear deformation is related to clay content (Hart 2017). Ductile shear deformation features are also encountered, including a wide variety of folds (open, isoclinal, recumbent, chevron, kink-band, and polyclinal) (Benn and Evans 2010). Only VDA deposits (among the other types of particle–fluid flows) can have evidence of similarly intense horizontal shear.

Material Consistency

Material consistency in tills can vary widely, depending on the type of till, its composition, and its compaction history. Lodgment tills are *overconsolidated*—they experienced higher overburden pressures from the weight of the ice than they would have experienced from the weight of the overlying deposits alone. They can be extremely densely packed and highly indurated (hardened), owing not only to the weight of overriding ice but also to the typically high clay content of such tills (Flint 1971; Benn and Evans 2010). Tills that have been subjected to the overburden pressures from hundreds or thousands of meters of glacier ice can be very firm to extremely firm when moist (very hard to extremely hard when dry)—approaching the degree of induration of solid rock. Tills subjected to less overburden pressure or to considerable shearing can be more dilated and looser. Clay-rich lodgment tills can be slightly to very sticky. Ablation tills generally are not clay rich and have consistencies that are loose to friable when moist (soft to slightly hard when dry).

4.2.3 Particle-Scale Characteristics of Tills

Particle Composition

Tills from alpine glaciers on and downvalley of volcanoes are necessarily composed largely of volcanic rock fragments, but the percentage of fresh volcaniclastic particles in a till depends on the availability of these particles for entrainment at the time of glacier advance (downvalley flow). Particle composition in such tills would likely be polymictic and consist of a variety of volcanic rock types. Tills from continental glaciers that had traversed nonvolcanic terrain before reaching a volcano might contain mainly various nonvolcanic rock types and few volcanic rock fragments.

In lodgment tills, hard lithologies are more likely to survive as clasts than soft rock types under the enormous weight and grinding action of large glaciers, but soft clast lithologies such as limestone are nevertheless common in tills (Kjær et al. 2003) (e.g., Fig. 35a–c). All rock types are likely to be preserved in ablation tills, which do not experience the crushing overburden pressures of lodgment tills.

Particle Roundness, Shape, Fracturing, and Surface Texture

A unique combination of roundness, shape, and surface texture of gravel-size clasts is acquired when they are transported by glaciers at the interface between sliding ice and stationary glacier beds. This combination of characteristics can be helpful in differentiating tills from other types of mass-flow and flood deposits at volcanoes. However, not all glacier-transported clasts in tills are moved along glacier beds; they ride instead on or within the flowing ice. The physical characteristics of those clasts would be inherited from their initial modes of clast formation and pre-entrainment environments (Flint 1971).

Lodgment tills contain a limited proportion of englacial clasts that experienced high pressure abrasion at the glacier bed. Some of these clasts are faceted, striated (usually longitudinally), and polished, and they typically have slightly to moderately rounded edges (Benn and Evans 2010) (Fig. 35). Others are simply irregularly shaped and subrounded or passed through with little effect (Hart 2017). Flow-parallel abrasion occurs on one side of a clast when it has been frozen into the base of the glacier or held in position beneath a sliding glacier for extended periods, or it can occur on two sides simultaneously when a clast maintains a quasi-stable orientation within a deforming till (Benn and Evans 2010). Multiple orientations of striae on a clast face reflect rotation of that clast while in contact with bedrock (Menzies et al. 2018). Longitudinally striated and faceted clasts, sometimes described as "bullet"- or "flatiron"-shaped, are unique to lodgment tills and diagnostic of emplacement by glaciers. Prolonged abrasion on only one side also can result in asymmetrical "stoss and lee" clast geometries, where a lodged clast is smoothed on the upglacier side but has a rough, fractured downvalley side (Benn and Evans 2010). Striations, shaping, and polishing of clasts locked in ice are most pronounced in clasts composed of softer rock types. However, many englacial clasts do not contact the glacier bed and others are not locked into the ice for prolonged periods, so faceting and striations are found only on a small proportion of the total clast population (Flint 1971).

Englacial and supraglacial clasts that were never dragged along the glacier bed (commonly found in end moraines and melt-out till) have different properties. These clasts are transported passively on or within glacier ice, where little or no abrasion occurs. Such clasts have roundness and shape inherited from past histories and rock properties, but in most cases, clasts are angular rockfall debris that landed on the glacier surface (Boulton and Deynoux 1981). A massive poorly sorted deposit containing only angular boulders

Fig. 35 Glacially smoothed and striated clasts. **a** Equant striated, polished cobble with multiple striation orientations and a crescentic gouge on limestone cobble, about 25 cm long, collected near Kongsvegen in NW Spitsbergen (Norway). Copyrighted photo by M. J. Hambrey, www.glaciers-online.net, used with permission. **b** In situ glacially striated clast of Carboniferous limestone found embedded in till from the British Irish ice sheet of the Last Glacial Maximum at Tunstall beach along the Holderness coast in East Yorkshire (UK). Clast is approximately 12 cm wide and shows striations with two principal orientations. Photo by J. Sutherland, used with permission. **c** Flatiron-shaped limestone clast from Swiss Alps that is faceted and striated on all sides. Coin is 24 mm in diameter. USGS photo by T. Pierson. **d** Bullet-shaped, faceted basalt clast that is highly polished, shows crescentic gouges, but is only faintly striated; collected from Exit Glacier near Seward, Alaska (USA). Coin is 24 mm in diameter. USGS photo by T. Pierson

is unlikely to be a VDF deposit because boulders in concentrated lahars are invariably crudely rounded during transport of more than a kilometer or two. VDA deposits have only angular boulders, however.

Many lodgment tills show evidence of considerable fracturing and crushing of clasts while being overridden (Benn and Evans 2010). Intense fracturing of clasts also occurs in debris avalanches, but pervasively fractured clasts in VDA deposits tend to remain relatively intact (fracture patterns termed *jigsaw fractures*), while intensively fractured clasts crushed by glaciers are unlikely to remain intact.

Particle-Size Distribution and Sedimentary Texture

The range of particle sizes in tills is highly variable and can range from mostly clay to mostly coarse clasts, depending on the lithology of the terrain overridden and scoured by ice (Flint 1971) and on the type of till (Mills 1977b). Size distributions are typically bimodal to polymodal (Boulton 1987); reported clay contents vary from <1 to 80 wt% (Flint 1971; Hart 2017)—clay contents above about 20 wt% far exceed clay contents of other volcanic wet flows or PDCs.

Lodgment tills can be richer in clay and finer grained overall than ablation tills (Vorren 1977). Highly indurated lodgment tills are fines-rich because they deform mainly by fracturing interlocked particles where shearing under high normal stress is prolonged (Hooke and Iverson 1995). This fits with the observation that lodgment till size distributions are extremely poorly sorted and coarsely skewed (Landim and Frakes 1968). There is little spatial variation in size distribution at outcrop scale (Boulton and Deynoux 1981).

Ablation tills have lower fines contents than most lodgment tills, and ablation tills of alpine glaciers are likely to have distributions reflective of mass-wasting from valley sides (Boulton and Deynoux 1981), with clay contents less than 10 wt% (Flint 1971) and probably well less than that in volcanic terrain. Sorting values (Folk and Ward 1957) for tills overall are very poor to extremely poor (almost all $>3\varphi$; many $>4.5\varphi$) (Landim and Frakes 1968). These values have considerable overlap with sorting values of VDF and VDA deposits.

Matrix Vesicles and Intergranular Voids

To our knowledge, vesicles have not been reported in lodgment tills and might only occur in some ablation tills that include supraglacial debris-flow deposits. However, relatively high void ratios are reported in some lodgment tills due to dilation during shear and in ablation tills due to minimal compaction during melting of debris-rich ice (Benn and Evans 2010).

4.2.4 Key Characteristics of Tills

Tills commonly occupy volcanic drainages where volcanic wet-flow deposits are also likely to be found, and multiple lines of evidence may be needed to distinguish them. The following features of tills are particularly useful in their identification. Because tills can so closely resemble VDF and VDA deposits, recognition of deposit architecture and erosional and depositional landforms associated with a deposit are often required for correct deposit identification. Use of deposit sedimentary textures in outcrop alone may not be sufficient. A more complete list of characteristic properties is found in Chap. "Deposit Identification: Key and Summary Tables". Statements in italicized font draw comparisons with the different volcanic wet-flow and PDC deposits.

- **Landscape-Scale Criteria**
  - Terminal moraines are characteristic landforms.
    Nothing similar produced by volcanic wet flows.
  - Lateral moraines are characteristic landforms.
    Can resemble flow levees produced by VDAs and large VDFs, but glacial moraines are typically much taller (tens of meters high) and more continuous (e.g., Fig. 29a*).*
  - Ground moraine deposits on volcanoes are commonly thin (several meters or less) and discontinuous.
    VDF and VDA deposits tend to be thicker.
  - Striated, smoothed, and sometimes polished and sculpted bedrock surfaces are common.
    Striations are also created by large flank-collapse landslides, VDAs, VDFs, and pyroclastic density currents (PDCs), but glacial striations are more widespread, often more continuous, and occur together with smoothly sculpted rock forms.
  - Trimlines occur high on valley sides.
    VDAs can also form trim lines, but glacier trimlines are generally higher up on valley sides and consistently parallel with valley floors (e.g., Fig. 29c*), whereas trimlines caused by rapid wet flows may vary in elevation above the bed due to momentum-driven superelevation.*
  - Crushed logs and vegetation may be found beneath or at margins of lodgment till if forest vegetation was overridden by ice.
    Logs and other woody debris are commonly pushed to the side or to the front of VDF and VDA deposits.

- Randomly oriented logs can be buried within stagnant ice melt-out deposits.
  Such logs can also be found in VDA deposits.
- Hummocky topography is found in melt-out till.
  Hummocky topography is commonly found in VDA deposits but is more pronounced than in hummocky till. Vertical relief in till deposits is meters to tens of meters, whereas in VDA deposits it is tens to hundreds of meters.
- Kettles may be present in ground moraine deposits.
  Similar ice meltout pits can occur in VDF, VHF, VWF, and VDA deposits.
  Similar phreatic explosion pit craters can occur in PDC deposits.

• **Outcrop-Scale Criteria**
- Extreme consolidation/induration occurs in lodgment tills.
  Does not apply to ablation tills.
  Lodgment tills can be harder than any other wet or dry flow deposits without secondary cementation.
- Lack of grading in most tills
- Shear deformation is common in lodgment tills but not in ablation tills.
  Shear deformation is locally present in some VDA deposits.

• **Particle-Scale Criteria**
- Striated gravel-size clasts are common (but not plentiful) in lodgment tills.
  Not present in ablation tills.
  Such clasts are not present in deposits of volcanic wet flows unless remobilized from till.
- Matrix in lodgment tills may exhibit fissility—a flaky, subhorizontal foliation.
  Not found in volcanic wet-flow deposits.
- Till boulders, not abraded on glacier bed, are angular unless remobilized from fluvial gravel.
  Angular boulders are found in both lodgment and ablation tills.
  Boulders in lahar deposits are typically crudely rounded after transport over short distances.
  Boulders in VDA deposits are usually angular.
- Unusually high clay contents occur in lodgment tills, especially those emplaced by large alpine glaciers and ice sheets.
  Significantly higher clay contents in some lodgment tills than in most VDF, VHF, and VDA deposits.
  Clay-size particles are largely absent in PDC deposits.
- Matrix vesicles and intergranular voids are absent in lodgment tills; possibly present in some ablation tills.
  Vesicles and marginally localized intergranular voids are common in VDF deposits.
  Intergranular voids are common in PDC and VWF deposits but absent in VDA and VHF deposits.

References

Allen S, Cas R (2001) Transport of pyroclastic flows across the sea during the explosive, rhyolitic eruption of the Kos Plateau Tuff, Greece. Bull Volcanol 62:441–456. https://doi.org/10.1007/s004450000107

Alloway B, McComb P, Neall V, Vucetich C, Gibb J, Sherburn S, Stirling M (2005) Stratigraphy, age, and correlation of voluminous debris-avalanche events from an ancestral Egmont volcano: implications for coastal plain construction and regional hazard assessment. J R Soc New Zeal 35:229–267

Allstadt KE, Shean DE, Campbell A, Fahnestock M, Malone SD (2015) Observations of seasonal and diurnal glacier velocities at Mount Rainier, Washington, using terrestrial radar interferometry. Cryosphere 9:2219–2235. https://doi.org/10.5194/tc-9-2219-2015

Andrews GDM, Branney MJ (2011) Emplacement and rheomorphic deformation of a large, lava-lake rhyolitic ignimbrite: Grey's Landing, southern Idaho. Geol Soc Am Bull 123:725–743

Antos JA, Zobel DB (2005) Plant responses in forests of the tephra-fall zone. In: Dale VH, Swanson FJ, Crisafulli CM (eds) Ecological responses to the 1980 eruption of Mount St. Helens. Springer, New York, pp 47–58

Aramaki S (1956) The 1783 activity of Asama volcano, Part I. Jpn J Geol Geog 27(1):189–229

Bacon CR (1983) Eruptive history of Mount Mazama and Crater Lake caldera, Cascade Range U.S.A. J Volcanol Geotherm Res 18:57–115. https://doi.org/10.1016/0377-0273(83)90004-5

Banks NG, Hoblitt RP (1996) Direct temperature measurements of deposits, Mount St. Helens, Washington,

1980–1981. U S Geol Surv Prof Pap 1387:1–76. https://doi.org/10.3133/pp1387

Belousov A, Voight B, Belosouva M (2007) Directed blasts and blast-generated pyroclastic density currents: a comparison of the Bezymianny 1956, Mount St Helens 1980, and Soufrière Hills, Montserrat 1997 eruptions and deposits. Bull Volcanol 69:701–740. https://doi.org/10.1007/s00445-006-0109-y

Belousov A, Belousova M, Hoblitt R, Patia H (2020) The 1951 eruption of Mount Lamington, Papua New Guinea: devastating directed blast triggered by small-scale edifice failure. J Volcanol Geotherm Res 401:106947. https://doi.org/10.1016/j.jvolgeores.2020.106947

Benage M, Dufek J, Degruytera W, Geis D, Harpp K, Rader E (2014) Tying textures of breadcrust bombs to their transport regime and cooling history. J Volcanol Geotherm Res 274:92–107

Benn DI (1995) Fabric signature of subglacial till deformation, Breidamerkurjökull, Iceland. Sedimentology 42:735–747

Benn DI, Evans DJA (2010) Glaciers and glaciation, 2nd edn. Hodder Education, London, p 734

Benn DI, Owen LA, Osmaston HA, Seltzer GO, Porter SC, Mark B (2005) Reconstruction of equilibrium-line altitudes for tropical and sub-tropical glaciers. Quat Int 138–139:8–21. https://doi.org/10.1016/j.quaint.2005.02.003

Bernard J, Kelfoun K, Le Pennec JL, Vallejo VS (2014) Pyroclastic flow erosion and bulking processes: comparing field-based vs. modeling results at Tungurahua volcano, Ecuador. Bull Volcanol 76:1–6. https://doi.org/10.1007/s00445-014-0858-y

Björnsson H (1975) Subglacial water reservoirs, jökulhlaups, and volcanic eruptions. Jökull 25:1–14

Blong R (1984) Volcanic hazards: a sourcebook on the effects of eruptions. Academic Press, Sydney, p 424

Bonadonna C, Costa A, Folch A, Koyaguchi T (2015) Tephra dispersal and sedimentation. In: Sigurdsson H, Houghton B, McNutt SR, Rymer H, Stix J (eds) The encyclopedia of volcanoes, 2nd edn. Elsevier, Amsterdam, pp 587–597

Boudon G, Camus G, Gourgaud A, Lajoie J (1993) The 1984 nuée-ardente deposits of Merapi volcano, central Java, Indonesia: stratigraphy, textural characteristics, and transport mechanisms. Bull Volcanol 55:327–342

Boulton GS (1987) A theory of drumlin formation by subglacial sediment deformation. In: Menzies J, Rose J (eds) Drumlin symposium. Balkema, Rotterdam, pp 25–80

Boulton GS, Deynoux M (1981) Sedimentation in glacial environments and the identification of tills and tillites in ancient sedimentary sequences. Precamb Res 15:397–422

Brand BD, Mackaman-Lofland C, Pollock NM, Bendaña S, Dawson B, Wichgers P (2014) Dynamics of pyroclastic density currents: conditions that promote substrate erosion and self-channelization—Mount St Helens, Washington (USA). J Volcanol Geotherm Res 276:189–214. https://doi.org/10.1016/j.jvolgeores.2014.01.007

Brand BD, Bendaña S, Self S, Pollock N (2016) Topographic controls on pyroclastic density current dynamics: insight from 18 May 1980 deposits at Mount St. Helens, Washington (USA). J Volcanol Geotherm Res 321:1–17. https://doi.org/10.1016/j.jvolgeores.2016.04.018

Branney MJ, Kokelaar P (1992) A reappraisal of ignimbrite emplacement: progressive aggradation and changes from particulate to non-particulate flow during emplacement of high-grade ignimbrite. Bull Volcanol 54:504–520

Branney MJ, Kokelaar P (2002) Pyroclastic density currents and the sedimentation of ignimbrites. Geol Soc Lond Mem 27:1–143

Branney MJ, Sparks RSJ (1990) Fiamme formed by diagenesis and burial-compaction in soils and subaqeuous sediments. J Geol Soc 147(6):919–922

Brantley SR, Waitt RB (1988) Interrelations among pyroclastic surge, pyroclastic flow, and lahars in Smith Creek valley during first minutes of 18 May 1980 eruption of Mount St. Helens, USA. Bull Volcanol 50:304–326. https://doi.org/10.1007/BF01073588

Breard ECP, Lube G, Cronin SJ, Fitzgerald R, Kennedy B, Scheu B, Montanaro C, White JDL, Tost M, Procter JN, Moebis A (2014) Using the spatial distribution and lithology of ballistic blocks to interpret eruption sequence and dynamics: August 6, 2012, Upper Te Maari eruption, New Zealand. J Volcanol Geotherm Res 286:373–386. https://doi.org/10.1016/j.jvolgeores.2014.03.006

Brosch E, Lube G (2020) Spatiotemporal sediment transport and deposition processes in experimental dilute pyroclastic density currents. J Volcanol Geotherm Res 401:106946. https://doi.org/10.1016/j.volgeores.2020.106946

Brown RJ, Andrews DM (2015) Deposits of pyroclastic density currents. In: Sigurdsson H, Houghton B, McNutt SR, Rymer H, Stix J (eds) The encyclopedia of volcanoes, 2nd edn. Elsevier, Amsterdam, pp 631–648

Brown RJ, Branney MJ (2004) Event-stratigraphy of a caldera-forming ignimbrite eruption on Tenerife: the 273 ka Poris Formation. Bull Volcanol 66:392–416. https://doi.org/10.1007/s00445-003-0321-y

Brown RJ, Branney MJ, Maher C, Dávila-Harris P (2010) Origin of accretionary lapilli within ground-hugging density currents: evidence from pyroclastic couplets on Tenerife. Geol Soc Am Bull 122(1–2):305–320. https://doi.org/10.1130/B26449.1

Brown RJ, Bonadonna C, Durant AJ (2012) A review of volcanic ash aggregation. Phys Chem Earth 45–46:65–78. https://doi.org/10.1016/j.pce.2011.11.001

Brugman MM, Meier MF (1981) Response of glaciers to the eruptions of Mount St. Helens. In: Lipman PW, Mullineaux DR (eds) The 1980 eruptions of Mount

St. Helens, Washington. U S Geol Surv Prof Pap 1250:743–756
Bryan SE, Cas RAF, Martí J (1998a) Lithic breccias in intermediate volume phonolitic ignimbrites, Tenerife (Canary Islands): constraints on pyroclastic flow depositional processes. J Volcanol Geotherm Res 81:269–296
Bryan SE, Martí J, Cas RAF (1998b) Stratigraphy of the Bandas del Sur Formation: an extracaldera record of Quaternary phonolitic explosive eruptions from the Las Cañadas edifice, Tenerife (Canary Islands). Geol Mag 135:605–636
Bryan SE, Cas RAF, Martí J (2000) The 0.57 Ma plinian eruption of the Granadilla Member, Tenerife (Canary Islands): an example of complexity in eruption dynamics and evolution. J Volcanol Geotherm Res 103(1–4):209–238. https://doi.org/10.1016/S0377-0273(00)00225-0
Bryan SE, Peate IU, Peate DW, Self S, Jerram DA, Mawby MR, Marsh JS, Miller JA (2010) The largest volcanic eruptions on earth. Earth-Sci Rev 102:207–229
Bull SW, Cas RAF (2000) Distinguishing base-surge deposits and volcaniclastic fluviatile sediments: an ancient example from the Lower Devonian Snowy River Volcanics, south-eastern Australia. Sedimentology 47:87–98
Burgisser A, Bergantz GW (2002) Reconciling pyroclastic flow and surge: the multiphase physics of pyroclastic density currents. Earth Planet Sci Lett 202(2):405–418. https://doi.org/10.1016/S0012-821X(02)00789-6
Bursik MI, Kurbatov AV, Sheridan MF, Woods AW (1998) Transport and deposition in the May 18, 1980, Mount St. Helens blast flow. Geology 26:155–158
Bursik M, Patra A, Pitman EB, Nichita C, Macias JL, Saucedo R, Girina O (2005) Advances in studies of dense volcanic granular flows. Rep Prog Phys 68(2):271–302. https://doi.org/10.1088/0034-4885/68/2/R01
Calder ES, Sparks RSJ, Gardeweg MC (2000) Erosion, transport and segregation of pumice and lithic clasts in pyroclastic flows inferred from ignimbrite at Lascar volcano, Chile. J Volcanol Geotherm Res 104:201–235
Capaccioni B, Cuccoli F (2005) Spatter and welded air fall deposits generated by fire-fountaining eruptions: cooling of pyroclasts during transport and deposition. J Volcanol Geotherm Res 145:263–280
Carey SN (1991) Transport and deposition of tephra by pyroclastic flows and surges. In: Fisher RV, Smith GA (eds) Sedimentation in volcanic settings. SEPM (Soc Sediment Geol) Spec Publ 45:39–57
Carey S, Bursik M (2000) Volcanic plumes. In: Sigurdsson H, Houghton BF, McNutt SR, Rymer H, Stix J (eds) Encyclopedia of volcanoes. Academic Press, San Diego, pp 527–544
Carey S, Bursik M (2015) Volcanic plumes. In: Sigurdsson H, Houghton B, McNutt SR, Rymer H, Stix J (eds) The encyclopedia of volcanoes, 2nd edn. Elsevier, Amsterdam, pp 571–585
Carey SN, Sigurdsson H (1982) Influence of particle aggregation on deposition of distal tephra from the May 18, 1980, eruption of Mount St. Helens volcano. J Geophys Res Solid Earth 87(B8):7061–7072. https://doi.org/10.1029/JB087iB08p07061
Carey S, Sigurdsson H, Mandeville C et al (1996) Pyroclastic flows and surges over water: an example from the 1883 Krakatau eruption. Bull Volcanol 57:493–511. https://doi.org/10.1007/BF00304435
Carn SA, Watts RB, Thompson G, Norton GE (2004) Anatomy of a lava dome collapse: the 20 March 2000 event at Soufrière Hills volcano, Montserrat. J Volcanol Geotherm Res 131:241–264
Cas RAF, Wright JV (1987) Volcanic successions modern and ancient. Allen and Unwin, London, p 528
Cashman KV, Sparks RSJ (2013) How volcanoes work: a 25-year perspective. Geol Soc Am Bull 125(5/6):664–690. https://doi.org/10.1130/B30720.1
Chapin CE, Elston WE (1979) Introduction. In: Chapin CE, Elston WE (eds) Ash-flow tuffs. Geol Soc Am Spec Pap 180:1–4
Charbonnier SJ, Gertisser R (2008) Field observations and surface characteristics of pristine block-and-ash flow deposits from the 2006 eruption of Merapi volcano, Java, Indonesia. J Volcanol Geotherm Res 177(4):971–982. https://doi.org/10.1016/j.jvolgeores.2008.07.008
Cioni R, Marianelli P, Santacroce R, Sbrana A (2000) Plinian and subplinian eruptions. In: Sigurdsson H, Houghton BF, McNutt SR, Rymer H, Stix J (eds) Encyclopedia of volcanoes. Academic Press, San Diego, pp 477–494
Cioni R, Pistolesi M, Rosi M (2015) Plinian and subplinian eruptions. In: Sigurdsson H, Houghton B, McNutt SR, Rymer H, Stix J (eds) The encyclopedia of volcanoes, 2nd edn. Elsevier, Amsterdam, pp 519–535
Clark PU (1991) Striated clast pavements: products of deforming subglacial sediment? Geology 19(5):530–533. https://doi.org/10.1130/0091-7613(1991)019<0530:SCPPOD>2.3.CO;2
Cole PD, Calder ES, Sparks RSJ, Clarke AB, Druitt TH, Young SR, Herd RA, Harford CL, Norton GE (2002) Deposits from dome-collapse and fountain-collapse pyroclastic flows at Soufrière Hills volcano, Montserrat. Geol Soc Lond Mem 21:231–262. https://doi.org/10.1144/GSL.MEM.2002.021.01.11
Cole PD, Neri A, Baxter PJ (2015) Hazards from pyroclastic density currents. In: Sigurdsson H, Houghton B, McNutt SR, Rymer H, Stix J (eds) The encyclopedia of volcanoes, 2nd edn. Elsevier, Amsterdam, pp 943–969
Cuffey KM, Paterson WSB (2010) The physics of glaciers, 4th edn. Elsevier, Amsterdam, 674 pp
Dade WB, Huppert HE (1996) Emplacement of the Taupo ignimbrite by a dilute turbulent flow. Nature 381:509–512
Dellino P, Dioguardi F, Doronzo DM, Mele D (2020) A discriminatory diagram of massive versus stratified deposits based on the sedimentation and bedload transportation rates. Experimental investigation

and application to pyroclastic density currents. Sedimentology 67(4):2013–2039
Dinehart RL (1998) Sediment transport at gaging stations near Mount St. Helens, Washington, 1980–1990: data collection and analysis. U S Geol Surv Prof Pap 1573:105 pp
Douillet GA (2021) The supercritical question for pyroclastic dune bedforms: an overview. Sedimentology 68(4):1698–1727
Douillet GA, Pacheco DA, Kueppers U, Letort J, Tsang-Hin-Sun E, Bustillos J, Hall M, Ramón P, Dingwell DB (2013) Dune bedforms produced by dilute pyroclastic density currents from the August 2006 eruption of Tungurahua volcano, Ecuador. Bull Volcanol 75:62. https://doi.org/10.1007/s00445-013-0762-x
Douillet GA, Taisne B, Tsang-Hin-Sun È, Müller SK, Kueppers U, Dingwell DB (2015) Syn-eruptive, soft-sediment deformation of deposits from dilute pyroclastic density current: triggers from granular shear, dynamic pore pressure, ballistic impacts and shock waves. Solid Earth 6:553–572. https://doi.org/10.5194/se-6-553-2015
Druitt TH (1995) Settling behaviour of concentrated dispersions and some volcanological applications. J Volcanol Geotherm Res 65(1–2):27–39. https://doi.org/10.1016/0377-0273(94)00090-4
Druitt TH (1998) Pyroclastic density currents. Geol Soc Lond Spec Publ 145:145–182. https://doi.org/10.1144/GSL.SP.1996.145.01.08
Druitt TH, Bacon CR (1986) Lithic breccia and ignimbrite erupted during the collapse of Crater Lake caldera, Oregon. J Volcanol Geotherm Res 29:1–32
Druitt TH, Calder ES, Cole PD, Hoblitt RP, Loughlin SC, Norton GE, Ritchie LJ, Sparks RSJ, Voight B (2002) Small-volume, highly mobile pyroclastic flows formed by rapid sedimentation from pyroclastic surges at Soufrière Hills volcano, Montserrat: an important volcanic hazard. In: Druitt TH, Kokelaar BP (eds) The eruption of Soufrière Hills volcano, Montserrat, from 1995 to 1999. Geol Soc Lond Mem 21:263–279
Dufek J, Ongaro TE, Roche O (2015) Pyroclastic density currents: processes and models. In: Sigurdsson H, Houghton B, McNutt SR, Rymer H, Stix J (eds) The encyclopedia of volcanoes, 2nd edn. Elsevier, Amsterdam, pp 617–629
Ehlers J (1996) Quaternary and glacial geology. Wiley, Chichester, p 578
Esposti Ongaro T, Clarke AB, Voight B, Neri A, Widiwijayanti C (2012) Multiphase flow dynamics of pyroclastic density currents during the May 18, 1980 lateral blast of Mount St. Helens. J Geophys Res 117:B06208. https://doi.org/10.1029/2011JB009081
Eychenne J, Le Pennec JL, Troncoso L, Gouhier M, Nedelec J-M (2012) Causes and consequences of bimodal grain-size distribution of tephra fall deposited during the August 2006 Tungurahua eruption (Ecuador). Bull Volcanol 74:187–205. https://doi.org/10.1007/s00445-011-0517-5
Eyles N, Eyles CH, Miall AD (1983) Lithofacies types and vertical profile models; an alternative approach to the description and environmental interpretation of glacial diamict and diamictite sequences. Sedimentology 30:393–410
Fay H (2002) Formation of kettle holes following a glacial outburst flood (jökulhlaup), Skeiðarársandur, southern Iceland. In: The extremes of the extremes: extraordinary floods. Proceedings of a symposium held at Reykjavik, Iceland, July 2000. IAHS Publ 271:205–210
Fierstein J, Hildreth W (1992) The plinian eruptions of 1912 at Novarupta, Katmai National Park, Alaska. Bull Volcanol 54:646–684. https://doi.org/10.1007/BF00430778
Fierstein J, Nathenson M (1992) Another look at the calculation of fallout tephra volumes. Bull Volcanol 54:156–167. https://doi.org/10.1007/BF00278005
Fisher RV (1961) A proposed classification of volcaniclastic sediments and rocks. Geol Soc Am Bull 72:1409–1414
Fisher RV (1963) Bubble-wall texture and its significance. J Sed Res 33(1):224–227
Fisher RV (1966a) Mechanism of deposition from pyroclastic flows. Am J Sci 264:350–363
Fisher RV (1966b) Rocks composed of volcanic fragments and their classification. Earth-Sci Rev 1:287–298
Fisher RV (1977) Erosion by volcanic base-surge density currents: U-shaped channels. Geol Soc Am Bull 88:1287–1297
Fisher RV (1979) Models for pyroclastic surges and pyroclastic flows. J Volcanol Geotherm Res 6:305–318
Fisher RV (1995) Decoupling of pyroclastic currents: hazards assessments. J Volcanol Geotherm Res 66(1–4):257–263. https://doi.org/10.1016/0377-0273(94)00075-R
Fisher RV, Schmincke H-U (1984) Pyroclastic rocks. Springer-Verlag, Berlin, p 472
Fisher RV, Waters AC (1970) Base surge bed forms in maar volcanoes. Am J Sci 268:157–180
Fitzgerald RH, Kennedy BM, Wilson TM, Leonard GS, Tsunematsu K, Keys H (2017) The communication and risk management of volcanic ballistic hazards. In: Fearnley CJ, Bird DK, Haynes K, McGuire WJ, Jolly G (eds) Observing the volcano world. Advances in volcanology. Springer, Cham. https://doi.org/10.1007/11157_2016_35
Flint RF (1971) Glacial and Quaternary geology. Wiley, New York, p 892
Folk RL (1974) Petrology of sedimentary rocks. Hemphill, Austin, 184 pp
Folk RL, Ward WC (1957) Brazos River bar: a study in the significance of grain size parameters. J Sediment Petrol 27:3–26
Francis PW (1993) Volcanoes—a planetary perspective. Clarendon Press (Oxford University Press), New York, p 443

Freundt A, Schmincke H-U (1992) Abrasion in pyroclastic flows. Geol Rundsch 81:383–389

Freundt A, Wilson CJN, Carey SN (2000) Ignimbrites and block-and-ash deposits. In: Sigurdsson H, Houghton BF, NcNutt SR, Rymer H, Stix J (eds) Encyclopedia of volcanoes. Academic Press, San Diego, pp 581–599

Gardner JE, Andrews BJ, Dennen R (2017) Liftoff of the 18 May 1980 surge of Mount St. Helens (USA) and the deposits left behind. Bull Volcanol 79:8. https://doi.org/10.1007/s00445-016-1095-3

Gardner JE, Nazworth C, Helper MA, Andrews BJ (2018) Inferring the nature of pyroclastic density currents from tree damage: the 18 May 1980 blast surge of Mount St. Helens, USA. Geology 46(9):95–798. https://doi.org/10.1130/G45353.1

Gehrels MJ, Newnham RM, Lowe DJ, Wynne S, Hazell ZJ, Caseldine C (2008) Towards rapid assay of cryptotephra in peat cores: review and evaluation of various methods. Quat Int 178:68–84. https://doi.org/10.1016/j.quaint.2006.10.014

Girolami L, Druitt TH, Roche O (2015) Towards a quantitative understanding of pyroclastic flows: effects of expansion on the dynamics of laboratory fluidized granular flows. J Volcanol Geotherm Res 296:31–39. https://doi.org/10.1016/j.jvolgeores.2015.03.008

Glicken H (1996) Rockslide debris avalanche of May 18, 1980, Mount St. Helens volcano, Washington. U S Geol Surv Open-File Rep 96-677, 90 pp and 5 plates

Gorshkov GS (1959) Gigantic eruption of the Volcano Bezymianny. Bull Volcanol 20:77–109. https://doi.org/10.1007/BF02596572

Gorshkov GS, Dubik YM (1970) Gigantic directed blast at Shiveluch volcano (Kamchatka). Bull Volcanol 34:261–288. https://doi.org/10.1007/BF02597790

Grunewald U, Sparks RSJ, Kearns JC, Komorowski J-C (2000) Friction marks on blocks from pyroclastic flows at the Soufriere Hills volcano, Montserrat: implications for flow mechanisms. Geology 28(9):827–830. https://doi.org/10.1130/0091-7613(2000)28<827:FMOBFP>2.0.CO;2

Guðmundsson MT, Sigmundsson F, Björnsson H (1997) Ice-volcano interaction of the 1996 Gjálp subglacial eruption, Vatnajökull, Iceland. Nature 389:954–957

Gudnason J, Thordarson T, Houghton BF, Larsen G (2017) The opening subplinian phase of the Hekla 1991 eruption: properties of the tephra fall deposit. Bull Volcanol 79:34. https://doi.org/10.1007/s00445-017-1118-8

Gudnason J, Thordarson T, Houghton BF, Larsen G (2018) The 1845 Hekla eruption: grain-size characteristics of a tephra layer. J Volcanol Geothem Res 350:33–46. https://doi.org/10.1016/j.jvolgeores.2017.11.025

Guthrie RH, Friele P, Allstadt K, Roberts N, Evans SG, Delaney KB, Roche D, Clague JJ, Jakob M (2012) The 6 August 2010 Mount Meager rock slide-debris flow, Coast Mountains, British Columbia: characteristics, dynamics, and implications for hazard and risk assessment. Nat Hazards Earth Syst Sci 12:1277–1294. https://doi.org/10.5194/nhess-12-1277-2012

Hart JK (2017) Subglacial till formation: microscale processes within the subglacial shear zone. Quat Sci Rev 170:26–44

Heiken G, Wohletz K (1991) Fragmentation processes in explosive volcanic eruptions. In: Fisher RV, Smith GA (eds) Sedimentation in volcanic settings. SEPM Spec Publ 45:19–26

Heming RF, Carmichael ISE (1973) High-temperature pumice flows from the Rabaul caldera, Papua New Guinea. Contrib Mineral Petrol 38:1–20

Herd RA, Edmonds M, Bass VA (2005) Catastrophic lava dome failure at Soufrière Hills volcano, Montserrat, 12–13 July 2003. J Volcanol Geotherm Res 148:234–252

Hildreth W, Fierstein J (2012) The Novarupta-Katmai eruption of 1912—largest eruption of the twentieth century: centennial perspectives. U S Geol Surv Prof Pap 1791:259 pp

Hoblitt RP (2000) Was the 18 May 1980 lateral blast at Mt St Helens the product of two explosions? Phil Trans R Soc A 358:1639–1661. https://doi.org/10.1098/rsta.2000.0608

Hoblitt RP, Kellogg KS (1979) Emplacement temperatures of unsorted and unstratified deposits of volcanic rock debris as determined by paleomagnetic techniques. Geol Soc Am Bull 90(1):633–642

Hoblitt RP, Miller CD, Vallance JW (1981) Origin and stratigraphy of the deposit produced by the May 18 directed blast. In: Lipman PW, Mullineaux DR (eds) The 1980 eruptions of Mount St. Helens, Washington. U S Geol Surv Prof Pap 1250:401–419

Holt EW, Taylor HP (1998) $^{18}O/^{16}O$ mapping and hydrogeology of a short-lived ($\approx$10 years) fumarolic (>500°C) meteoric-hydrothermal event in the upper part of the 0.76 Ma Bishop Tuff outflow sheet, California. J Volcanol Geotherm Res 83:115–139

Hooke RLeB, Iverson NR (1995) Grain-size distribution in deforming subglacial tills: role of grain fracture. Geology 23:57–60

Houghton B, Carey RJ (2015) Pyroclastic fall deposits. In: Sigurdsson H, Houghton B, McNutt SR, Rymer H, Stix J (eds) The encyclopedia of volcanoes, 2nd edn. Elsevier, Amsterdam, pp 599–616

Houghton BF, Smith RT (1993) Recycling of magmatic clasts during explosive eruptions: estimating the true juvenile content of phreatomagmatic volcanic deposits. Bull Volcanol 55:414–420

Houghton BF, Wilson CJN, Pyle DM (2000a) Pyroclastic fall deposits. In: Sigurdsson H, Houghton BF, McNutt SR, Rymer H, Stix J (eds) Encyclopedia of volcanoes. Academic Press, San Diego, pp 555–570

Houghton BF, Wilson CJN, Smith RT, Gilbert JS (2000b) Phreatoplinian eruptions. In: Sigurdsson H, Houghton BF, McNutt SR, Rymer H, Stix J (eds) Encyclopedia of volcanoes. Academic Press, San Diego, pp 513–525

Houghton B, White JD, Van Eaton AR (2015) Phreatomagmatic and related eruption styles. In: Sigurdsson H, Houghton B, McNutt SR, Rymer H, Stix J (eds) The encyclopedia of volcanoes, 2nd edn. Elsevier, Amsterdam, pp 537–552

Houghton BF, Swanson DA, Biass S, Fagents SA, Orr TR (2017) Partitioning of pyroclasts between ballistic transport and a convective plume: Kīlauea volcano, 19 March 2008. J Geophys Res Solid Earth 122(5):3379–3391

Huggel C, Ceballos JL, Pulgarín B, Ramírez J, Thouret J-C (2007) Review and reassessment of hazards owing to volcano–glacier interactions in Colombia. Annu Glaciol 45:128–136. https://doi.org/10.3189/172756407782282408

Hughes SR, Druitt TH (1998) Particle fabric in a small, type-2 ignimbrite flow unit (Laacher See, Germany) and implications for emplacement dynamics. Bull Volcanol 60:125–136

Inman DL (1952) Measures of describing the size distribution of sediments. J Sediment Petrol 22:125–145

Iverson NR (1990) Laboratory simulations of glacial abrasion: comparison with theory. J Glaciol 36:304–314

Iverson NR (2017) Determining glacier flow direction from till fabrics. Geomorphology 299:124–130. https://doi.org/10.1016/j.geomorph.2017.10.005

Iverson NR, Hooyer TS, Thomason JF, Graesch M, Shumway JR (2008) The experimental basis for interpreting particle and magnetic fabrics of sheared till. Earth Surf Process Landf 33:627–645

Kamata H, Mimura K (1983) Flow directions inferred from imbrication in the Handa pyroclastic flow deposit in Japan. Bull Volcanol 46:277–282

Kamb B, Raymond CF, Harrison WD, Engelhardt H, Echelmeyer KA, Humphrey N, Brugman MM, Pfeffer T (1985) Glacier surge mechanism. 1982–1983 surge of Variegated Glacier, Alaska. Science 227(4686):469–479. https://doi.org/10.1126/science.227.4686.469

Kieffer SW (1981a) Blast dynamics at Mount St. Helens on 18 May 1980. Nature 291:568–570. https://doi.org/10.1038/291568a0

Kieffer SW (1981b) Fluid dynamics of the May 18 blast at Mount St. Helens. In: Lipman PW, Mullineaux DR (eds) The 1980 eruptions of Mount St. Helens, Washington. U S Geol Surv Prof Pap 1250:379–400

Kieffer SW, Sturtevant B (1988) Erosional furrows formed during the lateral blast at Mount St. Helens, May 18, 1980. J Geophys Res 93(B12):14793–14816

Kieffer SW, Meiburg E, Best J, Austin J (2021) The mysterious grooves of Volcán Bárcena: a review of the role of streamwise counter-rotating vortices during erosion by dilute pyroclastic density currents. Bull Volcanol 83:26. https://doi.org/10.1007/s00445-021-01440-9

Kjær KH, Krüger J, van der Meer JJM (2003) What causes till thickness to change over distance? Answers from Myrdalsjökull, Iceland. Quat Sci Rev 22:1687–1700

Kocurek G, Hunter RE (1986) Origin of polygonal fractures in sand, uppermost Navajo and Page sandstones, Page, Arizona. J Sediment Res 56(6):895–904. https://doi.org/10.1306/212F8A7B-2B24-11D7-8648000102C1865D

Komorowski J-C, Jenkins S, Baxter PJ, Picquiout A, Lavigne F, Charbonnier S, Gertesser R, Preece K, Cholik N, Budi-Santoso A, Surono (2013) Paroxysmal dome explosion during the Merapi 2010 eruption: processes and facies relationships of associated high-energy pyroclastic density currents. J Volcanol Geotherm Res 261:260–294

Kuntz MA, Rowley PD, MacLeod NS, Reynolds RL, McBroome LA, Kaplan AM, Lidke DJ (1981) Petrography and particle-size distribution of pyroclastic-flow, ash-cloud, and surge deposits. In: Lipman PW, Mullineaux DR (eds) The 1980 eruptions of Mount St. Helens, Washington. U S Geol Surv Prof Pap 1250:525–539

Landim PMB, Frakes LA (1968) Distinction between tills and other diamictons based on textural characteristics. J Sediment Petrol 38:1213–1223

Lane CS, Cullen VL, White D, Bramham-Law CWF, Smith VC (2014) Cryptotephra as a dating and correlation tool in archaeology. J Archaeol Sci 42:42–50. https://doi.org/10.1016/j.jas.2013.10.033

Legros F, Martí J (2001) Formation of inversely graded basal layers in ignimbrites by progressive aggradation. J Volcanol Geotherm Res 111(1–4):25–33. https://doi.org/10.1016/S0377-0273(01)00208-6

Lipman PW, Mullineaux DR (eds) (1981) The 1980 eruptions of Mount St. Helens, Washington. U S Geol Surv Prof Pap 1250:844 pp. https://doi.org/10.3133/pp1250

Liu EJ, Cashman KV, Rust AC (2015) Optimizing shape analysis to quantify volcanic ash morphology. GeoResJ 8:14–30. https://doi.org/10.1016/j.grj.2015.09.001

Lorenz V (1974) Vesiculated tuffs and associated features. Sedimentology 21:273–291

Lube G, Cronin SJ, Platz T, Freundt A, Procter JN, Cl H, Sheridan MF (2007) Flow and deposition of pyroclastic granular flows: a type example from the 1975 Ngauruhoe eruption, New Zealand. J Volcanol Geotherm Res 161:165–186

Lube G, Breard ECP, Esposti-Ongaro T, Dufek J, Brand B (2020) Multiphase flow behaviour and hazard prediction of pyroclastic density currents. Nat Rev Earth Environ 1:348–365. https://doi.org/10.1038/s43017-020-0064-8

Macorps E, Charbonnier SJ, Varley NR, Capra L, Atlas Z, Cabré J (2018) Stratigraphy, sedimentology and inferred flow dynamics from the July 2015 block-and-ash flow deposits at Volcán de Colima, Mexico. J Volcanol Geotherm 349:99–116. https://doi.org/10.1016/j.jvolgeores.2017.09.025

Major JJ (2023) Subaerial volcaniclastic deposits—influences of initiation mechanisms and transport behaviour on characteristics and distributions. In: Di Capua A, De Rosa R, Kereszturi G, Le Pera E, Rosi M, Watt SFL (eds) Volcanic processes in the sedimentary record: when volcanoes meet the environment. Geol Soc Lond Spec Publ 520:29–100. First published online 12 July 2022. https://doi.org/10.1144/SP520-2021-142

Major JJ, Pierson TC, Hoblitt RP, Moreno H (2013) Pyroclastic density currents associated with the 2008–2009 eruption of Chaitén volcano (Chile)—forest disturbances, deposits, and dynamics. Andean Geol 40(2):324–358

Manga M, Patel A, Dufek J (2011) Rounding of pumice clasts during transport: field measurements and laboratory studies. Bull Volcanol 73:321–333. https://doi.org/10.1007/s00445-010-0411-6

Mark DM (1974) On the interpretation of till fabrics. Geology 2(2):101–104. https://doi.org/10.1130/0091-7613(1974)2<101:OTIOTF>2.0.CO;2

Mastin LG (2001) A simple calculator of ballistic trajectories for blocks ejected during volcanic eruptions. U S Geol Surv Open-File Rep 01-45, 16 pp. https://pubs.usgs.gov/of/2001/0045/

Mastin LG, Van Eaton AR, Lowenstern JB (2014) Modeling ash fall distribution from a Yellowstone supereruption. Geochem Geophys Geosyst 8:3459–3475. https://doi.org/10.1002/2014GC005469

McClelland E, Druitt TH (1989) Paleomagnetic estimates of emplacement temperatures of pyroclastic deposits on Santorini, Greece. Bull Volcanol 51:16–27

McPhie J, Doyle M, Allen R (1993) Volcanic textures—a guide to the interpretation of textures in volcanic rocks. Centre for Ore Deposit and Exploration Studies, University of Tasmania, Hobart, p 196

Mele D, Dioguardi F, Dellino P, Isaia R, Sulpizio R, Braia G (2015) Hazard of pyroclastic density currents at the Campi Flegrei Caldera (Southern Italy) as deduced from the combined use of facies architecture, physical modeling and statistics of the impact parameters. J Volcanol Geotherm Res 299:35–53

Mellors RA, Waitt RB, Swanson DA (1988) Generation of pyroclastic flows and surges by hot-rock avalanches from the dome of Mount St. Helens volcano, USA. Bull Volcanol 50:14–25

Menzies J, van der Meer JJM, Shilts WW (2018) Subglacial processes and sediments. In: Menzies J, van der Meer JJM (eds) Past glacial environments, 2nd edn. Elsevier, Amsterdam, pp 105–158

Miller JMG (1996) Glacial sediments. In: Reading HG (ed) Sedimentary environments: processes, facies and stratigraphy. Blackwell Science, London, pp 454–484

Mills HH (1977a) Basal till fabrics of modern alpine glaciers. Geol Soc Am Bull 88:824–828

Mills HH (1977b) Textural characteristics of drift from some representative Cordilleran glaciers. Geol Soc Am Bull 88:1135–1143

Mills HH (1991) Three-dimensional clast orientation in glacial and mass-movement sediments: compilation and preliminary analysis. U S Geol Surv Open-File Rep 90-128, 71 pp

Mimura K (1984) Imbrication, flow direction and possible source areas of the pumice-flow tuffs near Bend, Oregon, U.S.A. J Volcanol Geotherm Res 21:45–60

Miyabuchi Y (1999) Deposits associated with the 1990–1995 eruption of Unzen volcano, Japan. J Volcanol Geotherm Res 89:139–158

Moore JG (1967) Base surge in recent volcanic eruptions. Bull Volcanol 30:337–363

Moore JG, Melson WG (1969) Nuees ardentes of the 1968 eruption of Mayon volcano, Philippines. Bull Volcanol 33:600–620

Moore JG, Sisson TW (1981) Deposits and effects of the May 18 pyroclastic surge. In: Lipman PW, Mullineaux DR (eds) The 1980 eruptions of Mount St. Helens, Washington. U S Geol Surv Prof Pap 1250:421–438

Morrissey M, Zimanowski B, Wohletz K, Buettner R (2000) Phreatomagmatic fragmentation. In: Sigurdsson H, Houghton BF, McNutt SR, Rymer H, Stix J (eds) Encyclopedia of volcanoes. Academic Press, San Diego, pp 431–445

Neuendorf KKE, Mehl JP, Jackson JA (2005) Glossary of geology, 5th edn. American Geological Institute, Alexandria, VA, 779 pp

Newhall CG, Self S (1982) The volcanic explosivity index (VEI) an estimate of explosive magnitude for historical volcanism. J Geophys Res Oceans 87(C2):1231–1238. https://doi.org/10.1029/JC087iC02p01231

Pacheco-Hoyos JG, Aguirre-Díaz GJ, Dávila-Harris P (2020) Elutriation pipes in ignimbrites: an analysis of concepts based on the Huichapan ignimbrite, Mexico. J Volcanol Geotherm Res 403:107026. https://doi.org/10.1016/j.jvolgeores.2020.107026

Pierson TC (1985) Initiation and flow behavior of the 1980 Pine Creek and Muddy River lahars, Mount St. Helens, Washington. Geol Soc Am Bull 96:1056–1069. https://doi.org/10.1130/0016-7606(1985)96<1056:IAFBOT>2.0.CO;2

Pierson TC, Janda RJ, Thouret JC, Borrero CA (1990) Perturbation and melting of snow and ice by the 13 November 1985 eruption of Nevado del Ruiz, Colombia, and consequent mobilization, flow, and deposition of lahars. J Volcanol Geotherm Res 41:17–66

Pierson TC, Daag AS, Delos Reyes PJ, Regalado MTM, Solidum RU, Tubianosa BS (1996) Flow and deposition of posteruption hot lahars on the east side of Mount Pinatubo, July–October 1991. In: Newhall CG, Punongbayan RS (eds) Fire and mud: eruptions and lahars of Mount Pinatubo, Philippines. Philippine Institute of Volcanology and Seismology, Quezon City, University of Washington Press, Seattle, pp 921–950

Pierson TC, Major JJ, Amigo Á, Moreno H (2013) Acute sedimentation response to rainfall following the explosive phase of the 2008–2009 eruption of Chaitén volcano, Chile. Bull Volcanol 75:723. https://doi.org/10.1007/s00445-013-0723-4

Polacci M (2005) Constraining the dynamics of volcanic eruptions by characterization of pumice textures. Annu Geophys 48(4/5):731–738

Pollock NM, Brand BD, Rowley PJ, Sarocchi D, Sulpizio R (2019) Inferring pyroclastic density current flow conditions using syn-depositional sedimentary structures. Bull Volcanol 81:46. https://doi.org/10.1007/s00445-019-1303z

Porreca M, Mattei M, MacNiocaill C, Giordano G, McClelland E, Funiciello R (2008) Paleomagnetic evidence for low-temperature emplacement of the phreatomagmatic Peperino Albano ignimbrite (Colli Albani volcano, Central Italy). Bull Volcanol 70:877–893. https://doi.org/10.1007/s00445-007-0176-8

Porter SC, Carson RJ (1971) Problems of interpreting radiocarbon dates from dead-ice terrain, with an example from the Puget Lowland of Washington. Quat Res 1(3):410–414. https://doi.org/10.1016/0033-5894(71)90074-3

Post A, Streveler G (1976) The tilted forest: glaciological-geologic implications of vegetated neoglacial ice at Lituya Bay, Alaska. Quat Res 6(1):111–117. https://doi.org/10.1016/0033-5894(76)90043-0

Rader E, Geist D, Geissman J, Dufek J, Harpp K (2015) Hot clasts and cold blasts: thermal heterogeneity in boiling-over pyroclastic density currents. In: Ort MH, Porreca M, Geissman JW (eds) The use of palaeomagnetism and rock magnetism to understand volcanic processes. Geol Soc Lond Spec Publ 396:1–11. https://doi.org/10.1144/SP396.16

Risica G, Rosi M, Pistolesi M, Speranza F, Branney MJ (2022) Deposit-derived block-and-ash flows: the hazard posed by perched temporary tephra accumulations on volcanoes; 2018 Fuego disaster, Guatemala. J Geophys Res Solid Earth Solid Earth 127:e2021JB023699. https://doi.org/10.1029/2021JB023699

Robinson JE, Bacon CR, Major JJ, Wright HM, Vallance JW (2017) Surface morphology of caldera-forming eruption deposits revealed by lidar mapping of Crater Lake National Park, Oregon—implications for deposition and surface modification. J Volcanol Geotherm Res 342:61–78. https://doi.org/10.1016/j.jvolgeores.2017.02.012

Roche O (2012) Depositional processes and gas pore pressure in pyroclastic flows: an experimental perspective. Bull Volcanol 74:1807–1820

Rose WI, Durant AJ (2011) Fate of volcanic ash: aggregation and fallout. Geology 39(9):895–896. https://doi.org/10.1130/focus092011.1

Ross PS, White JDL (2005) Mafic, large-volume, pyroclastic density current deposits from phreatomagmatic eruptions in the Ferrar large igneous province, Antarctica. J Geol 113(6):627–649

Rowley PD, Kuntz MA, MacLeod NS (1981) Pyroclastic-flow deposits. In: Lipman PW, Mullineaux DR (eds) The 1980 eruptions of Mount St. Helens, Washington. U S Geol Surv Prof Pap 1250:489–512

Rowley PD, MacLeod NS, Kuntz MA, Kaplan AM (1985) Proximal bedded deposits related to pyroclastic flows of May 18, 1980, Mount St. Helens, Washington. Geol Soc Am Bull 96:1373–1383

Sarocchi D, Sulpizio R, Macías JL, Saucedo R (2011) The 17 July 1999 block-and-ash flow (BAF) at Colima Volcano: new insights on volcanic granular flows from textural analysis. J Volcanol Geotherm Res 204:40–56

Sato H, Fujii T, Nakada S (1992) Crumbling of dacite dome lava and generation of pyroclastic flows at Unzen volcano. Nature 360:664–666. https://doi.org/10.1038/360664a0

Saucedo R, Macías JL, Bursik M (2004) Pyroclastic flow deposits of the 1991 eruption of Volcán de Colima, Mexico. Bull Volcanol 66:291–306

Schmincke H-U (2004) Volcanism. Springer, Berlin, p 324

Schmincke H-U, Fisher RV, Waters AC (1973) Antidune and chute and pool structures in the base surge deposits of the Laacher See area, Germany. Sedimentology 20:553–574

Scott KM (1988) Origins, behavior, and sedimentology of lahars and lahar-runout flows in the Toutle-Cowlitz River system. U S Geol Surv Prof Pap 1447-A:74 pp

Scott WE, Gardner CA (2017) Field-trip guide to Mount Hood, Oregon, highlighting eruptive history and hazards. U S Geol Surv Sci Invest Rep 2017-5022-G:1–115. https://doi.org/10.3133/sir20175022g

Scott WE, Hoblitt RP, Torres RC, Self S, Martinez MML, Nillos T (1996) Pyroclastic flows of the June 15, 1991, climactic eruption of Mount Pinatubo. In: Newhall CG, Punongbayan RS (eds) Fire and mud: eruptions and lahars of Mount Pinatubo, Philippines. Philippine Institute of Volcanology and Seismology, Quezon City, University of Washington Press, Seattle, pp 545–557

Scott KM, Tucker DS, Riedel JL, Gardner CA, McGeehin JP (2020) Latest Pleistocene to present geology of Mount Baker volcano, northern Cascade Range, Washington. U S Geol Surv Prof Pap 1865:170 pp

Segerstrom K (1950) Erosion studies at Parícutin volcano, State of Michoacan, Mexico. In: Geologic investigations in the Parícutin area, Mexico. U S Geol Surv Bull 956-A:1–164

Segura G, Brubaker LB, Franklin JF, Hinckley TM, Maguire DA, Wright G (1994) Recent mortality and decline in mature *Abiesamabilis*: the interaction between site factors and tephra deposition from Mount St. Helens. Can J For Res 24:1112–1122

Segura G, Hinckley TM, Brubaker LB (1995) Variations in radial growth of declining old growth stands of *Abiesamabilis* after tephra deposition from Mount St. Helens. Can J For Res 25:1484–1492

Seymour VA, Hinckley TM, Morikawa Y, Franklin JF (1983) Foliage damage in coniferous trees following volcanic ashfall from Mt. St. Helens. Oecologia (Berlin) 59:339–343

Shawe DR, Snyder DB (1988) Ash-flow eruptive megabreccias of the Manhattan and Mount Jefferson calderas, Nye County, Nevada. United States Department of Energy Office of Scientific and Technical Information. https://www.osti.gov/biblio/6997858

Sheridan MF (1970) Fumarolic mounds and ridges of the Bishop Tuff, California. Bull Geol Soc Am 81:851–868. https://doi.org/10.1130/0016-7606(1970)81[851:FMAROT]2.0.CO;2

Sheridan MF (1979) Emplacement of pyroclastic flows: a review. In: Chapin CE, Elston WE (eds) Ash-flow tuffs. Geol Soc Am Spec Pap 180:125–136

Siebert L, Glicken H, Kienle J (1989) Debris avalanches and lateral blasts at Mount St Augustine volcano, Alaska. Nat Geogr Res 5(2):232–249

Siebert L, Simkin T, Kimberly P (2010) Volcanoes of the world, 3rd edn. University of California Press, Berkeley, p 568

Sigurdsson H, Houghton BF, McNutt SR, Rymer H, Stix J (eds) (2000) Encyclopedia of volcanoes. Academic Press, San Diego, p 1417

Sigurdsson H, Houghton B, McNutt SR, Rymer H, Stix J (eds) (2015) The encyclopedia of volcanoes, 2nd edn. Elsevier, Amsterdam, 1421 pp

Smith GA (1987) The influence of explosive volcanism on fluvial sedimentation: the Deschutes Formation (Neogene) in central Oregon. J Sediment Petrol 57:613–629

Smith RT, Houghton B (1995) Vent migration and changing eruptive style during the 1800a Taupo eruption: new evidence from the Hatepe and Rotongaio phreatoplinian ashes. Bull Volcanol 57:432–434

Smith GA, Katzman D (1991) Discrimination of eolian and pyroclastic-surge processes in the generation of cross-bedded tuffs, Jemez Mountains volcanic field, New Mexico. Geology 19:465–468

Smith GA, Lowe DR (1991) Lahars: volcano-hydrologic events and deposition in the debris flow—hyperconcentrated flow continuum. In: Fisher RV, Smith GA (eds) Sedimentation in volcanic settings. SEPM (Soc Sediment Geol) Spec Publ 45:59–70

Sparks RSJ (1976) Grain size variations in ignimbrites and implications for the transport of pyroclastic flows. Sedimentology 23:147–188

Sparks RSJ, Self S, Walker GPL (1973) Products of ignimbrite eruptions. Geology 1:115–118

Sparks RSJ, Sigurdsson H, Carey SN (1980) The entrance of pyroclastic flows into the sea, I. Oceanographic and geologic evidence from Dominica, Lesser Antilles. J Volcanol Geotherm Res 7:87–96

Sparks RSJ, Gardeweg MC, Calder ES, Matthews SJ (1997) Erosion by pyroclastic flows on Lascar volcano, Chile. Bull Volcanol 58:557–565

Stieglitz RD, Inden RF (1969) Development of cavernous sediment in a non-beach environment. J Sed Res 39(1):342–344

Sulpizio R (2005) Three empirical methods for the calculation of distal volume of tephra-fall deposits. J Volcanol Geotherm Res 145(3–4):315–336. https://doi.org/10.1016/j.jvolgeores.2005.03.001

Sulpizio R, Dellino P (2008) Sedimentology, depositional mechanisms and pulsating behaviour of pyroclastic density currents, chap 2 In: Gottsmann J et al (eds) Caldera volcanism: analysis, modelling and response. Dev Volcanol 10:57–96. https://doi.org/10.1016/S1871-644X(07)00002-2

Sulpizio R, Dellino P, Doronzo DM, Sarocchi D (2014) Pyroclastic density currents: state of the art and perspectives. J Volcanol Geotherm Res 283:36–65. https://doi.org/10.1016/j.jvolgeores.2014.06.014

Sumner JM (1998) Formation of clastogenic lava flows during figure eruption and scoria cone collapse; the 1986 eruption of Izu-Oshima volcano, eastern Japan. Bull Volcanol 60:195–212

Suzuki-Kamata K, Kusano T, Yamasaki K (2009) Fractal analysis of the fracture strength of lava dome material based on the grain size distribution of block-and-ash flow deposits at Unzen volcano, Japan. Sediment Geol 220:162–168

Swanson FJ, Jones JA, Crisafulli CM, Lara A (2013) Effects of volcanic and hydrologic processes on forest vegetation: Chaitén volcano, Chile. Andean Geol 40:359–391

Swanson FJ, Jones J, Crisafulli C, González ME, Lara A (2016) Puyehue-Cordón Caulle eruption of 2011: tephra fall and initial forest responses in the Chilean Andes. Bosque 37(1):85–96. https://doi.org/10.4067/S0717-92002016000100009

Taddeucci J, Edmonds M, Houghton B, James MR, Vergniolle S (2015) Hawaiian and Strombolian eruptions, chap 27. In: Sigurdsson H, Houghton B, McNutt SR, Rymer H, Stix J (eds) The encyclopedia of volcanoes, 2nd edn. Elsevier, Amsterdam, pp 485–503. https://doi.org/10.1016/B978-0-12-385938-9.00027-4

Taddeucci J, Alatorre-Ibargüengoitia MA, Cruz-Vázquez O, Del Bello E, Scarlato P, Ricci T (2017) In-flight dynamics of volcanic ballistic projectiles. Rev Geophys 55(3):675–718. https://doi.org/10.1002/2017RG000564

Tanguy J-C (1994) The 1902–1905 eruptions of Montagne Pelée, Martinique: anatomy and retrospection. J Volcanol Geotherm Res 60(2):87–107. https://doi.org/10.1016/0377-0273(94)90064-7

Thouret J-C (1990) Effects of the November 1985 eruption on the snowpack and ice cap of Nevado del Ruiz volcano, Colombia. J Volcanol Geotherm Res 41:177–201

Thouret J-C, Ramírez CJ, Gibert-Malengreau B, Vargas CA, Naranjo JL, Vandemeulebrouck J, Valla F, Funk M (2007) Volcano–glacier interactions on composite cones and lahar generation: Nevado del Ruiz,

Colombia, case study. Annu Glaciol 45:115–127. https://doi.org/10.3189/172756407782282589

Torres RC, Self S, Martinez MML (1996) Secondary pyroclastic flows from the June 15, 1991, ignimbrite of Mount Pinatubo. In: Newhall CG, Punongbayan RS (eds) Fire and mud: eruptions and lahars of Mount Pinatubo, Philippines. Philippine Institute of Volcanology and Seismology, Quezon City, University of Washington Press, Seattle, pp 665–678

Trabant DC, Waitt RB, Major JJ (1994) Disruption of Drift glacier and origin of floods during the 1989–1990 eruptions of Redoubt volcano, Alaska. J Volcanol Geotherm Res 62:369–385

Trofimovs J, Amy L, Boudon G, Deplus C, Doyle E, Fournier N, Hart MB, Komorowski J-C, Le Friant A, Lock EJ, Pudsey C, Ryan G, Sparks RSJ, Talling PJ (2006) Submarine pyroclastic deposits formed at the Soufrière Hills volcano, Montserrat (1995–2003): what happens when pyroclastic flows enter the ocean? Geology 34(7):549–552. https://doi.org/10.1130/G22424.1

Valentine GA, Fisher RV (2000) Pyroclastic surges and blasts. In: Sigurdsson H, Houghton BF, McNutt SR, Rymer H, Stix J (eds) Encyclopedia of volcanoes. Academic Press, San Diego, pp 571–580

Valentine GA, Harrington CD (2006) Clast size controls and longevity of Pleistocene desert pavements at Lathrop Wells and Red Cone volcanoes, southern Nevada. Geology 34(7):533–536

Van der Meer JJM, Kjaer KH, Krüger J (1999) Subglacial water-escape structures and till structures, Sléttjökull, Iceland. J Quat Sci 14(3):191–205. https://doi.org/10.1002/(SICI)1099-1417(199905)14:3<191::AID-JQS436>3.0.CO;2-%23

Van Eaton AR, Wilson CJN (2013) The nature, origins and distribution of ash aggregates in a large-scale wet eruption deposit: Oruanui, New Zealand. J Volcanol Geotherm Res 250:129–154. https://doi.org/10.1016/j.jvolgeores.2012.10.016

Van Eaton AR, Mastin LG, Herzog M, Schwaiger HF, Schneider DJ, Wallace KL, Clarke AB (2015) Hail formation triggers rapid ash aggregation in volcanic plumes. Nat Commun 6(1):7860

Vespermann D, Schmincke H-U (2000) Scoria cones and tuff rings. In: Sigurdsson H, Houghton BF, McNutt SR, Rymer H, Stix J (eds) Encyclopedia of volcanoes. Academic Press, San Diego, pp 683–694

Voight B, Davis MJ (2000) Emplacement temperatures of the November 22, 1994 nuée ardente deposits, Merapi volcano, Java. J Volcanol Geotherm Res 100:371–377

Vorren TO (1977) Grain-size distribution and grain-size parameters of different till types on Hardangervidda, south Norway. Boreas 6(2):219–227. https://doi.org/10.1111/j.1502-3885.1977.tb00351.x

Waitt RB (1981) Devastating pyroclastic density flow and attendant air fall of May 18—stratigraphy and sedimentology of deposits. In: Lipman PW, Mullineaux DR (eds) The 1980 eruptions of Mount St. Helens, Washington. U S Geol Surv Prof Pap 1250:439–458

Walder JS (2000a) Pyroclast/snow interactions and thermally driven slurry formation. Part 1: theory for monodisperse grain beds. Bull Volcanol 62:105–118

Walder JS (2000b) Pyroclast/snow interactions and thermally driven slurry formation. Part 2: experiments and theoretical extension to polydisperse tephra. Bull Volcanol 62:119–129

Wales N, Gomez-Velez JD, Newman BD, Wilson CJ, Dafflon B, Kneafsey TJ, Soom F, Wullschleger SD (2020) Understanding the relative importance of vertical and horizontal flow in ice-wedge polygons. Hydrol Earth Syst Sci 24:1109–1129. https://doi.org/10.5194/hess-24-1109-2020

Walker GPL (1971) Grain-size characteristics of pyroclastic deposits. J Geol 79:696–714

Walker GPL (1972) Crystal concentration in ignimbrites. Contrib Mineral Petrol 36:135–146. https://doi.org/10.1007/BF00371184

Waters AC, Fisher RV (1971) Base surges and their deposits: Capelinhos and Taal volcanoes. J Geophys Res 76:5596–5614

Watt SFL, Gilbert JS, Folch A, Phillips JC, Cai XM (2015) An example of enhanced tephra deposition driven by topographically induced atmospheric turbulence. Bull Volcanol 77:35. https://doi.org/10.1007/s00445-015-0927-x

Waythomas CF, Pierson TC, Major JJ, Scott WE (2013) Voluminous ice-rich lahars generated during the 2009 eruption of Redoubt volcano, Alaska. J Volcanol Geotherm Res 259:389–413

Whelley PL, Jay J, Calder ES, Pritchard ME, Cassidy NJ, Alcaraz S, Pavez A (2012) Post-depositional fracturing and subsidence of pumice flow deposits: Lascar volcano, Chile. Bull Volcanol 74:511–531. https://doi.org/10.1007/s00445-011-0545-1

White JDL, Houghton BF (2006) Primary volcaniclastic rocks. Geology 34:677–680

White JDL, Riggs NR (eds) (2001) Volcaniclastic sedimentation in lacustrine settings. Spec Publ No 30. International Association of Sedimentologists, Blackwell Scientific, Oxford, 309 pp

Wilson CJN (1980) The role of fluidization in the emplacement of pyroclastic flows: an experimental approach. J Volcanol Geotherm Res 8:231–249

Wilson CJN (1984) The role of fluidization in the emplacement of pyroclastic flow, 2. Experimental results and their interpretation. J Volcanol Geotherm Res 20:55–84

Wilson CJN (1985) The Taupo eruption, New Zealand. II. The Taupo ignimbrite. Phil Trans R Soc Lond Ser A Math Phys Sci 314(1529):229–310. https://doi.org/10.1098/rsta.1985.0020

Wilson CJN (2001) The 26.5 ka Oruanui eruption, New Zealand: an introduction and overview. J Volcanol Geotherm Res 112:133–174

Wilson L, Head JW (1981) Morphology and rheology of pyroclastic flows and their deposits, and guidelines for future observations. In: Lipman PW, Mullineaux DR (eds) The 1980 eruptions of Mount St. Helens, Washington. U S Geol Surv Prof Pap 1250:513–524

Wilson CJN, Hildreth W (1998) Hybrid fall deposits in the Bishop Tuff, California: a novel pyroclastic depositional mechanism. Geology 26(1):7–10

Wilson CJN, Houghton BF (2000) Pyroclast transport and deposition. In: Sigurdsson H, Houghton BF, McNutt SR, Rymer H, Stix J (eds) Encyclopedia of volcanoes. Academic Press, San Diego, pp 545–554

Wilson CJN, Walker GPL (1981) Violence in pyroclastic flow eruptions. In: Self S, Sparks RSJ (eds) Tephra studies. D. Reidel Publishing Co, Dortrecht, pp 441–448

Wilson CJN, Walker GPL (1982) Ignimbrite depositional facies: the anatomy of a pyroclastic flow. J Geol Soc 139(5):581–592

Wilson CJN, Houghton BF, Kampt PJJ, McWilliams MO (1995) An exceptionally widespread ignimbrite with implications for pyroclastic flow emplacement. Nature 378(6557):605–607

Winner WE, Casadevall TJ (1981) Fir leaves as thermometers during the May 18 eruption. In: Lipman PW, Mullineaux DR (eds) The 1980 eruptions of Mount St. Helens, Washington. U S Geol Surv Prof Pap 1250:315–320

Wright JV, Smith AL, Self S (1980) A working terminology of pyroclastic deposits. J Volcanol Geotherm Res 8:315–336

Wright JV, Self S, Fisher RV (1981) Towards a facies model for ignimbrite-forming eruptions. In: Self S, Sparks RSJ (eds) Tephra studies. D. Reidel Publishing Co, Dortrecht, pp 433–439

Wright HMN, Lesti C, Cas RAF, Porreca M, Viramonte JG, Folkes CB, Giordano G (2011) Columnar jointing in vapor-phase-altered, non-welded Cerro Galán ignimbrite, Paycuqui, Argentina. Bull Volcanol 73:1567–1582. https://doi.org/10.1007/s00445-011-0524-6

Yamamoto T, Takarada S, Suto S (1993) Pyroclastic flows from the 1991 eruption of Unzen volcano, Japan. Bull Volcanol 55:166–175

Zanella E, Gurioli L, Pareschi MT, Lanza R (2007) Influences of urban fabric on pyroclastic density currents at Pompeii (Italy): 2. Temperature of the deposits and hazard implications. J Geophys Res Solid Earth 112:B05214. https://doi.org/10.1029/2006JB004775

Zielinski T, van Loon AJ (1996) Characteristics and genesis of moraine-derived flowtill varieties. Sediment Geol 101:119–143

Zimanowski B, Büttner R, Dellino P, White JDL, Wohletz KH (2015) Magma–water interaction and phreatomagmatic fragmentation. In: Sigurdsson H, Houghton B, McNutt SR, Rymer H, Stix J (eds) The encyclopedia of volcanoes, 2nd edn. Elsevier, Amsterdam, pp 473–484

Deposit Identification: Key and Summary Tables

Abstract

This chapter presents two tools for use in the field to aid in differentiation and identification of volcanic deposits emplaced on dry land: a quick-reference key for deposit identification (Table 1) and a set of tables summarizing all the evaluation criteria for each deposit type, including each type of volcanic wet flow and the three other types of deposits at volcanoes containing volcanic particles (Tables 2, 3, 4, 5, 6, 7, and 8). Assessment of the physical characteristics of deposits should be carried out at different scales in the field, and field observations can benefit from additional particle-size analysis in the laboratory. If appropriate, field measurement of thermoremanent magnetic orientation of large in-situ clasts can provide additional insight. It is important to note that the identification criteria used in this book are characteristics best preserved in fresh (unaltered) deposits but may be preserved in many older, altered deposits, as well.

1 Overview

Hazardous volcanic wet flows emplace large volumes of volcanic sediment on the flanks of volcanoes and in river valleys farther downstream. A major question for a volcanologist in the field who is mapping deposits and assessing volcanic hazards in this terrain is, "What process emplaced this deposit?" We suggest that certain key criteria, summarized in this chapter, can greatly aid in the differentiation and identification of volcanic wet-flow deposits—deposits emplaced by lahars, including both volcanic debris flows (VDFs) and volcanic hyperconcentrated flows (VHFs), volcanic debris avalanches (VDAs), and volcanic water floods (VWFs)—from each other and from similar-looking pyroclastic density current (PDC), tephra fall (TF), and volcano-glacial deposits. We further suggest that assessment of multiple criteria is fundamental for accurate deposit differentiation and identification.

This chapter presents an identification key (Table 1) and tables of diagnostic criteria for each deposit type (Tables 2, 3, 4, 5, 6, 7, and 8) to aid in the task of deposit differentiation and identification. These criteria focus on field assessment of physical characteristics at different scales (summarized in Chaps. "Lahar Deposits—Debris-Flow Type", "Lahar Deposits—Hyperconcentrated-Flow Type", "Volcanic Debris-Avalanche Deposits", "Volcanic Water-Flood Deposits", and "Other Processes Depositing Volcanic Sediment"), augmented by laboratory particle-size analysis and, if needed, field measurement of thermoremanent magnetic orientation of in situ clasts. In addition,

T. C. Pierson et al., *Deposits of Volcanic Wet Flows*, Advances in Volcanology,
https://doi.org/10.1007/978-3-031-66574-5_12

Table 1 Identification key for fresh volcaniclastic deposits emplaced on land

Deposit characteristics	LAHAR (VDF)	LAHAR (VHF)	VDA	VWF	PDC (conc)	PDC (dilute)	TF	Till (LT)	Till (AT)
Landscape-Scale									
Relatively smooth planar surface (± large swales or convexities)	**P**	**P**	r (loc)	P	**P****	r	P	r	0
Isolated/clustered clasts studding surface	**P**	**P**	0	P	P (baf) r (ig)	0	r	0	P
Channelized surface (primary channels)	P	P	0	**P**	r (baf) P (ig)	r	0	0	0
Bars in channels (primary)	P	P	0	**P**	0	0	0	0	0
Flow levees	**P***	0	P	0	**P***	0	0	0	0
Striations on flow-path bedrock surfaces	P**	0	**P**	0	P**	r	0	**P**	0
Lobate margins concentrating coarse clasts	**P***	0	0	0	**P***	0	0	0	0
Valley-filling deposits (topography controlled)	**P****	**P****	**P****	**P****	**P****	r	0	0	P
Terrain-mantling deposits	0	0	0	0	r	P	**P**	P	0
OUTCROP-Scale	*Within individual deposits*								
Massive (unstratified)	**P**	P	**P**	r	**P**	P	P	**P**	**P**
Diffusely stratified	0	**P**	0	r	P	P	P	0	0
Well stratified	0	P	0	**P**	0	P	P	0	r
Bed forms	0	r	0	**P**	r	P	0	0	0
Strong fabric	r	0	0	P	r	0	0	P	0
Graded beds	P	r	r	r	P	P	P	0	0
Fluid-escape structures	r	P	0	0	P	r	0	0	0
Soft-sediment deformation (internal structures)	r	P	r	0	0	0?	0	0	r
Shear deformation (internal)	0	0	P	0	0	0	0	P	0
Evidence of heat	r	r	0	0	**P**	P	r	0	0

(continued)

Table 1 (continued)

Deposit characteristics	LAHAR (VDF)	LAHAR (VHF)	VDA	VWF	PDC (conc)	PDC (dilute)	TF	Till (LT)	Till (AT)
Loose/soft consistency	0	r	0	**P**	P	P	**P**	0	P
Moderate consistency	r	**P**	r	r	r	r	r	r	r
Hard/firm consistency	**P**	r	**P**	0	r	r	r	**P**	0
Particle-Scale	*Within individual deposits*								
Uniform clast composition	r	r	P (loc)	0	P	P	P	0	0
Uniform matrix color	**P**	**P**	r (loc)	**P**	**P**	**P**	r	P	P
Clast surfaces striated	0	0	0	0	0	0	0	P	0
Pumice clasts angular	0	0	0	0	r	r	**P**	0	r
Poorly to extremely poorly sorted	**P**	P	**P**	r	**P**	P	0	**P**	**P**
Pitted, crudely rounded boulder-size dense clasts	**P**	r	0	r	r	0	0	0	0
Cobble-size or larger clasts pervasively fractured but intact when deposited	0	0	**P**	0	0	0	r	0	0
Boulder-size dense clasts angular	r	P	**P**	P	r	**P**	**P**	P	**P**
Voids between gravel-size clasts (except in levees/fronts)	0	0	P	P	P	P	**P**	0	r
Bubble vesicles in silty sand matrix	**P**	r	0	0	0	r	0	0	r

P Yes; almost always present; r Rare; (loc) Locally present
P Yes; sometimes present 0 No; absent
* For relatively small flows only
** For relatively large flows only
See subsequent tables for more detail or see the relevant chapters for detailed description and discussion. Process abbreviations: VDF = volcanic debris flow, VHF = volcanic hyperconcentrated flow, VDA = volcanic debris avalanche, VWF = volcanic water flood, PDC = pyroclastic density current, TF = tephra fall, LT = lodgment till, and AT = ablation till. In PDCs, some characteristics differ between pumiceous PDFs (ignimbrites [ig]) and lithic PDFs (block-and-ash flows [baf]). Angularity of boulders is a function of travel distance as well as of process and particle composition

photographs of vertical outcrop faces can be taken in the field, and these can later be compared with the outcrop photos in Chap. "Annotated Outcrop Photographs: Volcanic Wet-Flow Deposits and Other Unconsolidated Volcanic Deposits".

It is important to note that the identification criteria used in this book are characteristics best preserved in fresh (unaltered) deposits. Furthermore, only deposits emplaced on dry land are considered. The terms *matrix* and *matrix material* refer to particles finer than 2 mm in diameter—the sedimentologist's "sand, silt, and clay," or the volcanologist's "ash." Sorting values (coefficients) reported in this book from volcanological studies and references are derived by the phi standard deviation method (Inman 1952), which was later modified as the inclusive graphic standard deviation method (Folk 1974). Sorting values reported in this book from sedimentological studies and references use Folk's inclusive standard deviation method. In the analysis of most volcaniclastic sediments, there would be only slight differences in calculated sorting coefficients, depending on the method used. But sedimentologists and volcanologists differ markedly in describing how "good" or "poor" the sorting of a deposit is, based on the same coefficients (see Sect. 4.3 and Table 6 in Chap. "Deposit Identification and Interpretation").

2 Deposit Identification Key

Table 1 is a quick-reference key that allows one to rapidly compare different characteristics of volcanic sediments at different scales. The summary tables in Sect. 3 provide supporting information for this key in greater detail, and the relevant chapters of the book provide the full background information and reference citations.

3 Summary Tables of Diagnostic Criteria

See Tables 2, 3, 4, 5, 6, 7, and 8.

Table 2 Summary table of volcanic debris flow (VDF) deposit characteristics

Process	*Volcanic debris flow (VDF)—concentrated lahar*	
Deposit type	Clay-rich VDF deposit	Clay-poor VDF deposit
Deposit architecture	Deposits are generally very to extremely large–derived from gravitational collapse of part of a volcanic edifice. Upper surfaces are relatively smooth and flat, except where incised by post-depositional erosion. Typically, few clasts protrude upward from deposit surface. Deposits form valley fills of variable thickness. Upstream of main depositional zone, deposits may form a relatively thin coating on existing valley topography. Flow levees composed of coarse clasts are rare or not present; marginal tangles of woody debris (crude levees) are common. Large debris blocks (appearing as hummocks) may be present (inherited from transformed VDAs)	Upper deposit surfaces vary from being smooth and flat to slightly convex upward, except where incised by post-depositional erosion. Surfaces are typically studded with clasts protruding out of deposit, arranged as single isolated clasts or clast clusters. Thin layers of fines (cap layers) are common in shallow surface depressions. Deposit margins of relatively small flows (where deposit is unconfined by channel or valley sides) are commonly confined by flow levees of cobbles and boulders; where levees do not form, margins are commonly lobate in plan view, with coarsest available clasts forming steep rims of partially openwork gravel. Deposits typically form terraces along valley margins after the more dilute lahar tails incise the fresh deposits
Lithofacies	Unstratified diamict (Dmm and Dmg). Deposit matrix is fines-rich with clay content > 5 wt%. Clay-rich VDFs seldom transform downstream to VHFs, and deposit characteristics are maintained for long distances away from source. Overall fining of gravel-size clasts does occur with distance from source, however	Unstratified diamict (Dcm and Dcg). Deposits commonly have two depositional facies that are defined by particle-size variations: the channel facies with abundant cobble and boulder clasts, and the overbank facies with largest clasts commonly only up to pebble-gravel size. Deposit matrix is relatively poor in fines with < 5 wt% clay. Thin silty partings (Fm, Fl) may be found between flow units. Downstream transition to VHF deposits is common
Erosion	Erosive, particularly at high velocities and on steep slopes, but less erosive than clay-poor VDFs. Nonerosive on gentle slopes	Powerfully erosive on steep slopes. Flows can striate bedrock and can carve relatively deep U-shaped channels in unconsolidated material. Nonerosive on gentle slopes
Effects on vegetation	Highly destructive to all vegetation along main flow path. Total leveling and burial of forests is possible. Tree trunks submerged in flows along valley sides (where flow velocities were low) can be encased in clay-rich mud coats several centimeters thick after flow passes. There is limited data on degrees of impact	Variable impacts on forest trees, depending on size of flow and location of trees relative to flow path. Impacts ranges from total destruction and removal of large trees to nondestructive coatings of muddy sand on tree trunks and branches. Flows commonly cause severe abrasion of tree trunks on upstream sides, extending from the base of the trunk to the upper mudline. Gravel clasts are commonly driven into standing tree trunks, embedded several centimeters into the bark or wood. Branches or small trees may be bent in downstream direction by flow and have branch tips tapered by abrasion to form sharp points

(continued)

Table 2 (continued)

Process	*Volcanic debris flow (VDF)—concentrated lahar*	
Deposit type	Clay-rich VDF deposit	Clay-poor VDF deposit
Stratification	Deposits are unstratified and can range up to 20 m thick along main valley axes, thinning to 1 m or less on valley sides	Deposits are unstratified. Individual deposits commonly form layers 0.5 to 2 m in thickness, but they are thicker where ponded behind obstructions or channel constrictions. Cap layers and sole layers within deposits can resemble bedding; these (along with abrupt changes in grading or particle size) can form stratigraphic breaks between stacked flow units (from multiple lahars or multiple surges within a lahar). Without these breaks, stacked units of similar age may not have recognizable depositional contacts, and multiple flow units may appear as one thick flow unit
Bed forms and structures	No bed forms or hydraulically formed structures have been identified in these flow deposits	Trains of symmetrical standing waves and hydraulic jumps have been observed in flowing clay-poor VDFs, so we presume that antidunes and chute-and-pool structures form in these flows; preservation of these structures in deposits is uncertain, however. Large bed forms and structures include openwork boulder bars downstream of flow obstructions, longitudinal whaleback bars, bouldery point bars, boulder clusters, and isolated outsized clasts
Gravel-size clast framework	Matrix-supported frameworks are typical but clast-supported frameworks are possible. Voids between clasts are rare	Clast-supported frameworks are typical, but matrix-supported frameworks are possible. Voids between clasts are absent in deposit interiors but common along deposit margins, especially in small deposits
Fabric	Random to weak fabrics (long-axes parallel to flow direction) and imbrication of platy clasts are common; long axes may be transverse to flow direction at deposit margins	Random to weak imbricate fabric (long-axes parallel to flow direction); may be transverse to flow direction at deposit margins. Imbrication may be strong in boulder bars behind flow obstructions
Grading	Deposits are typically normally graded in thicker sections	Variable grading. Deposits are commonly inversely graded, but normally graded and ungraded deposits occur as well. Some deposit cross-sections may display all three types–normal grading in tops of deposits, inverse grading in bases, and ungraded zones in central part
Internal deformation	Deformation of large fragile clasts is possible. Crushing of pebble-size clasts can occur in sheared sole layers of thick flows	Deformation of large fragile clasts is possible. Reverse-faulting of substrates has been noted
Fluid-escape structures	Absent	Generally absent, but rare dewatering pipes have been noted
Material consistency	Very firm to extremely firm when moist, hard to very hard when dry, and slightly sticky to sticky and slightly plastic to plastic when wet. Deposits generally cannot be dug with a shovel. Chunks of deposit break off when struck hard with a tool	Typically firm to very firm when moist and slightly hard to hard when dry, nonsticky and nonplastic to slightly plastic. Deposits usually can be dug with a shovel with moderate to high difficulty. Deposits break off in chunks when struck by a pick or shovel

(continued)

Table 2 (continued)

Process	*Volcanic debris flow (VDF)—concentrated lahar*	
Deposit type	Clay-rich VDF deposit	Clay-poor VDF deposit
Evidence of heat	Unlikely. Charred wood or prismatically jointed and bread-crust clasts may be present if lahar incorporated primary pyroclastic deposits	Typically absent, although direct triggering of lahars by PDCs may result in primary prismatic jointing or acquisition of thermoremanent magnetism in large clasts, localized charring of wood by hot boulders, or matrix oxidation in vicinity of large hot clasts. Incorporation of primary pyroclastic deposits by lahars may result in scattered evidence of high heat in deposits
Particle composition	Composed mainly of volcanic rock fragments of various lithologies (from units in edifice), likely including both fresh volcaniclastic and hydrothermally altered fragments (polymictic). Also likely to contain accidental lithic fragments entrained along flow path	Composed mainly of volcanic rock fragments, commonly of one rock type, but mixtures may contain significant amounts of admixed accidental lithic particles picked up along flow path (volcanic and/or nonvolcanic lithologies). Mixtures are typically polymictic
Particle roundness, shape, fracturing, and surface texture	Some rounding of source-rock gravel-size clasts has been observed; hydrothermally altered volcanic rock is softer and more susceptible to rounding than fresh hard rock. Sand grains are angular. Deposits may include rounded gravel clasts picked up from stream beds	Where source rock is fresh and hard, cobble- and boulder-size clasts are typically subangular to subrounded; sand is angular. Rounding from clast collisions increases with clast size and distance traveled. Boulder surfaces are typically rough and pitted. Deposits may include well rounded gravel clasts if river gravel is incorporated and any entrained pumice clasts are well rounded
Particle-size distribution and sedimentary texture	Deposits are typically composed of muddy sandy gravels or muddy gravelly sands; matrix material packs all spaces between gravel-size clasts. Particle-size distributions are commonly bimodal. Grain sizes vary from clay size to very large boulders. Clay content of matrix is >5 wt% and likely includes clay minerals. Sorting coefficients are typically $>4.0\ \varphi$	Deposits are mainly sandy gravels or gravelly sands (slightly muddy); matrix material packs all spaces between clasts except along deposit margins. Size distributions are commonly bimodal. Grain sizes typically vary from clay size to large boulders. Matrix clay contents are <5 wt%; clay minerals are rarely present. Sorting coefficients can vary from $2.0\ \varphi$ to $>4.0\ \varphi$)
Matrix vesicles and intergranular voids	Small bubble vesicles (~1 mm) are commonly visible in matrix. Otherwise, voids between clasts do not exist	Small bubble vesicles (~1 mm) are commonly visible in matrix. Other types of voids between clasts do not exist in the main body of the deposit where voids are packed with matrix material, but voids may be present in coarse marginal flow levees and flow fronts

Table 3 Summary table of volcanic hyperconcentrated flow (VHF) deposit characteristics

Process	*Volcanic hyperconcentrated flow—dilute lahar*
Deposit type	VHF deposit (previously called a lahar-runout deposit)
Deposit architecture	Aggradational assemblages form downstream-tapering sediment bodies at and below channel gradients of about 0.01 m/m. Surfaces are typically studded with outsized clasts and clast clusters. Deposition is commonly unconfined (forming broad sheets) on debris aprons and alluvial fans/plains. In confined valleys, VHF deposits emplaced by tails of large lahars are commonly inset as fills within recently incised channels and preserved as inset fill terraces. Sediments commonly coarsen toward channel axis. Smooth, flat-surfaced berms commonly form along channel margins near the level of maximum stage (level of peak flow). Channels characteristically have finely braided form with subtle low-amplitude, longitudinal bars; pronounced bar forms are absent
Lithofacies	Lithofacies of VHFs are among the most variable in volcanic sediments. Dmm, Dmg, Dms, Dcm, Dcg, and Dcs lithofacies are emplaced in channel-facies deposits by concentrated VHFs; Gmm, Gmo, Gcm, Gh, Sm, Sh, Ss, and Sl lithofacies are found in deposits emplaced by dilute VHFs (and in overbank facies of concentrated VHFs). Thin silty partings and layers (Fm, Fl) are common between sandy beds and on tops of deposits (consolidation laminations and cap layers)
Erosion	VHFs can be highly erosive in steep channels. In low-gradient channels, flows tend to aggrade their beds by continuous deposition, while at the same time shifting laterally and vigorously eroding channel banks. VHFs characteristically leave wide, shallow, rectangular channels when flowing in channels cut into unconsolidated material
Effects on vegetation	VHFs typically cause only moderate to minimal damage to trees and shrubs along flow paths, unless high-velocity flows are involved. Abrasion to tree trunks commonly is confined to within a meter of the bed (impacts from saltating bed load) and near peak flow level (impacts from floating debris). Only thin filmy coatings of mud and fine sand are found on tree trunks and other vertical surfaces
Stratification	Faint, diffuse, subhorizontal (plane-parallel) bedding is common in sandy deposits, but some sandy deposits are massive; sharp contacts may form between bedding planes when sediments of contrasting size and density are being transported by the same flow (e.g., small lithic-particle bed-load transport along with larger suspended/floating pumice). Deposits of relatively concentrated VHFs are commonly massive and transitional to VDF deposits. Thin (mm to cm), subhorizontal silty laminae spaced centimeters or tens of centimeters apart (cap layers and consolidation laminations) are a common type of pseudo-bedding found in VHF deposits
Bed forms and structures	Internal erosion surfaces may be produced by scour during lateral migration of channels; such truncation of bedding cuts across the highly uniform subhorizontal bedding produced during steady aggradation. Low-angle cross-bedding related to antidune formation is sometimes present; high-angle cross-bedding is not observed (although scour features can form steep contacts with adjacent beds). Gravel lenses, clast clusters, and single outsized clasts buried in sand (and sitting on deposit surfaces) are common. Deposits of single VDFs can have VHF deposits at their base (commonly faintly stratified); these sandy sole layers grade upward into the more poorly sorted, massive VDF units–mixing with river water produces the VHF at flow front, which is then overridden by the VDF

(continued)

Table 3 (continued)

Process	*Volcanic hyperconcentrated flow—dilute lahar*
Deposit type	VHF deposit (previously called a lahar-runout deposit)
Gravel-size clast framework	Clast-supported frameworks are normally found in channel facies deposits, although outsized individual clasts are supported by sandy matrix material
Fabric	Generally weak flow-parallel grain orientation occurs for smaller elongate clasts (A-axis < 10–15 cm) carried partly in suspension. Elongate clasts (A-axis > 10–15 cm) are transported by rolling along channel beds and generally have flow-transverse particle orientation. Clast imbrication is weak (in contrast to VWF deposits)
Grading	Grading is variable. Inverse grading is relatively common, reflecting incremental accretion by a dilute flow front and more concentrated main flow body. Normally graded and ungraded beds are also relatively common
Internal deformation	Shear deformation has not been observed, but soft-sediment deformation due to undrained loading of freshly emplaced deposits is common
Fluid-escape structures	Cap layers and consolidation laminations may be disrupted by vertically escaping pore water to form dish structures—a characteristic feature of VHF deposits. Dish structures, flame structures (injection of sediment into overlying beds), and dewatering (elutriation) pipes are relatively common. Convolute bedding produced by soft-sediment deformation from rapid loading and liquefaction of saturated deposits is relatively common. However, many outcrops do not display these structures
Material consistency	Consistency is typically slightly firm, although it can range from friable to firm when moist (soft to slightly hard when dry). These deposits are almost always firmer (harder) than water-flood deposits and looser (softer) than VDF deposits. Deposits are generally nonsticky and nonplastic when wet
Evidence of heat	No consistent evidence of high heat occurs in VHF deposits, although random pieces of charcoal or prismatically jointed clasts can be entrained from source PDC deposits
Particle composition	Volcanic sediment in deposits is composed of lithic and/or pumice rock fragments; particles may include primary volcaniclastic, secondary (reworked) volcaniclastic, and epiclastic particles. Deposits are generally polymictic. Segregation of low-density pumice particles in beds or lenses is common
Particle roundness, shape, fracturing, and surface texture	Fresh volcaniclastic gravel-size clasts are angular to subangular, but gravel recycled from older deposits can have variable degrees of rounding. Sand is angular. Particle shapes are commonly equant. Cobbles and boulders previously transported by VDFs may have rough surfaces and be surrounded by earlier, more intense abrasion; entrained fluvial gravel clasts may have smooth surfaces and be rounded or well rounded. Percentages of rounded clasts and degrees of rounding typically increase with distance traveled. Any pumice clasts are likely to be rounded to well rounded. Clasts are unlikely to be visibly fractured (fracturing during transport produces breakage into separate clasts)
Particle-size distribution and sedimentary texture	Bulk deposit mixtures are commonly slightly muddy gravelly sands or slightly muddy sandy gravels, with most gravel in the granule to fine pebble range. Cobbles and boulders occur as single outsized clasts or clast clusters that were rolled along channel bed and then buried by aggrading sand. Pebbles and granules are typically concentrated in beds or lenses. Particle-size distributions are usually unimodal or bimodal but can be polymodal in high-concentration VHF deposits. Dominant mode is sand. Total fines content of deposits is generally 3–10 wt% (only a small fraction of the fines in the original flow remains in VHF deposits). Sorting coefficients are commonly in the range 1.0–3.0 φ

(continued)

Table 3 (continued)

Process	*Volcanic hyperconcentrated flow—dilute lahar*
Deposit type	VHF deposit (previously called a lahar-runout deposit)
Matrix vesicles and intergranular voids	Neither bubble vesicles nor irregular voids between particles have been reported dispersed within deposits. However, merged and flattened bubble vesicles (several millimeters in diameter) have been found trapped beneath low-permeability cap layers, indicating the ease of upward air escape during deposit consolidation

Table 4 Summary table of volcanic debris avalanche (VDA) deposit characteristics

Process	*Volcanic debris avalanche (VDA)*
Deposit type	VDA deposits originating from huge deep-seated edifice collapses (involving summit) and smaller flank-collapse failures
Deposit architecture	Hummocks (surface mounds) and closed depressions often filled with small ponds or lakes are characteristic, although broad areas of flat-surfaced mixed- or lahar-facies material have few if any hummocks. Hummocks can range from a few meters to more than a kilometer in long dimension and from a meter up to about 200 m in height. Hummocks are smaller with distance from source and may be concentrated in lateral embayments where deposits are thin, where valley sides are laterally constricted, or where stranded against bedrock hills or buried bedrock ridges. Deposits form wedge-shaped valley fills where laterally confined or fan-shaped bodies when unconfined. Deposits are unusually thick (hundreds of meters) compared to other volcaniclastic deposits. Tall (tens of meters) and steep lateral margins and flow fronts (if preserved) are typical. Deposit surfaces commonly have prominent linear ridges parallel to flow direction, short and irregular compressional ridges transverse to flow direction, short ridges in a herring-bone pattern, and longitudinal color banding. Deposits from flank collapses may have transverse ridges near deposit margins
Lithofacies	Block facies and mixed facies are two end members of a broad range of depositional facies described for debris-avalanche deposits. Block-facies material contains relatively intact rock masses from the collapsed edifice (*debris-avalanche blocks*), which can be recognized by some preservation of original edifice structure and stratigraphy, commonly with abruptly juxtaposed colors and textures. Debris-avalanche blocks are highly fractured and can display jigsaw fractures down to the level of microscopic grains. Faulting of debris-avalanche blocks is common and may extend into surrounding mixed facies. Mixed-facies material dominantly contains well-mixed, typically polymictic, finer-grained material that is largely homogeneous in color and texture, with limited amounts of large clasts (from broken debris-avalanche blocks) suspended in the finer material. These large clasts are commonly clustered. Proportion of mixed-facies material increases distally due to progressive comminution of debris-avalanche blocks. Some deposits may occur together with clay-rich VDF deposits transformed from parts of the avalanche during emplacement
Erosion	Underlying bedrock surfaces can be eroded and striated. VDAs can erode tens of meters into underlying deposits and incorporate surficial vegetation and substrate bedrock and sediments
Effects on vegetation	Total destruction and removal of all vegetation in flow paths results from VDA motion. Relatively sharp boundaries between scoured valley sides and undisturbed vegetation above are common, although vegetation adjacent to the boundary may suffer damage from flying rock fragments

(continued)

Table 4 (continued)

Process	*Volcanic debris avalanche (VDA)*
Deposit type	VDA deposits originating from huge deep-seated edifice collapses (involving summit) and smaller flank-collapse failures
Stratification	Deposits form unstratified (massive) diamicts without primary bedding. Bedding of pyroclastic deposits and lavas within debris-avalanche blocks is commonly inherited from a failed edifice. Pseudo-bedding may occur in basal shear zones from multiple sub-parallel shears within a deposit
Bed forms and structures	No bed forms occur. Clastic dikes of matrix material are commonly injected into debris-avalanche blocks
Gravel-size clast framework	Mixed-facies deposits most commonly have matrix-supported frameworks. Clasts within block-facies deposits are typically clast supported. Voids between clasts are absent in the mixed facies but occur between intrablock clasts within the block facies. In small deposits resulting from flank collapses, intergranular voids may form
Fabric	Weak bed-parallel and flow-parallel alignment of elongate clasts is possible within basal shear zones but random fabric is typical elsewhere in deposits. Clast distribution in outcrop faces is not random; clustering of clasts in outcrop faces is common but usually localized
Grading	Matrix-facies deposits are typically ungraded, but they can show gross inverse grading at basal shear zones. Grading relationships can be complicated by overlapping lobes of retrogressive landslides
Internal deformation	Deformation within blocks may vary from little or none, to brittle deformation only (faulting), to highly localized shearing near the base of deposits (shear planes or narrow shear zones), to more broadly distributed plastic shear deformation of large clasts and debris-avalanche blocks. Small debris-avalanche blocks are more likely to exhibit shear deformation than large blocks. Large, sheared clasts of coarsely fractured hard rock break apart into smaller clasts that may be dispersed over short distances, forming clast clusters or lenses of angular clasts. Finely fractured or clay-rich debris-avalanche blocks can deform plastically and appear to "smear out." Soft-sediment deformation of underlying deposits (the VDA substrate) can be intense at the flow front, both beneath the VDA deposits or extending out beyond the flow front (the "bulldozer facies"). Parts of the substrate can be incorporated into VDA deposits and intensely deformed. Deformation structures are not usually present in deposits of shallow flank failures
Fluid-escape structures	Rare or absent
Material consistency	Highly variable. In clay-rich VDA deposits, deposit material is friable to very firm when moist and slightly hard to very hard when dry. Parts may be slightly plastic and slightly sticky when wet. In clay-poor VDA deposits (typical of flank collapses), deposits tend to be friable to firm when moist, without plasticity or stickiness when wet
Evidence of heat	Edifice collapse during an eruption include hot rock fragments in the avalanche debris (from magma ascending into the edifice). Entrained semi-molten fragments could develop primary prismatic jointing and breadcrusted clast surfaces (mainly silicic clasts) or cauliflower-textured surfaces (mainly mafic clasts). Some large deposits show evidence of high basal-frictional heating in the formation of melt-induced frictionites or pseudotachylites. Thermal effects are rare in shallow flank-collapse deposits

(continued)

Table 4 (continued)

Process	*Volcanic debris avalanche (VDA)*
Deposit type	VDA deposits originating from huge deep-seated edifice collapses (involving summit) and smaller flank-collapse failures
Particle composition	Relatively soft hydrothermally altered rock fragments, usually mixed with hard, unaltered volcanic rock fragments (all sizes) are typical in large VDA deposits originating from deep edifice collapses. Mostly or exclusively unaltered rock is found in smaller deposits originating from shallower flank collapses. All VDA deposit mixtures are grossly polymictic and can contain accidental nonvolcanic rock fragments, but intrablock zones of monomictic particles (clasts and matrix) are common. Colors of rock in debris-avalanche blocks in large VDA deposits are typically intense due to pre-collapse hydrothermal alteration, and contrasting colors are commonly juxtaposed to produce a mottling effect between and within debris-avalanche blocks. Boundaries between colors may be sharp or diffuse. Intrablock mixed-facies material typically has same color as blocks in which it is formed, but interblock mixed facies material is typically polymictic, having more neutral gray and brown colors
Particle roundness, shape, fracturing, and surface texture	Particles are typically angular to very angular, although minor rounding of mixed facies clasts can occur. Particle shapes reflect joint patterns in source rock. Intense fracturing of clasts within debris-avalanche blocks in large VDA deposits is ubiquitous and intrablock clasts are very angular, with or without clast deformation. Fractures may be open and intersecting with little relative movement between clast pieces, giving impression of a jigsaw puzzle ("jigsaw fractures"). Jigsaw fractures are not observed or reported at flank-collapse VDA deposits
Particle-size distribution and sedimentary texture	Size distributions are polymodal. Grain sizes in large deposits sourced from deep in the edifice vary from clay size to enormous debris-avalanche blocks hundreds of meters in diameter, but clay content is spatially variable within deposits and depends on amount and distribution of hydrothermally altered rock in the deposit. Sorting coefficients are typically >4.0 φ. Sediment textures depend on source: clay-rich deposits from deep-seated failures are typically muddy gravelly sand to muddy sandy gravel. Flank collapses mainly produce sandy gravel and gravelly sand, typically with low silt and clay contents. Note, however, that a pre-failure soil mantle containing clay formed by weathering on a volcano flank could add clay to such a VDA deposit
Matrix vesicles and intergranular voids	Bubble vesicles are absent in mixed-facies matrix material. Voids may be present in gaps along open fractures within and between debris-avalanche blocks in the block facies, forming zones of localized open-work structure In large clay-rich deposits, matrix packs all spaces between clasts except within debris-avalanche blocks, where open cracks may exist. In smaller clay-poor deposits, silt- and clay-size grains are largely absent, and deposits are not tightly compacted; thus, intergranular voids are common

Table 5 Summary table of volcanic water flood (VWF) deposit characteristics

Process	*Volcanic water flood (VWF)*
Deposit type	Muddy water-flood deposit (volcano-fluvial deposit)
Deposit architecture	Sediment bodies form downstream-tapering wedges, either narrowly confined to valleys or spread widely on alluvial planes. Later erosion may leave only remnants preserved as alluvial fill terraces. Braided channel patterns (multiple shallow anastomosing channels separated by low longitudinal bars) result from very high sediment loads and are nearly ubiquitous in volcanic settings. Ripple and dune bed forms are common on upper bounding surfaces. Juxtaposition of sharply bounded sediment patches (defined by abruptly different average particle sizes) is common on deposit surfaces. Thin, overlapping, sheetflood deposits are common on fans and aprons; thicker wedges of sediment are found in narrow valleys (up to about 30 m thick for single eruptions). Anabranching or single-thread channels become more common with lower posteruptive sediment-transport rates. Mid-channel longitudinal bars composed of pebble- to boulder-size gravel are common; point bars are also common on insides of channel bends. Berms composed mainly of sand and silt typically form in locations along channel margins where back eddies facilitate deposition of some bed load and suspended load
Lithofacies	A wide variety of gravelly and sandy lithofacies are found in VWF deposits, typically changing abruptly over short distances (vertically and laterally). Deposits are generally distinctly stratified. Lithofacies Gmo, Gcm, Gi, Gh, Sh, St, Sp, Ss are common. Sedimentary textures principally emcompass sand, gravel, sandy gravel, and gravelly sand. Facies associations define three generally recognized depositional facies: channel facies, overbank facies, and sheetflood facies. Gravel lenses are common in vertical sections of channel and sheetflood facies. Laminated fine sand and silt are common in overbank facies
Erosion	VWFs typically carry extreme sediment loads that cause channel beds to aggrade and channel banks to erode laterally. Channel shifting and avulsion are frequent during VWFs, leaving numerous abandoned, shallow, rectangular channels on braid plains. Channel width/depth ratios are high in channels formed in unconsolidated sediment
Effects on vegetation	Flows generally impart only moderate to light damage to woody vegetation. Damage to tree bark on upstream sides of trunks is usually limited to abrasion near the bed (from saltating bed-load particles) and near the peak stage level (from floating debris collisions). There is little to no coating of sediment on tree trunks. Some scouring of vegetation can occur along valley sides, but more commonly grass and small shrubs are only flattened and left pointing downstream
Stratification	Sharply defined bedding is typical unless sediment is highly uniform in grain size, in which case bedding may appear very faint, if visible at all (outcrop faces exposed to weather usually show such bedding better than freshly scraped exposures). Individual sandy beds are typically millimeters to centimeters thick. Gravel beds emplaced by moderate VWFs are typically centimeters to tens of centimeters. Large VWFs can emplace gravel beds that are meters thick. Steep cross-bedding (>15°) is common and characteristic of bed-load transport of sand- or gravel-size sediment. Plane-parallel bedding and both low- and high-angle cross-bedding are common in braided stream environments (upper flow regime). Bedding is often not laterally continuous for more than a few meters, due to scour from channel shifting
Bed forms and structures	Bed forms of medium and large scale (mesoforms and macroforms) are very common. These include ripples, dunes, plane-parallel (subhorizontal) bedding, antidunes, and chute-and -pool structures. Bed structures include various types of longitudinal and transverse bars, point bars, and finer-grained berms. No lateral flow levees composed of sediment are formed by VWFs
Gravel-size clast framework	Clast-supported frameworks are typical in gravel beds, bars, and lenses, but sand may fill gaps between clasts. Some coarse gravel deposits have voids between clasts—openwork structure

(continued)

Table 5 (continued)

Process	*Volcanic water flood (VWF)*
Deposit type	Muddy water-flood deposit (volcano-fluvial deposit)
Fabric	Elongate particles tend to align with flow direction. Imbrication of platy particles (upstream dipping "shingling" of tabular or platy clasts) is common but not always visible where particles are more equant in shape
Grading	Deposits are generally ungraded. Vertical changes in grain size tend to be abrupt and are associated with localized unconformities caused by channel shifting
Internal deformation	Any deformation is postdepositional, such as riverbank undercutting and slumping
Fluid-escape structures	Dewatering structures are generally absent. Pore fluid is normally able to escape easily upon deposition, without the buildup of high pore pressures that can concentrate escaping pore fluid into narrow zones
Material consistency	Sediment is loose to very friable when dry, and loose to soft when moist. Sandy sediment grains have virtually no coherence. Sand and fine gravel particles ravel easily from dry outcrop faces when disturbed by a tool (or even touched). Deposits are very easy to dig or scrape when an outcrop face is damp
Evidence of heat	Indicators of high heat are unlikely to be present, although incorporation of old primary pyroclastic deposits by a VWF may add some charred wood or prismatically jointed clasts to a deposit
Particle composition	Deposits are typically rich in volcaniclastic particles reworked from primary volcaniclastic deposits, but they also commonly contain some secondary volcaniclastic and epiclastic particles eroded and remobilized from channel and bank deposits. Particles include crystals, rock fragments, and glass fragments. Both normal-density rock fragments and pumice or scoria particles are found in VWF deposits, but they are commonly segregated in lenses or beds in outcrop. VWF deposits are generally polymictic
Particle roundness, shape, fracturing, and surface texture	Rounding of particles is a function of particle size and age (fresh or weathered), distance from source, and grain density and composition (weathered grains or fresh pumiceous rock particles are softer and round more easily). Fresh volcaniclastic normal-density sand and fine gravel particles are mainly angular in proximal deposits. Coarse gravel and cobbles are commonly subangular. Boulders are commonly sub-angular to sub-rounded. Average roundness of clasts increases with distance downstream, with in-stream time as bed load, with VWF flow velocity (collision forces), and with degree of entrainment of older alluvium. Particles typically have equant shapes; rod-shaped and disc-shaped gravel-size particles are less common Fresh volcaniclastic particles typically have rough surfaces; older river gravels have relatively smooth surfaces. Fracturing of clasts during transport occurs but usually results in breakage. Pumice or scoria clasts at almost any distance from source are generally well rounded
Particle-size distribution and sedimentary texture	VWF deposits mainly have bimodal or unimodal particle-size distributions. Dominant mode typically is sand in channel facies, although gravel can dominate in some bed forms; generally, only trace amounts of silt and clay are mixed in with the sand. Overbank facies are typically dominated by fine sand and silt. Sorting values are $< 1.0\ \varphi$ in sandy deposits. Sorting is poorer in gravel bars but better than in gravelly lahar deposits
Matrix vesicles and intergranular voids	Vesicles are absent but irregularly shaped voids between gravel-size clasts are common. Voids between coarse sand grains may also be seen with a hand lens due to typically low fines contents in channel-facies deposits

Table 6 Summary table of characteristics of pyroclastic density current (PDC) deposits emplaced on the flanks of stratovolcanoes and lava-dome complexes and in downstream river valleys

Process	*Pyroclastic density current (PDC)*		
Deposit type	Concentrated PDC deposit—lithic (Block-and-ash flow deposit)	Concentrated PDC deposit—pumiceous/scoriaceous	Dilute PDC deposit—pumiceous/scoriaceous or lithic
Deposit architecture	Deposits of individual flows typically form highly elongated, tabular-shaped, slightly downslope-thickening sediment bodies that are relatively simple in planform shape and convex upward in cross-section. Their abruptly terminating frontal and lateral margins are lobate in shape, sometimes branching, and commonly rimmed with openwork rims of coarse clasts. Lateral flow levees are common Multiple flows from repeated dome collapses form sediment accumulations in the form of downstream-tapering wedges—narrow when confined as valley fills or broad when spread out on debris fans. They have moderately rough, flat surfaces, typically studded with large clasts	Deposits of individual flows typically form highly elongated, tabular-shaped, slightly downslope-thickening sediment bodies that are relatively complex in planform shape and convex upward in cross-section. Their abruptly terminating frontal and lateral margins are highly and complexly branching and lobate (digitate). Margins are commonly rimmed with openwork rims of coarse clasts. Lateral flow levees are common. Flow deposit tops may display transverse ridges Multiple pumiceous PDCs produce downstream-tapering valley fills or broad fan deposits with moderately smooth upper surfaces that are flat or undulating. Surfaces are commonly scribed with the complex flow patterns of individual currents	Deposits of individual flows typically form thin, highly elongated, tabular-shaped, terrain-mantling, downcurrent-thinning sediment bodies that are commonly lobate in planform shape. Their frontal and lateral margins are diffuse Multiple dilute-current deposits would form stacked or overlapping deposits with properties similar to those of individual deposits. Deposits are commonly rich in broken pieces of wood, and other vegetal debris
Lithofacies	Deposits are massive unstratified diamicts, having lithofacies Dmm, Dmg, Dcm, Dcg, Gmm, Gcm, and Sm. Lithofacies can appear similar to VDF deposits and some glacial tills	Deposits are mainly massive or weakly stratified diamicts. Common lithofacies are Dmm, Dms, Dcm, Gm and Go (locally)	Deposits can be massive or stratified, sometimes well stratified. Overall, they are finer grained than deposits of most concentrated PDCs. Lithofacies commonly include those with ashy (sandy) textures with zones or lenses of gravel and fine plant debris. Lithofacies codes include Dmm, Dms, Gp, Gm, Gh, Sm, Sp, Sl

(continued)

Table 6 (continued)

Process	*Pyroclastic density current (PDC)*		
Deposit type	Concentrated PDC deposit—lithic (Block-and-ash flow deposit)	Concentrated PDC deposit—pumiceous/scoriaceous	Dilute PDC deposit—pumiceous/scoriaceous or lithic
Erosion	Highly erosive on steep slopes; minimally erosive on gentle slopes	Highly erosive on steep slopes; minimally erosive on gentle slopes. Flows can striate bedrock, scour soil, and carve deep channels in glacier ice. They entrain significant amounts of rock and ice fragments	High-velocity currents are significantly erosive, carving furrows in soil or ice
Effects on vegetation	Highly destructive to trees in flow paths–from physical impact and abrasion of wood and from prolonged heating and charring of wood. Trees may be flattened, toppled, buried, or entrained in flows	Concentrated flows are highly destructive to trees in or adjacent to flow paths, scorching or burning exposed trees and charring buried wood. Trees may be flattened, toppled, buried, or entrained in flows. Deposits cool relatively slowly, depending on deposit thickness and interaction with meteoric and ground water	High-velocity dilute flows can scorch, severely abrade, knock down, remove tops from, de-limb, or defoliate trees along their flow paths, with damage gradients decreasing with distance away from source. Temperature effects are intense but brief (minutes)
Stratification	Individual deposits are unstratified (massive)	Individual deposits are unstratified to faintly stratified	Individual deposits may be distinctly stratified, diffusely (faintly) stratified, or massive
Bed forms and structures	Bed forms are absent	Bed forms are absent except in rare cases of high velocity flows on steep slopes near vents where cross-bedding can be produced	Mainly plane-parallel-bedded but deposits commonly display low-angle (< 15°) cross-stratification and scour structures
Gravel-size clast framework	Clast-supported or matrix-supported frameworks	Clast-supported or matrix-supported frameworks	Matrix-supported frameworks
Fabric	Strong primary flow-parallel fabrics are commonly developed when elongate particles (or woody debris) are present. Secondary fabrics may develop during degassing, deflation, and compaction	Strong primary flow-parallel fabrics are commonly developed when elongate particles or woody debris are present. Secondary fabrics may develop during degassing, deflation, and compaction. Upstream-dipping imbrication of platy clasts is often reported	Isotropic fabric when deposits lack clasts large enough to readily orient in the flow, but any clasts present would likely have flow-parallel alignment

(continued)

Table 6 (continued)

Process	*Pyroclastic density current (PDC)*		
Deposit type	Concentrated PDC deposit—lithic (Block-and-ash flow deposit)	Concentrated PDC deposit—pumiceous/scoriaceous	Dilute PDC deposit—pumiceous/scoriaceous or lithic
Grading	Mainly inversely graded or ungraded (rarely normally graded)	Grading is highly variable–ungraded, normally graded, or inversely graded. Grading in these PDCs may be limited to coarse-tail grading and density-segregation grading of clasts (low-density pumice concentrated at top and dense lithic clasts concentrated at bottom of deposit cross-sections	Commonly normally graded; may encompass separate and progressively finer depositional units resulting from waning current energy. Typically capped by fine-grained airfall deposits
Internal deformation	Not reported	Rheomorphic flow structures may be found in welded zones of some thick deposits	Syneruptive, soft-sediment deformation of deposits from dilute PDCs has been reported at a few volcanoes
Fluid-escape structures	Gas-escape pipes and other structures are rare, occurring only if a deposit is emplaced on wet ground	Gas-escape (elutriation) pipes are common, usually developing in the upper parts of unwelded deposits	Gas-escape pipes are not present unless a dilute PDC deposit unit overlies another actively degassing deposit and escaping gas moves upward through the capping deposit
Material consistency	Unwelded deposits are typically loose to very friable when dry and loose to soft when moist. They are nonsticky and nonplastic when wet. Sediment grains have little coherence, due to the fact that most deposits are fines-depleted because of ash elutriation during flow. Sand and fine gravel particles ravel from dry outcrop faces when disturbed by a tool. Deposits are very easy to dig or scrape when outcrop face is moist. If partially welded or cemented, sediments have varying degrees of induration up to full lithification	Unwelded deposits are typically loose to very friable when dry, and loose to soft when moist. They are nonsticky and nonplastic when wet. Sediment grains have little coherence. Sand and fine gravel particles ravel from dry outcrop faces when disturbed by a tool (or even touched). Very easy to dig or scrape when outcrop face is moist. If partially welded or cemented, sediments have varying degrees of induration, up to full lithification	Dilute PDC deposits typically contain more fines than other PDC types. They are commonly friable to firm when moist; slightly hard to hard when dry

(continued)

Table 6 (continued)

Process	*Pyroclastic density current (PDC)*		
Deposit type	Concentrated PDC deposit—lithic (Block-and-ash flow deposit)	Concentrated PDC deposit—pumiceous/scoriaceous	Dilute PDC deposit—pumiceous/scoriaceous or lithic
Evidence of heat	Evidence is typically abundant–prismatically jointed (radially fractured) clasts; clasts with breadcrusted and cauliflower surface textures; and uniform thermoremanent magnetic orientation of large clasts. Commonly varying degrees of charring of buried tree trunks or logs	Evidence is typically abundant–charred wood if forest vegetation was present; pink coloration on tops of thick deposits, and uniform thermoremanent magnetic orientation of large clasts. Hot gas-escape pipes (fossil fumaroles) may also be present, with strong oxidation coloration of particles within the pipe throat; and vapor-phase mineralization and cementation of matrix material surrounding the pipe. Evidence may include phreatic explosion pits on the deposit surface	Typically rare due to low solids concentration in current and rapid cooling of turbulent current by ingested air
Particle composition	Deposits are monomictic or oligomictic; small fractions of accidental lithic fragments are commonly included in mixtures. Juvenile gravel-size clasts and matrix material typically have the same color. If lithic PDCs are sourced from different parts of a collapsed lava dome, particle mixtures may include lava rock from different eruptions with different colors and vesicularities	Dominantly monomictic—composed of the same juvenile pumice/scoria particles. Mixtures also usually contain small fractions of dense lithic fragments of similar or different composition. Juvenile clasts and matrix usually have the same color	Monomictic or oligomictic. Currents from magmatic eruptions composed of the same juvenile pumice/scoria particles. Currents from phreatic or phreatomagmatic eruptions primarily contain dense lithic fragments or a combination of lithics and pumice/scoria. Mixtures commonly contain small fractions of accidental lithic fragments, as well
Particle roundness, shape, fracturing, and surface texture	Pebble/cobble clasts are angular to subangular. Dense lithic particles are typically equant in shape. Clast surfaces are rough. Larger blocks may be sub-rounded from collisions with other hard rocks	Pumice and scoria clasts are rounded to well rounded, although some rounded clasts may have been broken during transport (some angular edges may be present). Cognate and accessory lithic clasts are typically angular (few collisions with other hard rock fragments); accidental lithics may be more rounded, reflecting clast history prior to entrainment. Particles are typically equant in shape	Pumice and scoria clasts are rounded to well rounded, although some rounded clasts may have been broken during transport (some angular edges may be present). Lithic clasts may be angular or more rounded, reflecting clast history prior to entrainment. Particles are typically equant in shape

(continued)

Table 6 (continued)

Process	*Pyroclastic density current (PDC)*		
Deposit type	Concentrated PDC deposit—lithic (Block-and-ash flow deposit)	Concentrated PDC deposit—pumiceous/scoriaceous	Dilute PDC deposit—pumiceous/scoriaceous or lithic
Particle-size distribution and sedimentary texture	Sandy gravels to gravelly sands (fines content variable but overall fines-depleted). Bimodal to polymodal grain-size distributions, ranging from fine ash to volcanic blocks meters in diameter. Published sorting values range from about 2.0–4.0 φ	Grain-size distributions are highly variable—clast-enriched, clast-depleted, fines-enriched, or fines-depleted. Textures are mainly sandy gravels to gravelly sands. Bimodal to polymodal size distributions, ranging from fine ash to volcanic blocks meters in diameter. Sorting values are highly variable (1.5–5.0 φ)	Deposits of highly dilute flows have particle sizes finer than in concentrated PDCs. Median diameters range from −3.0 φ to +4.0 φ, and size distributions lack the pronounced fine and coarse tails of size distributions in deposits emplaced by concentrated basal flows. Sorting values are commonly 1.5–4.5 φ, and deposits of phreatomagmatic eruptions are particularly rich in fine ash and ash aggregates
Matrix vesicles and intergranular voids	Vesicles are absent, but irregularly shaped voids between clasts are common, giving some gravelly units openwork frameworks on deposit margins and surfaces. Voids between sand grains may also be seen with a hand lens	Vesicles are absent, but irregularly shaped voids between clasts are common, giving some gravelly units openwork frameworks on deposit margins and surfaces. Voids between sand grains may also be seen with a hand lens	Vesicles are possible in wet dilute PDC (base-surge) deposits. Spaces between clasts tend to be well packed with finer matrix material

Table 7 Summary table of tephra fall (TF) deposit characteristics

Process	*Tephra fall (TF)*
Deposit type	Subaerial tephra-fall deposit; airfall tephra deposit
Deposit architecture	Sediment bodies are topography-mantling layered deposits that extend over broad areas downwind of the source vent. In simple cases, thickness decreases steadily with distance from the vent and also with orthogonal distance from main downwind axis of the deposit. Variation in wind speed and direction (at different altitudes) and aggregation of ash particles due to atmospheric conditions can produce complex areal distribution patterns. Deposits from low-energy eruptions can form thick rings of cinders or spatter around a vent. Unlike volcano-fluvial deposits, TF deposits are not confined to valleys or channels
Lithofacies	Lithofacies can be similar over broad areas but deposits commonly exhibit marked vertical differences in sediment texture within a single deposit. Textural transitions within single deposits tend to be gradational. Typically, all particles (blocks, lapilli, and ash) rest loosely in contact with one another with voids between them. Lithofacies types include: Gmo, Gcm, Gh, Sm, Sh, Sl, Fl, and Fm. On valley floors, TF deposits can closely resemble VHF and VWF deposits
Erosion	No direct erosional effect on landscapes
Effects on vegetation	Forest vegetation effects depend on proximity to vent, coarseness of tephra, and erupted tephra volume. Effects include heat injury, burial, and physical impacts. In heavily impacted proximal areas, severe abrasion and burning, as well as complete stripping of branches and leaves from trees is possible; few trees survive. Leaf kill and some branch breakage or deformation of small trees is common in medial zones, as well as burial of smaller plants. Distal zones show little physical damage to trees, but thin coatings of ash on leaves can cause extensive seasonal leafkill but not necessarily tree mortality
Stratification	Plane-parallel beds drape over gentle topography to form terrain-mantling, laterally extensive beds of locally uniform thickness; beds gradually thin with distance from source vent. Beds terminating against steep slopes may be thicker due to sloughing from the steeper hillsides. Single deposits may or may not contain internal stratification, but bedding contacts tend to be gradational unless there are abrupt changes in erupted tephra composition, eruption output (eruptive pulses), or wind conditions
Bed forms and structures	No primary bed forms or structures are present, unless high wind at the time of deposition (or in intervals between deposition of tephra beds) produces eolian bed forms and structures
Gravel-size clast framework	Mainly clast-supported frameworks
Fabric	No preferred fabric, except in slope-parallel layering of partially molten spatter
Grading	Grading can be normal or inverse in various combinations within tephra beds, depending on changes in eruption rate and style and the wind speed and direction. Downslope raveling of coarse particles can result in inverse grading at bottoms of slopes
Internal deformation	Rheomorphic flow structures may be found in welded zones of thick agglutinates
Fluid-escape structures	No gas-escape pipes or structures are present unless a tephra bed overlies a PDC deposit that is still degassing
Material consistency	Loose (when dry) and soft (when moist) consistencies, except where raindrop impacts have packed fines into pore spaces to form a crust

(continued)

Table 7 (continued)

Process	*Tephra fall (TF)*
Deposit type	Subaerial tephra-fall deposit; airfall tephra deposit
Evidence of heat	At the time of deposition, juvenile tephra particles can include rock fragments ranging from partially molten spatter and bombs to fully solidified, ambient temperature particles. Physical evidence of high heat at the time of deposition includes aerodynamic particle shapes (fusiform bombs), plastic deformation of bombs during flight or upon impact, welding or sintering of particles in the deposit, prismatic jointing of clasts, breadcrusting or cauliflower-texturizing of clast outer surfaces, and charring of buried vegetation. Thermoremenant magnetization of clasts also provides evidence of high heat at the time of deposition
Particle composition	Juvenile particles from magmatic eruptions form monomictic tephra mixtures, whereas phreatomagmatic and phreatic eruptions are more likely to produce oligomictic and polymictic mixtures, respectively. Individual particle types can include fragments of crystalline rock, individual crystals, broken shards of dense glass, vesicular glass fragments, dense glass droplets and strands, and aggregates of fine ash (ash pellets and accretionary lapilli)
Particle roundness, shape, fracturing, and surface texture	Particles are angular at all sizes unless solidified while molten or recycled in the vent during eruption. Clasts commonly fracture and break apart upon impact. Most volcaniclastic and lithic particles are equant in shape; spatter particles are flattened, contorted, or elongate in shape. Glassy ash particles have a variety of shapes and are commonly curviplanar and extremely angular. Particle surfaces can be rough (breakage surfaces), semi-smooth (aerodynamically shaped during fall), or extremely smooth (glass or crystal faces)
Particle-size distribution and sedimentary texture	Tephra mixtures generally have unimodal or bimodal distributions; they are strongly bimodal if mixtures combine vesiculated and non-vesiculated particles (large vesiculated clasts have aerodynamic equivalence to small dense particles), such that pumice clasts can be 5 to 10 times the diameter of the equivalent dense-rock particles. Median particle size is typically in fine gravel (fine-medium lapilli) range or coarser close to vents, but size decreases and sorting improves with distance from source. Sorting values can range from 0.2 φ to 3.0 φ overall, though most commonly 1.0–2.0 φ. Lithic ballistic blocks can exceed 1 m diameter at distances of several kilometers from the vent
Matrix vesicles and intergranular voids	Bubble vesicles are normally absent in TF deposits but may form in tephra from phreatomagmatic (wet) eruptions. Tephra crusts may contain secondary vesicles formed by air trapped by raindrops deforming and compacting fine ash layers. Intergranular voids are common in most tephra deposits, especially where coarse ash and lapilli tephra are relatively well sorted. Coarse tephra typically has a prominent openwork structure

Table 8 Summary table of glacial till characteristics

Process	*Glacier transport and deposition*	
Deposit type	Lodgment till (basal till)	Ablation till (meltout till)
Deposit architecture	Deposits form irregular sheet-like sediment bodies on low-gradient surfaces in the flow paths of glaciers (on valley floors, benches, and ridge crests); deposits are commonly discontinuous. Till sheets are thin relative to the area of the deposit, commonly centimeters to meters thick where emplaced by alpine glaciers and up to hundreds of meters thick where emplaced by continental and ice-cap glaciers. Lodgment till may be interbedded with glaciofluvial deposits along margins or on valley floors, and it is commonly sheared, folded, and/or faulted	Meltout tills form irregular, sometimes hummocky sheets or piles of coarse sediment (hummocky ground moraine) in areas where stagnant ice downwasted. Thickness is highly variable. Ablation till commonly forms distinct high ridges along sides of active glaciers (lateral moraines) or in front of glacier snouts (end moraines). Lateral and end moraines are typically well-defined, convex, boulder-rich ridges (may be sharp-crested or broad and rounded); they commonly contain thrust faults)
Lithofacies	Characteristic lithofacies are highly to extremely consolidated diamicts with little variation laterally or vertically. They tend to be richer in matrix material than other volcanic diamicts (VDF, VDA, and PDC deposits). Common lithofacies codes include Dmm, Dcm, Dms, Dcs	Characteristic lithofacies are poorly to moderately compacted diamicts with only moderate variation laterally or vertically. Deposits are typically rich in large clasts (cobble- and boulder-size)–similar to VDF, VDA, and PDC deposits. Common lithofacies codes include Dmm, Dcm, Dms, Dcs
Erosion	Characteristic small-scale erosional features on overridden bedrock that are associated with till emplacement (both lodgment and ablation tills) include striations, grooves, chatter marks, crescentic fractures, conchoidal fractures, lunate fractures, and overall smoothing and polishing of bedrock surfaces. Associated intermediate-scale features include flutes, furrows, roches moutonnées, whalebacks, rock drumlins, and crag-and-tail features. Associated large-scale features include U-shaped valleys, valley-side trimlines, glacial cirques at heads of valleys, valley-side terraces composed of silty or clayey lacustrine deposits, and proglacial meltwater channels	Characteristic small-scale erosional features on overridden bedrock that are associated with till emplacement (both lodgment and ablation tills) include striations, grooves, chatter marks, crescentic fractures, conchoidal fractures, lunate fractures, and overall smoothing and polishing of bedrock surfaces. Associated intermediate-scale features include flutes, furrows, roches moutonnées, whalebacks, rock drumlins, and crag-and-tail features. Associated large-scale features include U-shaped valleys, valley-side trimlines, glacial cirques at heads of valleys, valley-side terraces composed of silty or clayey lacustrine deposits, and proglacial meltwater channels
Effects on vegetation	Glaciers override and crush forests. Lodgment tills commonly overlie the flattened, crushed remains of trees or other vegetal debris	Rock debris flowing or falling off glacier margins can bury forests. Layers of logs and brush may be buried in ablation till at the former glacier margins. Debris layers on stagnant glacier ice can also host tree growth; as ice melts, supraglacial debris deforms to produce "drunken forests."

(continued)

Table 8 (continued)

Process	*Glacier transport and deposition*	
Deposit type	Lodgment till (basal till)	Ablation till (meltout till)
Stratification	Lodgment till is unstratified, except where interbedded with overridden glaciofluvial deposits. Sometimes till exhibits pseudo-stratification (closely spaced subhorizontal shear bands or planes) or fissility (crinkly subhorizontal foliation) that can resemble bedding	Ablation till is generally unstratified but may be crudely bedded where localized accretion of mass-flow debris sliding off glacier surfaces produces faint layering
Bed forms and structures	None	None
Gravel-size clast framework	Typically, matrix-supported frameworks	Typically, clast-supported frameworks
Fabric	Generally strong fabrics, with elongate clasts (including sand grains) having preferred orientations roughly parallel to flow direction; clasts may dip upglacier slightly (imbrication). Some transverse orientations may be found. Matrix particles may be strongly oriented parallel to horizontal micro-shear planes, lending a fissile structure to deposits	Random to weak fabric due to vertical sloughing or collapsing of wet debris on melting ice. If debris is saturated and flows as a debris flow on or adjacent to the ice, stronger fabrics can develop
Grading	Ungraded	Generally ungraded, though grading is possible where tills contain debris-flow or glaciofluvial depositional elements
Internal deformation	Brittle and ductile shear deformation are commonly present due to the shearing action of overriding ice. Brittle deformation forms include subhorizontal jointing with polished/striated joint surfaces, large-scale normal and thrust faulting, microshears, brecciation, boudins (lensoid structures), rotational structures, and crushed clasts and sand grains Ductile deformation includes a variety of types of folding	Possible deformation may occur related to slumping and flowage, particularly in meltout till
Fluid-escape structures	Generally absent but have been reported in till near overpressured subglacial meltwater conduits	Generally absent
Material consistency	Consistencies are very firm to extremely firm (when moist) and very to extremely hard (when dry) due to overconsolidation by the weight of overriding ice. Tills can be highly indurated, almost like solid rock. These tills are extremely difficult or impossible to dig with a shovel or pick in outcrops	Consistencies are generally loose to friable (when moist) and loose to slightly hard (when dry), although some ablation tills containing debris-flow deposits may have a firmer/harder consistency similar to that of VDF deposits
Evidence of heat	None	None
Particle composition	Alpine lodgment tills are dominantly polymictic mixtures of hard rock fragments from relatively nearby upvalley sources. In valleys on volcanoes, mainly volcanic rock types are present, although fragments of country rock can be entrained. Continental tills contain abundantly polymictic mixtures with fragments of many nonvolcanic rock types coming from far-distant terranes	Ablation tills on volcanoes are dominantly polymictic mixtures of hard volcanic rock fragments from relatively nearby upvalley sources

(continued)

Table 8 (continued)

Process	*Glacier transport and deposition*	
Deposit type	Lodgment till (basal till)	Ablation till (meltout till)
Particle roundness, shape, fracturing, and surface texture	Rounding is specific to certain types of clasts. Clasts in supraglacial and englacial debris are angular; they have not been dragged along the glacier bed. Clasts frozen into ice at the base of the glacier are forcefully ground against overridden bedrock; these clasts are subangular to subrounded with smoothed, striated, and faceted surfaces. Some sculpted clasts may be "bullet-shaped" or have asymmetric "stoss and lee" geometries (smoothed on one side but rough and fractured on the other. Shape is largely a function of jointing and fracturing in source rock on volcanoes (fragments of lava flows and domes, or tephra particles). In addition, clasts with previous erosional histories and degrees of roundness may be entrained by glaciers and deposited in tills	Shape is largely a function of jointing and fracturing in source rock on slopes above glaciers (fragments of lava flows and domes, or tephra particles). Very little rounding can occur to clasts in ablation tills–mostly only angular clasts from rockfall or snow avalanches can move onto glacier surfaces
Particle-size distribution and sedimentary texture	Particle sizes range from clay size to boulders, with size distributions commonly bimodal to polymodal. Lodgment tills (in contrast to VDF, VDA, and PDC deposits) are commonly enriched significantly in silt- and clay-size particles, reflecting subglacial rock crushing and abrasion. Deposits usually have matrix-supported internal structures. Sorting coefficients are $>3.0\ \varphi$. Matrix material may show abrupt localized variations in sedimentary texture	Particles range from clay to boulders but distributions are relatively fines-poor. Depending on valley-side rock types, distributions commonly polymodal and more gravelly than lodgment tills. Sorting coefficients are generally $>3.0\ \varphi$. Clay contents in alpine ablation tills are <10 wt%
Matrix vesicles and intergranular voids	Both are absent; matrix material packs all spaces between clasts	Vesicles are absent unless ablation tills contain debris-flow deposits. Internal voids are generally absent unless till is very loosely packed

References

Folk RL (1974) Petrology of sedimentary rocks. Hemphill, Austin p 184

Inman DL (1952) Measures of describing the size distribution of sediments. J Sediment Petrol 22:125–145

Annotated Outcrop Photographs: Volcanic Wet-Flow Deposits and Other Unconsolidated Volcanic Deposits

Abstract

Comparison of outcrop photographs is an important additional tool for use in the field to aid in the tasks of deposit differentiation and identification. The collection of outcrop photographs in this chapter provides a basis for visual comparison. Images, mainly of vertical outcrop faces, are presented for all the deposit types described in this book—the four types of volcanic wet-flow deposits and the pyroclastic density-current, tephra-fall, and glacial deposits that can closely resemble them. These photos illustrate that each of these types of deposits has characteristics and features that are recognizable in outcrops, and key identification features are noted in the figure captions. Laboratory analysis also may be needed to make a final determination, but preliminary identification of flow type is often possible from only characteristics and features observed in the field. These selected outcrop photographs are not intended to show the full range of sedimentary textures, structures, and features to be found in these types of deposits. Instead, photographs were selected to highlight the similarities, as well as the differences, between all the main categories of unconsolidated volcanic and volcaniclastic deposits. Additional outcrop photographs can be seen in the process-specific chapters in the main text.

Field identification of volcanic wet-flow deposits and their differentiation from pyroclastic and glacial deposits is challenging (Fig. 1). The collection of outcrop photographs in this chapter is intended to aid in that task. Images of all the types of deposits discussed in this book are included: those emplaced by volcanic debris flows (VDFs), volcanic hyperconcentrated flows (VHFs), volcanic debris avalanches (VDAs), volcanic water floods (VWFs), pyroclastic density currents (PDCs), tephra falls (TFs), and glaciers. These photos illustrate that each of these types of deposits has characteristics and features that are recognizable in outcrops, and key identification features are noted in the annotations. Laboratory analysis also may be needed to make a final determination, but identification of flow type is often possible from only characteristics and features observed in the field. Unless otherwise noted in the captions, deposit images in this chapter show vertical faces of outcrops.

The photographs included here of the deposits emplaced by pyroclastic density currents, tephra fall, and glaciers are limited. These selected examples are not intended to be a comprehensive representation of the broad range of sedimentary textures, structures, and features to be found in these deposits. Instead, photographs were selected to highlight the similarities, as well as the differences, between all the main categories of fragmental deposits composed of volcanic rock particles. Additional outcrop

T. C. Pierson et al., *Deposits of Volcanic Wet Flows*, Advances in Volcanology,
https://doi.org/10.1007/978-3-031-66574-5_13

Fig. 1 A typical field outcrop of volcanic wet-flow deposits in a forested river valley downstream of an active volcano. This vertical outcrop face in the upper Muddy Fork of the Sandy River downstream of Mount Hood (Oregon, USA) was scraped with a trenching shovel to reveal the true colors and sedimentary textures of the several deposits exposed here. It provides a good and realistic example of the complexity that can be encountered at outcrops in the field. When standing in front of such an outcrop for the first time, several initial questions can be posed: (a) How many separate depositional units are present?, (b) What are the characteristics of the unit boundaries or contacts? (c) What are the dominant sedimentary textures and clast compositions within each unit? (d) Are paleosols, datable organic materials, cultural debris, or other features present that could indicate disconformities and hiatus durations? Once the initial questions are answered, the more specific outcrop-scale and particle-scale deposit characteristics described in the book can be addressed. Finally, other outcrops can be exposed and examined upstream and downstream of this one, bulk samples can be collected, and landscape-scale features can be evaluated by hiking around on the valley floor and by examining aerial photo, lidar, or satellite imagery. USGS photo by T. Pierson

photographs can be seen in the process-specific chapters in the main text. For more comprehensive discussions, diagrams, and photographs of pyroclastic density–current and tephra-fall deposits, please refer to the following summary texts and to the relevant references they contain: Fisher and Schmincke (1984), Cas and Wright (1987), Francis (1993), McPhie et al. (1993), Branney and Kokelaar (2002), and Schmincke (2004). A comprehensive compilation of the types and features of glacial deposits may be found in Flint (1971), Ehlers (1996), Benn and Evans (2010), and Menzies et al. (2018).

1 Deposits of Lahars (Volcanic Debris Flows–VDFs)

See Figs. 2, 3, 4, 5, 6, 7, 8, 9, 10, 11, 12, 13, 14, 15 and 16.

Fig. 2 May 18, 1980, eruption-triggered, clay-poor, volcanic debris flow (VDF) deposit (above shovel blade) containing dense-rock fragments, located on left bank of Muddy River about 20 km downstream of vent at Mount St. Helens (Washington, USA) (Pierson 1985). The slightly darker vertical strip in center of photo was scraped clean. The lahar was triggered by rapid snowmelt from a lithic pyroclastic density current (boiling-over phase of the eruption), which scoured snow on the east flank of the volcano minutes after the 8:32 AM start of the eruption. Deposit overlies a buried soil (the 1980 forest floor) developed on an older lahar deposit. The remnant of a second May 18 lahar deposit is visible at the top of the deposit (above white line); it was emplaced by a pumiceous VDF that occurred in the afternoon of the same day during the Plinian phase of eruption (see closeup in Fig. 3). USGS photo by T. Pierson

Fig. 3 Remnant of volcanic debris flow (VDF) deposit composed mainly of rounded pumice clasts in Muddy River valley, about 20 km downstream of Mount St. Helens (Washington, USA). Lahar was triggered by snowmelt from column-collapse pumiceous pyroclastic density currents scouring the snow-covered outer flank of the volcano on the afternoon of May 18, 1980 (Pierson 1985), and deposit includes some dense-rock lithic fragments. Pumice clasts are mostly well rounded. USGS photo by T. Pierson

Fig. 4 Base of clay-rich North Fork Toutle River volcanic debris flow (VDF) deposit (overbank facies), sitting on largely undisturbed forest floor organic matter, about 30 km downstream of Mount St. Helens (Washington, USA) (Scott 1988a). Irregular depressions in the outcrop face are molds of gravel clasts removed during scraping of the face. This lahar deposit has few clasts larger than pebble size because the source material—blocks of volcanic rock from the volcano edifice—was so thoroughly crushed and sheared during emplacement of the May 18 volcanic debris avalanche (VDA) deposit prior to initiation of the lahar from part of that VDA deposit. Some larger clasts were entrained from riverbed gravels during lahar passage. USGS photo by T. Pierson

Fig. 5 Clay-rich, matrix-supported, prehistoric (~ 60–100 ka) volcanic debris flow (VDF) deposit from Mount Hood (Oregon, USA) that transformed from a volcanic debris avalanche (VDA) and flowed north to the Columbia River from Mount Hood (Vallance 1999). This outcrop is in the town of Hood River, Oregon, 40 km downstream of the volcano. Note the angularity of gravel-size clasts (common for VDAs and the clay-rich lahars they generate). This deposit is highly consolidated; this exposure was made by a power excavator, which cut through hard clasts and matrix alike. Field notebook (13 × 21 cm) for scale. USGS photo by T. Pierson

Fig. 6 Lower unit is top of relatively fine-grained, clay-poor volcanic debris flow (VDF) deposit on the Sandy River, about 70 km downstream of source on Mount Hood (Oregon, USA). It was emplaced on a channel-adjacent paleoterrace several meters high that was subsequently buried. Deposit is topped by a thin silty cap layer, which is overlain by fine-grained, laminated fluvial sand—the overbank facies deposit of the first volcanic water flood (VWF) to overtop the paleoterrace. The lahar marked the onset in 1781 of the intermittent twelve-year-long Old Maid eruptive period at Mount Hood (a dome-building eruption), which triggered 23 m of syneruptive channel aggradation by VWFs at this location (Pierson et al. 2011). Tool shaft is 5 mm wide. USGS photo by T. Pierson

Fig. 7 Massive clay-poor, and mainly fine-grained volcanic debris flow (VDF) deposit on a debris apron on the lower flank of Villarrica volcano (Chile), visible in background. VDF deposit is capped by a thinner stratified volcanic hyperconcentrated flow (VHF) deposit, probably from the tail of that lahar. Deposits were emplaced by the lahar during the volcano's March 2015 eruption (Johnson and Palma 2015). Tool shaft is about 5 mm wide. USGS photo by T. Pierson

Fig. 8 Deposit of the May 18, 1980, Pine Creek lahar—the gray, clay-poor, volcanic debris flow (VDF) unit at the top of the exposure. Deposit was emplaced on a low fluvial gravel terrace (topped by a soil) in the Pine Creek valley about 15 km downstream of Mount St. Helens (Washington, USA). Note the relative thinness of the deposit at this location as compared to flow depth (marked by mud coatings and abrasion of tree trunks in background). Lahar flow velocity was back-calculated as 15–20 m/s at this location (Pierson 1985). Shovel shaft above blade is 75 cm long. USGS photo by T. Pierson

Fig. 9 Prehistoric, relatively fine-grained volcanic debris flow (VDF) deposit about 25 km downstream of Mount St. Helens (Washington, USA). Deposit is polymictic and matrix supported. Note that fine pebbles are angular in VDF deposits unless rounded river gravel was entrained by the lahar. This long-exposed outcrop face was etched by rain wash that removed some matrix material and caused gravel clasts to stand out. Compare this with a fresh outcrop face where clasts are not etched (Fig. 5). Pen barrel is 1 cm in diameter. USGS photo by T. Pierson

Fig. 10 Contact between two prehistoric, relatively fine-grained volcanic debris flow (VDF) deposits about 25 km downstream of Mount St. Helens (Washington, USA). The abrupt change in average particle size visible here may sometimes be the only way to distinguish separate flow units, although this contact is also marked by a thin silty parting (a cap layer). Interval marks on the ruler are 10 cm long. USGS photo by T. Pierson

Fig. 11 Dark fragile clast on right (debris-avalanche block) is surrounded by matrix material of a clay-rich, prehistoric, volcanic debris flow (VDF) that transported it after transforming from a volcanic debris avalanche (VDA) (Vallance 1999). The VDA was triggered by an edifice collapse from Mount Hood (Oregon, USA). The purple color (indicator of hydrothermal alteration) and fractured clasts within the block are clues for a VDA origin. This outcrop, 37 km downstream of the volcano, exposes the same deposit shown in Fig. 5. Pen barrel 1 cm in diameter. USGS photo by T. Pierson

Fig. 12 Deposit of a volcanic debris flow (VDF) in White River valley, 8 km downstream of Mount Hood summit (Oregon, USA). Clasts were sourced either from dome-collapse block-and-ash flow deposits lining channel 3–8 km upstream (angular shape) or from the riverbed gravel derived earlier from those deposits (rounded shape). VDF triggered by prolonged heavy rainfall. Deposit is monomictic (clasts and matrix are the same color). Vertical particle-size changes suggest deposition by several surges. USGS photo by T. Pierson

Fig. 13 Contact between two fine-grained prehistoric volcanic debris flow (VDF) deposits, about 50 km downstream of Mount St. Helens (Washington, USA). Upper unit has a faintly stratified sand/silt sole layer (indicating hyperconcentrated flow at the lahar flow front). Lower unit has a normally graded top and a cap layer. Shaft of tool is 5 mm wide. USGS photo by T. Pierson

Fig. 14 Clay-poor oligomictic volcanic debris flow (VDF) deposit—the concentrated upper-valley phase of the 2006 Eliot Branch lahar at Mount Hood (Oregon, USA) (Poole 2015). The wide range of particle sizes, the inclusion of woody debris in the deposit, and some rounding of nearly all boulder-size clasts are visible. Deposit is massive and ungraded. Outcrop is 7 km downstream of source on north flank of the volcano. Small boulders are both smoothly rounded (entrained from streambed) and roughly rounded and pitted (rounded in part during flow). USGS photo by T. Pierson

Fig. 15 Clay-poor, monomictic volcanic debris flow (VDF) deposit, which is located about 2 km downstream of the sediment source—fresh, loose, lithic pyroclastic density-current (PDC) deposits in the Mizunashi River drainage (Unzen volcano, Japan). PDCs were associated with the 1990–1995 dome-building eruption (Miyabuchi 1999). This deposit was more consolidated than the similar-looking source block-and-ash flow deposits upstream. The source lava dome was composed of porphyritic dacite. Deposit is about 2 m thick at this location. Cobble-size clasts are still angular to very angular due to the short transport distance. Pen barrel 1 cm in diameter. USGS photo by T. Pierson

Fig. 16 Unusually thick, normally graded volcanic debris flow (VDF) deposit (middle unit) of the enormous prehistoric Chillos Valley lahar that originated from Cotopaxi volcano (Ecuador) (Mothes et al. 1998). Exposure is in a quarry wall. Deposit is highly weathered, so that blocks of originally hard rock are soft enough to be planed off by excavators. Note that boulders show rounding presumably acquired in transport. USGS photo by J. Vallance

2 Deposits of Lahars (Volcanic Hyperconcentrated Flows—VHFs)

See Figs. 17, 18, 19, 20, 21, 22, 23, 24, 25, 26, 27, 28, 29, 30, 31, 32 and 33.

Fig. 17 Deposit of the March 19–20, 1982, lahar at Mount St. Helens (Washington, USA). It was triggered by a lava-dome explosion and flowed initially as a volcanic water flood (VWF) in two main pulses, quickly transformed to a volcanic debris flow (VDF), and finally transformed to volcanic hyperconcentrated flow (VHF) (Pierson and Scott 1985; Dinehart 1999; Pierson 1999). The deposit here, about 30 km from source, shows two depositional units. The lower gray unit (from the first pulse) is faintly stratified in its lower half and was emplaced mainly as a VHF. The upper brownish gray unit (from the second pulse) shows characteristics of deposition as a VDF. A thin fine-grained layer at the top of the section was likely emplaced by the tail of the second pulse. About 20 km farther downstream, the lahar (both pulses) had fully transformed to VHF at that location. Flow was right to left. USGS photo by T. Pierson

Fig. 18 Upper part of a volcanic hyperconcentrated flow (VHF) deposit displaying very faint and diffuse bedding (in the lower half of the photo) in an outcrop about 80 km downstream of source on Mount Hood (Oregon, USA) along the Sandy River (Pierson et al. 2009). Upward coarsening and disappearance of distinct bedding in the upper half of this VHF unit suggests a progressive increase in solids concentration as the deposit was incrementally accreted on the streambed. Flow was obliquely left to right. USGS photo by T. Pierson

Fig. 19 Progressively coarsening volcanic hyperconcentrated flow (VHF) deposit in a lahar depositional unit in the lower Sandy River, 82 km downstream of Mount Hood along the Sandy River (Oregon, USA) (Pierson et al. 2009). Deposit shows discontinuous, subhorizontal, consolidation laminations of slightly different thicknesses and orientations. Fewer and thinner laminations occur in the more gravelly facies than in the sandier facies below it. Inverse grading (upward coarsening) is due to a progressive increase in sediment concentration as incremental accretion of sediment was occurring on the riverbed. Flow was left to right. USGS photo by T. Pierson

Fig. 20 Consolidation laminations in a sandy volcanic hyperconcentrated flow (VHF) facies at the base of the 1980 North Fork Toutle River lahar deposit in downstream of Mount St. Helens (Washington, USA). The sandy VHF facies here grades upward into volcanic debris flow (VDF) facies at the top of the photo. Thin consolidation laminations are subhorizontal and nearly parallel but discontinuous and irregular. Top lamination shows wavy soft-sediment deformation. Lamination just below center of photo terminates in chaotic pattern of branching laminations (closeup shown in Fig. 21). Dark spots are insect burrows. USGS photo by T. Pierson

Fig. 21 Closeup of lower left quarter of Fig. 20, showing distal terminations of the laterally injected laminations, some of which are wavy and more irregular at their ends. The uppermost lamination shows a branching pattern from forceful injection of muddy porewater (from right to left) into the surrounding sand during consolidation of the deposit. The preservation of insect burrows in the sand demonstrates how relatively coherent this deposit is due to its fines content. USGS photo by T. Pierson

Fig. 22 A wedge-shaped vertical column of relatively coherent, faintly stratified deposit, emplaced by a volcanic hyperconcentrated flow (VHF) and capped by flotsam on the original deposit surface, (pen resting on this surface) is supported and protected by a cluster of small tree stems. This unusually preserved remnant of an otherwise totally eroded (at this location) VHF deposit is fully suspended above the floodplain surface present at the time the photo was taken, six years after deposit emplacement. Postdepositional erosion by the tail of this lahar and by subsequent floods reestablished the floodplain at a level close to where it was prior to the VHF. This VHF transformed from a VDF farther upstream—the 2006 Eliot Branch lahar at Mount Hood (Oregon, USA) (Poole 2015). Coherence of the material attests to the silt and clay that is mixed with the sand. Shovel grip is 35 mm in diameter. View is looking downstream. USGS photo by T. Pierson

Fig. 23 Eruption-triggered (Apr 4, 2009) volcanic hyperconcentrated flow (VHF) deposit in Drift River Valley downstream of Redoubt Volcano (Alaska, USA) (Waythomas et al. 2013). Vertical exposure about 1 m high and about 30 km downstream of source. Note faint plane-parallel bedding, the vertical cut bank, and the relatively smooth undulating deposit surface with a non-continuous cap layer (lightest gray color)—all important VHF characteristics. Note that post-emplacement erosion of the deposit surface locally truncates the bedding within the deposit. USGS photo by T. Pierson

Fig. 24 Deposit of eruption-triggered (Apr 4, 2009) volcanic hyperconcentrated flow (VHF) in Drift River Valley downstream of Redoubt Volcano (Alaska, USA) (Waythomas et al. 2013). Beds of plane-parallel-bedded sand with two interbedded stringers of fine gravel mark initial deposition by the lahar. Sand beds grade upward to a more massive and coarser unit, emplaced by a more concentrated phase of the lahar. A cap of laminated fine sand and silt was emplaced on the surface, probably by a combination of dilute waning flow before channel fully incised and by porewater expulsion during consolidation. Tool shaft is 5 mm wide. USGS photo by T. Pierson

Fig. 25 Contact between two fine-grained lahar deposits of different ages (probable volcanic hyperconcentrated flow overbank facies) in upper Sandy River, about 20 km downstream of Mount Hood (Oregon, USA). Contact is marked by a thin organic layer and possible disrupted cap layer. Lower unit is ~ 1500-year-old lahar from the Timberline eruptive period, with moderately well-developed A and B soil horizons. Upper unit was emplaced by a lahar triggered at start of the Old Maid eruptive period in 1781. This is an example of a paleosol marking a disconformity. Scraper shaft is 5 mm wide. USGS photo by T. Pierson

Fig. 26 Unstratified sandy volcanic hyperconcentrated flow (VHF) overbank facies beds emplaced by the April 4, 2009, eruption-triggered lahar along the Drift River downstream of Redoubt Volcano (Alaska, USA). A bedding-parallel lamination surrounded by brownish stain (at point of scraper) separates two depositional subunits laid down by lahar surges; it is probably a consolidation lamination. Scraper shaft is 5 mm wide. USGS photo by T. Pierson

Fig. 27 Beds deposited by a single volcanic hyperconcentrated flow (VHF), emplaced on an aggrading floodplain (here preserved as a low terrace) of the West Fork Hood River following transformation of a volcanic debris flow (VDF) to VHF due to tributary inflow dilution in the 2006 Eliot Branch lahar at Mount Hood (Oregon, USA) (Poole 2015). The site is 24 km downstream of the lahar initiation point. The sequence here records deposition by two distinct lahar pulses that were deposited at least within an hour of each other and possibly within tens of minutes of each other (Poole 2015). The darker gray depositional unit below the sharp contact (marked by arrow) records the first pulse. Four centimeters above the arrow is a thin consolidation lamination that formed within the lighter gray depositional unit (emplaced by the second pulse) due to consolidation of the still-liquefied sediment. Both VHF units have (1) fine-grained sole layers (from the more dilute fronts of two separate flow pulses), (2) coarser massive central parts (reflecting passage of the most concentrated main bodies of the pulses), and (3) normally graded tops that reflect progressive dilution of the lahar pulse tails during deposition). The fine-grained sole layer of the second pulse unit provided the lowest-permeability layer within the sequence (Poole 2015), which forced escaping pore water to move laterally during consolidation to create the lamination. The very dark unit at the base of the section is a buried soil developed on the pre-emplacement floodplain surface. The pencil on the right side of the scraped section is 15 cm long. Flow was right to left. USGS photo by T. Pierson

Fig. 28 Closeup of basal contact of lower volcanic hyperconcentrated flow (VHF) pulse unit shown in Fig. 27, which shows the well-defined, inversely graded sole layer. The finest-grained lower part of the sole layer shows a wavy contact with the slightly coarser sediment above, reflecting undrained-loading-induced soft-sediment deformation. The upper part of the 5-mm-thick tool shaft is in the lower right corner of the photo for scale. The dark unit is the organic-rich soil that was developed on the top surface of the floodplain. Flow was right to left. USGS photo by T. Pierson

Fig. 29 Closeup of contact between lower and upper volcanic hyperconcentrated flow (VHF) depositional units and the consolidation lamination shown in Fig. 27. Pencil in lower right is 15 cm long. Flow was right to left. USGS photo by T. Pierson

Fig. 30 Closeup of complex soft-sediment deformation at the contact between the lower and upper volcanic hyperconcentrated flow (VHF) depositional units shown in Fig. 27. Still-liquefied sand was injected from the lower unit into the base of the upper unit during emplacement of the upper unit due to undrained loading of the still liquefied sediment within the lower unit. The consolidation lamination shown in Figs. 27 and 29 is highly deformed here. The scraper shaft is 5 mm wide. Flow was right to left. USGS photo by T. Pierson

Fig. 31 Soft-sediment deformation and water-escape structure at boundary between the two volcanic hyperconcentrated flow (VHF) depositional units shown in Fig. 27. Both flow pulses occurred within a single lahar—the 2006 Eliot Branch lahar at Mount Hood (Oregon, USA) (Poole 2015). The lower medium-grained sand (still liquefied at the time of loading) injected water and sand upward into the freshly emplaced coarse-sand/fine-gravel unit. The strength of the upper third of the section to overhang the lower part is attributed to stronger coherence resulting from higher fines content in upper unit. Thickness of the overhanging upper section is about 50 cm. Flow was right to left. USGS photo by T. Pierson

Fig. 32 Flat-surfaced volcanic hyperconcentrated flow (VHF) deposit in Drift River valley downstream of Redoubt Volcano (Alaska, USA). Surface is studded with small individual clasts and clast clusters. The exposed part of the largest clast in lower left quadrant is about 20 cm in diameter. The lahar was triggered by an explosive eruption on April 4, 2009. USGS photo by T. Pierson

Fig. 33 Plane-parallel beds, low-angle cross-beds, and scour structures in deposit emplaced by dilute volcanic hyperconcentrated flow (VHF) in the Chaitén (Blanco) River (Chile). Same outcrop as Fig. 32 but higher in section. Flow was approximately toward camera. Sequence is described in detail ("type B deposit") in (Pierson et al. 2013). Deposit appears to have been emplaced by progressively more dilute flow (from bottom to top of outcrop), interpreted by increased cross-bedding and channel scouring near the top of the section. Pencil is 15 mm long. USGS photo by T. Pierson

3 Deposits of Volcanic Debris Avalanches (VDAs)

See Figs. 34, 35, 36, 37, 38, 39, 40, 41, 42, 43, 44, 45, 46, 47, 48, 49, 50, 51, 52, 53, 54, 55, 56 and 57

Fig. 34 Volcanic debris avalanche (VDA) deposit emplaced in the upper North Fork Toutle River valley on May 18, 1980, at Mount St. Helens (Washington, USA). View is southeast toward the volcano, with the northward facing edifice-collapse failure scar clearly visible. River channel has cut into the VDA deposit. USGS photo by D. Wieprecht, Jun 16, 1996

Fig. 35 Side of hummock (exposed by erosion) on 1980 volcanic debris avalanche (VDA) deposit downslope of Mount St. Helens (Washington, USA). Person in lower left provides scale. Juxtaposed rock units of different colors here have gradational contacts. Note that it may be necessary to find a large enough outcrop to see color contrasts between adjacent debris blocks in VDA deposits. USGS photo by T. Pierson

Fig. 36 Side of hummock (exposed by erosion) on 1980 volcanic debris avalanche (VDA) deposit at Mount St. Helens (Washington, USA). Person in lower left provides scale. Contacts here are sharp between rock units of different color. Sharp contacts appear to be inherited from original stratigraphy within the failed edifice. USGS photo by T. Pierson

Fig. 37 Block-facies hummock of volcanic debris avalanche (VDA) deposit from Mount Shasta (California, USA) (Crandell 1989), with debris blocks containing pieces of the former edifice but transported about 20 km from source. Intrablock contacts are sharp. Deposit extends across much of valley floor in distance. Small bushes (~1 m tall) on the ground surface above outcrop face provide an approximate scale. Photo by L. Siebert

Fig. 38 Basal contact of an 18 ka volcanic debris avalanche (VDA) deposit from the Colima volcanic complex (Mexico), resting on underlying soil. The VDA had been transforming to a volcanic debris flow (VDF) when it deposited this debris. The VDA deposit dammed a river, and a breakout flood from the impounded water eroded and entrained much of the VDA deposit, producing the Naranjo Debris Flow that flowed 90 km farther downvalley (Capra and Macías 2002). At this outcrop (relatively near the lake-breakout VDF source), the VDA deposit (upper gray unit) was at least partially liquefied and ductile, and the large clasts within it are still angular. The VDF deformed and injected the underlying paleosol (black unit) with slurry. The large cylindrical cavity is a cast of a log that rotted away following burial. Pen below cavity is about 15 cm long. USGS photo by T. Pierson

Fig. 39 Block facies in Jurassic volcanic debris avalanche (VDA) deposit in Central Allan Hills, Antarctica (Reubi et al. 2005). Note the reddish jigsaw-fractured block (on the right, behind person) and the light-colored, stratified sandstone block (on the left) are partly suspended in gray mixed material. Photo by O. Reubi. used with permission

Fig. 40 Juxtaposed textures of a shattered black debris block above with more pervasively crushed and sheared brownish gray unit below in the block facies of a 7000 yr BP volcanic debris avalanche (VDA) deposit, Colima volcanic complex (Mexico) (Cortés et al. 2019). Scale on right side of card is in centimeters. Such abrupt changes in sedimentary texture are common in VDA deposits. Photo by T. Pierson

Fig. 41 Volcanic debris avalanche (VDA) deposit from Popocatépetl volcano (Mexico) (Siebe et al. 2017) with a fault offset of several meters in a wedge-shaped reddish debris-avalanche block. The fault demonstrates brittle deformation. Sheared-off tapered point of wedge is above the person. Photo by L. Siebert

Fig. 42 Units of variable color marking zones of shear within Barriles volcanic debris avalanche (VDA) deposit from Barú volcano (Panama). Shear bands go from upper left to lower right. (Herrick et al. 2013). Such shear zones are a common feature of VDA deposits. Note that only the reddish and gray areas had been scraped clean; brown areas are covered by slope wash. Scraper shaft is 5 mm wide. Photo by L. Siebert

Fig. 43 Side of hummock on 1980 volcanic debris avalanche (VDA) deposit at Mount St. Helens (Washington, USA). Rock units have different textures as well as different colors. Note that clasts are clustered and not randomly spaced—a characteristic common in VDA deposits. Clusters are typically formed by the fracturing of hard rock clasts but with only minimal shearing and displacement of the resulting smaller fragments during transport. Colored intervals on folding rule are 10 cm long. USGS photo by T. Pierson

Fig. 44 Block facies of 1980 volcanic debris avalanche (VDA) deposit in upper North Fork Toutle River valley, Mount St. Helens (Washington, USA). Note streaks of reddish color (arrows) from sheared soft clasts; harder gray clasts do not appear to be sheared. Internal shearing is common in matrix material of VDA deposits. Pen is 15 cm long. USGS photo by T. Pierson

Fig. 45 Block facies of 1980 volcanic debris avalanche (VDA) deposit in upper North Fork Toutle River, Mount St. Helens (Washington, USA). Note abrupt juxtaposition of different textures and colors, plus occurrence of clusters of closely spaced angular clasts that result from in situ breakage of larger clasts. Pen is 15 cm long. USGS photo by T. Pierson

Fig. 46 Eroded side of a hummock in the block facies of 1980 Mount St. Helens volcanic debris avalanche (VDA) deposit (Washington, USA). USGS photo by T. Pierson, 2006

Fig. 47 Aerial view of the distal part of the 1980 volcanic debris avalanche (VDA) deposit in upper North Fork Toutle River valley, Mount St. Helens (Washington, USA). Landscape reflects the effects of 32 years of erosion—muting of relief, rounding of hummocks, and incision by the river. USGS photo by K. Spicer, 2012

Fig. 48 Contact between two debris-avalanche blocks in the 1792 Mayu-yama volcanic debris avalanche (VDA) deposit, Unzen volcanic complex (Japan) (Ohta 1969). Red scalebar segments are 10-cm long. Note high degree of fracturing in rock within the gray unit. Contact between blocks here is relatively sharp, but gradational contacts between blocks in VDA deposits are also common. USGS photo by T. Pierson

Fig. 49 Mixed facies in a volcanic debris avalanche (VDA) deposit from Cofre de Perote volcano (Mexico) (Carrasco-Núñez et al. 2010). Clasts are polymictic, angular to subrounded; matrix material is well mixed and monochromatic. As is common in tropical settings, both clasts and matrix in deposits thousands of years old are deeply weathered and can be cut with a hand tool. Weathering occurred mostly after deposit emplacement (transport would have deformed large clasts). Scraper shaft is 6 mm wide. Photo by L. Siebert

Fig. 50 Interfingered contact between shattered debris-avalanche block (above) and finer-grained, more ductile interblock matrix (below) in volcanic debris avalanche (VDA) deposit at Colima volcanic complex (Mexico) (Cortés et al. 2019). Pen is 15 cm long. USGS photo by T. Pierson

Fig. 51 Light-colored mixed-facies matrix material injected through a debris-avalanche block composed of dark gray ignimbrite. Injection occurred near the contact between the ignimbrite (on left) and lava rock (on right), but it did not follow along the contact exactly. Outcrop is in the Pleistocene volcanic debris avalanche (VDA) deposit from Dittmar volcano within the Lassen Volcanic Center (California, USA) (Muffler and Clynne 2015). Scale bar is 2 m long. Photo by L. Siebert

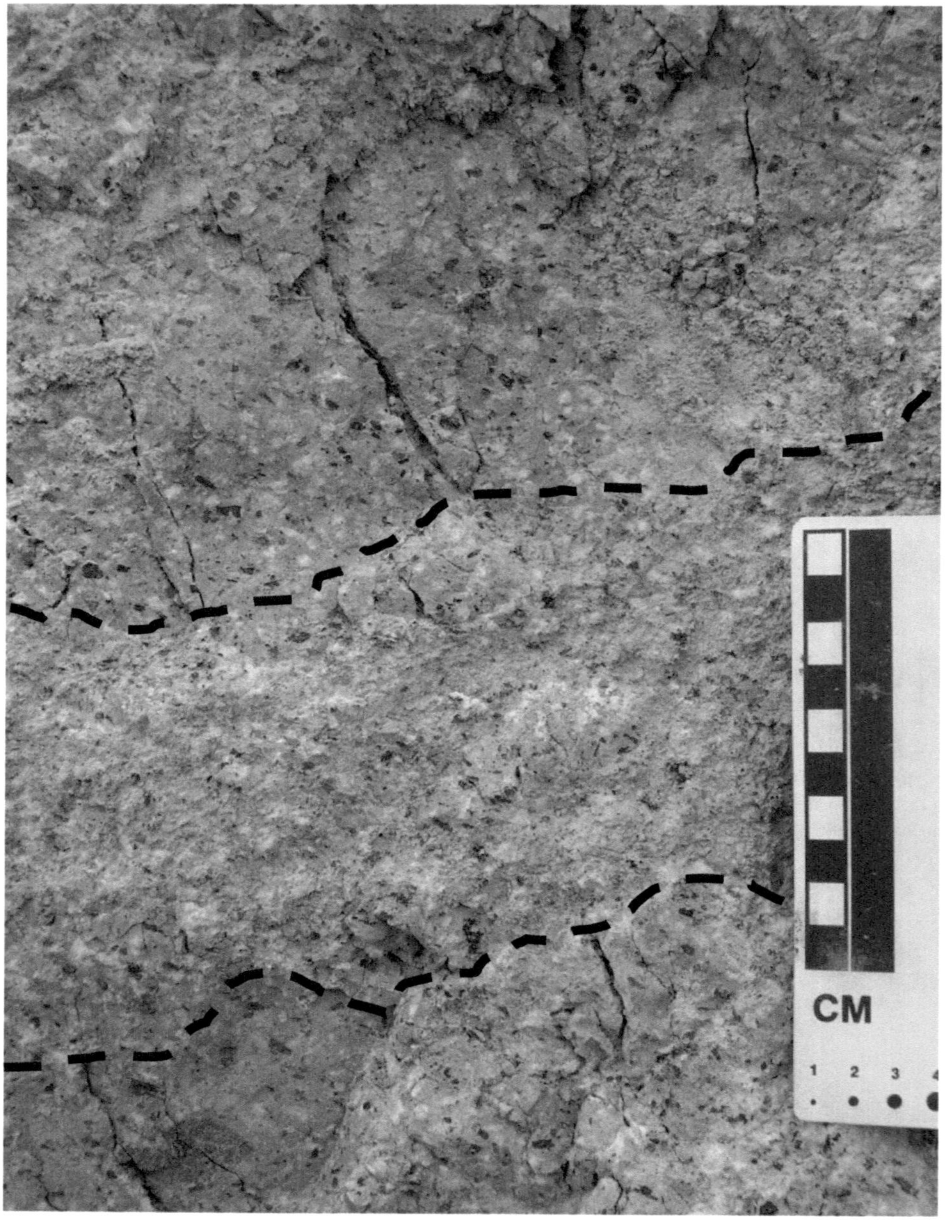

Fig. 52 Interior of large debris-avalanche block of fractured porphyritic dacite in the 1792 Mayu-yama volcanic debris avalanche (VDA) deposit, Unzen volcanic complex (Japan). The block appears to have been injected by similarly colored and textured crushed matrix material (outlined by dashes) during emplacement. Note truncation of fractures within the debris-avalanche block by the clastic dike. USGS photo by T. Pierson

Fig. 53 Jigsaw-fracture clast from an ancient, extremely large debris avalanche—the Oligocene Sevier megaslide in the Mrkagunt Plateau (Utah, USA) (Biek et al. 2019). Clast is 20-cm-wide. Note incipient fractures within clast fragments and separation of clast fragmens by injected mixed-facies matrix material. Photo by L. Siebert

Fig. 54 Jigsaw fractures in block of lava-dome porphyritic dacite in the 1792 Mayu-yama volcanic debris avalanche (VDA) deposit, Unzen volcanic complex (Japan). Note that little relative displacement of rock fragments occurred within this block during transport. Scraper shaft is 5 mm wide. USGS photo by T. Pierson

Fig. 55 Hummocky topography of the 1980 volcanic debris avalanche in the North Fork Toutle River valley about 15 km downstream of Mount St. Helens in 2006. Hummocks are partially vegetated but still actively eroding. USGS photo by T. Pierson

Fig. 56 Basal contact (at level of rock-hammer heads) of Riobamba volcanic debris avalanche (VDA) deposit from Chimborazo volcano (Ecuador) (Bernard et al. 2008). Note the truncated (planed off) dark clasts in substrate (to left of each hammer head) that were sheared off by the VDA. Dark elongated cavities in the outcrop are of unknown origin, possibly fossil animal burrows. Photo by L. Siebert

Fig. 57 Sharp basal contact of a Pleistocene volcanic debris avalanche (VDA) deposit from Dittmar volcano within the Lassen Volcanic Center (California, USA (Muffler and Clynne 2015). The 6-m-wide dark block of ignimbrite entrained in the VDA was planed off at the basal contact, indicating that shear was concentrated in a narrow shear zone at the base of the VDA and not distributed higher up in the avalanche. Photo by L. Siebert

4 Deposits of Volcanic Water Floods (VWFs)

See Figs. 58, 59, 60, 61, 62, 63, 64, 65, 66, 67, 68, 69, 70, 71, 72, 73 and 74.

Fig. 58 Plane-parallel-bedded, channel-facies coarse sand and fine gravel emplaced downstream of Mount Hood (Oregon, USA) by volcanic water flood (VWF) following the 1781–1793 eruption (Pierson et al. 2011). Riverbed aggradation due to excessive sediment loading of the Sandy River raised the riverbed about 23 m at this location, about 70 km downstream of the volcano. A debris fan fed by dome-collapse pyroclastic density currents and lahars provided loose, erodible sediment during heavy rain. Flow was right to left. Pocketknife is about 6 cm long. USGS photo by T. Pierson

Fig. 59 Volcanic water flood (VWF) deposit (channel facies) comprised of plane-parallel-bedded and cross-bedded coarse sand with interbedded coarse gravel lens. Deposit emplaced during prolonged channel aggradation during and following the 1781–1793 eruption of Mount Hood (Oregon, USA). Sediment source was about 70 km upstream (Pierson et al. 2011). Flow was left to right. Shovel shaft above blade is 3 cm in diameter. USGS photo by T. Pierson

Fig. 60 Finely bedded medium to coarse sand in a volcanic water flood (VWF) deposit in same outcrop exposure as in Fig. 59. Small dune bedforms and scour depressions are visible (lower regime flow). Flow was left to right. Historic accounts during this prolonged phase of posteruptive channel aggradation (Moulton and Dunlay 1990; Pierson et al. 2011) noted that flow was distributed among multiple braided channels and very shallow (10–20 cm). USGS photo by T. Pierson

Fig. 61 Finely bedded medium to coarse aggradational sand in a volcanic water flood (VWF) deposit in the same large outcrop as in Fig. 59 but at a different orientation to flow direction. Outcrop shows high-angle trough cross-bedding (created by three-dimensional dunes), with flow direction toward camera. Black pencil in center is 15 cm long. USGS photo by T. Pierson

Fig. 62 Two faintly bedded, medium- to coarse-grained sandy volcanic water flood (VWF) depositional units (channel facies) overlying a cobbly gravel bar in the Hood River, about 39 km downstream of Mount Hood (Oregon, USA). The sand units were emplaced by two sediment pulses within a single muddy flood that transformed from volcanic hyperconcentrated flow; the contact between them is marked by the arrow. This VWF's high sediment load was fed by the 2006 Eliot Branch lahar (Poole 2015), which was triggered by extremely heavy rainfall. A photo of the flood water surface at the approximate time and place this deposit was emplaced can be seen in Fig. 1c in chap "Volcanic Water Floods—Process Overview". The sand deposits show faint plane-parallel bedding in this freshly scraped outcrop (the sand is well sorted). Flow was toward the camera. Pencil is 15 cm long. USGS photo by T. Pierson

Fig. 63 Faintly plane-parallel-bedded, medium-grained sand in a volcanic water flood (VWF) deposit (channel facies) emplaced as a dilute lahar runout, following upstream transformation from debris flow to hyperconcentrated flow to water flood (Poole 2015). Location is about 40 km downstream of lahar source area on Mount Hood (Oregon, USA) and 0.7 km downstream of exposure in Fig. 62, on a midchannel sandbar in the Hood River. Flow was toward the camera. Pencil barrel is 8 mm wide. USGS photo by T. Pierson

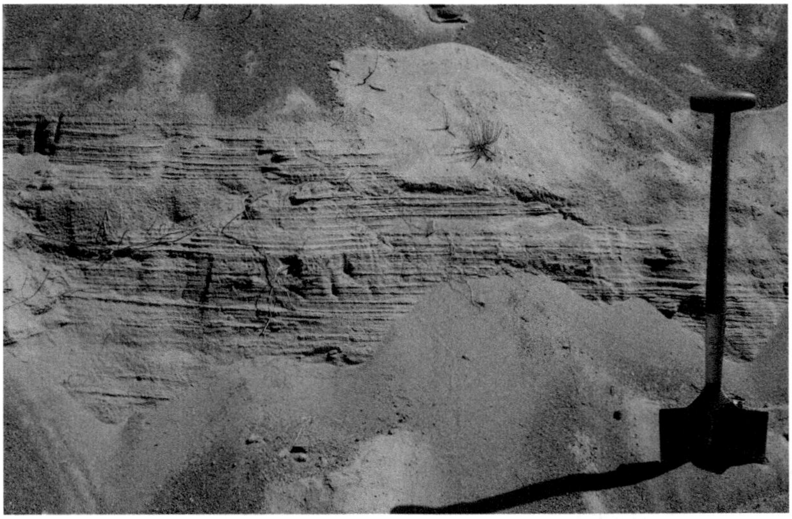

Fig. 64 Same volcanic water flood (VWF) sand beds as shown in Fig. 63, prior to scraping the outcrop face. The photo shows that weathered outcrop faces can reveal more bedding detail than cleanly scraped outcrops, especially in fluvial sands. Relief here between beds is due to slight variations in silt content among beds. Shovel blade is 17 cm wide. Flow was toward the camera. USGS photo by T. Pierson

Fig. 65 Well-stratified, loose/soft volcanic water flood (VWF) deposit (overbank facies) from the VWF tail of a dilute lahar that occurred immediately following the explosive phase of the 2008–2009 eruption of Chaitén volcano (Chile) (Pierson et al. 2013). Deposits emplaced by the hyperconcentrated-flow phase of the event aggraded and largely filled the nearby river channel, forcing the following muddy-flood phase of the lahar to overflow onto an alluvial fan surface where the town of Chaitén was built, depositing these sediments. An overhanging roof from the partly buried house in background prevented rainwash from eroding this deposit. Darkest material at bottom of pit is the pre-event soil surface. Shovel is 58 cm long. USGS photo by T. Pierson

Fig. 66 Well-stratified, loose/soft volcanic water flood (VWF) deposit (sheetflood facies) in a channel cut into a recently graded debris apron of lahar and VWF deposits on the lower west flank of Villarrica volcano (Chile). Deposit was emplaced during its 2015 eruption (Johnson and Palma 2015). Cut-and-fill channels (arrow) are common features that interrupt plane-parallel bedding in sheetflood-facies VWF deposits. Scraper blade about 12 cm wide. USGS photo by T. Pierson

Fig. 67 Well-rounded pebble-cobble fluvial gravel armoring the surface of a longitudinal bar in the Sandy River, about 70 km downstream of Mount Hood (Oregon, USA), about 200 years after this watershed experienced volcanic disturbance. View is vertically downward; pencil (15 cm long) lying on a clast is pointing right in flow direction. Imbricated elongated clasts are oriented transverse to flow direction (imbrication not visible). Finer-grained gravel and sand beneath a thin armor layer is common in rivers no longer experiencing high sediment loads. USGS photo by T. Pierson

Fig. 68 Plane-parallel- bedded coarse basaltic gravelly sand with outsized gravel clasts in volcanic water flood (VWF) deposit, emplaced by a huge eruption-triggered glacier outburst flood (jökulhlaup) from the Vatnajökull ice cap in southeast Iceland on November 5, 1996 (Guðmundsson et al. 1997). This deposit sits in a braid plain several kilometers wide. Pen is 15 cm long. USGS photo by T. Pierson

Fig. 69 Unstratified fluvial sandy gravel deposit emplaced by the volcanic water flood (VWF) resulting from a huge, eruption-triggered, glacier outburst flood (jökulhlaup) from the Vatnajökull ice cap in an extremely wide braided channel in southeast Iceland on November 5, 1996 (Guðmundsson et al. 1997). Matrix material was friable. USGS photo by T. Pierson

Fig. 70 Plane-parallel bedding and planar two-dimensional (2-D) cross-beds (formed by downstream-migrating 2-D dunes) in loose, medium- to coarse-grained sand deposits of a heavily sediment-laden syneruptive volcanic water flood (VWF) in the Negro River, downstream of Chaitén volcano (Chile) (Pierson et al. 2013). Flow was left to right. Pencil, 15-cm long, for scale. USGS photo by T. Pierson

Fig. 71 Cross-beded and plane-parallel-bedded coarse gravelly sand with interbedded lenses of fluvial gravel, emplaced by the volcanic water flood (VWF) phase of a syneruptive lahar in the Chaitén (Blanco) River downstream of Chaitén volcano (Chile). Such abrupt changes in the mean grain size of beds are relatively common in VWF deposits. Trenching shovel blade is 16 cm wide. USGS photo by J.J. Major

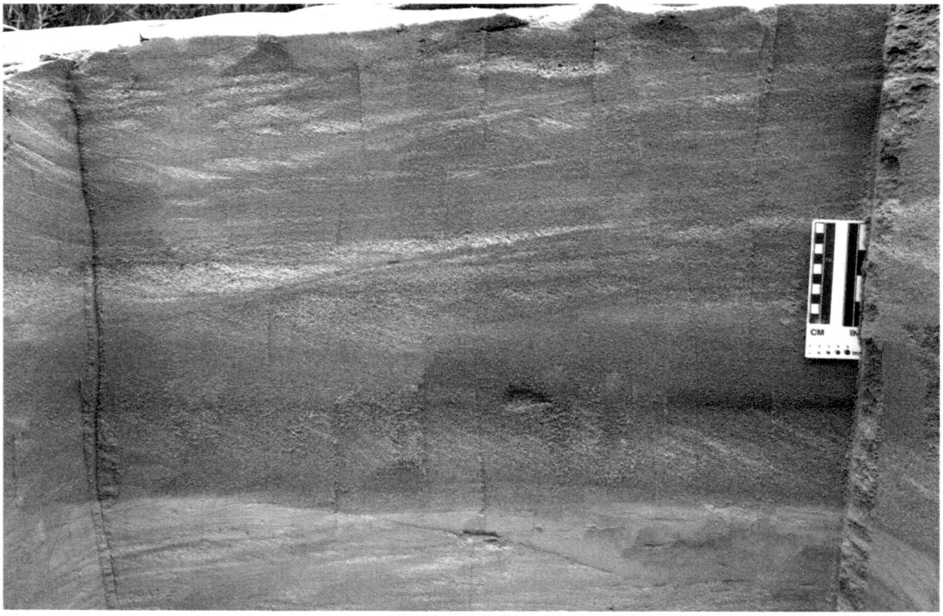

Fig. 72 Volcanic water flood (VWF) deposit (channel facies) of medium- to coarse-grained sand in the Negro River downstream of Chaitén volcano (Chile) (Pierson et al. 2013). Bed forms in upper part of outcrop face (above top of scale) exhibit plane-parallel bedding and climbing ripples. The middle of the face shows two-dimensional dune cross-bedding (high-angle cross-beds) and faint plane-parallel bedding. Flow was left to right. Outcrop was excavated in a flood berm along the left bank of the channel. USGS photo by T. Pierson

Fig. 73 Plane-parallel-bedded deposit of dense, glassy, rhyolite grains and pumice emplaced by a sustained, concentrated volcanic water flood (VWF) (channel facies), near its transition point to volcanic hyperconcentrated flow (VHF) in the rapidly aggrading Chaitén (Blanco) River, 10 km downstream of Chaitén volcano (Chile). This flood occurred immediately following widespread tephra deposition by the explosive phase of the 2008–2009 eruption. Flow was diagonally away (to right) from camera. The sequence is described in detail ("type C deposit") in (Pierson et al. 2013). Note density segregation of pumice grains at tops of individual beds. Similar looking VHF deposits were emplaced lower in this stratigraphic section (e.g., Figs. 32, and 33), indicating that a gradual transformation from VHF to VWF was recorded in this stratigraphic section. The VHF deposits are markedly firmer in consistency and more poorly sorted than these beds. Shovel shaft is about 3.5 cm in diameter. USGS photo by T. Pierson ▶

5

Fig. 74 Plane-parallel-bedded deposit of sand-size dense, glassy, rhyolite grains and gravel-size pumice clasts emplaced by a concentrated volcanic water flood (VWF) in the rapidly aggrading Chaitén (Blanco) River, 9 km downstream of Chaitén volcano (Chile) at the peak level of channel aggradation. Density segregation here is more pronounced than for the deposit in an outcrop 1 km downstream at a slightly lower stratigraphic level in Fig. 73. USGS photo by T. Pierson

5 Deposits of Pyroclastic Density Currents (PDCs)

See Figs. 75, 76, 77, 78, 79, 80, 81, 82, 83 and 84.

Fig. 75 Massive (unstratified) pumiceous pyroclastic density current (PDC) deposit—the Granadillo Ignimbrite, Tenerife (Canary Islands) (Bryan et al. 2000). It has a matrix-supported framework and is poor in lithic fragments. Particle-size distribution is strongly bimodal, but note that the small black lithic fragments (basalt) do not exhibit density-segregation grading with respect to the pumice clasts. Nor is there strong grading among pumice clasts alone. Pen is about 15 cm long. USGS photo by T. Pierson

Fig. 76 Unstratified lithic pyroclastic density current (PDC) deposit on lower flank of Colima volcano complex (Mexico), which was emplaced in 1818. It is a monomictic mixture and pebble-size clasts show slight rounding. Sedimentary texture closely resembles that of a volcanic debris flow (VDF) deposit, but the matrix material here is very friable (uncharacteristic of VDFs). USGS photo by T. Pierson

Fig. 77 Unstratified lithic pyroclastic density current (PDC) deposit (dome-collapse dacite block-and-ash flow) in White River valley, Mount Hood (Oregon, USA) (Scott and Gardner 2017). Nearest large block is about 1 m in diameter). Closeup of same deposit as shown in Fig. 10a in Chap "Other Processes Depositing Volcanic Sediment". This valley-confined deposit contains variably vesiculated dacite and is at least 10 m thick. Sedimentary texture resembles that of a volcanic debris-flow deposit, but the matrix material is friable. Photo by T. Pierson

Fig. 78 Diffusely plane-parallel-bedded scoriaceous pyroclastic density current (PDC) deposit on flank of a tuff cone, Halla volcano (South Korea) (Sohn 1996). This deposit, presumably emplaced by pulses of relatively small-volume dilute PDCs, closely resembles some VHF deposits. USGS photo by T. Pierson

Fig. 79 Low-angle cross-bedding in a sequence of distinctly stratified, scoriaceous, dilute pyroclastic density current (PDC) deposits emplaced on the flank of a tuff cone, Halla volcano (South Korea) (Sohn 1996). This deposit closely resembles some volcanic hyperconcentrated-flow and volcanic water-flood deposits. Note the truncation of bedding surfaces and bed forms by scour. USGS photo by T. Pierson

Fig. 80 Low-angle cross-bedding in distinctly stratified, scoriaceous, dilute pyroclastic density current (PDC) strata associated with plane-parallel- bedded deposits on a tuff cone on Halla volcano (South Korea) (Sohn 1996). This deposit closely resembles some volcanic hyperconcentrated-flow deposits. USGS photo by T. Pierson

Fig. 81 Base of the Poris Ignimbrite on Tenerife (Canary Islands) (Brown and Branney 2004). This is the light-colored unit in the upper two thirds of the image. The diffusely bedded and clast-depleted fine-grained basal layer grads upward into the ungraded main body of the ignimbrite. The ignimbrite overlies bedded tephra-fall pumice and accretionary lapilli tuffs. Pen is about 15 cm long. USGS photo by T. Pierson

Fig. 82 Proximal pyroclastic density current (PDC) deposit emplaced during the initial boiling-over phase of May 18, 1980, eruption of Mount St. Helens (Washington, USA) (Brand et al. 2016). Deposit is located on the upper northeast flank of the volcano. Outcrop shows steep regressive bed forms with stoss-side particle accretion forming a series of steeply dipping backset beds (flow direction was right to left). Scale bar is 52 cm long. Photo by B. Brand, used with permission

Fig. 83 Multiple bedded and massive proximal pumiceous pyroclastic density current (PDC) deposits on upper NE flank of Mount St. Helens (Washington, USA), emplaced on May 18, 1980 (Brand et al. 2014, 2016). The depositional surface was sloping ~ 15°. Both sharp and diffuse contacts separate the deposits. Bedding is both plane-parallel and cross-bedded. Featureless material at bottom of photo is ashy talus. Note the pronounced scour surface near the top of the outcrop. The trekking pole is about 1.1 m long. Photo by B. Brand, used with permission

Fig. 84 Thick ungraded, unstratified pyroclastic density current (PDC) deposit containing variably vesiculated basaltic andesite clasts, emplaced during the 2015 sub-Plinian eruptions of Calbuco volcano (Chile) (Castruccio et al. 2016). Note angularity of large clasts, the non-uniformity of large-clast spacing, and the vertical gas-escape pipes. Person is 1.75 m tall. Photo by B. Brand, used with permission

6 Deposits of Tephra Falls (TFs)

Figs. 85, 86, 87, 88, 89, 90, 91 and 92.

Fig. 85 Two unconformable sequences of bedded tephra fall (TF) deposits on the lower flank of Chimborazo volcano (Ecuador). Note the sharp contacts between depositional units and the diffuse bedding within units. Also note the uniformity of bed thickness and lithofacies at this scale, even where deposited on a slope. The main unconformity was caused by erosion from a late Pleistocene glacial advance. Photo by L. Siebert

Fig. 86 Part of stratigraphic section in Truful Truful River Gorge near Llaima volcano (Chile). Thick white layer (est. 3–4 m thick) is an 8.8 ka Plinian tephra fall (TF) deposit, which is overlain by a thicker, light-gray layer of dilute pyroclastic density current (PDC) deposits showing low-angle dune-type bed forms (Schindlbeck et al. 2014). Darker deposits above and below are volcano-fluvial and thin TF deposits. Photo by T. Pierson

Fig. 87 Thick, unstratified, ungraded, proximal basaltic tephra fall (TF) deposit from the 1783 Laki fissure eruption (Grimsvötn fissure system) (Thordarson et al. 2003). This deposit of mainly scoria fragments is quite poorly sorted for a TF deposit. USGS photo by T. Pierson

Fig. 88 Basaltic ribbon bombs (near-source spatter clasts) from a small rootless spatter cone developed on a large lava flow moving over swampy ground during the 934 CE eruption from Eldgja volcano, southeast Iceland. USGS photo by T. Pierson. These ribbon bombs did not have glassy rinds, but they were ejected while sufficiently molten to twist and deform while in flight but cooled and solidified quickly enough to be only slightly deformed or undeformed by landing impact

Fig. 89 Thin tephra fall (TF) ash beds on Whakaari/White Island volcano (Bay of Plenty, New Zealand) from small phreatomagmatic eruptions. Pocket knife case is about 8 cm long. USGS photo by T. Pierson, 1992

Fig. 90 Unstratified, ungraded, tephra fall (TF) deposit, dominantly rhyolite pumice, erupted from Hekla volcano (Iceland) during the Hekla 3 eruption, which occurred about 2,800 yr BP (Larsen et al. 2020). The light-colored pumice fragments are highly vesicular. Unit contains little ash matrix material between the lapilli. USGS photo by T. Pierson

Fig. 91 Section of proximal tephra fall (TF) deposits emplaced during the initial explosive phase of the 2008–2009 eruption of Chaitén volcano (Chile). End of measuring tape rests on pre-eruption ground surface and bends (at 21-inch mark) onto the tephra deposit surface. Note the leaves and twigs mixed into the lowest lapilli-rich unit—material stripped from forest trees during tephra fall. Topmost layers are composed mainly of very fine to extremely fine ash, which led to rapid rainfall runoff (due to low infiltration capacity) and a dramatic hydrogeomorphic response in the watershed to the eruption (Pierson et al. 2013). Another TF deposit emplaced by this same eruption but at another more distal location (Fig. 22a in Chap. "Other Processes Depositing Volcanic Sediment") appears quite different. Sheetwash erosion undoubtedly removed some of the top tephra layers at this location following deposition, based on evidence from other nearby tephra exposures. USGS photo by R. Hoblitt, Jan 24, 2010

Fig. 92 Small angular ballistic fragments (andesite) collected several kilometers from the vent of Sakurajima volcano (Japan) in 1986 following vent pressurization and explosive fragmentation and ejection of a solidified lava plug. The pen is 14 cm long. USGS photo by T. Pierson

7 Deposits of Glaciers

Figs. 93, 94, 95, 96, 97, 98, 99, 100 and 101.

Fig. 93 Dark gray Pleistocene lodgment till on lower flank of Mount Baker (Washington, USA). Note the uneven spatial distribution of boulder in the outcrop face and the smoothed, sculpted bedrock surface the deposit sits on. The till is overlain by a brownish orange volcanic debris-avalanche deposit. Gray color of river water is due to fines (glacial flour) suspended in the water. Outcrop face is approximately 10 m high. Photo by D.S. Tucker, used with permission

Fig. 94 Hand-held sample of overconsolidated (very hard), Pleistocene lodgment till on shoulder of Nisqually River valley at Mount Rainier (Washington, USA)—the same deposit as in Fig. 33b, c in Chap. "Other Processes Depositing Volcanic Sediment". This till matrix is rich in fines and is so highly compacted that it is nearly lithified; it was difficult to break with a hammer. USGS photo by T. Pierson

Fig. 95 The locally heterogeneous matrix of the overconsolidated Pleistocene lodgment till shown in Figs. 93 and 94, shown here in closeup, illustrates how tightly clay-rich till matrix material clings to clasts. It shows a clay rind caking a clast (outlined) and a sand pocket beneath the clast (arrow). In contrast, matrix material in lahar deposits is homogeneous and separates easily from clasts. The blade of the key is 7 mm wide. USGS photo by T. Pierson

Fig. 96 Lodgment till emplaced by Puget Lobe of the Cordilleran Ice Sheet at Double Bluff, Whidbey Island (Washington, USA). Note the combination of well-rounded and angular clasts that protrude from the fine-grained matrix. The ice sheet traversed a broad alluvial lowland and, as a result, the deposit is rich in clasts of well-rounded river gravel. The matrix material here is rich in silt and clay and the clasts are widely dispersed in the matrix, which is typical of lodgment tills emplaced by continental ice sheets. Hammer is about 35 cm long. Photo by D.S. Tucker, used with permission

Fig. 97 Overconsolidated, matrix-supported, Pleistocene lodgment till, emplaced by the Cordilleran Ice Sheet, has a fine-grained and slightly flaky texture and contains some exotic clasts. Outcrop is exposed in Baker River valley near mouth of Sulphur Creek on the lower southeast flank of Mount Baker (Washington, USA). Note that the gray color of this till is typical of the reducing conditions in many low-permeability, clay-rich tills in humid terrain. Knife handle is about 9 cm long. Photo by D.S. Tucker, used with permission

Fig. 98 Hummocky ground moraine on valley floor and a sharp-crested lateral moraine behind it downstream of the Tahoma Glacier in Mount Rainier National Park (Washington, USA). Glacially striated bedrock is in foreground in lower left. USGS photo by T. Pierson

Fig. 99 Ablation till (lower deposit) from late Pleistocene alpine glacier in Boulder Creek, lower east flank of Mount Baker volcano (Washington, USA) (Scott et al. 2020). Till is overlain by stratified deposits of the Boulder Creek Assemblage (lahar and volcano-fluvial deposits). Note angularity of boulder-size clasts (typical in alpine tills). A backpack (circled) at bottom center of photo gives the scale. If this were a VDF deposit, the boulder-size clasts would be crudely subrounded. Photo by D.S. Tucker, used with permission

Fig. 100 Ablation tills in a lateral moraine downstream of Deming Glacier in the Nooksack River valley on southwest flank of Mount Baker volcano (Washington, USA). Note two layers of buried trees (arrows), which mark the base of each till. Woody debris buried by glaciers in moraines typically occurs in discreet layers, whereas in lahars, woody debris is generally concentrated at flow margins but can be distributed throughout the deposit. Note people for scale (circled) near base of slope. Photo by D.S. Tucker, used with permission

Fig. 101 Pleistocene glacial outwash fan-delta foreset deposits emplaced by glacial meltwater streams emanating from a continental ice sheet. Outcrop is located near Squamish, British Columbia (Canada) and shows the characteristic coarse texture (commonly with an open-work framework) of gravelly outwash deposits. Bedding is marked by abrupt changes in average grain size, which is typical of most coarse-grained fluvial deposits. In valleys draining volcanoes, outwash deposits may be interbedded with volcanic water flood (VWF) deposits and appear very similar, but VWF deposits typically contain some angular/subangular gravel, due to mobilization of juvenile rock fragments. Gravel in glacial outwash deposits is typically rounded/well-rounded throughout. USGS photo by T. Pierson

References

Benn DI, Evans DJA (2010) Glaciers and Glaciation, 2nd edn. Hodder Education, London, p 734

Bernard, B, van Wyk de Vries B, Barba D, Leyrit H, Robin C, Alcaraz S, Samaniego P (2008) The Chimborazo sector collapse and debris avalanche: deposit characteristics as evidence of emplacement mechanisms. J Volcanol Geotherm Res 176(1):36–43. https://doi.org/10.1016/j.jvolgeores.2008.03.012

Biek RF, Rowley PD, Hacker DB (2019) The gigantic Markagunt and Sevier gravity slides resulting from mid-Cenozoic catastrophic mega-scale failure of the Marysvale volcanic field, Utah, USA. Geol Soc Am Field Guide 56:1–121

Brand BD, Mackaman-Lofland C, Pollock NM, Bendaña S, Dawson B, Wichgers P (2014) Dynamics of pyroclastic density currents: conditions that promote substrate erosion and self-channelization—Mount St Helens, Washington (USA). J Volcanol Geotherm Res 276:189–214. https://doi.org/10.1016/j.jvolgeores.2014.01.007

Brand BD, Bendaña S, Self S, Pollock N (2016) Topographic controls on pyroclastic density current dynamics: insight from 18 May 1980 deposits at Mount St. Helens, Washington (USA). J Volcanol Geotherm Res 321:1–17. https://doi.org/10.1016/j.jvolgeores.2016.04.018

Branney MJ, Kokelaar P (2002) Pyroclastic density currents and the sedimentation of ignimbrites. Geol Soc Lond Mem 27:1–143

Brown RJ, Branney MJ (2004) Event-stratigraphy of a caldera-forming ignimbrite eruption on Tenerife: the 273 ka Poris formation. Bull Volcanol 66:392–416. https://doi.org/10.1007/s00445-003-0321-y

Bryan SE, Cas RAF, Martí J (2000) The 0.57 Ma plinian eruption of the Granadilla Member, Tenerife (Canary Islands): an example of complexity in eruption dynamics and evolution. J Volcanol Geotherm Res 103(1–4):209–238. https://doi.org/10.1016/S0377-0273(00)00225-0

Capra L, Macías JL (2002) The cohesive Naranjo debris-flow deposit (10 km^3): a dam breakout flow derived from the Pleistocene debris-avalanche deposit of Nevado de Colima Volcano (México). J Volcanol Geotherm Res 117(1–2):213–235. https://doi.org/10.1016/S0377-0273(02)00245-7

Carrasco-Núñez G, Siebert L, Díaz-Castellón R, Vázquez-Selem L, Capra L (2010) Evolution and hazards of a long-quiescent compound shield-like volcano: Cofre de Perote, Eastern Trans-Mexican Volcanic Belt. J Volcanol Geotherm Res 197(1–4):209–224. https://doi.org/10.1016/j.jvolgeores.2009.08.010

Cas RAF, Wright JV (1987) Volcanic successions modern and ancient. Allen and Unwin, London, p 528

Castruccio A, Clavero J, Segura A, Samaniego P, Roche O, Le Pennec JL, Droguett B (2016) Eruptive parameters and dynamics of the April 2015 sub-Plinian eruptions of Calbuco volcano (southern Chile). Bull Volcanol 78:62. https://doi.org/10.1007/s00445-016-1058-8

Cortés A, Komorowski J-C, Macías J, Capra L, Layer P (2019) Late Pleistocene-Holocene debris avalanche deposits from Volcán de Colima, Mexico: portrait of a persistently hazardous volcano. https://doi.org/10.1007/978-3-642-25911-1_4

Crandell DR (1989) Gigantic debris avalanche of Pleistocene age from ancestral Mount Shasta volcano, California, and debris-avalanche hazard zonation, vol 1861. US Geology Survey Bulletin, 32 pp

Dinehart RL (1999) Sediment transport in the hyperconcentrated phase of the March 19, 1982, lahar. In: Pierson TC (ed) Hydrologic consequences of hot-rock/snowpack interactions at Mount St. Helens volcano, Washington, 1982–84, vol 1586. US Geology Survey Professional Paper, pp 37–52

Ehlers J (1996) Quaternary and glacial geology. John Wiley & Sons, Chichester, p 578

Fisher RV, Schmincke H-U (1984) Pyroclastic rocks. Springer, Berlin, p 472

Flint RF (1971) Glacial and Quaternary geology. John Wiley and Sons, New York, p 892

Francis PW (1993) Volcanoes – A Planetary Perspective. Clarendon Press (Oxford University Press), New York, p 443

Guðmundsson MT, Sigmundsson F, Björnsson H (1997) Ice-volcano interaction of the 1996 Gjálp subglacial eruption, Vatnajökull, Iceland. Nature 389:954–957

Herrick JA, Siebert L, Rose WI (2013) Large-volume Barriles and Caisán debris avalanche deposits from Volcán Barú, Panama. Geol Soc Am Spec Pap 498:1–22. https://doi.org/10.1130/2013.2498(09)

Johnson JB, Palma JL (2015) Lahar infrasound associated with Volcán Villarrica's 3 March 2015 eruption. Geophys Res Lett 42(15):6324–6331. https://doi.org/10.1002/2015GL065024

Larsen G, Róbertsdóttir BG, Óladóttir BA, Eiríksson J (2020) A shift in eruption mode of Hekla volcano, Iceland, 3000 years ago: two-coloured Hekla tephra series, characteristics, dispersal and age. J Quat Sci 35(1–2):143–154. https://doi.org/10.1002/jqs.3164

McPhie J, Doyle M, Allen R (1993) Volcanic textures – A guide to the interpretation of textures in volcanic rocks. Centre for Ore Deposit and Exploration Studies, University of Tasmania, Hobart, p 196

Menzies J, van der Meer JJM, Shilts WW (2018) Subglacial processes and sediments. In: Menzies J, van der Meer JJM (eds) Past glacial environments, 2nd edn. Elsevier, Amsterdam, pp 105–158

Miyabuchi Y (1999) Deposits associated with the 1990–1995 eruption of Unzen volcano, Japan. J Volcanol Geotherm Res 89:139–158

Mothes PA, Hall ML, Janda RJ (1998) The enormous Chillos Valley lahar: an ash-flow-generated debris flow from Cotopaxi volcano, Ecuador. Bull Volcanol 59:233–244

Moulton GE, Dunlay TW, eds. (1990) J Lewis Clark Expedition 6:12–14; (November 2, 1805–March 22, 1806) 7:48–49; (March 23, 1806–April 17, 1806). Lincoln, University of Nebraska Press

Muffler LJP, Clynne MA (2015) Geologic field-trip guide to Lassen Volcanic national park and vicinity, California. US Geol Surv Sci Invest Rep 2015–5067:67. https://doi.org/10.3133/sir20155067

Ohta K (1969) Study on the collapse of Mayu-Yama: on the mechanism for collapse. Fac Sci Kyushu Univ (in Jpn with Engl Abstr), Sci Shimabara Inst Volcanol Balneol

Pierson TC (1985) Initiation and flow behavior of the 1980 Pine Creek and Muddy River lahars, Mount St. Helens. Washington. Geol Soc Am Bull 96:1056–1069. https://doi.org/10.1130/0016-7606(1985)96<1056:IAFBOT>2.0.CO;2

Pierson TC, Scott KM (1985) Downstream dilution of a lahar: transition from debris flow to hyperconcentrated streamflow. Water Resour Res 21:1511–1524

Pierson TC, Pringle PT, Cameron KC (2011) Magnitude and timing of downstream channel aggradation in response to a dome-building eruption at Mount Hood, Oregon. Geol Soc Am Bull 123:3–20

Pierson TC, Major JJ, Amigo Á, Moreno H (2013) Acute sedimentation response to rainfall following the explosive phase of the 2008–2009 eruption of Chaitén volcano, Chile. Bull Volcanol 75:723. https://doi.org/10.1007/s00445-013-0723-4

Pierson TC (1999) Transformation of water flood to debris flow following the eruption-triggered transient-lake breakout from the crater on March 19, 1982. In: Pierson TC (ed) Hydrologic consequences of hot-rock/snowpack interactions at Mount St. Helens volcano, Washington 1982–84, vol 1586. US Geology Survey Professional Paper, pp 19–36

Pierson TC, Scott WE, Vallance JW, Pringle PT (2009) Eruption-related lahars and sedimentation response downstream of Mount Hood: field guide to volcaniclastic deposits along the Sandy River, Oregon. In: O'Connor JE, Dorsey RJ, Madin IP (eds) Volcanoes to vineyards: geologic field trips through the dynamic landscape of the Pacific Northwest, vol 15. Geology Society America Field Guide, pp 229–244

Poole M (2015) Sedimentologic changes in the deposits of an evolving lahar-flood, Hood River Basin, Mount Hood, Oregon. MSci Thesis, Portland State University, 130 pp

Reubi O, Ross P-S, White JDL (2005) Debris avalanche deposits associated with large igneous province volcanism: an example from the Mawson Formation, Central Allan Hills, Antarctica. Geol Soc Am Bull 117:1615–1628

Schindlbeck JC, Freundt A, Kutterolf S (2014) Major changes in the post-glacial evolution of magmatic compositions and pre-eruptive conditions of Llaima Volcano, Andean Southern Volcanic Zone, Chile. Bull Volcanol 76:830. https://doi.org/10.1007/s00445-014-0830-x

Schmincke H-U (2004) Volcanism. Springer, Berlin, p 324

Scott KM (1988a) Origins, behavior, and sedimentology of lahars and lahar-runout flows in the Toutle-Cowlitz River system, vol 1447-A. US Geology Survey Professional Paper, 74 pp

Scott WE, Gardner CA (2017) Field-trip guide to Mount Hood, Oregon, highlighting eruptive history and hazards. US Geol Surv Sci Invest Rep 2017–5022–G:1–115. https://doi.org/10.3133/sir20175022g

Scott KM, Tucker DS, Riedel JL, Gardner CA, McGeehin JP (2020) Latest Pleistocene to present geology of Mount Baker Volcano, northern Cascade Range, Washington, vol 1865. U S Geology Survey Professional Paper, 170 pp

Siebe C, Salinas S, Arana-Salinas L, Macías JL, Gardner J, Bonasia R (2017) The ~23,500 y 14C BP White Pumice Plinian eruption and associated debris avalanche and Tochimilco lava flow of Popocatépetl volcano, México. J Volcanol Geotherm Res 333–334:66–95. https://doi.org/10.1016/j.jvolgeores.2017.01.011

Sohn YK (1996) Hydrovolcanic processes forming basaltic tuff rings and cones on Cheju Island, Korea. Geol Soc Am Bull 108(10):1199–1211

Thordarson T, Larsen G, Steinþórsson S, Self S (2003) The 1783–1785 A.D. Laki-Grímsvötn eruptions II: appraisal based on contemporary accounts. Jökull 53:11–48

Vallance JW (1999) Postglacial lahars and potential hazards in the White Salmon River system on the southwest flank of Mount Adams, Washington, vol 2161. US Geology Survey Bulletin, 49 pp

Waythomas CF, Pierson TC, Major JJ, Scott WE (2013) Voluminous ice-rich lahars generated during the 2009 eruption of Redoubt Volcano, Alaska. J Volcanol Geotherm Res 259:389–413

Volcano Index

Locations, edifice types, and compositions of the volcanoes referenced in this book are listed alphabetically below; data are modified from Siebert et al. (2010) and Global Volcanism Program (2023). Most generic prefixes such as *Mount* or *Volcán* are not listed. Names of subsidiary vents referred to in the book are followed by the associated primary volcano name in parentheses, and coordinates of subsidiary features can be those of primary edifice. Volcano types are listed in order of the volcano names when subsidiary vents are listed with primary volcanoes. Names after a forward slash are synonyms, alternate names, or alternate spellings. The abbreviation "Is" is used for *island* or *islands*.

Rock types listed are the primary and secondary magma/lava types at each volcano, although other lava types may be present in minor proportions. Note that the rock type *andesite* in this table may include basaltic andesite.

Volcano name	Volcano type	Primary components	Location	Latitude longitude (degrees)
Adams	Stratovolcano	Andesite, basalt	Washington, USA	46.206 N 121.490 W
Antuco	Stratovolcano	Basalt, andesite	Central Chile	37.406 S 71.349 W
Asama	Stratovolcano	Andesite, dacite	Honshu, Japan	36.403 N 138.526 E
Augustine	Dome complex	Andesite, dacite	Alaska, USA	59.363 N 153.43 W
Baker	Stratovolcanoes	Andesite, dacite	Washington, USA	48.777 N 121.813 W
Bandai	Stratovolcanoes	Andesite	Honshu, Japan	37.598 N 140.076 E
Barú	Stratovolcano	Andesite	Panama	8.808 N 82.543 W
Bezymianny	Stratovolcano	Andesite	Kamchatka, Russia	55.972 N 160.595 E
Calbuco	Stratovolcano	Andesite	Southern Chile	41.326 S 72.614 W

T. C. Pierson et al., *Deposits of Volcanic Wet Flows*, Advances in Volcanology,
https://doi.org/10.1007/978-3-031-66574-5

Volcano name	Volcano type	Primary components	Location	Latitude longitude (degrees)
Casita (San Cristóbal)	Stratovolcanoes	Basalt, andesite	Nicaragua	12.702 N 87.004 W
Central Allan Hills	Large igneous province	Basalt	Antarctica	76.70 S 159.70 E (generalized loc.)
Chaitén	Caldera	Rhyolite	Southern Chile	42.833 S 72.646 W
Chichón, El	Dome complex	Trachyandesite	Mexico	17.360 N 93.228 W
Chimborazo	Stratovolcano	Andesite, dacite	Ecuador	1.464 S 78.815 W
Cofre de Perote	Shield volcano	Andesite, dacite	Mexico	19.492 N 97.15 W
Colima	Stratovolcano complex	Andesite	Mexico	19.514 N 103.62 W
Cotopaxi	Stratovolcano	Andesite, dacite	Ecuador	0.677 S 78.436 W
Crater Lake	Caldera	Dacite, andesite	Oregon, USA	42.93 N 122.12 W
Cumbre Vieja	Stratovolcanoes	Basinite, phonolite	La Palma, Canary Is, Spain	28.57 N 17.83 W
Dittmar	Stratovolcano	Andesite	California, USA	40.453 N 121.347 W
Fernandina	Stratovolcano	Basalt	Galápagos Is, Ecuador	0.37 S 91.55 W
Fuego–Acatenango	Stratovolcano pair	Basalt, andesite	Guatemala	14.473 N 90.880 W
Galunggung	Stratovolcano	Basalt, andesite	Java, Indonesia	7.25 S 108.058 E
Gjálp (Grímsvötn)	Fissure vents, caldera	Basalt	Iceland	64.42 N 17.33 W
Halla	Shield volcano, tuff cones	Basalt, trachyte	Jeju Is, South Korea	33.37 N 126.53 E
Hekla	Stratovolcano	Andesite, basalt	Iceland	63.98 N 19.70 W
Hood	Stratovolcano	Andesite	Oregon, USA	45.374 N 121.695 W
Iliamna	Stratovolcano	Andesite	Alaska, USA	60.032 N 153.090 W
Irazú	Stratovolcano	Andesite, basalt	Costa Rica	9.979 N 83.852 W

Volcano name	Volcano type	Primary components	Location	Latitude longitude (degrees)
Iriga	Stratovolcano	Andesite, basalt	Luzon, Philippines	13.457 N 123.457 E
Jefferson	Stratovolcano	Andesite, dacite	Oregon, USA	44.674 N 121.800 W
Jocotitlán	Stratovolcano	Dacite, andesite	Mexico	19.73 N 99.758 W
Kasatochi	Stratovolcano	Basalt, andesite	Aleutian Is, Alaska	52.177 N 175.508 W
Katla	Subglacial	Basalt, rhyolite	Iceland	63.63 N 19.05 W
Kelut	Stratovolcano	Andesite, basalt	Java, Indonesia	7.93 S 112.308 E
Kharimkotan/ Harimkotan	Stratovolcano	Andesite, dacite	Kuril Is, Russia	49.12 N 154.508 E
Komagatake	Stratovolcano	Andesite	Hokkaido, Japan	42.061 N 140.681 E
Laki (Grímsvötn)	Fissure vents, caldera	Basalt	Iceland	64.42 N 17.33 W
Lamington	Stratovolcano	Andesite, trachyandesite	Papua New Guinea	8.95 S 148.15 E
Láscar	Stratovolcano	Andesite, dacite	Northern Chile	23.37 S 67.73 W
Las Derrumbadas	Dome complex	Rhyolite	Mexico	19.30 N 97.47 W
Lassen Volcanic Center	Stratovolcano	Andesite, dacite	California, USA	40.492 N 121.508 W
Lastarria	Stratovolcano	Andesite, dacite	Northern Chile	25.17 S 68.50 W
Llaima	Stratovolcano	Basalt, andesite	Central Chile	38.692 S 71.729 W
Llullaillaco	Stratovolcano	Dacite	Northern Chile	24.72 S 68.53 W
Markagunt Plateau (Marysvale Volcanic field)	Volcanic field	Andesite	Southern Utah, USA	38.4 N 112.4 W (generalized location of VDA source)
Mayon	Stratovolcano	Andesite	Luzon, Philippines	13.257 N 123.685 E

Volcano name	Volcano type	Primary components	Location	Latitude longitude (degrees)
Meager	Volcanic complex	Dacite, andesite	British Columbia, Canada	50.63 N 123.50 W
Merapi	Stratovolcano	Andesite	Java, Indonesia	7.542 S 110.442 E
Michinmahuida/ Minchinmávida	Stratovolcano	Andesite, basalt	Southern Chile	42.793 S 72.439 W
Misti, El	Stratovolcano	Andesite, dacite	Peru	16.294 S 71.409 W
Miyakejima	Stratovolcano	Basalt, andesite	Izu Is, Japan	34.079 N 139.529 E
Mombacho	Stratovolcano	Andesite, basalt	Nicaragua	11.826 N 85.968 W
Monts Dore	Volcanic complex	Rhyolite, trachyandesite	France	45.35 N 2.49 E
Nevado de Colima (Colima)	Stratovolcano	Andesite	Mexico	19.514 N 103.62 W
Nevado del Huila	Stratovolcano	Andesite, dacite	Colombia	2.93 N 76.03 W
Nevado del Ruiz	Stratovolcano	Andesite, dacite	Colombia	4.859 N 75.322 W
Nevado de Toluca	Stratovolcano	Andesite, dacite	Mexico	19.108 N 99.758 W
Novarupta–Katmai	Complex volcano	Rhyolite, dacite	Alaska, USA	58.27 N 155.157 W
Ontake	Stratovolcano	Andesite, basalt	Honshu, Japan	35.890 N 137.48 E
Osorno	Stratovolcano	Basalt, andesite	Southern Chile	41.10 S 72.493 W
Overo, Cerro	Maar	Andesite	Northern Chile	23.52 S 67.67 W
Pacaya	Volcano complex	Basalt, dacite	Guatemala	14.381 N 90.601 W
Parícutin	Pyroclastic cone	Andesite	Mexico	19.493 N 102.251 W
Parinacota–Pomerape	Stratovolcano pair	Andesite, rhyodacite	Northern Chile	18.17 S 69.15 W
Pichu Pichu	Volcanic complex	Andesite	Peru	16.45 S 71.24 W
Pinatubo	Stratovolcano	Dacite, andesite	Luzon, Philippines	15.13 N 120.35 E

Volcano name	Volcano type	Primary components	Location	Latitude longitude (degrees)
Piton de Neiges (Piton de la Fournaise)	Shield volcanoes	Basalt	Reunion Is, France (Indian Ocean)	21.231 S 56.713 E
Popocatépetl	Stratovolcano	Andesite, dacite	Mexico	19.023 N 98.622 W
Rainier	Stratovolcano	Andesite	Washington, USA	46.853 N 121.760 W
Raung	Stratovolcano	Andesite	Java, Indonesia	8.125 S 114.042 E
Redoubt	Stratovolcano	Andesite, basalt	Alaska, USA	60.485 N 152.742 W
Rinjani	Stratovolcano	Andesite, basalt	Lombok Is, Indonesia	8.42 S 116.47 E
Ruapehu	Stratovolcano	Andesite	New Zealand	39.28 S 175.57 E
Sakurajima	Stratovolcano	Andesite, dacite	Kyushu, Japan	31.585 N 130.657 E
Sangay	Stratovolcano	Andesite	Ecuador	2.002 S 78.341 W
Santa Ana	Stratovolcano	Andesite	El Salvador	13.853 N 89.630 W
Santiaguito (Santa María)	Dome complex, stratovolcano	Dacite, andesite	Guatemala	14.756 N 91.552 W
San Vicente	Stratovolcano	Andesite, dacite	El Salvador	13.595 N 88.837 W
Semeru	Stratovolcano	Andesite, basalt	Java, Indonesia	8.108 S 112.92 E
Shasta	Stratovolcano	Andesite, dacite	California, USA	41.409 N 122.193 W
Shiveluch/Sheveluch	Stratovolcano	Andesite, basalt	Kamchatka, Russia	56.653 N 161.360 E
Socompa	Stratovolcano	Dacite, andesite	Chile/Argentina	24.40 S 68.25 W
St. Helens	Stratovolcano	Dacite, andesite	Washington, USA	46.20 N 122.18 W
Taranaki/Egmont	Stratovolcano	Andesite	New Zealand	39.30 S 174.07 E
Tashirodake	Stratovolcano	Andesite	Honshu, Japan	40.43N 140.41E
Taupo	Caldera	Rhyolite	New Zealand	38.82 S 176.00 E

Volcano name	Volcano type	Primary components	Location	Latitude longitude (degrees)
Tenerife	Stratovolcano	Phonolite	Canary Is, Spain	28.271 N 16.641W
Three Sisters	Stratovolcanoes	Andesite, basalt	Oregon, USA	41.17 N 121.77 W
Tokachidake	Stratovolcanoes	Andesite, basalt	Hokkaido, Japan	43.416 N 142.690 E
Tutupaca	Stratovolcano	Basalt, andesite	Peru	17.026 S 70.372 W
Unzen	Volcanic complex	Andesite, dacite	Kyushu, Japan	32.757 N 130.294 E
Usu	Stratovolcano	Basalt, andesite	Hokkaido, Japan	42.541 N 140.843 E
Villarrica	Stratovolcano	Basalt, andesite	Central Chile	39.42 S 71.93 W
Whakaari/White Is	Stratovolcanoes	Andesite, dacite	Bay of Plenty, New Zealand	37.52 S 177.18 E
Yakedake	Stratovolcanoes	Andesite, dacite	Honshu, Japan	36.224 N 137.590 E

References

Global Volcanism Program (2023) [Database] Volcanoes of the World (v 5.1.5; 15 Dec 2023). Distributed by Smithsonian Institution, compiled by Venzke E. 10.5479/si.GVP.VOTW5-2023.5.1

Siebert L, Simkin T, Kimberly P (2010) Volcanoes of the world, 3rd edn. University of California Press, Berkeley, 568 pp

Index

T. C. Pierson et al., *Deposits of Volcanic Wet Flows*, Advances in Volcanology,
https://doi.org/10.1007/978-3-031-66574-5

GPSR Compliance

The European Union's (EU) General Product Safety Regulation (GPSR) is a set of rules that requires consumer products to be safe and our obligations to ensure this.

If you have any concerns about our products, you can contact us on ProductSafety@springernature.com

In case Publisher is established outside the EU, the EU authorized representative is:

Springer Nature Customer Service Center GmbH
Europaplatz 3
69115 Heidelberg, Germany

Batch number: 10731393

Printed by Printforce, the Netherlands